Nucleosynthesis and Its Implications on Nuclear and Particle Physics

NATO ASI Series

Advanced Science Institutes Series

A series presenting the results of activities sponsored by the NATO Science Committee, which aims at the dissemination of advanced scientific and technological knowledge, with a view to strengthening links between scientific communities.

The series is published by an international board of publishers in conjunction with the NATO Scientific Affairs Division

A	Life Sciences	Plenum Publishing Corporation
B	Physics	London and New York
C	Mathematical and Physical Sciences	D. Reidel Publishing Company Dordrecht, Boston, Lancaster and Tokyo
D	Behavioural and Social Sciences	Martinus Nijhoff Publishers
E	Engineering and Materials Sciences	The Hague, Boston and Lancaster
F	Computer and Systems Sciences	Springer-Verlag
G	Ecological Sciences	Berlin, Heidelberg, New York and Tokyo

Series C: Mathematical and Physical Sciences Vol. 163

Nucleosynthesis and Its Implications on Nuclear and Particle Physics

edited by

Jean Audouze
Institut d'Astrophysique,
Paris, France

and

Nicole Mathieu
Laboratoire de l'Accélérateur,
Orsay, France

D. Reidel Publishing Company

Dordrecht / Boston / Lancaster / Tokyo

Published in cooperation with NATO Scientific Affairs Division

Proceedings of the NATO Advanced Research Workshop
(Fifth Moriond Astrophysics Meeting) on
Nucleosynthesis and Its Implications on Nuclear and Particle Physics
Les Arcs, France
March 17-23, 1985

Library of Congress Cataloging in Publication Data

NATO Advanced Study Institute on Nucleosynthesis and Its Implications on Nuclear
and Particle Physics (1985: Les Arcs, France)
    Nucleosynthesis and its implications on nuclear and particle physics.

    (NATO ASI series. Series C, Mathematical and physical sciences; vol. 163)
    Proceedings of the NATO Advanced Study Institute (Fifth Moriond Astrophysics
Meeting) on Nucleosynthesis and Its Implications on Nuclear Physics, Les Arcs, France,
March 17—23, 1985"—T.p. verso.
    "Published in cooperation with NATO Scientific Affairs Division."
    Includes index.
    1. Nucleosynthesis—Congresses. 2. Nuclear astrophysics—Congresses.
I. Audouze, Jean. II. Mathieu, Nicole, 1946-   . III. North Atlantic Treaty
Organization. Scientific Affairs Division. IV. Title. V. Series: NATO ASI series.
Series C, Mathematical and physical sciences; no. 163.
QB450.N38   1985   523.01'97   85-28159
ISBN 90-277-2173-4

Published by D. Reidel Publishing Company
P.O. Box 17, 3300 AA Dordrecht, Holland

Sold and distributed in the U.S.A. and Canada
by Kluwer Academic Publishers,
190 Old Derby Street, Hingham, MA 02043, U.S.A.

In all other countries, sold and distributed
by Kluwer Academic Publishers Group,
P.O. Box 322, 3300 AH Dordrecht, Holland

D. Reidel Publishing Company is a member of the Kluwer Academic Publishers Group

---

All Rights Reserved
© 1986 by D. Reidel Publishing Company, Dordrecht, Holland.
No part of the material protected by this copyright notice may be reproduced or utilized
in any form or by any means, electronic or mechanical, including photocopying, recording
or by any information storage and retrieval system, without written permission from the
copyright owner.

Printed in The Netherlands

CONTENTS

PREFACE xi

I − PRIMORDIAL NUCLEOSYNTHESIS 1

H. Reeves	Primordial nucleosynthesis in 1985	3
H. Reeves	The puzzle of lithium in evolved stars	13
H. Reeves	High energy particles in dark molecular clouds	23
P. Delbourgo-Salvador, G. Malinie, J. Audouze	Standard big bang nucleosynthesis and chemical evolution of D and ^{3}He	27
G. Steigman, D.S.P. Dearborn, D.N. Schramm	The survival of Helium-3 in stars	37
G. Steigman	How degenerate can we be ?	45
J. Audouze, D. Lindley, J. Silk	Early-photoproduction of D and ^{3}He and pregalactic nucleosynthesis of the light elements	57
R. Schaeffer, P. Delbourgo-Salvador, J. Audouze	Influence of quark nuggets on primordial nucleosynthesis	65
T.P. Walker, E.W. Kolb, M.S. Turner	Primordial nucleosynthesis with generic particles	71
D. N. Schramm	Dark matter and cosmological nucleosynthesis	79
B.J. Carr, W. Glatzel	Nucleosynthetic consequences of population III stars	87

II – EXPLOSIVE OBJECTS   95

J.W. Truran   Nucleosynthesis accompanying classical nova
                     outbursts   97

M. Wiescher, J. Görres, F.-K. Thielemann, H. Ritter
                     Reaction rates in the RP-process and
                     nucleosynthesis in novae   105

J.C. Wheeler   Type I supernovae   113

R. Canal, J. Isern, J. Labay, R. López
                     Nucleosynthesis and type I supernovae   121

F.-K. Thielemann, K. Nomoto, K. Yokoi
                     Explosive nucleosynthesis in carbon
                     deflagration models of type I supernovae   131

E. Müller, Y. Eriguchi
                     Differentially rotating equilibrium models
                     and the collapse of rotating degenerate
                     configurations   143

S.E. Woosley, T.A. Weaver
                     Theoretical models for type I and type II
                     supernova   145

W. Glatzel, M.F. El Eid, K.J. Fricke
                     Pair creation supernovae with rotation   167

N. Langer, M.F. El Eid, K.J. Fricke
                     Nucleosynthesis and massive star evolution   177

C. de Loore, N. Prantzos, M. Arnould, C. Doom
                     Evolution and nucleosynthesis of massive
                     stars with extended mixing   189

N. Prantzos, C. de Loore, C. Doom, M. Arnould
                     Nucleosynthesis in massive, mass losing,
                     stars   197

A. Maeder   Nucleosynthesis in massive stars : winds
                     from WR stars and isotopic anomalies in
                     cosmic rays   207

J.-P. Luminet   Explosive disruption of stars by big black
                     holes   215

B. Pichon   Nucleosynthesis in pancake stars   223

I.J. Danziger   Optical supernova remnants   233

S. Cahen, R. Schaeffer, M. Cassé
Light curves of exploding Wolf-Rayet stars ... 243

III - S PROCESS ... 251

F. Käppeler — s-process nucleosynthesis - Stellar aspects and the classical model ... 253

H. Beer — s-process nucleosynthesis below A=90 ... 263

W.M. Howard, G.J. Mathews, K. Takahashi, R.A. Ward
Pulsed-Neutron-Source models for the astrophysical s-process ... 271

G.J. Mathews, R.A. Ward, K. Takahashi, W.M. Howard
Stellar s-process diagnostics ... 277

K. Takahashi, G.J. Mathews, R.A. Ward, S.A. Becker
Production and survival of ^{99}Tc in He-shell recurrent thermal pulses ... 285

N. Prantzos, J.-P. Arcoragi, M. Arnould
Neutron capture nucleosynthesis in massive stars ... 293

A. Jorissen, M. Arnould
A parametrized study of the $^{13}C(\alpha,n)^{16}O$ neutron source ... 303

IV - CHEMICAL EVOLUTION AND CHRONOMETERS ... 313

F. Matteucci, L. Greggio
Type I SNe from binary systems : consequences on galactic chemical evolution ... 315

B. Barbuy — Magnesium isotopes and galactic evolution ... 325

Ph. Durouchoux — ^{26}Al experimental results ... 331

M. Cassé, N. Prantzos
^{26}Al produced by Wolf-Rayet stars and the 1.8 MeV line emission of the galaxy ... 339

H.P. Trautvetter — ^{26}Al-destruction in neutron rich environments ... 347

B.S. Meyer, D.N. Schramm
Certainties and uncertainties in long-lived chronometers ... 355

M. Arnould, N. Prantzos
More about nucleocosmochronology : the reliability of the long-lived chronometers, and the production of extinct radioactivities ... 363

F.-K. Thielemann, J.W. Truran
Chronometer studies with initial galactic enrichment ... 373

K.-L. Kratz, W. Hillebrandt, J. Krumlinde, P. Möller,
F.-K. Thielemann, M. Wiescher, W. Ziegert
A possible unified interpretation of the solar $^{48}Ca/^{46}Ca$ abundance ratio and Ca-Ti isotopic anomalies in meteorites ... 389

V - RELEVANT NUCLEAR PHYSICS PROBLEMS ... 397

R. Mochkovitch, K. Nomoto
The screening of photodisintegration reactions ... 399

R. Mochkovitch, M. Hernanz
Electron polarization in nuclear reactions at high density ... 407

G.J. Mathews, S.D. Bloom, K. Takahashi, G.M. Fuller,
R.F. Hausman Jr. Large-basis shell-model technology in nucleosynthesis and cosmology ... 413

R.N. Boyd
Properties of anomalous nuclei and their possible effects on stellar burning ... 423

U. Schröder, H.W. Becker, J. Görres, C. Rolfs,
H.P. Trautvetter, R.E. Azuma, J. King
Stellar reaction rate of $^{14}N(p,\gamma)^{15}O^+$ ... 431

H.P. Trautvetter, A. Redder, C. Rolfs
Recent progress in experimental determination of the $^{12}C(\alpha,\gamma)^{16}O$-reaction rate ... 435

R.N. Boyd
Radioactive ion beams : research motivation and methods of production ... 443

CONTENTS   ix

VI - NEUTRINOS AND MONOPOLES   451

    D. Vignaud   The gallium solar neutrino experiment Gallex   453

    G. Waysand   Real-time detection of low energy solar
                      neutrinos with indium and superheated
                      superconductivity   463

    A. Dar   Can neutrinos from Cygnus x-3 be seen by
                      proton decay detectors ?   477

    L. Gonzalez-Mestres, D. Perret-Gallix
                      Detection of magnetic monopoles with
                      metastable type I superconductors   487

LIST OF AUTHORS   495

LIST OF PARTICIPANTS   497

PREFACE

The different components of the Universe and this entity as a whole are and will be the source of many complex problems that the scientific community has to face. Among these problems, those concerning the origin and the evolution of the observed matter and its composition have received recently (at least in part) quite convincing solutions and answers. After the pioneering work of H. Bethe and C. Von Weiszacker (1939) who proposed that hydrogen is transformed into helium through the now classical carbon-nitrogen-oxygen cycle, a flourishing activity occured in the fifties. The culmination of this effort corresponds to the publications of the famous Burbidge, Burbidge, Fowler and Hoyle (1957) and Cameron (1957) papers. Those two papers have set up the basis of nucleosynthesis which is synonymous of nuclear astrophysics. In this book, we still use the classification of the different nucleosynthesis processes proposed at that time by these pionners.

Such progress has been made possible because nuclear physicists (especially experimentalists who determine probabilities of nuclear reactions) have been deeply involved in these astrophysical problems. Every graduate student knows now the importance of the 7 MeV excited state of ^{12}C in the occurence of the triple alpha reaction transforming ^{4}He into ^{12}C which is the first step of the synthesis of the element such that $A \geqslant 12$. We are living still today the "golden age" of nuclear astrophysics because of the advent of many developments which follow these first incisive contributions. Let us quote for instance (i) the primordial nucleosynthesis which takes place just after the Big Bang and which is responsible of the formation of the lightest elements (D, ^{3}He, ^{4}He and ^{7}Li) ; this research has been initiated by R.V. Wagoner in 1967, (ii) the interaction between the cosmic rays and the interstellar medium responsible for the formation of the light elements (Li, Be, B) and studied in France by H. Reeves and his group, (iii) the explosive nucleosynthesis appearing at the end of the sixties and mainly due to W.D. Arnett, J.W. Truran and their colleagues, (iv) the interpretation of many isotope anomalies showed by some specific isotope samples (especially those found in the meteorite Allende) undertaken by several nuclear astrophysicists including D.D. Clayton and H. Reeves (v) the determination by experimentalists, modelists and theorists of many nuclear cross sections work done in USA, GFR and France. The list of the achievements of nuclear astrophysics could be much longer. Its length contrasts with the very few number of texts or conference books devoted to this topic. Up to 1982 which correspond to the publication of "Essays in Nuclear Astrophysics", a book in honor of the W.A. Fowler seventieth birthday, the only books devoted to these problems are those written (or edited) by Fowler (in 1967), Reeves (in 1967), Clayton (in 1968), Arnett and Schramm (1973), Audouze and Vauclair (in 1980) and Wilkinson (in 1981). We should add to this list the excellent review of Trimble (1975).

Given the importance of the topic and the activity of the scientists involved in such problems, it has appeared most timely to organize (with the help of NATO who is very gratefully acknowledged), a workshop on this subject. I am extremely pleased to report that the scientific content and excellent oral presentations are reflected by the outstanding quality of the papers rassembled in this book. This workshop which has been attended by the large majority of the researchers active in the field, has been most profitable to the participants (and even the organizers !). I really hope that the reader will share part of my enthusiasm.

My optimism concerning the value of this book comes from the fact that all the major questions dealing with nucleosynthewsis and its relation with microphysics have been touched upon here : the first part of the book is devoted to primordial nucleosynthesis and constitutes an echo of the exciting debate which took place concerning the values and possibly the limitations on the baryonic density and the number of neutrino (and lepton) families. Many interesting contributions deal with a search of non baryonic dark matter component (gravitinos, generic particles, quark nuggets, ...) consistent with the primordial abundances of the very light elements.

PREFACE

The second part of the book provides the reader with an account of the most outstanding progress achieved in explosive nucleosynthesis : explosion of novae - the building and consideration of models of type I and type II supernovae - the actual nucleosynthetic role of massive (especially Wolf-Rayet) stars - the possible existence of "pancake stars" which could suffer huge gravitational effects of black holes, etc... Related problems like the light curves and the observed abundances of novae and supernovae are also reviewed.

In part III almost all the specialists of the s-process nucleosynthesis (coming from the slow absorption of neutrons during the helium burning phases) have given an account of their most recent results.

Part IV is concerned with some aspects of chemical evolution especially those which concern the galactic effect of type I supernovae and which attempt to explain the very recent observations concerning stellar Mg isotopic ratios. The implications of the recent detections of the gamma-ray line produced by the ^{26}Al decay are presented such as several thorough discussions of current models of nucleo-cosmo-chronology.

At the beginning of the book (part I) the relation of nucleosynthesis with particle physics is obvious. Its relation with nuclear physics is of course conspicuous. A few recent major nuclear physics experiments are reported in Part IV such as some theoretical work dealing with electron screening effects and progress in shell models and also new perspectives regarding the operation of radioactive ion beams.

The book ends up with a report on the two new european projects which plan to attempt to detect solar neutrinos by using the gallium or the Indium technique. Finally there were two problems which deserve to be considered in this context : the search of energetic neutrinos coming from our sources as mysterious and may be elusive as Cygnus X3 and the question of magnetic monopole(s).

I really hope that the reader will have as much pleasure in using this book as I had to organize this workshop with my friends and to edit this book with Nicole Mathieu. In my "Avant-Propos" written in french I have attempted to thank all the individuals responsible for the success of this workshop. Let me thank again the NATO organization, J. Tran Thanh Van, the members of the scientific organization committee and all the authors and participants for having made this meeting and the outcome of it i.e. this book as useful, pleasant and profitable. I would like to thank also Professor Pierre Papon, Directeur Général du CNRS and my colleagues, members of the Groupe de Réflexion "Cosmologie, Physique Nucléaire et Physique des Particules" for their continuous support.

Jean AUDOUZE
Institut d'Astrophysique du CNRS

REFERENCES

Audouze J. and Vauclair S. (1980) : An introduction to nuclear astrophysics - Reidel, Dordrecht
Barnes C.A., Clayton D.D., and Schramm D.N., (eds), 1982, Essays in nuclear astrophysics, Cambridge University Press- Cambridge
Burbidge E.M., Burbidge G.R., Fowler W.A. and Hoyle F., 1957, Rev. Mod. Phys., $\underline{29}$, 547
Cameron A.G.W., 1957, Chalk River Report CRL-41
Clayton D.D., 1968, Principles of stellar evolution and nucleosynthesis, Mc Graw Hill - New York
Fowler W.A., 1967, Nuclear astrophysics, American Philosophical Society - Phyladelphia
Reeves H., 1967, Stellar evolution and nucleosynthesis, Gordon and Breach - New York
Schramm D.N. and Arnett W.D., (eds), Explosive nucleosynthesis, University of Texas - Austin
Trimble V., 1975, Rev. Mod. Phys., $\underline{47}$, 877
Wilkinson D. (ed.), 1981, Nuclear astrophysics Pergamon Press - Oxford

AVANT PROPOS

Cet ouvrage, le cinquième de la série des réunions d'Astrophysique de Moriond, rassemble les contributions présentées en mars 1985 aux Arcs lors de l'atelier subventionné par l'OTAN et consacré à la nucléosynthèse des éléments chimiques. Je tiens à souligner le rôle essentiel tenu par Jean Tran Thanh Van dans l'organisation matérielle de ces rencontres et je remercie Dr M. di Lullo, responsable des programmes scientifiques de l'OTAN, pour son aide et sa compréhension. Je remercie mes amis du comité d'organisation, Catherine Césarsky, Philippe Crane, Tom Gaisser, Dennis Hegyi et Jim Truran. Je tiens à exprimer ma gratitude à Nicole Mathieu et Marie-Claude Pantalacci qui ont collaboré avec charme et compétence à l'organisation scientifique de l'atelier. Je remercie enfin l'ensemble des participants qui nous ont beaucoup aidé en nous faisant parvenir dans les délais leurs très intéressantes contributions.

Jean AUDOUZE

Je dédie ces compte rendus à tous ceux qui ont apporté des contributions majeures en Nucléosynthèse et qui n'ont pu se joindre à nous : C.A.Barnes, H.Bethe, E.M. et G.R.Burbidge, A.G.W.Cameron, D.D.Clayton, W.A.Fowler, F.Hoyle et E.E.Salpeter.

Jean AUDOUZE

COLLECTION OF NICE SENTENCES UNINTENTIONALLY PRONOUNCED BY THE SPEAKERS

"I am not God ; I would like to be. Maybe I'll convince you when I am not in front of you."
                                        Audouze

"(as chairman, to close a long discussion between two participants) : Have several beers together !"
                                        Arnett

"(about supernovae) : If you have two of them, you have a class."
                                        Wheeler

"A tunnel has two entrances"
                                        Pomanski

"By the way, we are here for speculating"
                                        Luminet

"Nature has not been kind with us"
                                        Danziger

"I am here on the strength of the weak interaction with nucleo-synthesis. Some of the stuff I serve you may be half-baked"
                                        Shapiro

"We don't want to destroy the Sun"
                                        Dearborn

"The subject is not hot because we cooperate with both Americans and Russians. But separately !"
                                        Durouchoux

"We have, of course, calculated something"
                                        Cassé

"Sometimes, the motivations are so big that we just make something to satisfy the users"
                                        Takahashi

"I left out some points which did not fit the curve"
                                        X

"(about some model maker) : He would only use reaction rates blessed by the pope"
                                        Trautvetter

"(about a frequently shown figure) : It is your last chance to understand it"
                                        Arnould

I – PRIMORDIAL NUCLEOSYNTHESIS

PRIMORDIAL NUCLEOSYNTHESIS IN 1985

Hubert Reeves
Service d'Astrophysique, Centre d'Etudes Nucléaires, Saclay, France
Institut d'Astrophysique de Paris, France

Several reviews of the subject have been presented in the recent years (Yang et al, 1984)(Audouze, 1984) (Beaudet and Reeves, 1984). I plan here to discuss some recent developments, to present an overall assessment and also to mention a few interesting new fields of application of primordial nucleosynthesis as a research tool in the physics of the cosmos.

1. THE LITHIUM- 7 AS A COSMOLOGICAL PARAMETER

Before discussing the recent measurements of the Spite (Spite and Spite 1982; Spite, Maillard and Spite 1984), I want to review the historical situation (Boesgaard 1976; Duncan 1981). Fig.1 shows the abundance of Li in the stars of two clusters; a very young one (the Pleiades; 5×10^7 y) and a slightly older one (the Hyades; 6×10^8 y). In both cases, the abundance curve is flat, at a value of Li/H $\sim 10^{-9}$ on the L.H.S. of the curve, but decreases progressively on the R.H.S. This is understood through the fact that, at lower surface temperature, stars have increasingly deep surface convective zone, and increasingly hot bottom for this convective zone, where lithium is gradually burned by thermonuclear (p,α) reactions. It is assumed that all these stars have received, at birth, from the interstellar gas, the same original abundance of Li/H $\sim 10^{-9}$. We, consequently, understand why the depletion is more important for the older cluster ($\sim 10^9$ y) than for the younger one ($\sim 10^8$ y).

This same effect is held to be responsible for the strong depletion of Li in the solar surface (Li/H $\sim 10^{-11}$) (although a convincing theoretical explanation is still lacking here). In fact the lithium meteoritic ratio (which, when appropriately

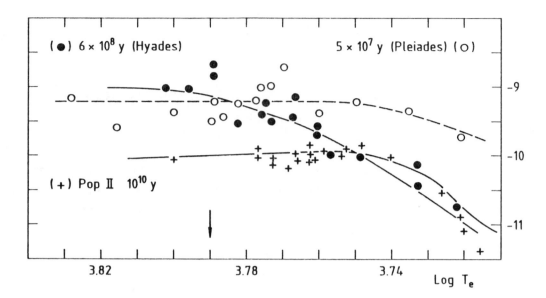

Fig.1 : Lithium abundance curves in the Pleiades (O); in the Hyades (●) and in old Pop II stars (+). In all cases the abundances are approximately constant in the left part of the curve, and decrease gradually on the right, due to the gradual depletion at the bottom of the convective zone.

normalized to silicon, gives also an equivalent value of Li/H$\sim 10^{-9}$) is taken as a proof that the lithium abundance has not varied significantly in the last 5×10^9y.

This explains why the results of the Spite came as a surprise. A naive extrapolation of the previous argument would foresee a complete depletion of lithium in old Pop II stars ($\sim 10^{10}$y). But their data give Li/H$\sim 10^{-10}$, only one tenth of the later value. Further, the Spite, in view of the fact that the points on the L.H.S. of the abundance curve (fig 1) all gathered around the value 10^{-10} while the points on the R.H.S. were falling (just as for the young clusters), argued that the flat curve value ($\sim 10^{-10}$) could be considered as an undepleted value, reflecting the Big Bang yield.

The strength of their argument is based on statistics. At first, the number of lithium-measured stars was rather small. But, as time went on, more and more stars were added to their list, strengthening their case. The reader will make up his own mind by looking at the graph. The accuracy of each measurement is claimed to be better than a factor of two.

The theoretical situation with respect to the plausibility of keeping the Li abundance in a Pop II star undepleted for 10^{10} years has been investigated by Michaud, Fontaine and Beaudet (1984). These authors conclude that the constancy of the Li abundance in the range $5500 < T_{eff} \lesssim 6200$ poses a serious problem to evolutionnary models. The minimum reduction, if α , the ratio of the mixing length to the scale height, is a constant in this range, is a factor of four, with important variations along the temperature scale. Only an highly artificial of adjustment of α could be invoked to explain an abundance variation of less than a factor of two.

The theoretical difficulties and fine tuning of parameters needed to account for the data have also been studied by Cayrel et al (1984), and Cayrel and Däppen (1984), including also the effect of turbulent diffusion.

Clearly the situation needs clarification. At the present time it seems wise to taken a prudent path. I will use, as the cosmological parameter, the value Li/H=10^{-10} but keep in mind that these stars may have suffered some depletion. The value 10^{-9} should then be used as an upper limit.

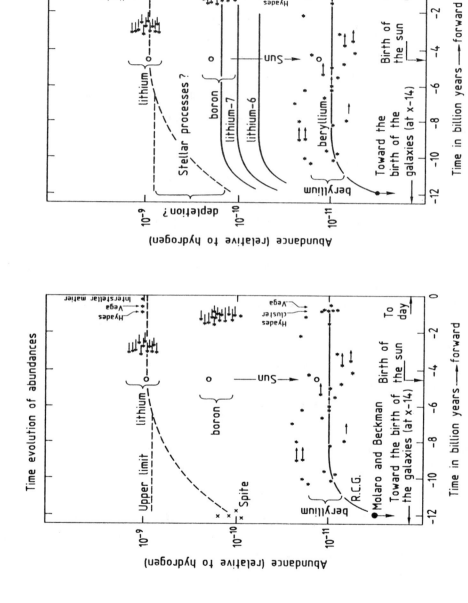

Fig.2 and 3: Evolutionary abundance curves for lithium beryllium and boron (from Reeves and Meyer 1978) with added points. The dashed curve for lithium represents the variation of lithium with time if the Li value of the Pop II is undepleted.

The situation is presented in a graph of evolutionnary abundance curve as a function of the galactic life (from Meyer and Reeves, 1978) (fig 2). In this respect it is interesting to mention the search of ^{9}Be in Pop II stars by Molaro and Beckman (1984). They obtain an upper limit of Be/H < 3×10^{-12} three times smaller than the average stellar values. This fits with the idea of GCR formation of Be (Meneguzzi, Audouze, Reeves, 1971).

The fig 3 shows the abundance of the other L- elements, produced by cosmic rays, as predicted by the fact that the cross-sections and cross-section ratios are, by now, well known (see paper II). It appears that the boron, the lithium-6 are accounted for, but in the Spite hypothesis, a stellar contribution of ^{7}Li is required. Many candidates have been put forward in the litterature (Cameron and Fowler 1971).

2. HELIUM-4 ABUNDANCE

The best piece of evidence is the Kundt and Sargent (1983) measurement on compact galaxies (the low O/H testifies that the stellar contribution to ^{4}He is likely to be very small). I present it in the fig 4 for every one to see the uncertainties. I will use here $X_4 = 0.245 \pm 0.01$ as the cosmological value of He-4.

3. DEUTERIUM

There is a new determination of the interstellar D/H ratio in dark clouds (Dalgarno and Lepp 1984). The data is obtained from an analysis of the DCO^+/HCO^+, involving the various molecular exchange reactions such as :

$$HD + H_3^+ \longleftrightarrow H_2 + H_2D^+$$

$$H_2D^+ + CO \longleftrightarrow DCO^+ + H_2$$

$$\longrightarrow HCO^+ + HD$$

together with

$$D + HCO^+ \longleftrightarrow H + DCO^+$$

The quoted result, $3 \times 10^{-6} <$ D/H $< 4 \times 10^{-5}$ is in agreement with the atomic determination in thin clouds by (Bruston et al 1981; Vidal Madjar et al 1983) with

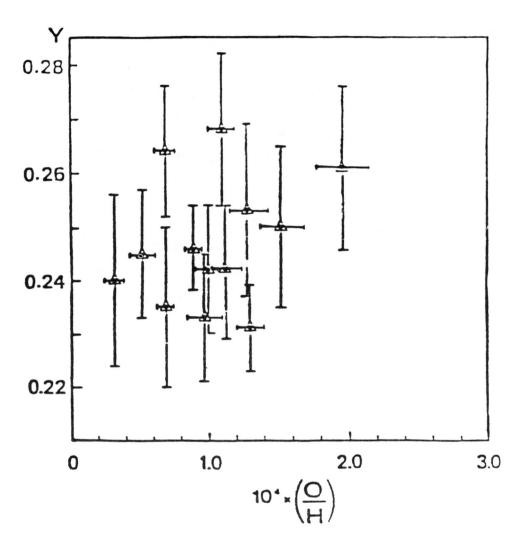

Fig. 4: Helium abundance in compact galaxies (with small level of stellar activities) as a function of the oxygen abundance (Kunth and Sargent 1983).

the value indirectly derived from the solar wind helium isotopic composition (Geiss and Reeves 1972) (Geiss and Bochsler 1979).

4. HE^3 DETERMINATION

The recent data on interstellar He^3 by Rood, Bania, and Wilson (1984) indicates that, because of important stellar production, He^3 may not be a very useful cosmological observable. The data range from $He^3/H \sim 2 \times 10^{-4}$ to $\sim 2 \times 10^{-5}$. Since this last number is not in disagreement with the solar system estimate from the solar wind (Geiss and Reeves 1972) (Geiss and Bochsler 1979) we shall keep it for our best estimate.

5. COSMOLOGICAL YIELDS

Fig 5 shows the Big Bang yields (Beaudet and Reeves, 1984). Although done with a different code (Yahil and Beaudet 1976), the result is in agreement with the calculations of Yang et al (1984). The uncertainties in the nuclear rates are small, except for lithium-7 (a factor of two). The boxes include the corresponding integrated uncertainties.

The theoretical curve are labelled by the parameter g, the coefficient of the relativistic density term ($\rho = g(\pi^2/30) T^4$). This parameter describes the number of particle species and their statistical factor (g = $\Sigma g_b + (7/8) \Sigma g_f$ (b for bosons, f for fermions). The number quoted in fig 5 refers to the standard picture of light neutrinos (N_ν = 2,3 or 4). In recent years, in the spirit of unified theories, it has become fashionable to introduce many new light particles, whose cosmological effect can also be integrated in g.

From fig 5 it appears that the data is compatible with a density range ρ_b between 3 and 6 x 10^{-31}, leading to Ω_b = 0.06 to 0.12 if Ho = 50km/sec/Mpc, or Ω_b = 0.015 to 0.03 if Ho = 100 km/sec/Mpc, if the pristine D/H is within a factor of two or three of its solar system value (Audouze and Tinsley 1974; Delbourgo-Salvator et al 1985), and if $9 < g < 12$, (two or three light ($M_b \lesssim 100$ keV) neutrinos).

Thus it appears that primordial nucleosynthesis fits the observed data on the light element abundances. In my opinion the agreement is as good as one can expect it to be...Furthermore it leaves little space for the existence of new light particles as implied by many unified schemes, in particular those incorporating the

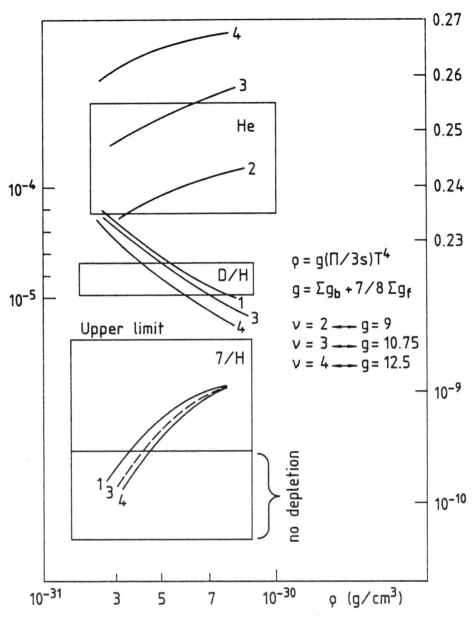

Fig.5 : Big Bang yields as a function of present cosmic baryonic density (from Beaudet and Reeves 1984). The curves g= 9, 10.75, 12.5 gives the assumed value of the statistical factor g. These values correspond for instance to the hypothesis of 2,3 or 4 light neutrinos.

existence of shadow particles (interacting only through the gravitational field (Kolb et al 1985a)).

6. STABILITY OF THE LAWS OF PHYSICS

Several attempts have been made to check whether the laws of physics change with time. The result from the Oklo natural reactor about 1.5×10^9 years ago (Irvine 1983) and from the study of quasars several billion years ago (Pagel 1983) have shown no detectable change on the values of the constants of physics. Recently (Kolb et al 1985; Kolb 1985b) have pushed the investigation to the first minutes of the universe, by asking "how much change can we introduce in the gravitational constant G, in the Fermi constant G_f, and in the proton-neutron mass difference, before spoiling the compatibility of the light element yields with standard nucleosynthesis"? It appears, from this study, that the variation must be less than about one percent in all cases, confirming, in a dramatic way, the constancy of the laws of nature, through a period which has witnessed a change of 10^9 in temperature and 10^{27} in the number densities! This constancy, which can be seen as a "kindness" of nature for the theoretical physicist, carries its own interrogation: how can these numbers remain so constant when everything else change by so large factors, in an apparently totally "anti-machian" way?

REFERENCES

Audouze, J. 1984, in Primordial Helium, ed. by P.A.Shaver et al, ESO p.3
Audouze, J. and Tinsley, B.M., 1974, Astrophys.J. 192, 487.
Beaudet, G. and Reeves, H., 1984, Astron. Astrophys. 134, 240.
Boesgaard, A.M., 1976, Publ.Astron.Soc.Pacific, 88, 353.
Bruston, P., Audouze, J., Vidal-Madjar, A. and Beaudet, G., 1981, Ap.J. 243,161.
Cameron, A.G.W., and Fowler, W.A. 1971, Ap.J. 164, 111.
Cayrel, R., Cayrel, G., Campbell, B., Däppen, W., 1984 Astroph.J.(in press)
Cayrel, R., Däppen, W., 1984, in preparation
Dalgarno, A., and Lepps, S., 1984, Ap.J. 287, L47.
Delbourgo-Salvador, P., Gry, Malinie, G. and Audouze, J., 1985, Astron.Astrophys. (in press).
Duncan, D.K., 1981, Ap.J. 248, 651.
Geiss, J., and Bochsler, P. 1979, Proc. 14th Solar Wind Conf., Berlin
Geiss, J., and Reeves, H., 1972, Astron.Astrophys. 18, 126.
Irvine, J.M. 1983, in the Constants of Physics. The Royal Society of London,p.29.
Kolb, E.W., Seckel, D., and Turner, M.S. 1985a preprint
Kolb, E.W., Perry, M.J. and Walker, T.P. 1985b, in preparation
Kolb, E.W., 1985, preprint Fermi Lab "Cosmology in theories with extra dimension".
Kunth, D., and Sargent, W.L.W. 1983, Ap.J. 273, 81.

Meneguzzi, M., Audouze, J., and Reeves, H., 1971, Astron.Astrophys. 15, 337.
Michaud, G., Fontaine, G., and Beaudet, G., 1984, Astrophys.J. 282, 206.
Molaro, P. and Beckman J., 1984, Astron.Astrophys. 139, 394.
Pagel, B.E., 1983, in "the Constants of Physics", the Royal Society of London, p.35.
Rood, R.T., Bania, T.M. and Wilson, T.L., 1984, Ap.J. 280, 629.
Spite, F., and Spite, M., 1982, Astron.Astrophys. 115, 357.
Spite, M., Maillard, J.P., and Spite, F. 1984, Astron.Astrophys. 141, 56.
Vidal-Madjar, A., Laurent, C., Gry, C., Bruston, P., Ferlet, R., and York, D.G., 1983, Astro. Ap 120, 589.
Yahil, A., and Beaudet, G., 1976, Astrophys.J. 206, 26.
Yang, J., Turner, M.S., Steigman,G., Schramm, D.N. and Olive K.A., 1984, Ap.J. 281, 493.

THE PUZZLE OF LITHIUM IN EVOLVED STARS

Hubert Reeves
Service d'Astrophysique, Centre Nucléaires de Saclay, France
Institut d'Astrophysique, Paris, France

The recent reports on the possible presence of lithium-6 in stars (some above the M.S.), (Andersen, Gustafsson and Lambert 1984) has renewed the interest in the hypothesis of spallation reactions in the surfaces of red giants (Canal, Isern and Sanahuja 1980). Indeed no mechanism has been proposed to generate lithium-6 except high energy reactions.

Spallation implies the acceleration of fast particles which, in turn, generate X-rays throught ionization of the ambient medium. The amount of energy spent in X-rays, for each lithium produced, depends on the energy spectrum of the fast particles. A reasonable estimate (which is also a lower limit) should be 3 ergs per lithium atom (Ryter et al 1970). The main mechanisms contributing to the generation of lithium-6 and 7 are (with comparable yields in normal stellar abundances) (p + C, O $\rightarrow$ ^{6}Li, ^{7}Li) and (α + α $\rightarrow$ ^{6}Li, ^{7}Li).

The excitation functions are presented in fig 1,2 and 3, from a summary of recent data. In each case, the isotopes are generated in roughly equal abundances (within a factor of two); in other words we should expect n (^{6}Li)/n(^{7}Li) $\sim$ 1.

STELLAR DATA

Through recent satellite observations (mostly the Einstein satellite) we have some data (or upper limits) on the X-ray activity of red giants. Typical values of the ratio of X-ray to optical luminosity are 10^{-6} to 10^{-7} (Vaiana 1981) for the bright red giants, with an X-ray luminosity of $\sim 10^{29}$ erg s^{-1}. This can be used as an upper limit to the X-ray flux generated by fast particle deceleration.

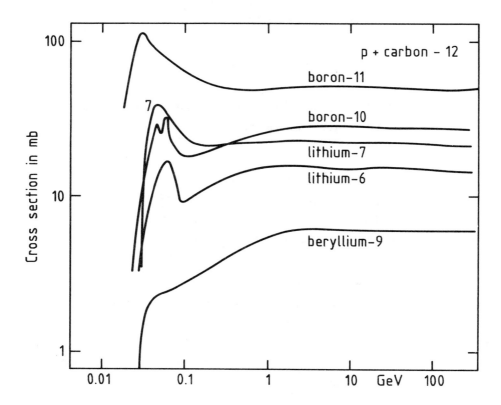

Fig.1: Excitation cross sections for the bombardment of proton on carbon-12. The curve includes all decay products.

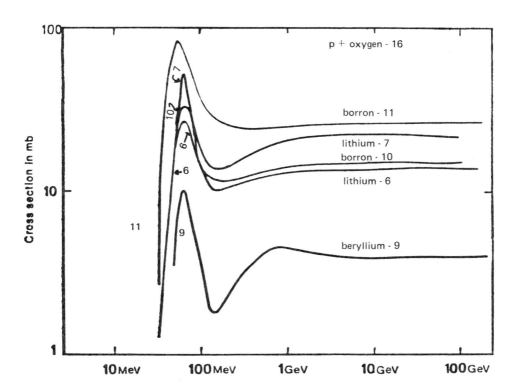

Fig.2: Excitation cross sections for the bombardment of proton on oxygen-16. The curve includes all decay products.

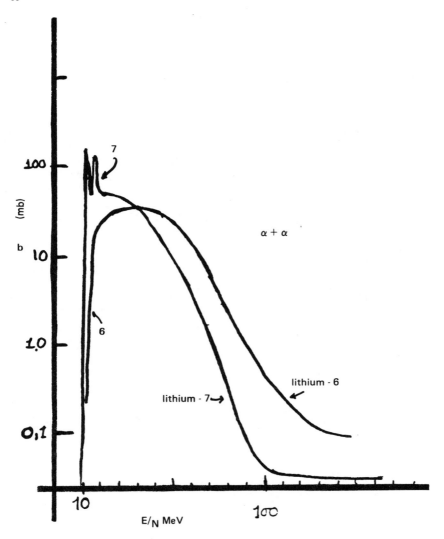

Fig.3: Excitation cross sections for the bombardment of helium-4 on helium-4.. The curves includes all decays products. The abscissa is the laboratory energy in MeV per nucleon (the laboratory energy is four time larger).

THE PUZZLE OF LITHIUM IN ENVOLVED STARS

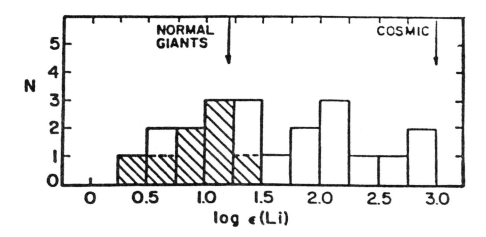

Fig.4: Number of weak G band stars with given lithium to hydrogen abundance ratio as a function of the abundance. The value of 3.0 corresponds to Li/H= 10^{-9}; the value of 0 corresponds to Li/H=10^{-12}. Dashed areas represent upper limits (from Lambert and Sawyer 1984).

The optical observations are summarized in Lambert and Sawyer (1984). The weak G band stars are characterised by important carbon deficiencies, low carbon isotopic ratios and nitrogen enhancements, , unmistakably showing that the surface material has been processed by the CNO cycle. It is usually believed that this material, formerly deeply buried in the star, has been "dredged-up" to the surface during the stellar evolution toward the red giant region.

Another characteristic of these stars is the presence of lithium with unusually high abundance for evolved stars. The data is presented in fig.4, where the number of stars with given lithium abundance is plotted as a function of abundance. Lambert et Sawyer rightly point out that the highest Li/H coincide with the cosmic (interstellar) value of Li/H = 10^{-9}, with no values higher than this value.

In fig 5, the recent data on the lithium isotopic ratio in stars (some above the M.S.) is presented (Andersen et al 1984). An upper limit of ^{6}Li/^{7}Li <0.2 can be estimated, while an average value of 0.1 can be obtained, comparable with the solar system (1/12.5). The uncertainties are difficult to assess. It is not clear that lithium-6 has really been detected. Another study by Spite (1984) does not seem to confirm the presence of lithium-6 in red giant atmospheres.

PRODUCTION BY SPALLATIVE PROCESSES

With the upper limit of X-ray emission by fast particles ($<10^{29}$ erg/sec) and an upper limit for the duration of the red giant stage ($\lesssim 10^{16}$ sec), at the price of 3 ergs per Li atom, we can generate 3×10^{44} Li atoms, or fill a stellar mass of 3×10^{-4} M with the cosmic ratio of Li/H = 10^{-9}. This is far smaller than the estimated mass of the surface convective zone of red giants ($\sim 10^{-1} M_\odot$) in which we could plausibly believe that all the lithium made in the flaring activity should be mixed.

One could argue that, somehow, the mixing of the upper (active) layers with the deep convective zone is very inefficient. However the upper limit to the isotopic ratio Li^6/Li^7 $\lesssim 0.2$ could only be explained by <u>assuming</u> nuclear destruction of ^{6}Li in the convective zone (recall that the thermonuclear destruction rate of ^{6}Li is about 100 faster than for ^{7}Li). Thus the spallation mechanism appears to be inadequate when the <u>isotopic formation ratio</u> (^{6}Li/^{7}Li $\sim$ 1) and the <u>X-ray luminosity</u> are taken into account.

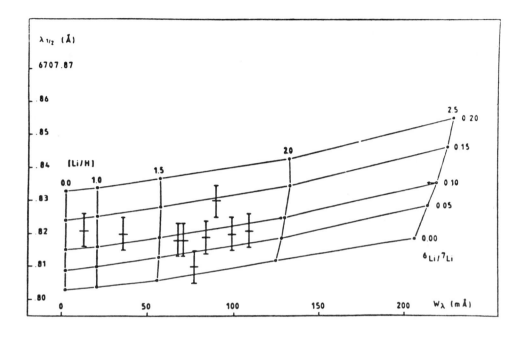

Fig.5: Measurements of the lithium isotopic ratio in stars. In ordinate, the observed wavelength; in abscissa, its equivalent width. The lines in the graph are the loci of equal isotopic ratios (from Andersen et al 1984).

OTHER SOLUTIONS

The presence of Li in evolved stars is custumarily ascribed to a mechanism of (^{3}He + ^{4}He $\rightarrow$ ^{7}Be $\rightarrow$ ^{7}Li) burning in the ashes of H-burning zone, as proposed by Cameron and Fowler (1971). This mechanism cannot account for the presence of ^{6}Li (if confirmed), nor for the cut-off of Li/H at the cosmic value (if statistically significant).

In view of this cut off, and of the apparently solar lithium isotopic ratio, it seems more natural to look not for a production mechanism of lithium but rather from a preservation of material inherited from interstellar matter. Lambert and Sawyer suggest a scheme involving diffusion in stellar surfaces, along the lines advocated by Michaud (1980) for Am and Ap stars. Diffusion, however, while (generally) respecting isotopic ratios, would give rise to local enhancements of elements which would seem to have no reason to respect the cosmic limit.

The problem is : how to preserve the cosmic abundance of lithium (or to deplete it by less than one or two orders of magnitude) in stellar surfaces with unmistakable sign of mixing with CNO processed layers...

Consider, for instance, a fresh addition of matter with local galactic abundances (Meyer 1985), not previously heated to typical stellar temperature. For instance, imagine that, while inflating their sizes, these giant stars gradually eat a collection of Jupiter - like planets. The addition of matter with (Li/H=10^{-9}) would respect the cosmic limit and the isotopic ratio (if no prior gas- grain separation has taken place). This hypothesis would predict that the highest Li stars would show the least depleted abundances of other elements. The data reported in Lambert and Sawyer (their table 2) does not fulfill this prediction. This difficulty applies to other types of addition mechanisms of undepleted (unheated) matter, such as accretion while crossing thick interstellar clouds. (A further test would be the detection of deuterium in the CH bands)..

SUMMARY

The following observations are reported for some evolved stars.

a) CNO abundances and isotopes, implying CNO cycle alteration b) Li abundances reaching, but not above, the interstellar value (Li/H $\leq 10^{-9}$) c) Lithium isotopic ratio as in the solar system ($Li^6/Li^7 \sim 0.1$ with an upper limit at < 0.2)). d) X-ray fluxes of 10^{29} erg/sec

Neither spallation processes in the stellar surfaces, nor ($He^3 + He^4$) reaction after completion of H burning, nor mixing with interstellar material can account for all these features at the same time. This problem is open for a solution.

REFERENCES

Andersen, J., Gustafson, B. and Lambert, D.L., 1984, Astron.Ap. 136, 65.

Barbuy, B., 1978, Astron.Astrophys. 67, 339.

Cameron, A.G.W., and Fowler, W.A. 1971, Ap.J., 163, 111

Canal, R., Isern, J., and Sanahuya, B., 1980, Ap.J. 235, 504.

Lambert, D.L., and Sawyer, S.R., 1984, Astrophys.J. 283, 192.

Maurice, E., Spite, F., and Spite, M., 1984, Astr.Ap. (in press).

Meyer, J.P., 1985, Ap.J. Supp. 57, 151 and 57, 173.

Michaud, G., 1980, Ap.J. 85, 589.

Ryter, C., Reeves, H., Gradsztajn, E., and Audouze, J., 1970, Astron.Astrophys. 8-, 389.

Spite, M. (private communication)

Vaiana, J., 1981, Ap.J. 245, 163.

HIGH ENERGY PARTICLES IN DARK MOLECULAR CLOUDS

Hubert Reeves
Service d'Astrophysique, Centre d'Etudes Nucléaires de Saclay, France
Institut d'Astrophysique de Paris, France

INTRODUCTION

The role of galactic cosmic rays in generating the light elements lithium, beryllium and boron by bombardment of interstellar matter has been elucidated some years ago (Meneguzzi, Audouze and Reeves (MAR) 1971) (Reeves, 1974), (Reeves and Meyer 1978). It appears that this mechanism is responsible for the production of the isotopes ^{6}Li, ^{9}Be, ^{10}B and ^{11}B and for a small part of the ^{7}Li. The remaining part of this isotope has been generated in the Big Bang and perhaps also in stellar processes. I have reviewed the situation in my paper "Primordial Nucleosynthesis in 1985" in this Moriond 1985 volume (paper I).

Ferlet and Dennefeld (1984) have reported the detection of lithium in clouds of interstellar matter of the Ophiuchus region. The isotopic ratio appears to vary from place to place. In one case, a ratio of ^{6}Li/^{7}Li $\sim$ 0.025 is reported, about three times smaller than the solar system value of 0.1. These variations, if confirmed, are difficult to account for in the standard scenarios. Various ideas have been proposed, largely unconvincing. Here we study quantitavely the implications of local irradiations in OB associations, as discussed previously by (Meyer 1978). It is generally believed that solar type stars are born in stellar associations, together with massive short-lived stars which explode in supernovae before the dissociation of the association (Reeves 1978). These events could give rise to local fast particle fluences and to appreciable amount of spallation-generated lithium-6. The local enhancement of the isotopic ratio (up to 0.10) would later be diluted in the pool of galactic matter but would be retained in any planetary system in formation. This hypothesis can be tested in the light of our present knowledge of the electromagnetic fluxes accompanying these fast particles. Here we shall study, in turn, the various windows where these fluxes could be observed.

ENERGETICS

Typical masses of dark clouds associated with OB associations are $\sim 3 \times 10^4$ $M_\odot$ or 3×10^{61} nucleons. The local enrichments of lithium-6 needed in the present discussion are $Li^6/H \sim 10^{-10}$ or 3×10^{51} Li^6 atoms. The discussion of Ryter et al (1970) summarized in my paper "High Energy Reactions in Red Giants" of this volume (paper II), implies a cost of ~ 3 erg per ^6Li or a total of 3×10^{51} ergs. Normalized to the cloud, the energy requirement is ~ 200 eV per nucleon.

During its life as a young group, an OB association typically witnesses about ten SN explosions (Reeves 1978) of about 10^{51} ergs each. The energetic requirement could be met if one third of this energy is emitted in fast particles, not an unlikely possibility.

GAMMA RAY OBSERVATIONS (HIGH ENERGY)

The study of MAR has shown that the present mean flux of galactic cosmic ray $\phi_G \sim 10$ p $cm^{-2} sec^{-1}$ will produce an amount of $Li^6/H \sim 10^{-10}$ during the whole galactic life T_G ($\sim 10^{10}$ y). To generate an equivalent amount during the life of the OB association ($\sim 10^7$ y) would require a flux of $\phi_{OB} \sim 10^4$ p $cm^{-2} sec^{-1}$ (some 10^3 times larger).

Through the observations of the satellite Cos B (Cassé and Paul 1980), we have good coverage of the gamma ray flux above 100 MeV. There is certainly no sign of strong local enhancement of fast particles which would result in gamma rays by the $p+p \rightarrow \pi_o \rightarrow \gamma$ ($\gtrsim 100$ MeV). The interpretation of gamma ray sources implies, at best, an increase of a factor of five ($\phi_{OB}/\phi_p \leq 5$) for the flux of all particles above a few hundred MeV. Furthermore, since the gamma ray flux is proportional to the product of the fast particle flux and the local density, the sources may be reflecting mostly the matter density enhancement.

GAMMA RAY OBSERVATIONS (LOW ENERGY)

Fluxes of energetic particles with $E < 300$ MeV would not generate gamma rays observable by CosB. However they could be detected through gamma ray lines from excited states of heavy nuclei (Meneguzzi and Reeves 1975). From comparison with measured cross sections, it appears that the collision of fast protons on interstellar carbon should produce about ten gamma rays of 4.4 MeV (first excited state) for each ^{6}Li generated. From the previous discussion, this means about $3\times10^{52}\gamma$ (4.4).

The mean flux in 10^7 years should be $\phi_\gamma = 10^{38} \gamma$ (4.4) sec^{-1}.

The HEAO-3 satellite has searched for gamma ray lines in various young associations (Durouchoux et al 1981). For instance, in the Ophiucus group, at 500 light-years, an upper limit of $7\times10^{-4}\gamma$ cm^{-2} sec^{-1} is reported, some twenty times larger than the flux predicted from the present estimate ($3\times10^{-5}\gamma$ cm^{-2} sec^{-1}). Thus the gamma ray line upper limits can not, by themselves, rule out our hypothesis.

MATTER IONIZATION IN CLOUDS

The observations of the DCO$^+$/HCO$^+$ ratio in dark clouds can potentially lead to information on the ionization state of dark clouds. The flux of low energy cosmic rays needed to generate the local enhancement of ^{6}Li would results in an ionization rate $\int H \sim 10^{-14}$ sec^{-1}, probably too large to be reconcile with this isotopic ratio, altough according to Dalgarno and Lepp (1980), the case may not be as clear cut as previously thought (Guelin et al 1977).

OBSERVATIONS OF X-RAYS

The deceleration of fast protons results in an X ray production of $\sim$ 3ergs per lithium, as discussed before (paper II). The mean flux of $L_x \sim 10^{52}$ erg $/10^7$ y = 3×10^{37} erg/sec, needed in the context of our hypothesis, is $\sim 10^5$ larger than the mean X-ray flux from typical dark cloud ($L_x \sim 2\times10^{32}$ erg s^{-1}) (Montmerle et al 1983). The fact that most of the X-ray emission in Ophiucus occurs from a ring may imply that most of the X-ray radiation is emitted and reabsorbed inside a thick region. However it appears very unlikely that this burial effect could account for the factor of 10^5 missing here.

INFRARED LUMINOSITY

Thick clouds or thin clouds, the energy should come out somewhere. The infrared luminosity of the typical clouds ($\sim 10^5 L_\odot$) is large enough to accomodate the hypothesis under discussion if the X-rays accompanying the formation of Li are all absorbed and later reemitted in this spectral region.

CONCLUSIONS

The present status of the art, including a) the CosB data on the high energy gamma rays, b) the HCO^+ isotopic ratio, and c) the X-ray fluxes from Einstein, makes it <u>most unlikely</u> that local irradiation in OB associations could explain the reported variations of lithium isotopic ratio. The last word on this question could eventually come from the gamma ray lines since the clouds are very unlikely to be opaque to this radiation.

REFERENCES

Cassé, M. and Paul, J.A., 1980, Ap.J. <u>237</u>, 236.
Dalgarno and Lepp, S. 1984, Ap.J., <u>287</u>, L47.
Durouchoux, Ph., Montmerle, T., Jacobson, A., Ling, J., Mahoney, W., Riegler, G., Wheaton, W., 1981, IRCC, XG 3-4, 74
Ferlet, R., and Dennefeld, M., 1984, Astron.Astrophys. <u>138</u>, 303.
Guelin, M., Langer, W.D., Sell, R.L., Wootten, H.A., 1977, Ap.J. <u>217</u>, L165.
Meneguzzi, M., Audouze, J., Reeves, H., 1971, Astron.Astrophys. <u>15</u>, 337.
Meneguzzi, M., and Reeves, H., 1975 Astr.Ap. 40, 99
    1975, Astr.Ap. 40, 91.
Montmerle, T., Koch-Miramond, L., Falgarone, E., and Grindlay, J.E. 1983, Ap.J. <u>269</u>, 182.
Reeves, H., 1974, Ann.Rev.Astron.Ap. <u>12</u>, 437.
Reeves, H., 1978, in Protostars and Planets, ed.T.Gehrels, University of Arizona Press, p.399.
Reeves, H., Meyer, J.P., 1978, Ap.J. <u>226</u>, 613.
Ryter, C., Reeves, H., Gradsytajn, E., and Audouze, J., 1970, Astron.and Astroph. <u>8</u>, 389.

STANDARD BIG BANG NUCLEOSYNTHESIS AND CHEMICAL EVOLUTION OF D AND ^{3}HE

P. DELBOURGO-SALVADOR[1], G. MALINIE[2], J. AUDOUZE[1,3]
1 Institut d'Astrophysique du CNRS, 75014 Paris - France
2 Centre d'étude de Limeil, Villeneuve Saint Georges - France
3 Laboratoire René Bernas, Orsay - France.

ABSTRACT: Effect of stellar and galactic evolution on the abundances of D and ^{3}He are examined in an attempt to relate the observations of these two light elements with their primordial abundances.
It is shown that in order to reconcile the baryonic densities deduced respectively from ^{4}He and D in the frame of Standard Big Bang nucleosynthesis models, the ratio between the primordial and the present abundance of D should be as large as 7-10. This can be achieved only by specific models of galactic evolution discussed here.

I. INTRODUCTION

The production of the very light elements (D, ^{3}He, ^{4}He and ^{7}Li) predicted in the frame of the canonical Big Bang nucleosynthesis constitute presently one of the major cosmological tools: the comparison between these predictions and the observed abundances of these elements leads to quite important constraints on the present baryonic density of the Universe such that $\Omega_B \leq 0.02$ (and therefore on its overall dynamical evolution) and on the maximum number of neutrinos (and leptons) families (see e.g. H. Reeves, this conference, Yang <u>et al.</u> 1984, Boesgaard and Steigman, 1985).
Given the importance of such implications, one should analyse them very carefully especially in relation with stellar and galactic evolution processes which affect the abundances of these elements over the galactic history since their early formation. In that respect, the relation between the observed D and ^{3}He abundances which can only be determined presently either in the Solar System or in the interstellar medium (ISM), (Vidal-Madjar <u>et al.</u>, 1983) and their primordial values, depends strongly on these evolution processes. D is indeed destroyed in any stellar zone for which $T \geq 10^5$K and immediately transformed into ^{3}He. The situation is not the same for ^{4}He and

^{7}Li which have been observed respectively in "lazy" blue compact galaxies (Kunth and Sargent, 1984) and in halo stars (Spite and Spite, 1982). Moreover, when one makes use of the predictions of the Standard Big Bang nucleosynthesis together with the simplest galactic evolution models (Audouze and Tinsley, 1974) which predict a ratio $D_{primordial}/D_{now}\sim 2$, one notices (following Vidal-Madjar and Gry, 1984) a conspicuous discrepancy between the baryonic density deduced from ^{4}He and D respectively. This discrepancy can only be avoided if $D_{primordial}/D_{now}\sim 7-10$. This is why in this contribution we analyse different plausible hypotheses regarding the stellar and galactic evolution of D and ^{3}He and their effect on the $D_{primordial}/D_{now}$ and $^3He_{primordial}/^3He_{now}$ ratio.

II. STELLAR PRODUCTION AND DESTRUCTION OF D AND ^{3}He

For the purpose of this analysis, one should determine carefully all the production and destruction sites (stellar and not) concerning D and ^{3}He.

D is totally destroyed in stellar interiors and as a matter of fact, its destruction by the reaction $D(p,\gamma)^3He$ occurs during the pre main sequence phase of stars.

If stars lose large amount of mass by stellar winds during this period (Hartman, 1984), the matter ejected is D free and ^{3}He rich. This effect has not been taken into account in standard model of chemical evolution and because the mass loss rates are not known precisely.

^{3}He is a less fragile species and can even be produced in low mass stars by the incomplete p-p cycle. So in low mass stars at the beginning of the main sequence : $X(^3He)=X(^3He)_{initial}+X(D)_{initial}$ (because D is entirely transformed into ^{3}He). During the Red Giant Branch (RGB), part of ^{3}He initial had been destroyed and a certain amount had been produced depending on the mass of the star. According to Iben and Truran (1978) : $(^3He/H)_{RGB}=K(M/M_\odot)^{-2}+0.7(^3He/H)_{initial}$ where K is a production coefficient.

During this phase, significant mass loss M takes place, according to Fusi-Pecci and Renzini (1976) : $M \propto M^{-2}$.

The fate of ^{3}He during the subsequent phases of stellar evolution is quite uncertain. During the Asymptotic Giant Branch, ^{3}He can be destroyed by $^3He(\alpha,\gamma)^7Be$ if the temperature at the base of the convective envelope is high enough (T>a few 10^7 K). Therefore during this phase, low and intermediate mass stars should eject material with less ^{3}He than during the RGB.

The galactic enrichment of ^{3}He comes mainly from low mass stars, because it requires long period of time to be achieved. This excludes (together with the fact that ^{3}He is more destroyed in high mass stars) that high and intermediate mass stars release significant amounts of ^{3}He into the ISM.

III. CHEMICAL EVOLUTION OF D AND ^{3}He

The equations describing the evolution of the abundance by mass of an element i is :

$$\frac{d\sigma X_i}{dt} = -\nu\,\sigma(t)X_i + \int_{M_1}^{M_2} E_i(M)\phi(M)\nu\,\sigma(t-\tau_M)dM + \delta X_{inf} \qquad (1)$$

where σ is the gas density, ν the astration rate, $E_i(M)$ the mass fraction of isotope i released by stars of mass M to the interstellar medium, δ the infall rate, X_{inf} the abundance of element i in the infalling material and ϕ is the initial mass function.

The first term of the r.h.s. of this equation is the star formation rate which is proportional to σ the gas density, the second is the fraction returned to the ISM, the third term represents the infalling material.

For D, this equation is very simple : because D is only destroyed in stars, the second term of r.h.s. is equal to zero (Audouze et al., 1976)

$$\frac{d\sigma X_D}{dt} = -\nu\sigma X_D + \delta X_{Dinf} \qquad (2)$$

The present observed gas density allows to establish a simple relation between the infall rate δ and the astration rate ν :

$$\delta = \nu(1-R)\,\frac{\sigma_{now} - \exp\{-\nu(1-R)\,t_{now}\}}{1 - \exp\{-\nu(1-R)\,t_{now}\}} \qquad (3)$$

where R is the fraction of initial gas processed in the hypothesis of instantaneous recycling and σ_{now} the present gas density (we adopt t_{now}=12.5 Gyr and σ_{now}=0.05).

We now consider two reasonable hypotheses which effects are to increase the ratio $D_{primordial}/D_{now}$:
(i) infall of processed material
(ii) mass loss during the pre main sequence phase.

The two relations (2) and (3) are sufficient to treat the deuterium case. We can calculate the ratio D_p/D_n as a function of the infall rate δ in the case (i). We obtain the classical factor 2 without infall. Fig. I shows the results of this calculation.

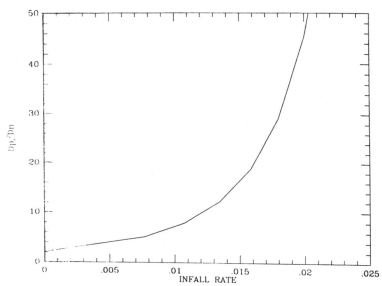

Fig. I
Case (i) : Infall of processed material. The ratio $D_{primordial}/D_{now}$ is plotted as a function of the infall rate δ (% of the total mass of the Galaxy).

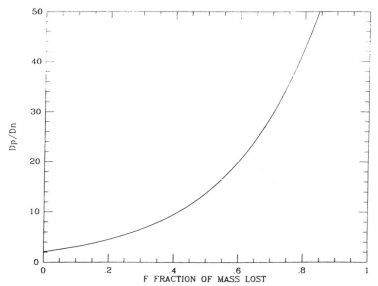

Fig. II
Case (ii) : Mass loss during the pre main sequence phase. The ratio $D_{primordial}/D_{now}$ is plotted as a function of f the fraction of mass lost.

Fig. II shows the ratio D_p/D_n as a function of f the fraction of the star lost in the case (ii). Equation (2) becomes :

$$\frac{d\sigma X_D}{dt} = -\nu(1 + f)\sigma X_D + \delta X_{inf}$$

This figure corresponds to a case without infall.
But we must keep in mind that primordial D is transformed into ^{3}He, so a too important value for primordial deuterium might produce overabundance of ^{3}He, inconsistent with the observations.
Chemical evolution of ^{3}He is not as simple as for D and the prescriptions coming from stellar evolution (see § II) are :
. ^{3}He is produced by incomplete p-p cycle in low mass stars, the production is proportional to $(M/M_\odot)^{-4}$ (for M<5 M☉).
. The fraction of ^{3}He of the star ejected in the ISM are :
70 % if M<2 M☉
25 % if 2M☉<M<5M☉
No ^{3}He is ejected if M>5 M☉.

The results of chemical evolution of D and ^{3}He are shown on Figure III and IV for model (i) in which there is an infall of processed material and on figure V for model (ii) in which significant mass loss of D free and ^{3}He rich material takes place during the pre main sequence phase.
In case (i), the outcome is that with an infall of processed material $\delta=0.015$ (1.5% of the total mass of the galaxy), one obtain a ratio $D_p/D_{now}=15$ and the primordial abundances are $(X_D)_p=10^{-4}$, $(X^3He)_p=5 \cdot 10^{-5}$, the protosolar abundances are $(X_D)_\odot=3 \cdot 10^{-5}$, $(X^3He)_\odot= 5 \cdot 10^{-4}$ (which are compatible with the Solar System observations), the interstellar abundances now are $(X_b)=7 \cdot 10^{-6}$, $(X^3He)= 6 \cdot 10^{-5} - 1.3 \cdot 10^{-4}$ depending on K.
One can note that the abundance of ^{3}He in the ISM now is very sensitive to the value of K, the production coefficient but the solar value is not affected by a change of this coefficient because ^{3}He is produced and ejected by low mass stars with a long life time so ^{3}He appears in the ISM after the birth of the Sun. This could explain the dispersion of the values of ^{3}He observed by Rood et al. 1984 in regions with more low mass stars but it seems in contradiction with the fact that high value of ^{3}He are found at large galactocentric distance.
In case (2), the conclusions are that with a mass loss of 20 % of the total mass (the determinations of this mass loss are still very uncertain (Hartman 1984)), one obtains :

$(X_D)_p = 10^{-4}$ $(X^3He)_p = 5 \cdot 10^{-5}$

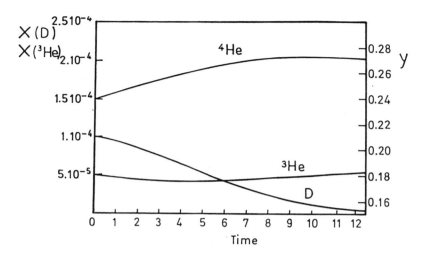

Fig. III
Results of the chemical evolution model with infall of processed material. The abundances of D, ^{3}He, ^{4}He are plotted as a function of time in Gyr. The infall rate is $\delta=0.012$. The production rate of ^{3}He is 5.10^{-5} $(M/M_\odot)^{-4}$.

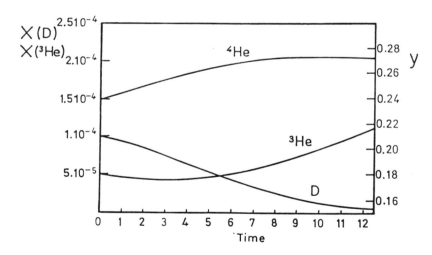

Fig. IV
Same as Fig. III but with a production rate for ^{3}He of 5.10^{-4} $(M/M_\odot)^{-4}$.

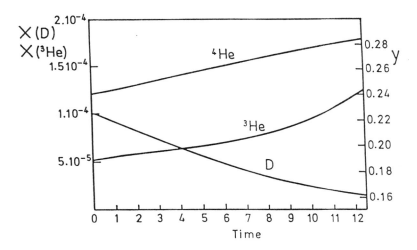

Fig. V
Results of the chemical evolution model with mass loss during the pre main sequence phase and no infall. The fraction of mass lost is f=20%, the production rate of ^{3}He is 5.10^{-4} $(M/M_\odot)^{-4}$. The abundances of D, ^{3}He and ^{4}He are plotted as a function of time.

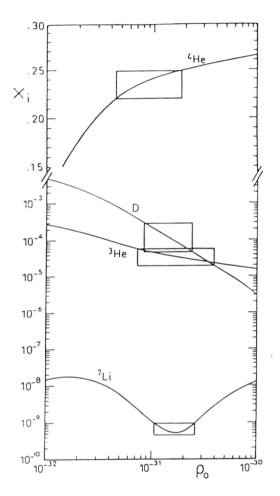

Fig. VI

Comparison between the light elements abundances computed and the observed abundances (boxes) taking into account chemical evolution models with $D_p/D_n \cong 10$.

$(X_D)_0 = 2.10^{-5}$ $(X^3He)_0 = 6.10^{-5}$
$(X_D)_{ISM} = 10^{-5}$ $(X^3He)_{ISM} = 8.10^{-5} - 1.3 \ 10^{-4}$.

The effect of the production factor K is the same as in the case (1). But one could explain the dispersion of ^{3}He measurement by the presence of young stars ejecting D free and ^{3}He rich and then avoid the difficulties outlined in case (i).

CONCLUSION

From this investigation, one can conclude that specific (but quite reasonable !) models of galactic evolution implying either the infall (or inflow) of processed interstellar material or preferably significant stellar mass losses occuring during the pre main sequence phase lead to D_p/D_n ratios consistent both with the ^{3}He and ^{4}He abundances in the frame of the classical Big Bang nucleosynthesis (figure VI).
This analysis shows convincingly that in order to reconcile the observations regarding D and ^{4}He with their primordial abundances deduced from the canonical Big Bang models one should invoke such galactic models. It is the price one has to pay to deduce from the early nucleosynthesis cosmological constraints as important as the maximum baryon density and the maximum number of neutrino and lepton families.

REFERENCES

- Audouze J., Lequeux J., Reeves H., Vigroux L. ; 1976, Astrophys. J., 208, L51
- Audouze J., Tinsley B.M. ; 1974, Astrophys. J. 192, 487
- Boesgaard A.M., Steigman G., Ann. Rev. Astron. Astroph.,1985
- Delbourgo-Salvador P., Vangioni-Flam E., Malinie G., Audouze J., 1984, in "The Big Bang and Georges Lemaitre", Ed. A. Berger, Reidel, Dordrecht
- Fusi-Pecci F., Renzini A., 1976, Astron. Astrophys. 39, 413
- Hartman L., 1984 - Comments astrophys. 10, 97
- Iben I., Truran J.W., 1978, Astrophys. J. 220, 980
- Kunth D., Sargent W.L.W., 1984, Astrophys. J., 273, 81
- Michaud G., Fontaine G., Beaudet G., 1984, Astrophys. J., 282, 206.
- Reeves, H., these proceedings.
- Rood R.T., Bania T.M., Wilson T.L., 1984 Astrophys. J. 280,629
- Spite F., Spite M., 1982, Astron. Astrophys. 115, 357
- Vidal-Madjar A., Gry C., 1984 Astron. Astrophys. in press.
- Vidal-Madjar A., Laurent C., Gry C., Bruston P., Ferlet R. and York D.G., 1983, Astron. Astrophys., 120, 58
- Yang J., Turner M.S., Steigman G., Schramm D.N., Olive K.A., 1984, Astrophys. J. 281, 493.

THE SURVIVAL OF HELIUM-3 IN STARS

G. Steigman[1], D. S. P. Dearborn[2] and D. N. Schramm[3]

[1]Bartol Research Foundation, University of Delaware
Newark, Delaware 19716, USA

[2]Lawrence Livermore National Laboratory, P. O. Box 808
Livermore, California 94550, USA

[3]Astronomy and Astrophysics Center, University of Chicago
Chicago, Illinois 60637, USA

ABSTRACT. If the observed abundance of ^{3}He is to be used – in concert with the predictions of Primordial Nucleosynthesis – to bound from below the universal density of nucleons, the survival of ^{3}He in the material processed through stars must be estimated. The results of detailed stellar evolution calculations which follow the destruction of ^{3}He are outlined. The survival of ^{3}He as a function of stellar mass, composition and mass loss is presented as are estimates of the primordial abundances of D and ^{3}He; the cosmological consequences are summarized.

1. INTRODUCTION

If the production and destruction of ^{3}He in the course of galactic evolution were known, the presently observed interstellar and/or solar system abundances could be extrapolated back to pregalactic epochs. Since any deuterium processed through stars is converted to ^{3}He via radiative proton capture, the evolution of ^{3}He may permit bounds to be set on the primordial abundance of D plus ^{3}He (Yang et al. 1984). If the destruction of ^{3}He can be constrained, significant lower limits to the nucleon density – critical to questions of dark matter and the number of light neutrinos (or, other "inos") – may be derived (Yang et al. 1984). In addition to the cosmological motivation, it is important to study the evolution of ^{3}He during recent epochs (Rood, Steigman and Tinsley 1976) to interpret the comparison of solar system determinations with those of the interstellar medium (Rood, Bania and Wilson 1984).

We have studied the survival of ^{3}He in stars covering a range of masses ($M/M_\odot = 8, 15, 25, 50, 100$) and compositions (Pop. I, Pop. II), with and without mass loss (Dearborn, Schramm and Steigman 1985, DSS). In Section 2 the physics of the survival of ^{3}He in stars is outlined and our (DSS) results summarized. The trends in ^{3}He survival as a function of the stellar mass, composition and mass loss will be described. Then, in Section 3, we use our results to estimate the average survival of ^{3}He integrated over a stellar generation, assuming an initial mass function. Finally, we use the observed abundances of ^{3}He and D (Boesgaard and Steigman 1985) to estimate the primordial abundances and to derive bounds to the nucleon-to-photon ratio $\eta = N/\gamma$.

2. SURVIVAL OF HELIUM-3

In stars, ^{3}He is produced by the burning of deuterium: $D(p,\gamma)^3$He; this reaction, which has a low Coulomb barrier, is rapid at temperatures in excess of 6×10^5K. The destruction of ^{3}He: ^{3}He(^{3}He, $2p$)^{4}He or ^{3}He(α,γ)^{7}Be, is inhibited by a more significant Coulomb barrier; temperatures in excess of 7×10^6K are required. For hydrogen burning stars, "new" ^{3}He is produced, independent of the prestellar abundances of D and ^{3}He. In the cooler, outer layers, ^{3}He will survive (Iben 1967; Rood 1972). Indeed, stars of $M \lesssim 2M_\odot$ are <u>net producers</u> of ^{3}He (Iben 1967; Rood 1972). In more massive stars, however, sufficiently high temperatures are achieved - throughout most (but not all) of the star-to burn ^{3}He to ^{4}He and beyond. We have (DSS), therefore, concerned ourselves with the destruction of ^{3}He in the more massive stars ($M \geq 8M_\odot$). These results are summarized below. Later, we will supplement our work with that of Iben and Truran (1978), to attempt to account for $M \leq 8M_\odot$.

2.1 Trends In The Survival of ^{3}He

The fraction of ^{3}He which survives stellar processing has been calculated (DSS) for models with Pop. I abundances (Y=0.28, Z=0.02) and Pop. II abundances ($0.22 \leq Y \leq 0.30$, $0.0004 \leq Z \leq 0.004$), with mass loss (Dearborn et al. 1978) and without. For each model we computed the fraction of ^{3}He which survives in the material returned from stars, g_3, for $M/M_\odot = 8, 15, 25, 50, 100$. A selection (the essence) of our results are shown in Figure 1 where the results for g_3 versus M are displayed for three models.

The trends in g_3 illustrated in Figure 1 are easy to understand. Consider the results for the model with mass loss (Dearborn et al. 1978). Mass loss removes the outer - cooler - layers of the star before the ^{3}He can be destroyed. As a result, more ^{3}He survives compared with the model of the same composition but with no mass loss ($g_3(\dot{M}\neq 0) \geq g_3(\dot{M}=0)$). This effect is more pronounced for the more massive stars ($M \gtrsim 25M_\odot$); for less massive stars the nuclear burning timescale is comparable to, or less than, the mass loss timescale. The most massive stars ($M \gtrsim 50M_\odot$) lose mass so rapidly that, roughly independent of their initial masses, they evolve as if $M \simeq 50M_\odot$. Thus, for the mass loss rates of Dearborn et al (1978), for $M \geq 50M_\odot$, $g_3(\dot{M}\neq 0) \simeq g_3(M=50M_\odot; \dot{M}=0)$. As a result, for Pop. I abundances, $g_3 \geq 0.23$.

The dependence of g_3 on chemical composition is also consistent with expectation. For example, DSS find that for Z fixed, if Y decreases, then g_3 increases. This is because such models have a lower core temperature, a lower central luminosity and a smaller convective zone; more of the star is sufficiently cool for ^{3}He to survive. In contrast, if for fixed Y, Z decreases, less ^{3}He survives. This is because such models have higher core temperatures, higher luminosity and larger convective zones; more of the star is hot enough to destroy ^{3}He.

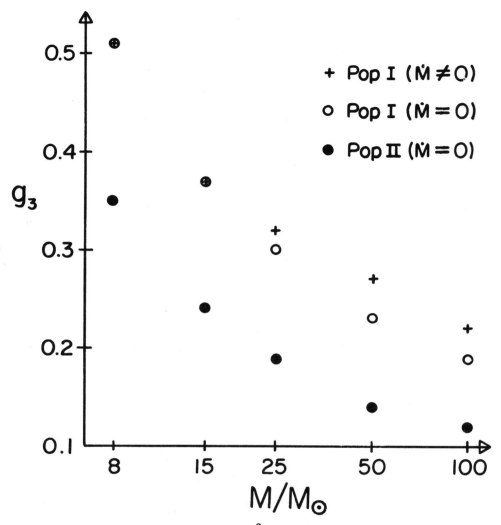

Figure 1. The survival fraction of ^{3}He, g_3, is shown as a function of stellar mass for three models. The model indicated by (+) has Pop. I abundance (Y=0.28, Z=0.02) and a mass loss rate from Dearborn et al. (1978). The open circles describe the same Pop. I model without mass loss. The filled circles are for a non-mass losing, Pop. II (Y=0.25, Z=0.0004) model.

3. THE SURVIVAL OF HELIUM-3 IN A STELLAR GENERATION

If the distribution of masses in a stellar generation is described by an initial mass function f(M), then the average survival of ^{3}He — integrated over the IMF — is

$$<g_3> = [\int f(M)dM]^{-1} \int g_3(M)f(M)dM. \quad (1)$$

For a Salpeter (1955) mass function,

$$f(M) = (M-\mu)/M^{2.35}, \quad (2)$$

where $\mu \simeq 1.0$–1.4 is the remnant mass. If we restrict our attention to the high mass stars ($M \gtrsim 8M_\odot$), then we may use the results (DSS) summarized in Figure 1 to derive $<g_3>_8$,

Pop. I, $\dot{M} \neq 0$: $<g_3>_8 = 0.34 \pm 0.03$, (3a)
Pop. I, $\dot{M} = 0$: $<g_3>_8 = 0.33 \pm 0.05$, (3b)
Pop. II, $\dot{M} = 0$: $<g_3>_8 = 0.22 \pm 0.03$. (3c)

The "uncertainties" in (3) are due to the range in remnant masses and the interpolation between masses in our models. As expected, the greatest survival for ^{3}He is for the massing losing, Pop. I model. However, since the mass loss is relatively unimportant for the lower mass stars, which dominate the IMF, the difference between the two Pop. I models is negligible. The "hotter" Pop. II models destroy more ^{3}He. Since the Pop. II result (3c) is for a model without mass loss, this should provide a firm <u>lower</u> bound to $<g_3>_8$.

Clearly, in integrating over a generation of stars, the lower mass stars are crucial. Iben and Truran (1978) have considered the survival of ^{3}He in stars of 3–8$M_\odot$. For this range they find,

$$g_3(M) = 0.7 + \frac{18}{7}(10^4 y_{23i})^{-1}(\frac{M_\odot}{M})^2, \quad (4)$$

where $y_{23i} = [(D + {}^3He)/H]_i$ is the sum of the initial abundances of D and ^{3}He. Using (2) and (4) in (1) – with $\mu=1$ – we find,

$$<g_3>_3 = 0.26 + 0.63 <g_3>_8 + 0.05 \, (10^4 y_{23i})^{-1}. \quad (5)$$

Utilizing our previous results, we now find,

Pop. I : $<g_3>_3 = 0.47$, (6a)

Pop. II: $<g_3>_3 = 0.40$. (6b)

In an attempt to extend our results down to the lowest mass stars which could have evolved in the lifetime of the galaxy, we assumed for $0.8 < M/M_\odot < 3$, that $g_3=1$; for such stars we took as an estimate of the remnant mass, $\mu=0.7$. With these assumptions, we find,

$$<g_3>_{0.8} = 0.49 + 0.43 <g_3>_8, \quad (7a)$$

Pop. I : $<g_3>_{0.8} = 0.63$ (7b)

Pop. II: $<g_3>_{0.8} = 0.59$ (7c)

Since low mass stars are likely net producers of ^{3}He, these estimates may well be too conservative (i.e.: g_3 may even exceed ~0.6).

4. CONSTRAINTS FROM BOUNDS TO THE PRIMORDIAL ABUNDANCES OF D AND HELIUM-3

Observations of ^{3}He in the interstellar gas (via the hyperfine line of ^{3}He$^+$) are very difficult and the few results obtained to date (Rood, Bania and Wilson 1984) are confused. According to Rood, Bania and Wilson (1984), interstellar abundances range from solar (or, possibly, less) to an order of magnitude higher. Indeed, it should be remembered that Rood, Steigman and Tinsley (1976) noted that, due to net production in low mass stars, the interstellar abundance of ^{3}He should exceed the solar system abundance. However, to build a bridge back to pregalactic abundances, it seems safest to start with the solar system abundances (Boesgaard and Steigman 1985).

$$y_{23\odot} = \left(\frac{D+^3He}{H}\right)_\odot = 3.6 \times 10^{-5}, \tag{8a}$$

$$y_{3\odot} = (^3He/H)_\odot = 1.4 \times 10^{-5}. \tag{8b}$$

If, instead of abundance by number, we wish to use the mass fraction, then for $Y_\odot = 0.25$ we have,

$$X_{23\odot} = 6.0 \times 10^{-5}, \quad X_{2\odot} = 3.1 \times 10^{-5}. \tag{9}$$

Yang et al (1984) were the first to note that if g_3 could be estimated, the primordial abundances of D and ^{3}He could be constrained. Using a "one-cycle" approximation, they found an upper bound to the primordial abundance of D plus ^{3}He,

$$y_{23P} < y_{23\odot} + (g_3^{-1} - 1)y_{3\odot}. \tag{10}$$

Using the abundances in (8) in (10), we find,

$$10^5 y_{23P} < 2.4 + 1.8 g_3^{-1}. \tag{10'}$$

If $g_3 > 0.2$ (see (3c)), then $10^5 y_{23P} \lesssim 11$. According to Yang et al (1984), this constraint on y_{23P} provides a lower bound to the nucleon-to-proton ratio: $\eta \geq 3 \times 10^{-10}$. For $\eta \geq 3 \times 10^{-10}$ and 3 flavors of light (<<1MeV), 2-component neutrinos, the primordial abundance of ^{4}He exceeds $Y_P = 0.240$ (for a neutron half-life of 10.4 min.; Yang et al. (1984)). If, indeed, $g_3 > 0.4(0.6)$, then $10^5 y_{23P} \leq 6.9(5.4)$ and, $10^{10} \eta \geq 4(5)$. In these cases, the primordial mass fraction of ^{4}He is predicted to be even larger: $Y_P \geq 0.243(0.246)$. If these estimates are reliable, there is less and less room for an "extra" neutrino flavor (provided that $Y_p \leq 0.25$).

ACKNOWLEDGMENTS

We have profited from valuable discussions with Jim Truran and Bob Rood. This work is supported at Bartol by DOE grant DE-AC02-78ER-05007 at Chicago by DE-AC02-80ER-10773 A004 and by the DOE at Lawrence Livermore Lab.

REFERENCES

Boesgaard, A. M. and Steigman, G. 1985, Ann. Rev. Astron. Astrophys. 23, 319.
Dearborn, D. S. P., Blake, J. B.., Hainebach, K. C. and Schramm, D. N. 1978, Ap. J. 223, 552.
Dearborn, D. S. P., Schramm, D. N. and Steigman, G. 1985, Ap. J. (Submitted, May 1985).
Iben, I. 1967, Ap. J. 147, 624.
Iben, I. and Truran, J. W. 1972, Ap. J. 220, 980.
Rood, R. T. 1972, Ap. J. 177, 681.
Rood, R. T., Steigman, G. and Tinsley, B. M. 1976, Ap. J. (Lett.) 207, L57.
Rood, R. T., Bania, T. M. and Wilson, T. L. 1984, Ap. J. 280, 629.
Salpeter, E. E. 1955, Ap. J. 121, 161.
Yang, J., Turner, M. S., Steigman, G., Schramm, D. N. and Olive, K. A. 1984, Ap. J. 281, 493.

HOW DEGENERATE CAN WE BE?

Gary Steigman
Bartol Research Foundation, University of Delaware
Newark, DE 19716, USA

ABSTRACT. Since relic neutrinos are virtually unobservable, a large – but undetectable – lepton asymmetry is possible. Cosmological constraints can provide bounds to such a universal neutrino degeneracy. The weakest constraint is from considerations of the age and density of the present Universe. More significant constraints follow from the formation of structure in the expanding Universe. The most stringent bounds to neutrino degeneracy follow from comparisons of the predictions of Big Bang Nucleosynthesis with the observed abundances of the light elements. All these approaches are reviewed and constaints to neutrino degeneracy derived. It is shown that neutrino degeneracy will <u>not</u> permit the Universe to be "closed" by nucleons.

1. INTRODUCTION

The Universe is not symmetric between particles and antiparticles (Steigman 1976). There are, for example, virtually no antibaryons; there is a Universal Baryon Asymmetry. To quantify the Baryon Asymmetry, we may compare the baryon number (the difference between baryons and antibaryons) in a comoving volume to the number of photons in the same volume.

$$B = (N_B - N_{\bar{B}})/N_\gamma \approx n_B/n_\gamma \equiv \eta. \qquad (1)$$

The nucleon-to-photon ratio, η, is known from Big Bang Nucleosynthesis to be very small: $\eta \lesssim 10^{-9}$ (Yang et al. 1984). The Baryon Asymmetry, therefore, is very small.

What of the Lepton Asymmetry? There are two types of leptons: charged leptons ($l = e^{\pm}, \mu^{\pm}, \tau^{\pm}, \ldots$) and neutral leptons ($\nu \equiv \nu_e, \nu_\mu, \nu_\tau, \ldots$). The lepton asymmetry in charged leptons is easily constrained. Electrons and positrons are the only stable charged leptons; $e^{\pm}$s are the only relics from the Big Bang which could be present today. But, during the early evolution of the Universe electron-positron pairs would have annihilated when the temperature dropped below the electron mass. There were, however, an excess of electrons which survived in an abundance which insures the charge neutrality of the Universe: $n_e = n_p \approx n_B$. The total lepton number (defined similarly to the baryon number above) is then that due to neutral leptons plus the contribution from electrons.

$$L = L_\nu + L_l \approx L_\nu + O(B) \qquad (2)$$

In most Grand Unified Theories (GUTS) it is "natural" for L to be of order B (Dimopoulos and Feinberg 1979; Schramm and Steigman 1979; Nanopoulos, Sutherland and Yildiz 1980; Turner 1981). In this case, $L \approx O(B)$ and $L_\nu \approx O(B) \approx \eta \lesssim 10^{-9}$ so that any neutrino degeneracy would be very small. In this case, for each neutrino flavor, the ratio of relic neutrinos and antineutrinos to relic photons is the "usual" value (see, for example, Steigman 1979).

$$n_\nu = (n_\nu + n_{\bar\nu})/n_\gamma = 3/4 (T_\nu/T_\gamma)_0^3 = 3/11. \qquad (3)$$

However, as Harvey and Kolb (1981) have shown, it is possible to find GUTS where it can be arranged that $|L| \gg 1$ while $B \ll 1$. For such models, $|L| \approx |L_\nu|$ and $n_\nu \gg 1$.

Since we are unable to observe the relic neutrinos and determine the asymmetry directly, it is of value to look elsewhere for constraints to L_ν. To this end, the present Universe and its evolutionary history provide an ideal laboratory. If the relic neutrinos are "too" degenerate they will dominate the present universal mass density (Weinberg 1962). Massless or massive degenerate neutrinos may influence the evolution of structure in the Universe (Freese, Kolb and Turner 1983). Most significantly, neutrino degeneracy will affect the abundances of the light elements produced during Big Bang Nucleosynthesis (Wagoner, Fowler and Hoyle 1967; Beaudet and Goret 1976; Yahil and Beaudet 1976; Beaudet and Yahil 1977; David and Reeves 1980; Fry and Hogan 1982; Scherrer 1983).

The above cited, cosmological approaches to constraining a universal relic neutrino degeneracy will be reviewed and the best – most constraining – bounds will be derived. In Section 2 some useful formulae are presented. The age/density constraint is discussed in Section 3. Constraints from the evolution of structure in the expanding Universe are derived in Section 4. In Section 5 the most restrictive bounds to neutrino degeneracy are derived from Big Bang Nucleosynthesis (BBN). In particular, in Section 5, it is shown that neutrino degeneracy does not permit the Universe to be "closed" by nucleons. The results are summarized in Section 6.

2. SOME USEFUL FORMULAE

As the Universe expands and cools, the ratio of the neutrino chemical potential (μ_ν) to the thermal energy (T_ν) remains constant. This ratio, $X = \mu_\nu/T_\nu$, the degeneracy parameter, provides a measure of the degeneracy. In terms of X, the neutrino lepton number is,

$$L_\nu(X) = \frac{\pi^2}{12\zeta(3)}(T_\nu/T_\gamma)^3 \, X[1+(\frac{X}{\pi})^2]. \tag{4}$$

The energy density in massless or extremely relativistic (ER) degenerate neutrinos may be compared to that of the relic photons.

$$\rho_{ER}(X)/\rho_\gamma = 7/8(T_\nu/T_\gamma)^4 N_\nu(X), \tag{5a}$$

$$N_\nu(X) = 1 + 30/7(X/\pi)^2 + 15/7(X/\pi)^4. \tag{5b}$$

$N_\nu(X)$ is the "effective" number of equivalent, light neutrinos.

For massive, nonrelativistic (NR) degenerate neutrinos, the energy density is,

$$\rho_{NR}(X) = m_\nu n_\nu(X) n_\gamma, \tag{6}$$

where $n_\nu(X)$ is the ratio of the <u>sum</u> of neutrinos and antineutrinos to photons; $n_\nu(X) \geq n_\nu(0) = 3/11$. The case of massive degenerate neutrinos differs significantly from that of massless neutrinos. Langacker, Segre and Soni (1983) have shown that for a hierarchy of masses ($m_1 > m_2 > \ldots$), an initial asymmetry would be reduced to negligible levels for the heavier neutrinos ($X_1 < X_2 < \ldots$). They also show that the asymmetry of the lightest neutrino species would be reduced to a fixed point of order unity: $X_{MAX} = X(m_{MIN}) \approx O(1)$. For massive neutrinos, then, it is to be expected that $X_{NR} \leq O(1)$; a priori there are no such constraints for X_{ER}.

3. THE AGE/DENSITY CONSTRAINT

Neutrino degeneracy means that "extra" neutrinos (or antineutrinos) are present compared with the nondegenerate case. Such extra neutrinos will increase the neutrino contribution to the total energy density. Constraints to neutrino degeneracy may be found, therefore, by requiring that the degenerate neutrinos not contribute "too much" density (Weinberg 1962).

Since "high" density universes expand faster, they are - at present (when $T_{\gamma 0} \approx 3K$) - younger than low density universes. As a result, constraints to the present age of the Universe provide the best bounds to the present density. If we write the present density (ρ_0) in terms of the "critical" density ($\rho_c = 3H_0^2/8\pi G \approx 10^4 h_0^2$ eVcm^{-3}; $H_0 = 100 h_0$ kms^{-1}Mpc^{-1}) and the density parameter ($\Omega_0 = \rho_0/\rho_c$), then the age of the Universe may be related to the Hubble age ($H_0^{-1} = 9.8 h_0^{-1}$ Gyr),

$$t_o = H_o^{-1} f(\Omega_o) \; ; \; f(\Omega_o) < 1. \tag{7}$$

For ER neutrinos, if $\Omega_\nu = \Omega_o$, $f = (1 + \Omega_o^{1/2})^{-1}$. A lower limit to t_o leads to an upper limit on Ω_o and, hence, to $\rho_{ER}(X)$.

$$t_o \geq 13 \text{ Gyr} : N_\nu(X) \leq 1.2 \times 10^4, \; X_{ER} \leq 27, \tag{8a}$$

$$t_o \geq 10 \text{ Gyr} : N_\nu(X) \leq 4.6 \times 10^4, \; X_{ER} \leq 38. \tag{8b}$$

Thus, a substantial degeneracy could still exist among relic ER neutrinos without violating the age/density constraint.

The density contributed by NR neutrinos depends on two parameters: the mass (m_ν) and the neutrino-to-photon ratio ($n_\nu(X)$); the age/density constraint can only bound the product of the two.

$$t_o \geq 13 \text{ Gyr} : m_\nu n_\nu(X) \leq 6 \text{eV}, \tag{9a}$$

$$t_o \geq 10 \text{ Gyr} : m_\nu n_\nu(X) \leq 10 \text{eV}. \tag{9b}$$

To further constrain neutrino degeneracy we must turn to considerations of galaxy formation.

4. GALAXY FORMATION AND NEUTRINO DEGENERACY

Structure can only evolve in an expanding Universe during those epochs when the Universe is "Matter Dominated" (MD) - that is, when the density is dominated by NR particles (Meszaros 1974). If the present Universe is MD, the early universe was dominated by ER, degenerate neutrinos for epochs whose redshift exceeds z_{eq} where $1 + z_{eq} = \rho_o/\rho_{ER}(X)$. So that structure could have evolved from perturbations whose amplitude was sufficiently small to be consistent with the absence of anisotropies in the microwave radiation background, we must require (Steigman and Turner 1985) that $z_{eq} \geq 10^3$. Using t_o to constrain ρ_o, we find:

$$t_o \geq 13 \text{ Gyr} : N_\nu(X) \leq 44 , X_{ER} \leq 6, \qquad (10a)$$

$$t_o \geq 10 \text{ Gyr} : N_\nu(X) \leq 70 , X_{ER} \leq 7. \qquad (10b)$$

The requirement of a sufficiently long epoch of MD results in a significantly tighter bound on the degeneracy of ER neutrinos.

The situation is even more tightly constrained for NR degenerate neutrinos. "Ordinary" massive, nondegenerate neutrinos will damp existing perturbations by free-streaming and mixing up regions of high and low density (Bond, Efstathiou and Silk 1980). For nondegenerate NR neutrinos, the free-streaming scale is,

$$\lambda_\nu \text{ (Mpc)} \simeq 12(100\text{eV}/m_\nu). \qquad (11)$$

If we utilize the most conservative (least constraining) bound from the age of the Universe (eq. 9b) then,

$$\lambda_\nu \text{ (Mpc)} \gtrsim 120\ n_\nu(X). \qquad (12)$$

For $X=0$, $n_\nu = 3/11$ and $\lambda_\nu > 33$Mpc. Even for nondegenerate, massive neutrinos the free streaming scale is much too large (White, Frenk and Davis 1983). If the neutrino contribution to the present mass density is kept fixed, degenerate neutrinos are lighter than nondegenerate neutrinos and, therefore, they free-stream further, damping perturbations on even larger scales. If, indeed, relic neutrinos are degenerate, they must either be massless or, be so light as to be relativisitic even at present.

5. DEGENERATE NEUTRINOS AND BIG BANG NUCLEOSYNTHESIS

Nucleosynthesis in the context of the standard, hot Big Bang Model with three flavors of nondegenerate, light neutrinos yields primordial abundances of the light elements (D, ^{3}He, ^{4}He, ^{7}Li) in excellent agreement with current observational data (Yang et al. 1984). A lower bound to the nucleon-to-photon ratio (η) is provided by the observed abundances of D and ^{3}He (Yang et al. 1984): $\eta \geq \eta_{MIN} \approx 3\text{-}4 \times 10^{-10}$. From the observed abundances of D, ^{4}He and ^{7}Li an upper bound to η may be derived: $\eta \leq \eta_{MAX} \approx 7\text{-}10 \times 10^{-10}$. If, in contrast with the "standard" model, the neutrinos are degenerate, the predicted abundances will change and the current consistency may be destroyed. If the observed abundances are to be accounted for, neutrino degeneracy must be bounded.

5.1 Non-e Neutrinos

The effect of degeneracy in μ-, τ-, ... neutrinos is to increase the total energy density at the time of nucleosynthesis. Degeneracy in these neutrinos increases the expansion rate; the Universe cools to a fixed temperature earlier: $t \to t' = S^{-1}t$. The "speed-up" factor depends on the degeneracy.

$$S(X) = [1+\tfrac{7}{43}\Delta N_\nu(X)]^{1/2}, \qquad (13)$$

Where $\Delta N_\nu(X) = N_\nu(X)-1$ (see eq. 5). For X different from zero (and fixed η), the faster expansion results in increased abundances of D, ^{3}He and ^{4}He. To preserve consistency with the observed abundances requires η_{MIN} to increase (to reduce D and ^{3}He) and η_{MAX} to decrease (to reduce ^{4}He). There will be a critical value, X_c, such that for $|X| > X_c$, $\eta_{MIN} > \eta_{MAX}$ and NO consistency remains. If we adopt (Boesgaard and Steigman 1985) D/H $\leq 10^{-4}$ and $Y_p \leq 0.25$, then $X_c = 1.4$.

5.2 e-Neutrinos

For degenerate e-neutrinos, there are two effects. As for the non-e-neutrinos, the universe will expand more rapidly ($t \to t' = S^{-1}t$; $S = S(X_e)$). A much more important effect of e-neutrino degeneracy is to influence the neutron-to-proton ratio. In equilibrium,

$$\frac{n}{p} = \exp[-\Delta m/T - X_e]. \tag{14}$$

For $X_e < 0$, there is an excess of antineutrinos over neutrinos which causes the equilibrium abundance of neutrons to increase. The higher neutron abundance results in the production of larger amounts of all the light elements. To preserve the consistency with the observed abundances of D and ^{4}He requires: $X_e \geq -0.05$.

In contrast, for $X_e > 0$, the neutrino excess drives the neutron abundance down. The dearth of neutrons inhibits nucleosynthesis, decreasing the abundances of the light elements. If we insist that the primordial abundance of deuterium exceed that observed today, $D/H \geq 10^{-5}$ (Boesgaard and Steigman 1985), and that $Y_p \geq 0.23$, then we require $X_e \leq 0.10$.

5.3 Neutrino Degeneracy And The Nucleon Density

The present density of nucleons may be expressed in terms of the nucleon-to-photon ratio and the present density of relic photons ($n_N = \eta n_\gamma$). For a relic photon temperature $T_{\gamma 0} \leq 3K$ and a Hubble parameter $H_0 \geq 50$ kms^{-1} Mpc^{-1}, the nucleon density parameter (the ratio of the nucleon density, ρ_N, to the critical density ρ_c) is bounded by,

$$\Omega_N \leq 0.019 \eta_{10}, \quad \eta_{10} = 10^{10}\eta. \tag{15}$$

As we have already noted, consistency between the predictions of BBN and the observed abundances requires that $\eta_{10} \leq 7$–10 so that, $\Omega_N \leq 0.14$ – 0.19. A nucleon dominated Universe fails to be closed by a factor which exceeds ~ 5–7. Indeed, were $\Omega_N = 1$, then $\eta_{10} \geq 52(H_0/50)^2$. But, for $\eta_{10} \geq 50$, in the standard model, the primordial abundance of deuterium would be far too small ($D/H \leq 5 \times 10^{-8}$) and those of ^{4}He and ^{7}Li, far too large ($Y_p \geq 0.27$, $^7Li/H \geq 6 \times 10^{-9}$).

The question arises, though, whether in the presence of neutrino degeneracy, a large nucleon density might still yield – through BBN –

the light elements in their observed abundances (Yahil and Beaudet 1976; David and Reeves 1980). For $\Omega_N = 1$ ($n_{10} \gtrsim 50$), we must avoid the <u>overproduction</u> of ^{4}He and ^{7}Li. This can be achieved through an electron-neutrino degeneracy ($X_e > 0$) which will reduce the neutron abundance and, hence, the yields of the light elements. However, the yield of deuterium will be far too small. If the non-e-neutrinos are degenerate, the Universe will expand more rapidly and tend to leave behind more D (which hasn't been burned to ^{3}He, ^{4}He, etc.). With two (or more) flavors of degenerate neutrinos, BBN acquires two adjustable parameters: the e-neutrino degeneracy (X_e) and the overall speed-up factor (S). Through suitable fine tuning of η, X_e and S, it may be possible to recover the agreement between the primordial and observed abundances (David and Reeves 1980). There is, however, a serious problem which, heretofore, has gone unnoticed.

In the presence of degenerate, ER neutrinos, the Universe becomes MD <u>later</u>; there is less time available for structure to evolve. For $\Omega_N = 1$ and $S \geq 1$,

$$1 + z_{eq} \approx 150 n_{10}(0.3 + 1.4 S^2)^{-1} \qquad (16)$$

Since, to recover agreement with the light element abundances requires $S^2 \gg 1$, $z_{eq} \approx 10^2 n_{10}/S^2$. But, for structure to evolve, $z_{eq} \lesssim 10^3$, so that $S^2 \lesssim 0.1 n_{10}$.

For example, for $n_{10} \approx 50$, we see that $S^2 \lesssim 5$. But, for $n_{10} \approx 50$, agreement with the light element abundances would require $S^2 \gtrsim 50$; in this case, $z_{eq} \lesssim 10^2$. Indeed, even if we permit $H_0 = 38$ so that $\Omega_N = 1$ is achieved for $n_{10} = 30$, we see that $S^2 \lesssim 3$ compared to the value $S^2 \gtrsim 15$ required for BBN.

6. SUMMARY

Relic neutrinos are virtually impossible to observe directly and the only constraints to a universal neutrino degeneracy (asymmetry) follow from cosmological considerations.

Stable, massive neutrinos must be light ($m_\nu \lesssim 24\text{-}42\text{eV}$);

degenerate massive neutrinos even lighter. However, such light neutrinos have such large free-streaming scales that they damp perturbations on scales relevant for galaxy formation; this problem is worse for degenerate neutrinos. Thus, if relic neutrinos are degenerate, they will be massless or ER at present.

For massless or ER neutrinos, the weakest constraint on the degeneracy comes from the age/density constraint (Sec. 3): $X_{ER} \lesssim 27\text{-}38$. A much more restrictive bound comes from considerations of the growth of structure in the expanding Universe. The requirement that the Universe have been MD for at least a factor of 10^3 in expansion factor limits the degeneracy to $X_{ER} \lesssim 6\text{-}7$.

Any neutrino degeneracy with $|X| \approx O(1)$ will have serious effects on BBN. For non-e-neutrinos, the effect of their degeneracy is to speed-up the expansion of the Universe, leading to the synthesis of -in particular - more D and more ^{4}He. Consistency with observed abundances is retained only for: $|X| \lesssim 1.4$. Degenerate e-neutrinos (or antineutrinos) have a direct effect on the neutron abundance during BBN and, even a small degeneracy can have large effects on the resulting abundances of the light elements. Consistency with the observed abundances requires: $-0.05 \lesssim X_e \lesssim 0.10$.

Finally, we have seen that no combination of neutrino degeneracies (i.e., choices of X_e and S) will permit $\Omega_N = 1$ <u>and</u> $z_{eq} \gtrsim 10^3$. Galaxy formation and BBN conspire - even in the presence of neutrino degeneracy - to forbid a Universe closed by nucleons.

In answer to the question posed by the title: Not very!

ACKNOWLEDGMENTS

In the course of this work I have profitted from valuable discussions with many colleagues; particular thanks go to Hubert Reeves and Amos Yahil. The degenerate nucleosynthesis calculations described here were performed while I was a visitor at the University of Chicago and at Fermilab. I am grateful to Rocky Kolb, Bob Scherrer and Mike Turner for their assistance. This work is supported at Bartol by DOE grant DE-AC02-78ER-05007.

REFERENCES

Beaudet, G. and Goret, P. 1976, Astron. Astrophys. $\underline{49}$, 415.
Beaudet, G. and Yahil A. 1977, Ap. J. $\underline{218}$, 253.
Boesgaard, A. M. and Steigman, G. 1985, Ann. Rev. Astron. Astrophys. $\underline{23}$, 319.
Bond, J. R., Efstathiou, G. and Silk, J. 1980, Phys. Rev. Lett. $\underline{45}$, 1980.
David, Y. and Reeves, H. 1980, Phil. Trans. Roy. Soc. $\underline{A296}$, 415.
Dimopoulos, S. and Feinberg, G. 1979, Phys. Rev. $\underline{D20}$, 1283.
Freese, K., Kolb, E. W. and Turner, M. S. 1983, Phys. Rev. $\underline{D27}$, 1689.
Fry, J. N. and Hogan, C. J. 1982, Phys. Rev. Lett. $\underline{49}$, 1873.
Harvey, J. A. and Kolb, E. W. 1981, Phys. Rev. $\underline{D24}$, 2090.
Langacker, P., Segre, G. and Soni, S. 1982, Phys. Rev. $\underline{D26}$, 3425.
Meszaros, P. 1974, Astron. Astrophys. $\underline{37}$, 225.
Nanopoulos, D. V., Sutherland, D. and Yildiz, A. 1980, Lett. Nuovo Cim. $\underline{28}$, 205.
Scherrer, R. J. 1983, M.N.R.A.S. $\underline{205}$, 683.
Schramm, D. N. and Steigman, G. 1979, Phys. Lett. $\underline{B87}$, 14.
Steigman, G. 1976, Ann. Rev. Astron. Astrophys. $\underline{14}$, 339.
Steigman, G. 1979, Ann Rev. Nucl. Part. Phys. $\underline{29}$, 313.
Steigman, G. and Turner, M. S. 1985, Nucl. Phys. $\underline{B253}$, 375.
Turner, M. S. 1981, Phys. Lett. $\underline{B98}$, 145.
Wagoner, R. V., Fowler, W. A. and Hoyle, F. 1967, Ap. J. $\underline{148}$, 3.
Weinberg, S. 1962, Phys. Rev. $\underline{128}$, 1457.
White, S., Frenk, C. and Davis, M. 1983, Ap. J. (Lett.) $\underline{274}$, L1.
Yahil, A. and Beaudet, G. 1976, Ap. J. $\underline{206}$, 26.
Yang, J., Turner, M. S., Steigman, G., Schramm, D. N. and Olive, K. A. 1984, Ap. J. $\underline{281}$, 493.

EARLY-PHOTOPRODUCTION OF D AND ^{3}He AND PREGALACTIC
NUCLEOSYNTHESIS OF THE LIGHT ELEMENTS

Jean AUDOUZE[1,2], David LINDLEY[3] and Joseph SILK[4]

1   Institut d'Astrophysique du CNRS, Paris - France.
2   Laboratoire René Bernas, Orsay - France.
3   Theoretical Astrophysics, Fermilab, Batavia - USA.
4   Department of Astronomy, University of California, Berkeley
    USA.

ABSTRACT: Two unconventional scenarios concerning the early nucleosynthesis of the lightest elements (D, ^{3}He, ^{4}He, ^{7}Li) which attempt to alleviate the strong predictions of the Standard Big Bang nucleosynthesis are briefly reviewed here. The first one consists in a "Big Bang photosynthesis" of D and ^{3}He induced on ^{4}He and ^{7}Li by energetic photons coming from the decay of hypothetical massive neutrinos and gravitinos : in this case the present observed D abundance could be consistent with a closed universe or such that $\Omega=1$. The second one consists in a pregalactic formation of D, ^{3}He and ^{7}Li by the interaction of energetic ^{4}He nuclei produced by a first generation of massive stars with the primordial gas made only of pure hydrogen.

I. INTRODUCTION

This contribution presents two possible scenarios which have been designed to challenge the current "Canonical Big Bang nucleosynthesis" models. These models which are thoroughly described in this book (see eg the contributions of Reeves, Steigman and Delbourgo-Salvador et al.) lead to very stringent predictions on the present baryonic density (such that $\Omega_B \sim 0.01$), the overall dynamical evolution of the Universe (which should be expanding for ever) and the number of different families of leptons (which is 3 according to these models). Given the cosmological importance of such predictions which are not always considered in agreement with the current observed abundances (see eg Vidal-Madjar and Gry, 1984) and which are at variance with the present inflationary models which predict $\Omega=1$, it is very important to examine all the possible alternatives.
Another chapter of this book (Schaeffer et al.) shows that the existence of <u>stable</u> quark nuggets with atomic mass A around

$10^{16}-10^{18}$ constitutes one way to reconcile the present observed abundances of the very light elements with a large value of Ω. Here two other "non standard" scenarios are briefly reviewed. The first consists in the secondary effect on ^{4}He and ^{7}Li induced by energetic photons coming from the decay of gravitinos and/or massive neutrinos. The second one which consists in the possible effect of pregalactic cosmic rays on the primordial pure H gas has already been presented in this proceeding series (Audouze and Silk, 1983) and therefore will only be summarized here.

2. PHOTOSYNTHESIS OF D AND ^{3}He INDUCED BY MASSIVE NEUTRINO/GRAVITINO DECAY

In this scenario we assume the existence of unstable gravitinos and/or massive neutrinos which by decay produce photons energetic enough to cause the photofission of ^{4}He just after the outcome of the primordial nucleosynthesis. As suggested by Lindley (1979) and Hut and White (1984), one can imagine the existence of massive neutrinos ($m_\nu \sim 500$ MeV) with a lifetime $\tau_\nu \sim 10^4-10^5$ sec which could decay by releasing such energetic photons. This is also the case for the decaying gravitinos predicted by supersymmetric models (Fayet 1984). The mass of these still hypothetical particles range from 20 GeV to 1 TeV while their lifetime is $\tau_{grav} \sim 10^8 (100 \text{ GeV}/m_{grav})^{-3}$ sec.

The high energy photons which are produced by the decay of such particles not only induce photofission on ^{4}He (and ^{7}Li) but also are slowed down by the two following processes (i) production of $e^+ e^-$ pairs by interaction with the thermal photons above a critical photon energy E_γ such that : $E_\gamma kT = 1/50 \text{ MeV}^2$, (ii) Compton scattering or pair production in the presence of nuclei below this critical photon energy. Since the threshold for the ^{4}He photofission is about 20 MeV, the thermal temperature should be below 10^{-3} MeV. It is not until the temperature is lower than $2 \cdot 10^{-4}$ MeV that double photon pair production falls below threshold. When the temperature T of the Universe is such that $2 \cdot 10^{-4} < T < 10^{-3}$ MeV, the initial photons produce pairs which then undergo inverse Compton scattering and provide a spectrum of secondary photons. Some of these secondary photons which have an energy above the ^{4}He threshold contribute also to the photofission of the light elements. Therefore in evaluating the destruction of ^{4}He, ^{3}He and D, one should take into account this secondary production (Lindley, 1985).

The relative abundances of ^{4}He, ^{3}He and D are then given by a quite simple reaction network :

$$d\ N_4 = -\Sigma_4\ N_4\ dn_E/n_e$$

$$d\ N_3 = (-\Sigma_3\ N_3 + f_{43}\ \Sigma_4\ N_4)\ dn_E/n_e$$

$$d\ N_D = (-\Sigma_D\ N_D + f_{3D}\ \Sigma_3\ N_3 + f_{4D}\ \Sigma_4\ N_4)\ dn_E/n_e$$

where N_4, N_3 and N_D are the abundances (by number) of ^{4}He, ^{3}He and D, Σ_4 Σ_3 and Σ_D the corresponding destruction rates; the f are the branching ratios ; n_E is the number per unit volume of energetic decay electrons and positrons and n_e the thermal electron density.

In the case of gravitinos, since they do not probably decay until after the early nucleosynthesis the effects of their presence and decay on the abundances of the light elements might be important. If they exist, they should be as numerous as photons and since they are massive, they might easily dominate the density of the Universe during these early epochs, effect which should be avoided such as any distortion of the microwave background induced by decay products.

Nevertheless within these constraints, one could inquire whether gravitinos dominated nucleosynthesis followed by photoreactions induced by the decay of these particles can lead to light element abundances consistent with the observations. If the mass of the gravitinos is such that $m_{grav} > 50$ MeV, the Universe is dominated by non relativistic gravitinos when the n/p ratio freezes out. The expansion accelerates in such a way that the freeze out temperature is increased significantly such as the n/p ratio which is then close to 1. In this case after the primordial nucleosynthesis, there is nothing but ^{4}He (N_4 = n(He)/(A=4) = 0.25).

Figure 1 shows the partial transformation of ^{4}He to ^{3}He and D as a function of the gravitino lifetime : if the lifetime is too short nothing happens while if it is too long D and ^{3}He are also destroyed.

After having been dominated by the massive gravitinos, the Universe is again dominated by radiation after their decay. The calculations have been performed in the case where $n_B/n_\gamma \sim 5\ 10^{-10}$ at the time when gravitinos have significantly decayed. The results depend not only on the lifetime (and therefore the mass) of the gravitino but also on the electron energies : the 50 MeV electrons induce in fact more photofission than the 200 MeV ones. This is because the photoreaction cross sections have strong peaks at energies of some tens of MeV or less ; therefore 50 MeV electrons produce more inverse Compton photons than the 200 MeV electrons. For electron of higher energies (i.e. of 100 GeV), this trend is reversed and the number of photoreactions induced per

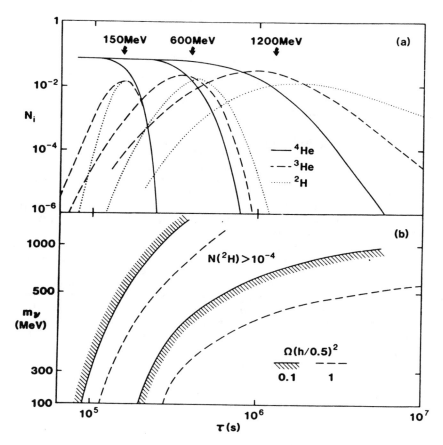

Figure 1   Effect of decaying gravitinos on light element abundances. Initial abundances were 0.25 ^{4}He by number (100% by mass) for the upper panel, and 0.06 (24% by mass) for the lower panel, with no ^{3}He or ^{2}H in either case. Massive gravitinos rapidly produce electron-positron pairs, which then cause photodissociation through inverse Compton photons. The two sets of curves are for 50 MeV and 200 MeV electrons. Abundances are plotted against gravitino lifetime.

gravitinos increase approximately in proportion to the mass of the gravitino (Lindley, 1985).
This increase is however compensated by the fact that in order to fix n_B/n_γ, the ratio n_{grav}/n_γ should decrease with the mass of the gravitino. Within the uncertainties of the calculation, the results are independent of the gravitino mass. In the results showed in Fig. 1, there is a regime where $n(^3He)>n(D)$ because the threshold of $^4He + \gamma \to {}^3He$ is lower than that of $^4He + \gamma \to D$. The n(D) abundance is higher than 10^{-5} within ranges of about a factor 5 in the gravitino lifetime.

Massive neutrinos can also be a source of energetic photons after they decay into energetic electrons. In this case, the number of neutrinos relative to that of photons falls off with their mass because the freeze out of neutrinos occurs at non relativistic temperatures, see e.g. Dicus et al. (1978). If neutrinos with mass > 1 MeV decay onto e^+ e^- pairs plus light neutrinos then the typical electron energy will be about 1/3 of the neutrinos mass. Figure 2a shows the resulting D, ^3He and ^4He abundances for three different electron energies. The results are quite similar to those obtained with gravitinos. In Figure 2b the lifetime of neutrinos such that $D/H \sim 10^{-4}$ has been plotted as a function of their mass (the results have been given for $n_B/n_\gamma = 5 \cdot 10^{-10}$ and $5 \cdot 10^{-9}$ respectively). The neutrino abundances decrease when their mass increase and then a wider range of lifetime is consistent with $D/H > 10^{-4}$.

If one of these two hypothesis (i.e. the existence of gravitinos or massive neutrinos) is established, primordial nucleosynthesis does not provide any more any cosmological constraint either on the baryon density (relative to that of photons) or on the number of neutrino families.

3. PREGALACTIC SYNTHESIS OF THE LIGHT ELEMENTS

As said in the introduction, we have already presented in these Moriond Rencontres (Audouze and Silk, 1983) the scenario by which one assumes that after the Big Bang the primordial gas is made only of hydrogen and then is enriched into the light elements D, ^3He, ^4He and ^7Li by the joint effect of contamination induced by very massive population III stars and of the spallation reactions triggered by pregalactic cosmic rays. This scenario is fairly contrived not by the energetic requirements but by the fact that in order to avoid any ^7Li overabundace relative to D, the energetic particles forming the pregalactic cosmic rays should have a miminim energy of at least 300 MeV/N.

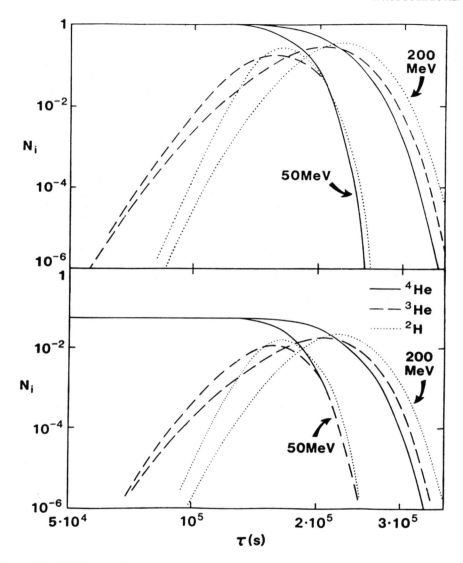

Figure 2 Effect of massive unstable neutrinos on light element abundances. The upper panel shows the resulting abundances for three different neutrino masses, as a function of lifetime, from an initial ^{4}He abundance of 0.07 by number (28% by mass). The lower panel shows the region in mass-lifetime parameter space in which an abundance of ^{2}H greater than 10^{-4} is synthesized.

4. CONCLUSION

The partial photofission of ^{4}He (and induced by the decay of gravitinos and/or massive neutrinos just after the early nucleosynthesis occuring in a closed (large baryonic density) universe can lead in principle to primordial abundances consistent with the observations. This type of scenario such as the pregalactic synthesis induced by cosmic rays on a pure hydrogen primordial gas constitute an alternative to the standard Big Bang nucleosynthesis and its cosmological consequences on the baryonic density and the number of neutrinos (and lepton families). Another alternative to this very important problem might also be found in this book proposed by Schaeffer et al. who analyse the effect induced by the existence of quark nuggets.

REFERENCES

1 Audouze J. and Silk J., 1983, in Formation and Evolution of galaxies and large structures in the Universe, Eds J. Audouze and J. Tran Thanh Van, NATO-ASI n° 117, Reidel Dordrecht p.267
2 Dicus, D.A., Kolb E.W. and Teplitz V.L., 1978, Ap.J., 221, 327.
3 Fayet P., 1984, in Proc. First ESO-CERN Symposium "Large Scale of the Universe, Cosmology and Fundamental Physics, Eds G. Setti and L. Van Hove, CERN Geneva p.35.
4 Hut P. and White S.D.M., 1984, Nature, 310, 637.
5 Lindley D., 1979, M.N.R.A.S., 188, 15 p.
6 Lindley D., 1985, Ap.J., to be published.

INFLUENCE OF QUARK NUGGETS ON PRIMORDIAL NUCLEOSYNTHESIS

R. Schaeffer[1], P. Delbourgo-Salvador[2], J. Audouze[2,3]

1   Service de Physique Théorique, CEN-Saclay,
    91191 Gif-sur-Yvette   cedex   France

2   Institut d'Astrophysique du CNRS, 98 bis Bd. Arago
    75014 Paris   France

3   Laboratoire René Bernas, 91405 Orsay   France

There are many indications that the baryonic content of the universe is rather low. The observation of the luminous matter in galaxies leads to a ratio Ω_b of the baryon density over the critical density of about 0.05. Similar values are obtained from the models of primordial nucleosynthesis that predict abundances of D, ^{3}He and ^{7}Li that favour unambiguously an open universe with values of Ω_b ranging from $5\ 10^{-3}$ to 0.1. On the other hand, the observation of the mass to light ratio at the cluster scales as well as the large scale velocity field of galaxies leads to $\Omega > 0.2$. To explain this discrepancy, the existence of dark matter has been postulated. From the results of nucleosynthesis calculations, one is unavoidably led to postulate that this dark matter must be non-baryonic, that it must have no strong interaction with baryons and nuclei. Although several hypothetical particles could be good candidates for this non-baryonic matter (massive neutrinos, "warm" or "cold" particles) none of these has yet been convincingly shown to exist.

It has been suggested[1] that small droplets, "nuggets", of quark matter could exist ad be stable or at least metastable with respect to their decay into ordinary nucleons or nuclei. These nuggets could be present in the universe and provide an alternate explanation for the dark matter. This possibility has many attractive features. It allows for an $\Omega = 1$ universe as is likely in many of the theories for the very early universe, with species of particles that all are of baryonic nature and can thus be transformed into one another. The ratio Ω_b/Ω of the baryon density over the total matter density is thus a natural outcome of the theory rather than a number that must be chosen a priori. Moreover, whereas the density fluctuations of baryons undergo Silk damping at the smaller scales, this is not the case[2] for quark nuggets due to their very low electric charge . Fluctuations at the galaxy scales are thus allowed and galaxies can be made easily. The question, however, must be raised whether the strong interactions of the quark nuggets will not spoil the agreement of calculated and observed abundances of the light elements since, as is well known, the early nucleosynthesis calculation produce very strong constraints on the various parameters that describe the universe.

The physical properties of the quark nuggets are quite uncertain[1,3] and several cases must be considered in a calculation of the early nucleosynthesis in presence of nuggets : the latter could be more stable than nuclei, metastable with a long lifetime or unstable with respect to their decay into baryons and nuclei. Nuggets will emit and absorb nucleons, according to the reactions

$$n + Q \rightleftarrows Q \qquad p + Q \rightleftarrows Q \quad . \qquad (1)$$

This changes their baryon number by one unit. We assume that subsequent reactions will change the number of nuggets and finally produce a population of nuggets with an average baryon number A equal to be most probable value. For the baryon absorption, we use the geometrical cross-section of the nugget plus a coulomb barrier for the protons. The baryon emission rate is treated in a way that is standard for the fission of ordinary nuclei[4]: it is proportional to the frequency of attempts for the emitted particle to go out of the nugget, multiplied by the relevant energy barrier factors[5]. The numerical calculations including the reactions (1) are performed using a computer code[6] with all standard nuclear reactions.

At $T \approx 100$ MeV where the calculation is started, the reactions (1) are in equilibrium and the relative abundance of baryons and nuggets is determined by the Saha equation. In the case of stable or metastable nuggets, the baryon emission is rapidly suppressed by the corresponding energy barriers as the temperature gets lower. Baryon absorption also gets suppressed, even in the case of stable nuggets : due to the high $n \rightleftarrows p$ and $n + p \rightleftarrows d$ reaction rates most neutrons are captured into nuclei that are prevented from being absorbed by the nuggets due to the Coulomb barrier. A simple model that reproduces the results of the more complete numerical calculations can be obtained as follows. The time evolution of the ratio Ω_b/Ω is given by

$$\frac{d}{dt}\frac{\Omega_b}{\Omega} = \frac{\Omega_{nug}}{\Omega}\lambda_{em} - \frac{\Omega_b}{\Omega}\lambda_{abs} \qquad (2)$$

with reaction rates

$$\lambda_{em} \sim \eta \frac{c}{R} \cdot \frac{1}{A} \sim n_Q A^{-4/3} \qquad (3)$$

$$\lambda_{abs} \sim \pi R^2 n_{nug} \langle v \rangle \sim \Omega_{nug} n_c A^{-1/3} \qquad (4)$$

here, r is the radius of a nugget, the factor $\eta c/R$ is the frequency of attempts baryon emission, and the factor $1/A$ is due to the fact that the emission rate is proportional to the number of nuggets that, for a given value of Ω_{nug}, goes as $1/A$. The factor n_Q in eq. 3 is the baryon number density <u>within a nugget</u> ($\sim 10^{39} \text{cm}^{-3}$), whereas n_c is the critical density of the universe ($\sim 10^{23} \text{cm}^{-3}$ at the time of nucleosynthesis). Due to this large difference of scale, the equilibrium (1) is strongly displaced in favour of the baryons even in the case of stable quark nuggets provided A is not too large. The value of A is given by the laws of physics. It is however quite uncertain and we consider it as a parameter. In case A

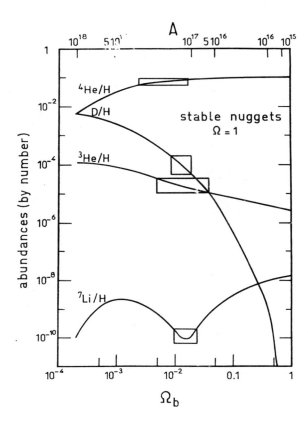

Figure 1 : Results of primordial nucleosynthesis in the case of stables nuggets with an Ω = 1 universe. The fraction of baryonic matter in the form of nucleons Ω_b (lower scale) is solely determined by the size A (upper scale) of the quark nuggets (eq. 4 to 6). The abundances by number of the light elements, which depend on the parameter A, are thus also a function of Ω_b. The boxes indicate the area allowed by observations for each of the species. All the observed abundances can be obtained simultaneously when our unique parameter A is chosen to be about 10^{17}. This leads to the prediction that $\Omega_b \sim 0.02$ and $\Omega_{nug} \sim 0.98$.

is smaller than $\sim 10^{16}$, only nucleons are present, whereas for $A > 10^{16}$ nucleosynthesis occurs in the presence of baryons only. The actual value of the ratio Ω_b/Ω can be obtained by asuming that every baryon emitted by a nugget becomes a proton or is embedded in a nucleus of D, ^{3}He or ^{7}Li:

$$\Omega_b/\Omega = 1 - e^{-X} \qquad (5)$$

with

$$X = \int_0^\infty \lambda_{em} dt = \left(\frac{A}{A_{tran}}\right)^{-4/3} \qquad (6)$$

where

$$A_{tran} = 1.6 \; 10^{16} \left(\frac{\varepsilon}{10 MeV}\right)^{-3/2} \eta^{3/4} \qquad (7)$$

is now a better estimate than the value (10^{16}) obtained from the ratio n_Q/n_C. The eq. 5 shows that the value of Ω_b/Ω - that is a result of the nucleosynthesis calculation - depends merely on the value of the parameter A. For a given value of Ω, the calculated abundances (Fig. 1) thus are a function of A, but can equivalently be plotted as a function of Ω_b. The observed values of the abundances correspond to $\Omega_b \sim 0.02$ and $A \sim 2 \; 10^{15}$ (for ε = 10MeV and $\eta \sim 0.1$).

The hypothesis that the universe is populated by stable or metastable quark nuggets thus is consistent with the observed abundances of light elements, despite the strong interactions of baryons and quark nuggets, provided the average baryon number of these nuggets is quite large (A $\sim 2 \; 10^{15}$). It leads to a universe with 98% of the matter in the form of nuggets and 2% in the form of baryons.

Astonishingly enough, unstable nuggets would not provide the observed rates[5], unless they decay at extremely late an epoch so as to insure that the decay products are not trapped into the galaxies (where the abundance observations are made).

References

1. E. Witten, Phys. Rev. D 30, 272 (1984)
2. R. Schaeffer, to be published
3. E. Farhi abd R.L. Jaffe, Phys. Rev. D 30, 2379 (1984)
4. D.D. Clayton, Prnciples of Stellar Evolution and Nucleosynthesis , Mc Graw Hill (1968)
5. R. Schaeffer, P. Delbourgo-Salvador, J. Audouze, to be published
6. P. Delbourgo-Salvador, C. Gry, G. Malinie and J. Audouze, Astron. Astroph. (1985), to be published

PRIMORDIAL NUCLEOSYNTHESIS WITH GENERIC PARTICLES

T.P. Walker, E.W. Kolb, and M.S. Turner
NASA/Fermilab Astrophysics Group
Fermi National Accelerator Laboratory
P.O. Box 500
Batavia, IL 60510
USA

ABSTRACT. We discuss a revision of the standard model for big bang nucleosynthesis which allows for the presence of generic particle species. The primordial production of ^{4}He and D + ^{3}He is calculated as a function of the mass, spin degrees of freedom, and spin statistics of the generic particle for masses in the range $10^{-2} \lesssim m/m_e \lesssim 10^2$. The particular case of the Gelmini and Roncadelli majoron model for massive neutrinos is discussed.

1. INTRODUCTION

The standard model of big-bang nucleosynthesis[1] has proved an invaluable tool for placing constraints on various aspects of particle physics models.[2] With this in mind, we have developed a method which treats the effects of a generic particle species on primordial nucleosynthesis. In section 2 we review the pertinent features of the standard model and continue in section 3 to discuss the alterations to the standard model which are necessary to include generic particles. Section 4 contains the results for some generic particles and section 5 presents a discussion of a specific model for massive neutrinos.

2. A REVIEW OF THE STANDARD MODEL

The physics of big-bang nucleosynthesis consists of two distinct phases.[3] During the first phase, characterized by T > 1 MeV (throughout we use units so that h = c = k = 1), the rate of weak interactions, $\Gamma_{wk} \propto G_F^2 T^5$, is larger than the expansion rate, $\Gamma_{exp} \propto (G_N \rho)^{1/2} \propto G_N^{1/2} T^2$, and thus light neutrinos maintain thermal equilibrium with photons via $e^-e^+ \leftrightarrow \bar{\nu}_i \nu_i$. Nucleons maintain chemical equilibrium through $p\nu_e \leftrightarrow ne$, $n \leftrightarrow pe\bar{\nu}_e$, and $pe \leftrightarrow n\nu_e$ interactions and the neutron-to-proton ratio follows an equilibrium value,

$$(\frac{n}{p}) = \exp(-\frac{Q}{T}) \; ; \; Q = m_n - m_p = 1.293 \text{ MeV}. \tag{1}$$

Neutrons and protons remain in equilibrium until weak interactions freeze out (i.e. $\Gamma_{wk} \approx \Gamma_{exp}$) at a temperature $T_F \approx 0.7$ MeV. From this point, (n/p) slowly decreases due to non-equilibrium neutron decay until the onset of the second phase: nucleosynthesis. This second phase, characterized by $T \lesssim 0.1$ MeV, involves various nuclear interactions which proceed until $\Gamma_{nuclear} \lesssim \Gamma_{exp}$. For $T \gtrsim 0.1$ MeV, the formation of deuterium via np $\rightarrow$ dγ is suppressed by photodissociation due to the relatively low binding energy of the deuteron (2.2 MeV). The relative abundance of deuterons can be expressed

$$\frac{n_d}{n_N} \propto \eta \exp(\frac{2.2 \text{MeV}}{T}), \tag{2}$$

where η is the nucleon-to-photon ratio. At $T \sim 0.1$ MeV, this bottleneck is circumvented and the various n,p,d reactions create D, ^{3}He, ^{3}H, ^{4}He, and ^{7}Li. The lack of stable mass 5 and 8 nuclei, Coulomb barriers, and the relatively large binding energy of ^{4}He result in most of the available neutrons being transformed into ^{4}He.

Primordial helium-4 production's dependence on these phases is described by three parameters: the nucleon-to-photon ratio η, the neutron half-life $\tau_{1/2}$, and the number of effective radiative degrees of freedom (usually parameterized as the number of light neutrino species N_ν). The amount of ^{4}He produced in the big-bang is dependent upon (n/p) at freeze out (Y_p = mass fraction of ^{4}He $\approx 2(n/p)_F/[1+(n/p)_F]$). The earlier weak interactions freeze out, the greater $(n/p)_F$ and the larger Y_p. Thus the increase in Y_p with increasing $N_\nu (\sim \Gamma_{exp}^{1/2})$ and $\tau_{1/2} (\propto \Gamma_{wk}^{-1/2})$. Y_p also depends upon the efficiency of nuclear interactions at converting neutrons into ^{4}He. As the number density of photons decreases, the deuterium bottleneck breaks at higher temperatures, thus resulting in the increase of Y_p with increasing η.

The primordial production of D and ^{3}He also depends on the efficiency of ^{4}He production. Nearly all D and ^{3}He are processed into ^{4}He and thus their abundance depends upon the nuclear interaction rate - expansion rate interplay. Higher nucleon abundances (η) mean faster D-^{3}He depletion while a faster expansion rate ($N^{1/2}$) leads to an earlier freeze out of nuclear interactions, when D-^{3}He abundances are larger. Therefore, D and ^{3}He decrease with increasing η or $N_\nu^{-1/2}$.

3. ALTERATIONS FOR GENERIC PARTICLES

We have altered the original computer code of Wagoner[4] to correctly treat the additional physics of a generic particle species. By generic particle, we mean a particle which maintains good thermal contact with either photons or light neutrinos throughout nucleosynthesis. As such, the ith generic particle can be described by a temperature T and has energy density and pressure

$$\rho_i(T) = \left[\frac{g_i}{2\pi^2}\int_{y_i}^{t\infty}\frac{\xi^2(\xi^2-y_i^2)^{1/2}d\xi}{e^{\xi+\theta_i}}\right]T^4 \quad (3)$$

$$P_i(T) = \left[\frac{g_i}{6\pi^2}\int_{y_i}^{t\infty}\frac{(\xi^2-y_i^2)^{3/2}d\xi}{e^{\xi+\theta_i}}\right]T^4 \quad (4)$$

where g_i is the number of spin states, (e.g. $g_e = 2$, $g_\nu = 1$, $g_\gamma = 2$, ...), $y_i = m_i/T$, and θ_i is $+(-)1$ for fermions (bosons) (in natural units $a = \pi^2/15$).

For $T \simeq 10$ MeV, the universe is dominated by electrons, neutrinos, photons, and the generic particles. The entropy in a volume $R^3(t)$ ($R(t)$ is the FRW scale factor[5]) can be decomposed into two parts:

$$S_\gamma = \frac{R^3}{T_\gamma}[2\times(\rho_e+P_e) + \rho_\gamma + P_\gamma + \sum_i{}^{(\gamma)}(\rho_i+P_i)] \quad (5)$$

and

$$S_\nu = \frac{R^3}{T_\nu}[2\times N_\nu(\rho_\nu + P_\nu) + \sum_j{}^{(\nu)}(\rho_\nu + P_\nu)]. \quad (6)$$

The factor 2 results from consideration of particles and anti-particles, and the sums are over generic species which maintain good thermal contact with photons or light neutrinos. Up until $T \sim 3$ MeV[6], light neutrinos and photons maintain thermal equilibrium via neutral and charged current weak interactions and are described by the same temperature. Until this decoupling we have

$$S = S_\gamma + S_\nu = \text{constant} \quad (T \geq T_{dec} \simeq 3 \text{ MeV}) \quad (7)$$

Below T_{dec}, the decoupled light neutrinos maintain a thermal distribution described by T_ν and we have

$$\begin{aligned}S_\gamma &= \text{constant} = S_\gamma(T_\gamma=T_\nu=T_{dec}) \\ S_\nu &= \text{constant} = S_\nu(T_\nu=T_\gamma=T_{dec})\end{aligned} \quad (T \leq T_{dec}) \quad (8)$$

Using the above formalism we can describe the dynamical evolution of the universe through primordial nucleosynthesis as a function of T_γ[7]. The scale factor, $R(T_\gamma)$ can be expressed in terms of combinations of integrals like those of equations (3) and (4), as can the neutrino temperature $T_\nu(T_\gamma)$. The expansion rate as a function of T_γ follows from the field equation

$$\left(\frac{\dot{R}}{R}\right)^2 = \Gamma_{exp}^2 = \frac{8\pi G}{3}\rho \quad (9)$$

or

$$\int dt = \sqrt{\frac{3}{8\pi G}} \int \frac{dR}{R(T_\gamma)\sqrt{\rho(T_\gamma)}}, \qquad (10)$$

The weak interaction physics is also a function of T_γ and T_ν through the phase space dependence of the initial and final states of the interactions $p\nu \leftrightarrow ne$, $pe \leftrightarrow n\nu$, and $n \leftrightarrow pe\nu$. We have supplemented Wagoner's code by numerically calculating the rates of these interactions, including radiative and Coulomb corrections[8], for given values of (T_γ, T_ν).

4. RESULTS FOR GENERIC PARTICLES

By assumption, a given generic particle species maintains good thermal contact through nucleosynthesis with either photons or light neutrinos and thus the physics of each generic species is completely specified by its mass, spin degrees of freedom, spin statistics, and annihilation mode. In figure 1, we show the ^{4}He (a) and D+^{3}He (b) yields for primordial nucleosynthesis with generic particles coupled either to photons or light neutrinos, having $g = 1, 2$, and in the mass range $10^{-2} \lesssim m/m_e \lesssim 10^2$. All calculations were done with $n_{10} = 3$, $\tau_{1/2} = 10.6$ min, and $N_\nu = 3.0$. Horizontal dashed lines indicate standard model yields with the indicated number of light neutrino species.

The plots of Y_p and D+^{3}He as a function of generic particle mass are most easily understood by considering three distinct regions characterized by m/m_e:

(I) $m/m_e \gtrsim 10^{1.75}$ ($m \gtrsim 30$ MeV)

(II) $1 \lesssim m/m_e \lesssim 10^{1.25}$ (0.5 MeV $\lesssim m \lesssim$ 10 MeV)

(III) $m/m_e \lesssim 10^{-1}$ ($m \lesssim 50$ keV)

These regions are indicated in figure 1(a) and 1(b).

In all the graphs, we recover the standard model results for $m \gtrsim 30$ MeV, since a particle of this mass will be present only in trivial abundances at the time of primordial nucleosynthesis. In addition, generic particles coupled to photons and having $m \lesssim 50$ keV overproduce ^{4}He and underproduce D+^{3}He because these particles act as "pseudo-light neutrinos" increasing ρ_{TOT}, and because they dump most of their entropy after nucleosynthesis, necessitating a larger value of η during the bottleneck phase.[9] Generic particles coupled to neutrinos do not alter η and thus these particles with masses below 50 keV act only as extra light neutrino degrees of freedom (the $g = 1$ particle is a boson and contributes slightly more to the energy density than its fermion counterpart).

We need to consider the energy density as a function of T_γ to understand the Y_p and D+^{3}He production for generic particles in the mass range 0.5 MeV - 10 MeV. The energy density for a universe containing a generic X particle which annihilates to photons can be written:

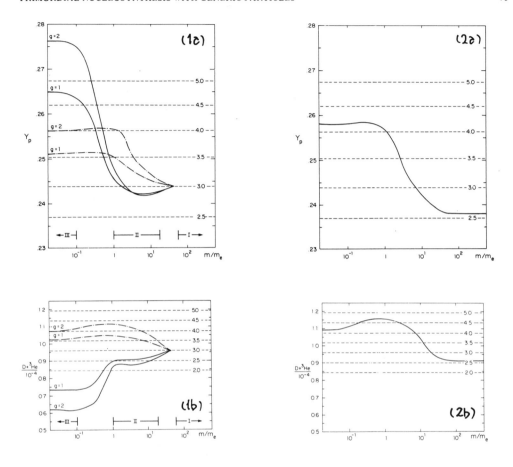

Figures 1(a) and 1(b) indicate primordial yields of ^{4}He and D+^{3}He, respectively, for generic particles coupled to photons (solid lines) or light neutrinos (dash-dot lines) as a function of particle mass. Spin degrees of freedom are labeled by g.

Figures 2(a) and 2(b) show primordial yields of ^{4}He and D+^{3}He, respectively, for the majoron model of Gelmini and Roncadelli. Here m represents the mass of the heavy neutrino and the light higgs.

In all figures, horizontal dashed lines represent standard model yields for the indicated number of light neutrino species.

$$\rho_{X \to \gamma's} = a_1 T_\gamma^4 + a_2 \left(\frac{T_\nu}{T_\gamma}\right)_{X \to \gamma's}^4 T_\gamma^4 + f(y_X^\gamma) T_\gamma^4, \tag{11}$$

where the first term accounts for electrons and photons, the second for light neutrinos, and the third for the X particles with $y_X^\gamma = m_X/T_\gamma$ and $f(y_X^\gamma)$ defined as T_γ^{-4} times equation (3). The energy density of the standard model is just

$$\rho_{STD} = a_1 T_\gamma^4 + a_2 \left(\frac{T_\nu}{T_\gamma}\right)_{STD}^4 T_\gamma^4 . \tag{12}$$

We see that for a given T_γ we can have $\rho_{X \to \gamma's} < \rho_{STD}$, even though we have an extra X particle, provided

$$a_2 \left(\frac{T_\nu}{T_\gamma}\right)_{X \to \gamma's}^4 + f(y_X^\gamma) < a_2 \left(\frac{T_\nu}{T_\gamma}\right)_{STD}^4 . \tag{13}$$

in this case, $(T_\nu/T_\gamma)_{X \to \gamma's} \leq (T_\nu/T_\gamma)_{STD}$, and thus inequality (13) holds provided

$$f(y_X^\gamma) < a_2 \left[\left(\frac{T_\nu}{T_\gamma}\right)_{STD}^4 - \left(\frac{T_\nu}{T_\gamma}\right)_{X \to \gamma's}^4 \right] . \tag{14}$$

It is possible for particles in the mass range 0.5 MeV - 10 MeV to satisfy this inequality and the resulting $\rho_{X \to \gamma's} < \rho_{STD}$ causes an underproduction of both ^{4}He and D+^{3}He.

In a similar way, it is possible to have a generic particle coupled to neutrinos which has a mass so that $\rho_{X \to \nu's} > (\rho_{STD}$ + energy density of the massless equivalent of the X). For these generic particles, we observe overproduction of ^{4}He and D+^{3}He relative to their extra light neutrino counterparts.

5. A SPECIFIC APPLICATION: HEAVY NEUTRINOS WITH MAJORONS

As a specific application of the generic particle approach, we have considered the effects of a heavy neutrino, like that described in majoron models[10], on primordial nucleosynthesis. In these models, neutrino masses are generated through the spontaneous breaking of a global B-L symmetry, and as such they contain a Goldstone boson, the majoron. The symmetry breaking is accomplished by expanding the Higgs sector to contain a triplet which couples to pairs of fermion doublet fields and acquires a non-zero vacuum expectation value (v.e.v.). The inclusion of the Higgs triplet introduces 6 massive scalars to the theory, 5 of which have masses on the order of the doublet v.e.v. (250 GeV) times a coupling constant. The remaining scalar picks up a mass similar to the heavy neutrino mass which in turn is proportional to the triplet v.e.v. The triplet v.e.v. is constrained by limits on majoron cooling of various astrophysical objects to be less than ~1 MeV.[11]

In majoron models, the dominant interaction is the majoron mediated annihilations of heavy to light neutrinos which has a cross section[12]

$$\sigma_{ann} = \frac{1}{32\pi} g_H^2 g_L^2 (\beta s)^{-1} , \qquad (15)$$

where $g_{H(L)}$ is the heavy (light) neutrino-majoron coupling, β the c.m. neutrino velocity, and s the total c.m. energy squared. It can be shown[13] that such interconversions are sufficiently fast to keep the heavy neutrino in equilibrium with light neutrinos until after nucleosynthesis.

Thus we can treat these heavy neutrinos as generic neutrinos coupled to light neutrinos in order to determine their effect on primordial nucleosynthesis. For simplicity we take light Higgs boson and heavy neutrino to have the same mass and take $N_\nu = 2$, $\eta_{10} = 3$, and $\tau_{1/2} = 10.6$ min. The results are shown in figures 2(a) and 2(b). Noting that neutrinos less massive than about 1 MeV overproduce ^{4}He, one might try to decrease η to bring $Y_p \simeq 25\%$, the accepted upper limit,[14] but we can't do this because that would make too much D+^{3}He.[15] Thus we conclude that majoron-type models with neutrino masses less than 1 MeV cannot reproduce observed primordial abundances. We also note that astrophysical limits on the triplet v.e.v. imply that $m_\nu \leq 1$ MeV and thus it appears that astrophysical/cosmological arguments can put severe constraints on the acceptable neutrino masses (i.e., $m_{\nu 6} \gtrsim 1$ MeV only) in majoron-type models, if not rule them out completely.

6. ACKNOWLEDGEMENTS

T.P.W. wishes to thank D. Seckel and G. Steigman for useful comments and discussions and the DOE and Indiana University Graduate Research and Development Fund for support.

REFERENCES

(1) Wagoner, R. V., Fowler, W. A., and Hoyle, F., Ap. J. **148**, 3 (1967).
(2) For a review see, Yang, J., Turner, M. S., Steigman, G., Schramm, D. N., and Olive, K. A., Ap. J. **281**, 493 (1984).
(3) For current reviews of the standard model see Steigman, G. and Boesgaard, A. M. in Ann. Rev. Astron. Astrophys. (1985); Yang, J., etal. see Ref. 2.
(4) Wagoner, R. V., Ap. J. **179**, 343 (1973).
(5) For a discussion of the physics of the "early" universe, see, Weinberg, S., Gravitation and Cosmology, New York: Wiley and Sons, Inc. (1972).
(6) The actual neutrino decoupling temperature is different for each species. See, Dicus, D. A., Kolb, E. W., Gleeson, A. M., Sudarshan, E. C. G., Teplitz, V. L., and Turner, M. S., Phys. Rev. **D26**, 2694 (1982).

(7) Details of this technique can be found in Kolb, E. W., Turner, M. S., and Walker, T. P. (in preparation).
(8) See Ref. 6.
(9) In reality, the effect which generic particles have on η during nucleosynthesis must vanish for masses $\lesssim$ 100's of eV, since photons dumped later than this cannot thermalize.
(10) Chikashige, Y., Mohapatra, R. N., and Peccei, R. D., Phys. Lett. **98B**, 265 (1981); Gelmini, G. B. and Roncadelli, M., Phys. Lett. **99B**, 411 (1981); Georgi, H. M., Glashow, S. L., and Nussinov, S., Nucl. Phys. **B193**, 297 (1981).
(11) Fukugita, M., Watamura, S., and Yoshimura, Phys. Rev. Lett. **48**, 1522 (1982); Glashow, S. L. and Manohar, A., Harvard Preprint HUTP-85/A031.
(12) Georgi, H. M., etal., see Ref. 9.
(13) See Ref. 7.
(14) Yang, J., etal., see Ref. 2.
(15) Ibid.
(16) Walker, T. P. (in preparation).

Dark Matter and Cosmological Nucleosynthesis

David N. Schramm
University of Chicago
and
Fermilab

ABSTRACT

It is shown that big bang nucleosynthesis requires at least some of the baryons in the universe must be dark. All current solutions seem to require at least two ad hoc assumptions. Of these more complex but complete solutions, the ability to produce large scale free structures where these dark baryons lie may be the best way to resolve the different proposed solutions to the several different dark matter problems.

This paper will first look at the various arguments regarding dark matter. In so doing, it will be emphasized that there are several cosmological dark matter problems[1,2], not all of which require exotic (non-baryonic) stuff. [This paper closely follows ref 1] In fact, it will be shown that some dark baryons are necessary, and how these dark baryons are distributed may eventually prove to be the way for resolving the nature of the bulk of the matter of the universe. In particular, we will emphasize the fact that no simple solution works which uses only an exotic particle with no additional assumptions.

It will then be shown that the additional constraint, that the large scale structure of the universe be scale-free[3,1] may severely constrain the allowable dark matter solutions even if we allow violations of Turner's law[4], that "the Tooth Fairy may only be invoked once in a scientific paper". Such a scale-free solution tells us that clusters are laid out in a fractal; since the fractal is close to dimension one, it may point in the direction of cosmic strings or other linear phenomena. The conclusion will be that the correct, "best", solutions may tell us something about phase transitions in the early universe. It will also be shown that the best way to resolve the various contradictory models may be to find the dark matter baryons which end up in different place in the different scenarios.

The Dark Matter Problems

As emphasized in ref. 1 and 2, there are at least three classes of dark matter problems:
1. Dynamics
2. Galaxy Formation
3. Inflation

The first is the well established[5] fact that the light emitting regions of the Universe have relatively low mass-to-light ratios so that if there were only that much mass in the Universe, the cosmological density parameter $\Omega \equiv \rho/\rho_{crit}$ would be ~ 0.007, a very open universe. However, if those same light emitting galaxies are observed interacting with more distant galaxies in binaries and small groups, then the implied mass of each galaxy is about a factor of ten larger and the implied Ω is ~ 0.07. this tells us that galaxies have massive halos which are dark. (As Schramm and Steigman[6] emphasize, it is the *light*, not the mass, which is missing.) If the dynamics of large clusters of galaxies are observed, then the implied mass per galaxy is even larger. If all galaxies have that much mass, then Ω is from 0.1 to ≤ 0.4. It is important to emphasize that *no system* implies an $\Omega > 0.4$.

To put these numbers in perspective, remember that Big Bang nucleosynthesis[7] puts an upper limit on the density of baryons in the universe of $\Omega_b < 0.15$ for a present background temperature of $T_0 \leq 2.8K$. Thus, within the uncertainties, all of the dark halo material could be baryonic, although the higher super-cluster masses may require something additional.

Hegyi and Olive[8] have shown that if the halo material is non-baryonic, then it must either be in blackholes (that were baryonic at the time of Big Bang nucleosynthesis) or low mass, astronomical objects (Jupiters or low mass, dim stars). This latter class would require a very stron peak at low mass in the initial mass function.

Freese and Schramm[9] have combined the nucleosynthesis arguments with age arguments to conclude that $\Omega_b \geq 0.03$. In particular, they show that the lower limit on globular cluster ages of 13×10^9 years implies that $h_0 \leq 0.7$. Therefore, the lower bound on h_b/n_γ from Yang et al.[7] yields a lower bound on Ω_b of 0.03. Thus, it is argued that *the bulk of the baryons must be dark*. As we will see, it is not clear whether these dark baryons make up the dark halos or whether they are distributed elsewhere. However, it should not be forgotten that the dynamical and the nucleosynthetic arguments are still consistent with $\Omega \sim 0.1$ and everything being baryonic[10].

The arguments for non-baryonic matter center on two additional dark matter arguments: galaxy formation and "inflation"; inflation being the most theoretical, it is perhaps the hardest to avoid unless we live in a special epoch in the history of the universe. As we will see, the galaxy formation arguments may have some intriguing loopholes.

The inflation argument comes from the proposal of Guth[11]. Although the detailed mechanism has been modified considerably (see review in ref. 12 and references therein), the basic paradigm is still there. Inflation is the name given to the process which sets the initial conditions of the universe by a rapid, de Sitter phase expansion. Such a phase will solve the horizon problem, the flatness problem, the monopole problem and it will produce fluctuations which might eventually yield galaxies. The detailed particle models which produce inflations consistent with observations still appear very ad hoc[12]. However, the flatness problem alone tells us that something like inflation must have occurred, or else we live in a very special epoch in the history of the universe. In particular, since Ω varies with time on a gravitational time scale, we can conclude that, since we still don't know whether Ω is significantly different from unity after $\sim 15 \times 10^9$ years, at the Planck time (10^{-43} seconds), Ω had to be fine tuned to be unity to ~ 60 decimal places. Inflation, or anything like it, solves this problem by making $\Omega = 1$ to a far greater accuracy and thus avoids forcing us to live in the special epoch where Ω first deviates from unity. Since $\Omega = 1$ is in excess of Ω_b, this tells us that the bulk of the matter of the universe is non-baryonic. Note also that since $\Omega_{clustered} < 0.4$, inflation tells us that the bulk of the matter of the universe is not clustered with the light emitting stuff.

The above sequence of arguments tended to favor massive ($\sim 10eV$) neutrinos as dark matter since they would be non-baryonic and would not strongly cluster with the light emitting stuff. Then the galaxy formation arguments came along and the shift from neutrinos to "cold" matter occurred. One point to be emphasized is: the galaxy formation arguments may have a significant loophole and the large scale structure arguments may force us to go through the loophole.

The galaxy formation arguments[13] center on the following points:
1. Quasars exist at redshifts[14] $z > 3$.
2. Anisotropies in the microwave background are small[15], $\delta T/T \lesssim 2 \times 10^{-5}$.
3. Baryonsynthesis requires primordial, adiabatic fluctuations[16].
4. Fluctuations, $\delta\rho/\rho$, grow linearly with the expansion until $\delta\rho/\rho \sim 1$ or until redshift $z \sim 1/\Omega$ for $\Omega < 1$.
5. Fluctuation growth does not start until matter domination.

If the universe were baryon dominated, then from arguments 2, 3, 4 and 5, $\delta\rho/\rho$ would only reach $\sim 2 \times 10^{-2}$ today since the background radiation decoupled at redshift $z \sim 3000$. But we know that $\delta\rho/\rho$ exceeds 1 today. Thus, we need something in addition to baryons. If the additional material were in the form of neutrinos or other particles which are relativistic, "hot", until just before galaxy formation, then condensed objects wouldn't form until $z \lesssim 1$, violating argument 1. The best candidate thus appears to be "cold" matter which is moving slowly and can begin clustering rapidly on galactic scales. It has been shown[13] that argument 4 also implies that Ω must exceed ~ 0.4 in order to have sufficient growht. This latter point appears to violate the dynamic argument that matter in clusters has $\Omega < 0.4$ since cold matter will cluster on smaller scales than superclusters. Thus, superclusters should contain all cold matter *if* light emitting stuff goes along with it.

Initial Solutions and Loopholes

From the above discussion, it is apparent that in the simple picture, no single dark-matter candidate works: hot particles don't form galaxies fast enough and cold particles put too much mass on small scales where it is not seen. Even hybrids of hot and cold particles fail[17,18]. Two classes of solution have been able to work: one is to have light, not trace mass[19], the other is to have cold particles decay to hot[20].

If light does not trace mass, then many (most) clumps of cold matter and baryons do not turn on their baryons and fail for some reason to shine. Such a failure to universally ignite requires a toothfairy, albeit a thermodynamic one (rather than a particle physics one). This thermodynamic toothfairy is added to the particle physics toothfairy (the one which produced the cold particles: axions, photinos, heavy leptons, planetary mass black holes, quark nuggets, etc.). The decaying scenario requires two toothfairies as well, but both are of the particle variety.

Some people have preferences for one or another variety of toothfairy. Thermodynamic ones might, in principle, be eliminated since they only require classical physics, but remember, *we still can't predict the weather*. On the other hand, particle ones might, in principle, be eliminated by future accelerator experiments.

Another way out is to duck the galaxy formation argument by having something other than matter carry the primordial fluctuation[21]. Under such a scenario, the fluctuations would not be smoothed by the free streaming of the matter and radiation. They would be sitting, waiting for the baryons to fall in once the baryons decouple from the radiation and the universe is no longer matter dominated. Such fluctuations behave like the old isothermal fluctuation model. Cosmic strings[22] are a fine example of such fluctuations if energy density in strings ρ_s, relative to the total matter energy density ρ, is $\rho_s/\rho \sim 10^{-3}$, then fluctuation growth in ρ to $\delta\rho/\rho \sim 1$ can occur for the baryons and rapid galaxy formation is no problem. Strings would have mass scales smaller than $\sim 10^{12} M_\odot$ damped by gravitational radiation, but otherwise behave like isothermals with, however, an equal power on all mass scales, Harrison-Zeldovich spectrum. Strings are also capable of producing non-random phases[23] which may be valuable for explaining large clusters and voids[24]

While baryosynthesis argues against primordial isothermals on scales larger than the horizon, it does not prevent subsequent phase transitions from producing isothermal-like fluctuations on scales smaller than the horizon at these phase transitions. An example is the possible generation of planetary mass black holes[25] or quark nuggets[26] at the quark-hadron-chiral transition. Such objects can serve as small scale isothermal seeds that baryons can condense upon[27,28] after recombination. Such objects can serve as the seeds for the Ostriker-Cowie[29] explosive galaxy formation scenarios.

Thus, galaxy formation either requires cold matter or isothermal-like fluctuations which require some phase transition in the early universe, either generating strings or some other remnant.

Large Scale Structure

The distribution of matter in space has been described in detail by Peebles[30] using the two-point correlation function $\xi(r)$. This is the excess probability over random that two objects are separated by a distance r. The correlation function for galaxies, ξ_{gg}, has been shown to have the form $\xi_{gg}(r) = \alpha_{cc} r^{-1.8}$. Similarly, it has been shown[31,32,33] that the correlation function of rich clusters ξ_{cc} also has the form $\xi_{cc} = \alpha_{cc} r^{-1.8}$; but for Abell $R \geq 1$ clusters $\alpha_{cc} \approx 20 \alpha_{gg}$ and α_{cc} is even larger for Abell $R \geq 2$ clusters.

At first glance this appears strange since the larger separated clusters would not have had as much gravitational clumping as the more closely separated galaxies. However, Kaiser[34] and Fry[35] pointed out that if clusters are merely the 3σ peaks in the galaxy distribution function, then one would obtain a factor of ~ 20 enhancement since the less clumped galaxies would dilute the amplitude for ξ. Such biasing would mean that $\xi_{gg}(r) \propto \xi_{cc}(r)$ so that when ξ_{gg} goes negative, ξ_{cc} should also be negative, only more so. Current data[36] indicates that ξ_{gg} deviates from the $r^{-1.8}$ power law and goes negative at $r \approx 20 h_0^{-1}$ Mpc (where $h_0 = H_0/100$ km/sec/Mpc), but ξ_{cc} is positive out to $r \gtrsim 100 h_0^{-1}$ Mpc. Thus, simple biasing seems in trouble. Of course, the negative correlations may be due to unfortunate statistical fluctuations in our neighborhood, but it should be remembered that essentially all models of galaxy formation with a Harrison-Zeldovich type fluctuation spectrum yield ξ_{gg} going negative for some r which is well below 100 Mpc. Therefore, a positive ξ_{cc} at 100 Mpc is in conflict with the biasing interpretation for all the standard galaxy formation scenarios.

This problem has been reexamined using a dimensionless approach[3] where all distances are measured in units of $L \equiv n^{-1.3}$ (where n is the density of objects in the catalogue). With this procedure, $\xi(r) = \beta(L)(hr/L)^{-1.8}$. This is analogous to the renormalization approach that is used in condensed matter physics to analyze scale-free transitions. In particular, if $\beta(L)$ were found to be constant for all, this would imply that $\xi(r)$ is scale free. It is interesting that $\xi(L)$ is approximately the same for both $R \geq 1$ and $R \geq 2$ as well as the Schectman[37] catalogue. However, $\beta(L)$ for galaxies is about a factor of 3 larger. Thus, this dimensionless approach has eliminated the dependence on cluster richness and reduced the entire dynamic range to a factor of ~ 3. That all the various cluster-cluster functions have the same $\beta(L)$ seems to imply that there is a scale-free process operating from scales of Mpc to 100's of Mpc. The slight enhancement of β_{gg} is simply implying that galaxies have had their clustering enhanced by gravity a factor of ~ 3 over the scale-free process which produced the $r^{-1.8}$ behavior.

A scale-free process of this type is known as a fractal[38]. The $r^{-1.8}$ power law tells us that the dimension D of the fractal is $D = 3 - 1.8 = 1.2$. This is pretty close to a linear, $D = 1$, fractal.

There are several possible scale-free processes which might yield a $D = 1.2$ fractal for laying out galaxies and clusters in the universe.

1. **Biasing** This can yield a scale-free process if the primordial fluctuation spectrum is scale-free. However, as mentioned above, the lack of negativity at large r in ξ_{cc} seems to argue against this mode.

2. **Strings** If the primordial fluctuations are created by cosmic strings, they would be laid out along a pattern left by the primordial string before it fragmented. The fractal dimension of this pattern depends on how the strings move and fragment, but a near linear $D \geq 1$ fractal is not unreasonable.

3. **Percolation** If galaxy formation is related to explosions of primordial seeds, and if the density of such seeds is sufficiently high, percolation can occur and a scale-free pattern can form. The fractal dimension of such a pattern depends on the geometry of the region where the seeds are dense(neutrino pancakes?).

Conclusions

Combining the need for a scale-free, large-scale structure mechanism with $\Omega = 1$ and with the requirement that galaxies form rapidly with $\delta T/T$ small, restricts us to:

1. biasing and cold dark matter
2. strings and hot dark matter
3. percolated explosions and hot dark matter.

If we relax the scale-free requirement then a fourth solution is also allowed:

4. decaying particles.

In the "bias and cold" case, the dark baryons are in non-shining clumps of cold matter unrelated to clusters of galaxies. Such a solution also requires ξ_{cc} to be negative at the same scales that ξ_{gg} goes negative. As Bardeen, Kaiser and Szalay[39] point out, at present, no cold biasing solution seems simultaneously consistent with $\Omega_{dynamic}$ and with ξ_{cc} vs ξ_{gg}.

The "string and hot" case requires the dark baryons to be in the halos of galaxies. The hot x-ray gas in rich clusters[6] implies that galaxies in clusters had significant baryonic matter associated with them which fell into the deep potential well of the cluster and was heated to x-ray temperatures. If all galaxies has this much baryonic material, it would yield dark baryonic halos. A strong (recent) plus for the string model comes from the work on superstrings which is the current best candidate for the Theory Of Everything (T.O.E.). While ten dimensional superstrings are themselves not directly related to the cosmic strings needed for galaxy formation, Witten[40] has shown that the two superstring models O(32) and E(8) x E(8), which are anomaly-free, finite and have chiral fermions, naturally lead to cosmic strings. These cosmic strings form when the supertheory breaks at the GUT epoch.

The "percolation and hot" case also has the dark baryons in halos, but this model uses baryons in early explosions so that the early universe is quite active which should lead to observational consequences. This model requires early universe phase transition activity since the seeds can only be sufficiently plentiful if isothermal-like fluctuations are generated on scales that eventually produce seeds. The quark-hadron transition is the best candidate for such a transition. However, while current estimates imply a first-order transition, it is still quite difficult[4] to get long-lived quark nuggets or planetary-mass black holes.

The fourth solution, like the first, already requires that one of our assumptions about current observations is wrong. It also requires that a very fine tuned particle physics situation exists with a massive particle decaying to light one just after the initiation of galaxy formation. In this case, the dark baryons would also be in the form of halo material.

In summary, the way to distinguish possibilities is to find the dark baryons. While the best current estimate might lean towards strings and hot stuff, there is clearly much more work to be done!

Acknowledgments

I would like to thank my recent collaborators Katy Freese, Keith Olive, Gary Steigman, Alex Szalay, Michael Turner and Nicola Vittorio.

This work was supported by NSF grant AST-8313128 and DOE grant DE-AC02-80ER10773 A004 at the University of Chicago and by DOE and NASA at Fermilab.

References

1. Schramm, D.N. *Il Nuovo Cimento* in press (1985).
2. Schramm, D.N. *Nuc.Phys.* **B252**, 53 (1985).
3. Szalay, A. and Schramm, D.N. *Nature* **314**, (1985).
4. Turner, M.S. *Ferminews* in press (1985).
5. Faber, S. and Gallagher, J. *Ann.Rev.Astron&Astrophys.* **17**, 135 (1977).
   Peebles, P.J.E., in *Physical Cosmology* (eds Audouze, J., Baliar, R., Schramm, D.) 1980).
6. Schramm, D.N. and Steigman, G. *Ap.J* **241**, 1 (1981).
7. Yang, J., Turner, M., Steigman, G., Schramm, D.N. and Olive, K. *Ap.J.* **281**, 493 and references therein (1984).
8. Hegyi, D. and Olive, K. *Phys.Lett.B* **126**, 28 (1982).
9. Freese, K. and Schramm, D.N. *Nuc.Phys.B.* **233**, 167 (1984).
10. Gott, J.R., Gunn, J., Schramm, D.N. and Tinsley, B. *Ap.J.* **194**, 543 (1974).
11. Guth, A. *Phys.Rev.D.* **23**, 347 (1981).
12. Olive, K. and Schramm, D.N. *Comments Nuclear and Particle Physics.* in press (1985).
13. Vittorio, N. and Silk, J. *Proc.Inner Space-Outer Space* in press (1985). Bond R. and Efstathiou, G. *Proc.Inner Space-Outer Space* in press (1985).
14. Frenk, C., White, S. and Davis, M. *Ap.J.* **271**, 417 (1983).
15. Usson, J. and Wilkinson, D. *Proc.Inner Space-Outer space* in press (1984).
16. Turner, M. and Schramm, D.N. *Nature* **279**, 301 (1979).
    Press, W. *Physica Scripta* **21**, 702 (1980).
17. Davis, M., LeCar, M., Pryor, C. and Witten, E. *Ap.J.* **250**, 423 (1978).
18. Occhionero F. (1985) to be published

19. Davis, M., Efstathiou, G., Frenk, C. and White, S. *Ap.J.Supp.* in press (1985).
20. Olive, K., Seckel, D. and Vishniac, E. *Phys.Lett.B.* in press (1984).
    Olive, K., Schramm, D.N. and Srednicki, M. *Phys.Lett.* in press and references therein (1985).
21. Vittorio, N. and Schramm, D.N. *Comments Nuclear and Particle Physics* in press (1985).
22. Vilenkin, A. *Phys.Rep.*, in press and references therein (1985).
23. Turock, N. and Schramm, D.N. 1984: *Nature*, 312 598 (1984).
24. Peebles, P.J.E. *Ap.J.* 277 470 (1984).
25. Crawford, M. and Schramm, D.N. *Nature* 298 538 (1982).
26. Witten, E. *Physics Rev.D.* 30 272 (1984).
27. Freese, K., Schramm, D.N. and Price, R. *Ap.J.* 275 405 (1983).
28. Raphaelli, N. in this volume (1986).
29. Ostricker, J. and Cowie, L. *Ap.J.* 243 L127 (1980).
30. Peebles, P.J.E. *The Large Scale Structure of the Universe* (Princeton Univ. Press, Princeton, 1981).
31. Bahcall, N. and Soniera, R. *Ap.J.* 270 20 (1983).
32. Klypin, A. and Kopylov, A. *Sov.Astra.Lett.* 9 41 (1983).
33. Hauser, M. and Peebles, P.J.E. *Ap.J.* 185 757 (1973).
34. Kaiser, N. *Ap.J.* in press (1984).
35. Fry, J. private communication (1983).
36. Davis, M. *Proc.Inner-Outer Space* loc.cit. (1984).
37. Schectman, S.A. *Ap.J.* submitted (1984).
38. Mandelbrot, B. *Fractal Symmetry of Nature* Freeman, San Francisco (1977).
39. Bardeen, J., Kaiser, N. and Szalay, A. Fermilab Preprint (1985).
40. Witten, E. *Phys.Lett.* in press (1985).
41. Applegate, J. and Hogan, C. Caltech Preprint (1985).

NUCLEOSYNTHETIC CONSEQUENCES OF POPULATION III STARS

B.J. Carr
Queen Mary College, University of London, England

W. Glatzel
Institute of Astronomy, Cambridge, England

Abstract: It is argued that the Population III stars with the most explanatory power are Very Massive Objects (VMOs) in the mass range $10^2 < M/M_\odot < 10^5$. Their evolution is reviewed, with particular emphasis on their nucleosynthetic consequences, and this leads to constraints on the mass spectrum of Population III stars. The possible relevance of Population III stars for some elemental anomalies and the cosmic helium abundance is discussed.

1. INTRODUCTION

The study of a pregalactic population of stars is motivated by their many possible cosmological consequences. Their remnants could account for the dark matter in galactic halos (White and Rees 1978), their light could produce distortions in the 3K background (Rowan-Robinson et al 1979) or even the 3K background itself (Rees 1978), their heat could provide a mechanism to reionize the Universe (Hartquist and Cameron 1977), their explosions and mass loss could produce a burst of initial enrichment (Truran and Cameron 1971), perhaps explaining certain elemental anomalies and contributing to the cosmic helium abundance (Talbot and Arnett 1971), and they might even play an important role in galaxy formation (Ostriker and Cowie 1981). These sorts of consequences have been studied in detail by Carr, Bond and Arnett (1984, hereafter CBA).

Of particular interest is the question of which stars could account for these phenomena without contradicting other observations. If most of the Universe is processed through Population III stars, then the requirement that they do not produce too much background light restricts their masses to $M > 10$ $M_\odot$ or $M < 0.1$ $M_\odot$. Stars in the mass range $10 < M/M_\odot < 100$ generate a lot of heavy elements (Weaver and Woosley 1980), so it would be difficult to understand how their remnants could provide the dark matter in galactic halos without producing too much enrichment. On the other hand, stars at earlier epochs might have been larger than 100 $M_\odot$ due to the influence of the 3K background (Kashlinsky and Rees

1983) and the absence of metals (Silk 1977). Such VMOs would be expected to collapse to black holes with high efficiency (without metal ejection) providing their mass exceeded a critical value $M_c \cong 200\ M_\odot$. Their remnants would therefore be plausible candidates for solving the dark matter problem (CBA) and they would avoid making too much background light providing they burnt out at a redshift $z \gtrsim 10$.

An important constraint on the mass range of black holes in the galactic halo comes from the requirement that their passage through the galactic disk does not heat up the disk stars too much. This condition is satisfied only for holes smaller than $10^6 M_\odot$ (Lacey 1984). We conclude that the most interesting candidates for Population III stars are VMOs in the mass range $10^2 < M/M_\odot < 10^5$. Their evolution is reviewed in Section 2 and their nucleosynthetic consequences will be discussed in Section 3. The dynamical constraints would also permit the dark matter to be contained in the remnants of SMOs in the mass range $10^5 \lesssim M/M_\odot < 10^6$. However, since SMOs collapse directly (without burning their nuclear fuel) due to relativistic instabilities if they have no initial metallicity (Fricke 1973), such stars would not be able to explain the other cosmological features mentioned above. We therefore emphasize the VMO scenario here, although SMOs would be more attractive if one wanted to perturb the standard picture as little as possible.

2. EVOLUTION OF VMOs

2.1 Hydrogen core burning

Because of their high masses, VMOs radiate near the Eddington limit. Initially this luminosity will derive from the pp-chain; CNO burning cannot take place due to zero initial metallicity. However, CNO burning takes over once the necessary catalysts have been created by the triple-alpha reaction (El Eid, Fricke and Ober 1983; hereafter EFO). This automatically generates a CNO abundance of about 10^{-9} (Bond, Arnett and Carr 1984; hereafter BAC). If, in addition, their is no initial helium, a slight amount must first be created via pp-reactions in order to trigger triple-alpha burning.

Luminosities close to the Eddington value cause radiation pressure driven mass loss. A rigorous theory of mass loss - sometimes also described in terms of a nuclear pulsational instability - does not exist since the detailed nonlinear mechanism of mass shedding is not very well understood. Consequently its treatment varies among different authors (BAC, EFO, Woosley and Weaver 1982). EFO assume a semi-empirical relation for the mass loss, whereas BAC derive an analytic relation between the total fraction of mass lost and the corresponding helium yield which is independent of the particular mechanism. This relation will be discussed later.

2.2 Hydrogen shell burning

When hydrogen is exhausted in the convective core, the helium core contracts and hydrogen ignites in a shell. BAC have run a 500 $M_\odot$ Population I model using an implicit hydrocode and treating convection time-dependently. In this case, 2.5 $M_\odot$ are burnt in the shell, releasing 3.5×10^{52} ergs on a local thermal timescale. This leads to a supersonic expansion of the envelope, since the gravitational binding energy of the shell is only 0.5×10^{52} ergs, and the resulting wind is expected to eject the envelope entirely. BAC conclude that the effect may be similar for Population III stars if CNO catalysts are convectively dredged up into the hydrogen-burning shell.

The calculations of EFO for Population III stars indicate that effective shell burning starts later than for Population I stars. In addition, they show that mass loss can substantially reduce the size of the shell source. They find that the shell propagates outwards, increasing the size of the helium core. They obtain evidence for complete envelope ejection only when M exceeds 500 $M_\odot$.

The Woosley-Weaver calculations for Population III stars also exhibit a density decrease in the envelope but the shell does not become unbound in their models. Since the results of BAC and EFO are quite similar, the discrepancy between the different authors is likely to be due to the different treatments of convection and mass loss rather than the different assumptions for the initial composition.

2.3 Helium core burning

Mass loss must also be expected in the helium burning phase. However, the pulsational instability is usually assumed to be damped by the extended envelopes surrounding the cores. Helium burning leads to carbon-oxygen cores consisting mainly ($\approx 90\%$) of oxygen. EFO found by numerical calculations the following linear relations between the various masses:

$$M_{CO} \cong 0.89\, M_{HX} \;;\quad M_{CO} \cong 1.09\, M_\alpha \;;\quad M_\alpha \cong 0.56\, M_i$$

where M_i is the initial mass, M_{HX} is the total mass at hydrogen exhaustion, M_α is the helium mass at hydrogen exhaustion, and M_{CO} is the carbon-oxygen core mass. The relation $M_\alpha = M_i/(2-Y_i)$, where Y_i is the initial helium abundance, as derived by BAC analytically, agrees remarkably well with the numerical result.

2.4 Dynamical phase

Carbon-oxygen cores in the mass range $30 < M_{CO}/M_\odot < 4 \times 10^5$ become dynamically unstable due to electron-positron pair creation. (More massive cores encounter a relativistical instability first). They therefore collapse and undergo explosive nuclear burning. Their ultimate fate – oscillation, explosion or collapse to a black hole – is determined by their mass and angular momentum. Table 1 shows the different mass ranges for zero angular momentum according to various authors.

TABLE 1: The fate of non-rotating carbon-oxygen cores

Mass Range ($M_\odot$)	Stability	Oscillation	Explosion	Collapse
Woosley & Weaver (1982)	---	---	60-100	>100
Bond et al (1984)	<32	~32	32-110	>110
Ober et al (1983)	<50	---	50-112	>112
Glatzel et al (1985)	<22	22-27	27-86	>86

For low masses the energy released by nuclear burning is not sufficient to disrupt the star, so it undergoes oscillations, eventually accompanied by explosive mass loss (Ober et al 1983). Below a critical mass $\hat{M}_{co}$, higher mass cores explode due to nuclear burning without leaving any remnant. Cores with mass above $\hat{M}_{co}$ proceed to such high temperatures that endothermic photodisintegration of nuclei takes place; this reinforces the collapse and ultimately leads to black hole formation. We note that the critical mass M_c of the initial hydrogen star is related to the critical core mass by $\hat{M}_c \simeq 1.64 \hat{M}_{co}$, providing the mass loss during hydrogen-burning is not too large.

The explosions are almost entirely due to explosive oxygen burning. However, incomplete silicon burning occurs for masses near $\hat{M}_{co}$ and, in these cores, a tiny fraction of iron group elements is synthesized (5% at maximum). The surroundings are therefore enriched mainly with oxygen but also with intermediate nuclei from sulphur to calcium.

FIGURE 1: The fate of rotating carbon-oxygen cores

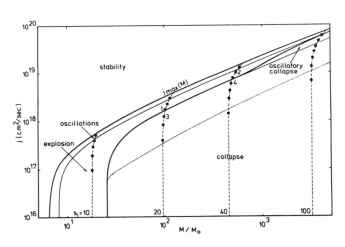

The stabilizing effect of rotation extends the mass range in which explosions can occur. For suitably chosen angular momentum, explosions are obtained for core masses up to $10^4 M_\odot$ corresponding to a hydrogen mass of $\sim 2 \times 10^4 M_\odot$ (Glatzel et al 1985). Above this mass, neutrino losses preclude any explosion. The nucleosynthesis in the rotating models is similar to the nonrotating ones. Figure 1 shows the different regimes of stability, oscillations, explosion and collapse in terms of specific angular momentum and mass for an n=0 polytrope. To obtain the masses and angular momenta for an n=3 polytrope (as is appropriate for $M > 10^2 M_\odot$), the numbers given have to be multiplied by factors of 3.36 and 5.86, respectively. We note that a VMO may fragment into a binary system if the angular momentum is too large. This only happens in the "stable" region of Figure 1, so the limiting mass for which explosions can occur is unaffected. However, black hole remnants could conceivably arise in the "stable" region if the angular momentum and mass of one of the fragments was in the collapse domain.

3. NUCLEOSYNTHETIC CONSEQUENCES

3.1 Constraints on the spectrum of Population III stars

Of all constraints which can be imposed on the spectrum of Population III stars, the nucleosynthetic ones are the most stringent for $4 M_\odot < M < M_c$. Since there exist Population I stars with metallicities as low as 10^{-3}, the Population III enrichment cannot exceed this value. Furthermore, if one assumes that Population II stars form after Population III stars, the maximum enrichment Z_{max} is reduced to 10^{-5},

FIGURE 2: Upper limit on the density of Population III stars

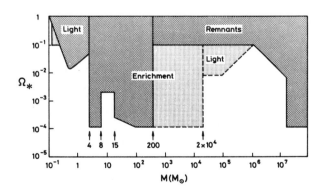

since this is the lowest metallicity found in Population II stars (Bond 1981). The associated limit on $\Omega_*(M)$, the density of the Population III stars in units of the critical density, is shown in Figure 2 on the assumption that $Z_{max} = 10^{-3}$ and that the metal yield of the stars is

$Z_{ej} \simeq 0.2$ for $4 < M/M_\odot < 8$, $Z_{ej} \simeq 0.01$ for $8 < M/M_\odot < 15$, $Z_{ej} \simeq 0.5 - (6.3 M_\odot/M)$ for $15 < M/M_\odot < 10^2$, and $Z_{ej} \simeq 0.5$ for $10^2 M_\odot < M < M_c$. The limit scales with the initial gas density, which is taken to be $\Omega_g = 0.1$.

Figure 2 also indicates the limits associated with background light and black hole remnant constraints. The light constraint is important in the mass range above $10^2 M_\odot$ only if the stars burn at a low redshift. However, as indicated earlier, if one allows for the effect of rotation, the enrichment limit can be extended up to a mass of $2 \times 10^4 M_\odot$, in which case the upper limit on $\Omega_*(M)$ would be as indicated by the dashed line.

Assuming a standard IMF,

$$n(M)dM \sim M^{-\alpha}dM \; ; \; M_{min} < M < M_{max},$$

one avoids overenrichment providing

$$M_{max} > M_c \left(\frac{Z_{ej} f_*}{Z_{max}}\right)^{1/(2-\alpha)} \quad \text{for } \alpha < 2$$

or

$$M_{min} < 4 M_\odot \left(\frac{Z_{ej} f_*}{Z_{max}}\right)^{1/(2-\alpha)} \quad \text{for } \alpha > 2$$

where f_* is the total fraction of the Universe going into the stars. If we want the Population III remnants to provide the dark matter, f_* must be close to 1. In this case, we can obtain a prescribed pregalactic enrichment by a suitable choice of the parameters α, M_{min}, and M_{max}. However, it requires rather fine-tuning to get an enrichment of $10^{-3}-10^{-5}$ (as would be required to explain the G-dwarf problem or spectral distortions in the 3K background) and - if $M_{max} < 0.1 M_\odot$ or $M_{min} > M_c$ - one gets no enrichment at all.

An alternative solution is to postulate two generations of stars, one containing a large fraction of the mass of the Universe and producing the dark matter, the other containing a small fraction and exploding to produce a small enrichment. This could come about rather naturally if the exploding stars form first. For one could envisage the characteristic star mass automatically shifting out of the exploding range, either because it is decreased by the cooling effects of the metals or because it is increased by the rise in the Jeans mass induced by the stars' heating effect. In either case, the enrichment produced by the first stars would be self-limited to a value in the range $10^{-3}-10^{-5}$ (CBA).

3.2 Abundance anomalies

Stars with [Fe/H] < -1 are found to have three times the solar oxygen-to-iron ratio and the ratio decreases with increasing metallicity (Sneden et al 1979). High [O/Fe] is observed both in field metal-poor stars and globular clusters. Furthermore, extragalactic HII regions exhibit a sulphur-to-oxygen ratio which decreases with increasing oxygen abundance

(French 1980) and field metal-poor stars also have enhancements in other oxygen burning products. Since exploding VMOs create an enrichment in oxygen and oxygen-burning products without ejecting iron group elements, they might be able to explain these anomalies.

Pagel and Edmunds (1981) have reviewed arguments indicating that some nitrogen may be primary. The mechanism of VMO hydrogen shell ejection discussed above provides a source of primary nitrogen if carbon and oxygen can be dredged up into the hydrogen-burning shell. If, on the other hand, VMOs with $M>M_c$ can eject metals by this mechanism, a strong limit could be placed on the number of the corresponding black hole remnants (Tarbet and Rowan-Robinson 1982).

3.3 Helium

If VMO remnants are to provide the missing mass, their helium enrichment is crucial. A relation between the helium enrichment ΔY of the surroundings and the total fraction ϕ_L of mass lost, which is independent of the particular details of mass loss, was derived by BAC. For high mass loss, the helium yield is low, since the mass is lost before helium is produced significantly. If the mass loss is sufficiently low that the boundary of the star shrinks more slowly than the hydrogen burning convective core $(\phi_L < (1-Y_i)/(2-Y_i))$, the helium yield is found to be:

$$\Delta Y = (1 - Y_i/2) \phi_L^2$$

The maximum yield ΔY_{max} is obtained if the surface just follows the edge of the shrinking convective core and, in this case, we obtain

$$\Delta Y_{max} = 0.25(1-Y_i)^2/(1-Y_i/2) = \begin{cases} 0.17 & \text{for } Y_i = 0.22 \\ 0.25 & \text{for } Y_i = 0 \end{cases}$$

This result is in remarkable agreement with the numerical calculations of EFO. These authors obtained ΔY between 0.126 and 0.177 over all mass ranges. We note that, in the case of envelope ejection during hydrogen-shell burning, the helium enrichment is always maximal.

Since we find an enrichment of $\Delta Y=0.17$ for a primordial helium abundance of $Y_i=0.22$, we would inevitably overproduce helium if a large fraction of dark matter were in VMO black holes. In this case only SMOs could provide the dark matter, since they collapse without any helium production. However, we could also adopt the point of view that the cosmic helium abundance is actually produced by Population III VMOs. The fraction F of the universe which is processed through VMOs and the resulting helium abundance Y are related by (Bond et al 1983):

$$F = \int_0^Y \frac{2(2-Y)}{(1-Y)^2} \exp\left(\frac{2Y}{Y-1}\right) dY$$

For small Y, this just gives $F \cong 4Y$, so one naturally gets the sort of value required if F is close to 1. More precisely, for $0.2<Y<0.25$ we need $0.9<F<0.95$.

Acknowledgements: This report is based on work done with our colleagues, W.D. Arnett, J.R.Bond, M.F.El Eid, K.J. Fricke and W.W. Ober. W.G. gratefully acknowledges support by the "Deutsche Forschungsgemeinschaft" under grant Gl 127/1-1 and the hospitality of the IOA, Cambridge.

REFERENCES

Bond,H.E.:1981, Ap.J. 248, 606.
Bond,J.R., Carr,B.J., Arnett,W.D.:1983, Nature 304, 514.
Bond,J.R., Arnett,W.D., Carr,B.J.:1984, Ap.J. 280, 825.
Carr,B.J., Bond,J.R., Arnett,W.D.:1984, Ap.J. 277, 445.
El Eid,M.F., Fricke,K.J., Ober,W.W.:1983, Astr.Ap. 119, 54.
French,H.B.:1980, Ap.J. 240, 41.
Fricke,K.J.:1973, Ap.J. 183, 941.
Glatzel,W., El Eid,M.F., Fricke,K.J.:1985, Astr.Ap., in press.
Hartquist,T.W., Cameron,A.G.W.:1977, Ap. Space Sci. 48, 145.
Kashlinsky,A., Rees,M.J.:1983, M.N.R.A.S. 205, 955.
Lacey,C.G.:1984, in Formation and Evolution of Galaxies and Large Structures in the Universe, ed. J. Audouze and J. Tran Thanh Van (Dordrecht: Reidel).
Ober,W.W., El Eid,M.F., Fricke,K.J.:1983, Astr.Ap. 119, 61.
Ostriker,J.P., Cowie,L.L.:1981, Ap.J. Lett. 243, L127.
Pagel,B.E.J., Edmunds,M.G.:1981, Ann. Rev. Astr. Ap. 19, 77.
Rees,M.J.:1978, Nature 275, 35.
Rowan-Robinson,M., Negroponte,J., Silk,J.:1979, Nature 281, 635.
Silk,J.:1977, Ap.J. 211, 638.
Sneden,C., Lambert,D., Whitaker,R.W.:1979, Ap.J. 234, 964.
Talbot,R.J., Arnett,W.D.:1971, Nature 229, 250.
Tarbet,P.W., Rowan-Robinson,M.:1982, Nature 298, 711.
Truran,J.W., Cameron,A.G.W.:1971, Ap. Space Sci. 14, 179.
Weaver,T.A., Woosley,S.E.:1980, Ann. NY Acad. Sci. 336, 335.
White,S.D.M., Rees,M.J.:1978, M.N.R.A.S. 183, 341.
Woosley,S.E., Weaver,T.A.:1982, in Supernovae: A Survey of Current Research, ed. M.J. Rees and R.J. Stoneham (Dordrecht: Reidel).

II – EXPLOSIVE OBJECTS

NUCLEOSYNTHESIS ACCOMPANYING CLASSICAL NOVA OUTBURSTS

James W. Truran
Department of Astronomy, University of Illinois
1011 W. Springfield, Urbana, IL 61801

ABSTRACT. Theoretical studies of thermonuclear runaways in degenerate hydrogen-rich matter leading to classical nova outbursts predict that nova ejecta can be characterized by substantial overabundances of the elements carbon, nitrogen, and oxygen as well as of other interesting nuclear species. Observational determinations of abundances in nova ejecta have now confirmed both that anomalous CNO enrichments do indeed occur and that the elements neon, sodium, magnesium and aluminum can also be present in significant amounts. This paper reviews the abundance patterns observed in novae and discusses these within the context of the theoretical models. We then examine the possible implications of these abundance enrichments for galactic nucleosynthesis. The possible role of nova in the production of ^{22}Na and ^{26}Al is also discussed.

1. INTRODUCTION

Theoretical studies have now confirmed the view that the outbursts of the classical novae are a consequence of thermonuclear runaways proceeding in accreted hydrogen shells on the white dwarf components of these close binary systems. For the conditions which are expected to obtain in this environment, nuclear energy generation is provided by the carbon-nitrogen-oxygen (CNO) cycle hydrogen-burning reaction sequences. Peak temperatures ~ 200-300 million K are achieved in explosive hydrogen burning on a hydrodynamic time scale. The natural limits imposed on the rate of nuclear energy generation in the operation of these reactions at high temperatures holds important implications for hydrodynamic models of nova outbursts (Truran 1982) and has led to the prediction that enhanced CNO abundances may be required to explain the observed characteristics of novae in outburst (Starrfield et al 1972). Hydrodynamic calculations which further confirm these conclusions have since been performed by a number of researchers (Sparks, Starrfield and Truran 1978; Starrfield, Truran and Sparks 1978; Prialnik, Shara and Shaviv 1978, 1979; Nariai, Nomoto and

Sugimoto 1980; Prialnik et al 1982; Starrfield, Sparks and Truran 1985).

My aim in this paper is to identify the contributions which novae may be expected to make to nucleosynthesis in the galaxy. In the following section, we first briefly review existing data on heavy element abundances in the ejecta of classical novae. Specific predictions concerning the formation of heavy nuclei accompanying thermonuclear runaways in nova outbursts are then elaborated. Discussion of the implications of nuclear processes in novae for galactic nucleosynthesis then follows.

2. COMPOSITION OF NOVA EJECTA

Recent observational investigations of the compositions of nova nebular remnants have provided critical support for and constraints upon the thermonuclear runaway model for the classical novae. Abundance determinations are available, in particular, for the ejecta of the novae V1500 Cyg (Ferland and Shields, 1978), HR Del (Tylenda 1978), DQ Her (Williams et al 1978), RR Pic (Williams and Gallagher 1979), T Aur (Gallagher et al 1980), V1668 Cyg (Stickland et al 1981), V 693 CrA (Williams et al 1985), and Nova Aquila 1982 (Snijders, Seaton and Blades 1982). The mass fractions in the form of heavy elements determined for the first six of these average ~ 0.23, and range from 0.039 for RR Pic to 0.56 for DQ Her. The bulk of this matter is in the form of carbon, nitrogen and oxygen. This is consistent with theoretical calculations of thermonuclear runaways leading to nova outbursts. Information concerning the isotopic composition of this material would also be extremely useful. Existing models predict that the $^{12}C/^{13}C$, $^{14}N/^{15}N$, and $^{16}O/^{17}O$ ratios characterizing nova ejecta will differ significantly from those of solar system matter.

Two recent novae, Nova Corona Austrina 1981 and Nova Aquila 1982, both show evidence of substantial enrichments of heavy nuclei other than carbon, nitrogen and oxygen. In particular, Nova Corona Austrina 1981 is enriched in C, N, O, Ne, Na, Mg, and Al and Nova Aquila 1982 is enriched in C, N, O, Ne, and Mg.

The presence of these increased concentrations of nuclei from neon to aluminum presents an interesting challenge to theories of nova outbursts. We note that significant increases in the concentrations of such heavy nuclei are not expected to accompany nova outbursts: the controlling CNO reaction sequences operating at temperatures $\sim$ 150-300 million degrees act largely to rearrange existing CNO isotopic abundances but not to increase the total number of CNO nuclei. The high CNO abundances must therefore reflect the envelope composition prior to the ignition of hydrogen. Breakout of the CNO cycle hydrogen burning sequences via $^{15}O(\alpha,\gamma)^{19}Ne(p,\gamma)^{20}Na$ can occur at high temperatures ($T \gtrsim 4 \times 10^8$ K), yielding neon and heavier nuclei (Wiescher et al 1985). Such conditions may be realized in thermonuclear runaways on neutron stars (Wallace and Woosley 1981; Starrfield et al 1982), but it seems unlikely that the neon, sodium, magnesium and alumnium enrichments seen in novae can be understood on this basis, since burning temperatures are typically not high enough for breakout to occur. Recent calculations of runaways on white dwarfs of masses approaching the Chandrasekhar limit predicted peak temperatures in the shell source approaching 3.5×10^8 K (Starrfield, Sparks and Truran 1985).

It would appear, rather, that the outward mixing of white dwarf core material must provide the enrichment mechanism. In particular, it has been found that shear-induced turbulent mixing between the white dwarf and the accreted material can lead to envelope CNO enrichments (Kippenhahn and Thomas 1978; MacDonald 1984). For the recent novae which show high Ne-Na-Mg-Al concentrations, such mixing up of core matter can explain the abundance peculiarities if one assumes an underlying O-Ne-Mg white dwarf (Law and Ritter 1983; Truran 1985; Williams et al 1985).

It is clear from this brief review of observational determinations of abundances in nova ejecta that significant abundance enrichments can occur, in support of the prediction of theoretical studies.

3. OVERVIEW OF NOVA NUCLEOSYNTHESIS

We now examine the predictions of theoretical models of novae with regard to the synthesis of a variety of nuclear species.

^{7}Li. The production of lithium in novae as ^{7}Li can result from hydrogen burning reaction sequences (Arnould and Nørgaard 1975; Starrfield et al 1978). As in the case of red giants, the ^{7}Be formed via ^{3}He(^{4}He,γ)^{7}Be must be transported outward by convection to cooler regions of the envelope faster than it can be converted to ^{7}Li by ^{7}Be(e$^-$,ν)^{7}Li and destroyed via ^{7}Li(p,γ)^{8}Be(2^{4}He). The temperature and convective histories are thus critical factors which determine the uncertainties in predictions of ^{7}Li production. The initial ^{3}He concentration of the envelope matter is also of interest, since much of the ^{3}He present may be converted readily to ^{7}Be. Model calculations by Starrfield et al (1978) predict that ^{7}Li enrichments of factors up to $\gtrsim 10^2$ may be achieved in matter subsequently ejected by novae. Unfortunately, only upper limits on lithium abundances in novae are currently available from observations (Friedjung 1979).

^{12}C and ^{16}O. The concentrations of these two dominant isotopes of carbon and oxygen represent the products of helium burning in intermediate mass and massive stars. The factors by which these two isotopes can be enriched in the ejecta of novae are too small to be important to galactic nucleosynthesis. The anomalously low ^{12}C/^{13}C and ^{16}O/^{17}O isotopic ratios predicted for nova ejecta further confirm that novae cannot represent significant sources of ^{12}C and ^{16}O.

^{14}N. It is well known that ^{14}N represents the dominant product of CNO-cycle hydrogen burning for hydrostatic burning conditions in stellar interiors. Under these circumstances, extremely high N/O and N/C ratios can be achieved as well as a high isotopic ratio ^{15}N/^{14}N. Theoretical calculations predict nitrogen excesses and observations reveal nitrogen enrichments up to factors ~ 100. Williams (1982) argues that novae may represent an important source of ^{14}N in the galaxy while Peimbert and Sarmiento (1984) argue against ^{14}N production by novae. The critical consideration is simply the fact that the characteristic ^{15}N/^{14}N ratio expected for nova ejecta is far too high with respect to solar for ^{14}N production to have galactic implications. Determinations of isotopic ratios for matter ejected by novae would provide important information regarding this matter.

^{13}C, ^{15}N, and ^{17}O. These isotopes, particularly ^{15}N, can be overproduced by large factors in nova explosions (Sparks, Starrfield and Truran 1976; Wiescher et al 1985). ^{15}N and ^{17}O overabundances of factors $\gtrsim 10^3$ are predicted for some models. Novae might in fact be able to account for the observed abundance levels of these two isotopes in galactic matter. Certainly, the ejecta of novae may be expected to be characterized by profoundly non-solar isotopic abundance patterns for carbon, nitrogen and oxygen. Detailed predictions will of course be sensitive to the temperature and convective histories of the ejected material.

^{19}F. Recent studies (Wiescher et al 1985) indicate that small overabundances of ^{19}F might also be achieved in nova ejecta. The predicted abundances are extremely temperature sensitive, and the required temperatures are on the high side of those typical of nova runaways. Based upon existing studies of the nuclear processing of matter in novae, it seems unlikely that novae represent the main site of ^{19}F production in the galaxy. It should be noted, however, that no other studied nucleosynthesis site has yet provided a sufficient level of production of ^{19}F to meet galactic requirements.

^{22}Na and ^{26}Al. Synthesis of the astrophysically important radioactive isotopes ^{22}Na ($\tau_{1/2}$ = 2.6 y) and ^{26}Al ($\tau_{1/2}$ = 7.2 x 10^5 y) can also be expected to occur under the explosive hydrogen burning conditions that are achieved in classical nova outbursts. The formation of significant concentrations of ^{22}Na provides an attractive possible explanation (Clayton 1975) for the highly ^{22}Ne-enriched Ne-E anomaly (Eberhardt et al 1979; Lewis et al 1979) identified in meteorites. Questions concerning ^{26}Al production are of course important both to consideration of the source of the ^{26}Mg anomalies found in meteorites, that correlate with Al/Mg ratios (Lee, Papanastassiou, and Wasserburg 1977), and to the interpretation of the diffuse galactic gamma-ray line emission from ^{26}Al (Mahoney et al 1984).

Numerical calculations of ^{22}Na and/or ^{26}Al production under explosive hydrogen-burning conditions have been performed by Arnould and Nørgaard (1978), Lazareff et al (1979), Arnould et al (1980),

Hillebrandt and Thielemann (1982) and Wiescher et al (1985). While
the situation here is complicated and there remain considerable uncertainties associated with predictions of nucleosynthesis in novae, it
is nevertheless possible to state several general conclusions. The
calculations of nucleosynthesis in novae by Hillebrandt and Thielemann
(1982) and Wiescher et al (1985) predict concentrations of ^{20}Ne and
^{21}Ne formed relative to ^{22}Ne + ^{22}Na which are too large to explain the
meteoritic Ne-E component. However, the ^{22}Na/^{23}Na production ratios
they obtain are consistent with the formation of a Ne-E component as a
consequence of ^{22}Na decay in grains formed in the expansion and
cooling of nova ejecta. We note that ^{22}Na decay provides an extremely
attractive mechanism for the formation of ^{22}Ne-enriched neon isotopic
anomalies, though the bulk of the ^{22}Ne in galactic matter was likely
formed rather as ^{22}Ne (Truran and Hillebrandt 1985).

The extent to which novae are responsible for the ^{26}Al present in
galactic matter is also of considerable interest. Clayton (1984) has
recently argued that novae represent the likely source of the ^{26}Al
detected in the interstellar medium by Mahoney et al (1984). His
estimate was based upon the numerical prediction of ^{26}Al production by
Hillebrandt and Thielemann (1982). The calculations by Wiescher et al
(1985), which include substantially revised thermonuclear reaction
rates, indicate a level of ^{26}Al production lower by a factor of
approximately ten than that of the previous work. Given their prediction of ^{26}Al production and reasonable estimates of nova frequencies
and mass contributions, it now seems difficult to understand how novae
can be a major galactic source of ^{26}Al. Further calculations are
necessary to clarify this issue.

4. NOVAE AND GALACTIC NUCLEOSYNTHESIS

In reviewing the role of novae in nucleosynthesis, it is important to
keep in mind both that the contributions of novae are small, relative
to the masses released into the interstellar medium by either planetary nebulae or supernova ejection, and that predictions regarding
novae are highly uncertain. The overall contribution of novae to the
abundance of a specific nucleus in the Galaxy is dependent upon the
frequency of nova outbursts, the average mass ejected per nova event,
and the fractional mass of the ejecta in the form of the nucleus of
interest. The rate of nova events for the Andromeda galaxy has been
determined to be 38 yr^{-1} (Arp 1956); we adopt a value 40 yr^{-1} for the
current rate of nova events in the Galaxy. (Any estimate of the time
dependence of the rate of nova activity over the history of the galaxy
is rendered uncertain by the fact that the progenitors of the classical novae have not been unambiguously determined.) Determinations of
the masses ejected by novae are also quite uncertain. Theoretical and
observational considerations suggest a "typical" mass ejected of order
10^{-4} M$_0$. Peimbert and Sarmiento (1984) argue for 10^{-5} M$_0$ ejected per
event on the basis of recent studies of Nova Cygni 1978 and Nova
Aquilae 1982, but both of these are quite fast novae which are distin-

guished by high levels of heavy element enrichment and thus may not reflect the behavior of the "average" classical nova. We adopt a value 10^{-4} $M_\odot$. We thus arrive at a rate of mass ejected per year for novae of (40 yr^{-1}) $(10^{-4} M_\odot/\text{event}) \sim 0.004$ $M_\odot \text{ yr}^{-1}$. This is a factor of ten or more below similar rough estimates for the contributions for planetary nebulae and supernovae.

In light of the fact that novae contribute only a small fraction of the mean mass of stellar ejecta returned to the interstellar medium, it is clear that only nuclei which are overproduced by significant factors are of interest with respect to galactic nucleosynthesis. The rarer isotopes of carbon, nitrogen and oxygen are particularly noteworthy in this regard. Determinations of isotopic-abundance patterns for nova ejecta, if possible, would provide critical information regarding this matter. Novae may also prove important contributors to the abundances of ^{7}Li and of the unstable nuclei ^{22}Na and ^{26}Al.

This research was supported in part by the National Science Foundation under grant AST 83-14415.

5. REFERENCES

Arnould, M., and Nørgaard, H. 1975, Astron. Astrophys. 42, 55.
Arnould, M., and Nørgaard, H. 1978, Astron. Astrophys. 64, 195.
Arnould, M., Nørgaard, H., Thielemann, F.-K., and Hillebrandt, W. 1980, Astrophys. J. 237, 931.
Arp, H.C. 1956, Astron. J. 61, 15.
Clayton, D.D. 1975, Nature 257, 36.
Clayton, D.D. 1984, Astrophys. J. 280, 144.
Eberhardt, D., Jungck, M.H.A., Meier, F.O., and Niederer, F. 1979, Astrophys. J. (Letters) 234, L169.
Ferland, G.J., and Shields, G.A. 1978, Astrophys. J. 226, 172.
Friedjung, M. 1979, Astron. Astrophys. 77, 357.
Gallagher, J.S., Hege, E.K., Kopriva, D.A., Williams, R.E., and Butcher, H.R. 1980, Astrophys. J. 237, 55.
Hillebrandt, W., and Thielemann, F.-K. 1982, Astrophys. J. 225, 617.
Kippenhahn, R., and Thomas, H.-C. 1978, Astron. Astrophys. 63, 265.
Law, W.Y., and Ritter, H. 1983, Astron. Astrophys. 123, 33.
Lazareff, B., Audouze, J., Starrfield, S., and Truran, J.W. 1979, Astrophys. J. 228, 875.
Lee, T., Papanastassiou, D.A., and Wasserburg, G.J. 1977, Astrophys. J. (Letters) 211, L107.
Lewis, R.S., Alaerts, L., Matsuda, J.-I., and Anders, E. 1979, Astrophys. J. (Letters) 234, L165.
MacDonald, J. 1984, Astrophys. J. 283, 241.
Mahoney, W.A., Ling, J.C., Wheaton, W.A., and Jacobson, A.S. 1984, Astrophys. J. 286, 578.
Nariai, K., Nomoto, K., and Sugimoto, D. 1980, Publ. Astr. Soc. Japan 32, 473.
Peimbert, M., and Sarmiento, A. 1984, Astron. Express 1, 97.

Prialnik, D., Livio, M., Shaviv, G., and Kovetz, A. 1982, Astrophys. J. 257, 312.
Prialnik, D., Shara, M.M., and Shaviv, G. 1978, Astron. Astrophys. 62, 339.
Prialnik, D., Shara, M.M., and Shaviv, G. 1979, Astron. Astrophys. 72, 192.
Snijders, M.A.J., Seaton, M.J., and Blades, J.C. 1982, in Proceedings of the Third European IAU Conference (Madrid), P. 177.
Sparks, W.M., Starrfield, S., and Truran, J.W. 1976, Astrophys. J. 208, 819.
Sparks, W.M., Starrfield, S., and Truran, J.W. 1978, Astrophys. J. 220, 1063.
Starrfield, W.M., Kenyon, S.J., Sparks, W.M., and Truran, J.W. 1982, Astrophys. J. 258, 683.
Starrfield, S., Sparks, W.M., and Truran, J.W. 1985, Astrophys. J. 291, 136.
Starrfield, S., Truran, J.W., Sparks, W.M. 1978, Astrophys. J. 226, 186.
Starrfield, S., Truran, J.W., Sparks, W.M., and Arnould, M. 1978, Astrophys. J. 222, 600.
Starrfield, S., Truran, J.W., Sparks, W.M., and Kutter, G.S. 1972, Astrophys. J. 176, 169.
Strickland, D.J., Penn, C.J., Seaton, M.J., Snijders, M.A.J., and Storey, P.J. 1981, M.N.R.A.S. 197, 107.
Truran, J.W. 1982, in C.A. Barnes, D.D. Clayton, and D.N. Schramm (EDS.), Essays in Nuclear Astrophysics, Cambridge University Press, Cambridge, P. 467.
Truran, J.W. 1985, in W.D. Arnett and J.W. Truran (EDS.). Nucleosynthesis: Challenges and New Developments, University of Chicago Press, Chicago.
Truran, J.W., and Hillebrandt, W. 1985, Preprint.
Tylenda, R. 1978, Acta Astron. 28, 333.
Wallace, R.K., and Woosley, S.E. 1981 Astrophys. J. Suppl. 45, 389.
Wiescher, M., Gorres, J., Thielemann, F.-K., and Ritter, H. 1985, Preprint.
Williams, R.E. 1972, Astrophys. J. (Letters) 261, L77.
Williams, R.E., and Gallagher, J.S. 1979, Astrophys. J. 228, 482.
Williams, R.E., Sparks, W.M., Starrfield, S., Ney, E.P., Truran, J.W., and Wyckoff, S. 1985, M.N.R.A.S. 212, 753.
Williams, R.E., Woolf, N.J., Hege, E.K., Moore, R.L., and Kopriva, D.A. 1978, Astrophys. J. 224, 171

REACTION RATES IN THE RP-PROCESS AND NUCLEOSYNTHESIS IN NOVAE

M. Wiescher
Institut für Kernchemie
Johannes Gutenberg Universität
Mainz
Federal Republic of Germany

J. Görres
Dept. of Physics
University of Pennsylvania
Philadelphia, PA 19104
USA

F.-K. Thielemann
Max-Planck-Institut für Physik und Astrophysik
Garching b. München
Federal Republic of Germany

H. Ritter
Universitätssternwarte
Maximilians Universität
München
Federal Republic of Germany

ABSTRACT. Nuclear structure information available on proton rich unstable nuclei were used to evaluate thermonuclear reaction rates in the rp-process in explosive hydrogen burning. These rates were applied in a systematic analysis for a variety of temperature conditions, appropriate to nova explosions, to study nucleosynthesis for isotopes of Ne Na Mg Al Si. The results are discussed in comparison with recent observations of elemental abundances in nova ejecta.

Recent observations of elemental abundances in novae outbursts have shown strong enhancements in Ne, Na, Mg, Al (Williams et al., 1983) and in Ne, Mg, Si, S (Snijders et al. 1984) for nova CrA 1981 and nova Aql 1982, respectively (Fig. 1). Nova events are interpreted as a thermonuclear runaway in an accreting hydrogen shell on top of a degenerate C-O-white dwarf in a stellar binary system. This results in temperatures during the outburst in the range of $1 \cdot 10^8 K < T < 3 \cdot 10^8 K$ (Starrfield et al.,1978). However, nucleosynthesis in hydrogen burning at such conditions does not lead to the observed elemental abundances (Arnould et al.,1980; Hillebrandt, Thielemann, 1982).

This raises the following questions, relevant for the general understanding of the nova mechanism:

1. Is the present knowledge of nuclear reactions in explosive hydrogen burning complete?

2. Can nova explosions result in higher temperatures, which enable a break-out from the hot CNO-cycle?

3. Can a nova explosion on a Ne-O-Mg-white dwarf explain the observed abundances?

In the following we attempt to provide answers to these questions, which may help to solve the present puzzle of abundances in nova ejecta.

For investigating these questions, reaction network calculations for hot hydrogen burning at different temperatures and densities were performed. The reaction network used contained 117 nuclei up to mass $A = 42$ linked by 473 beta-decay, proton, alpha and neutron reactions and their reverse reactions. The reaction rates were obtained, whenever possible, from the compilations of Fowler et al. (1975), Harris et al. (1983), Woosley et al. (1978). The rates for proton capture reactions on short-lived nuclei A<30, however, which are important in hot hydrogen burning, are rather uncertain. Wallace and Woosley (1981) estimated the rates for these reactions on the assumption that only one or two resonances make significant contributions to the reaction rates. Because additional structure data about the particular compound nuclei are available now we recalculated the rates taking into account all possible resonances as well as nonresonant reaction mechanism likely to contribute to the stellar reaction rate.

The energies of the possible resonances were calculated from the excitation energies of known proton unbound states in the compound nuclei. If the analogue nucleus does show a higher level density, additional states were assumed in that excitation range. Possible energy shifts between these analogue levels were taken into consideration. Corrections for the Thomas-Ehrman shifts were applied in particular for the level energies of the resonant states in ^{19}Ne$(p,\gamma)^{20}$Na. These shifts were calculated using standard formula (Thomas, 1952). Spin and parities for most

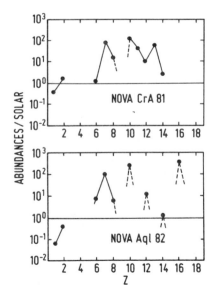

Figure 1. Observed elemental abundances in novae CrA81 and Aql 82 relative to solar abundances.

resonances were taken from the analogue states of the corresponding isospin multipletts. The resonance strengths which will determine the rate, were calculated by the well known relation:

$$\omega\gamma = \frac{2J+1}{(2j_T+1)(2j_p+1)} \frac{\Gamma_p \Gamma_\gamma}{\Gamma_{tot}} \qquad (1)$$

with J, j_T, j_r as spin of the resonance state, the target nucleus and the projectile, respectively. The total width Γ_{tot} and the partial widths Γ_p and Γ_γ for the entrance and exit channel of the particular resonance level determine the strength. The proton partial width is proportional to the Coulomb penetrability for a charged particle, which was calculated with standard codes for a nuclear channel radius $R = 1.20(A_T^{1/3} + A_r^{1/3})$, with A_T, A_p as target and projectile mass, respectively. The partial width is also proportional to the spectroscopic proton factor Θ_p^2, reflecting the single particle structure of the particular state. They were derived from the known spectroscopic neutron factors Θ_n^2 of the corresponding analogue levels (Endt, van der Leun, 1978). If no information was available a 'typical' value $\Theta_p^2 = 0.1$ has been assumed. If the analogue states of the resonances are bound, the γ-partial width was determined from the known lifetimes of these states. Otherwise, the single particle γ-width was calculated assuming predominantly decay to the ground-state or the first excited states, which were favoured by high energy and low l-transfer (E1, M1, E2 transitions). The results were weighted by the mean value of the distributions of known γ-ray transition strengths in the same mass region (Endt, 1979).

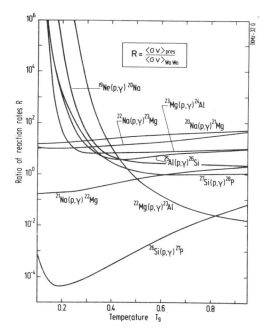

Figure 2. Comparison of the calculated reaction rates with previous estimates.

The direct capture (DC) is usually the dominant nonresonant term in (p,γ)-reactions and may therefore contribute considerably to the reaction rate. The cross section for possible DC-transitions to bound states were calculated in terms of the DC-model described by Rolfs, (1973). Usually the cross section is determined by the transitions to the groundstate and the first excited bound states. Only such transitions are therefore considered for the calculations. The spectroscopic factors for the final states were derived from the experimentally known ones of the analogue states (Ajzenberg-Selove, 1983; Endt, van der Leun, 1978) or from shell model predictions (Cole et al., 1976).

Figure 2 shows the comparion between the rates from Wallace and Woosley (1981) and the new ones described above. For the proton capture on the isotopes ^{20}Na, ^{21}Na, ^{22}Na the rates agree within one order of magnitude. However, the comparison of the different rates for the proton capture on ^{19}Ne, ^{21}Mg, ^{23}Mg, ^{25}Al and ^{27}Si shows a disagreement of several orders of magnitude in the temperature range below $T = 3 \cdot 10^8$ K. This results mainly from the addition of several new resonances as well as from the nonresonant direct capture contributions to the reaction rates. The new rate for ^{26}Si$(p,\gamma)^{27}$P is considerably smaller than quoted by Wallace and Woosley. The former rate was based mainly on the influence of a d-wave resonance at $E = 250$ keV. The existence of this proposed resonance could not be confirmed. The reaction rate at low temperatures is therefore mainly determined by a weak d-wave resonance at $E = 80$ keV and the direct capture process only. This results in the relatively small reaction rate.

The network calculations were performed for constant temperatures at $T = 1 \cdot 10^8$, $1.5 \cdot 10^8$, $2 \cdot 10^8$, $3 \cdot 10^8$, $4 \cdot 10^8$ and $5 \cdot 10^8$ K with a typical density of $\rho = 5 \cdot 10$ g/cm^3. The envelope composition was assumed to be 90% solar with a 10% admixture of C-O-white dwarf matter respective O-Ne-Mg-white dwarf matter. The calculations were stopped when a typical nova energy of $3 \cdot 10^{46}$ erg was reached.

The results of the calculations are displayed in terms of a reaction flow analysis in Figure 3. Shown are the time integrated reaction flows F_{ij} between the nuclei i and j:

$$F_{ij} = \int \left[\lambda_i Y_i + \sum_l N_A \rho \langle \sigma v \rangle_{il} Y_i Y_l - \sum_j \lambda_j Y_j - \sum_{jk} N_A \rho \langle \sigma v \rangle_{jk} Y_j Y_k \right] dt \ . \quad (2)$$

Y_{ijkl} are the nuclear abundances, λ_{ij} are the decay rates for β-decay or photodisintegration and $\langle \sigma v \rangle$ are the two particle interaction rates. Thick solid lines represent flows of $F > 10^{-3}$, dashed lines flows $10^{-3} > F > 10^{-4}$ and thin solid lines flows $10^{-4} > F > 10^{-6}$. It is easily to observe that at low temperatures $T < 1.6 \cdot 10^8$ K the dominant burning sequence is the hot CNO-multicycle, while in the temperature range $1.6 \cdot 10^8$ K $< T < 2 \cdot 10^8$ K also the hot NeNa- as well as the hot MgAl-cycle will show up. At higher temperatures $2 \cdot 10^8$ K $< T < 3.5 \cdot 10^8$ K the CNO-cycles are still closed, no leak reaction occurs between the CNO region and the NeNa region, but the reaction path in the mass region beyond Ne already runs far off line of stability and is determined by the capture rates on the particular radioactive nuclei involved. At temperatures above $T = 3.5 \cdot 10$ K CNO material will also leak to the NeNaMgAl region, by the reaction sequence ^{15}O$(\alpha,\gamma)^{19}$Ne$(p,\gamma)^{20}$Na. Beyond Ne, as already for temperatures in excess of $2 \cdot 10^8$ K, a continous flow-pattern emerges close to the proton drip line underlining the change from a series of burning cycles to a rapid proton capture process.

The reaction network was also applied to existing hydrodynamic nova models by Starrfield et al. (1978). The calculations were performed for temperature and density profiles of model 1, 2, 4, 8 assuming solar abundances with a slight admixture of

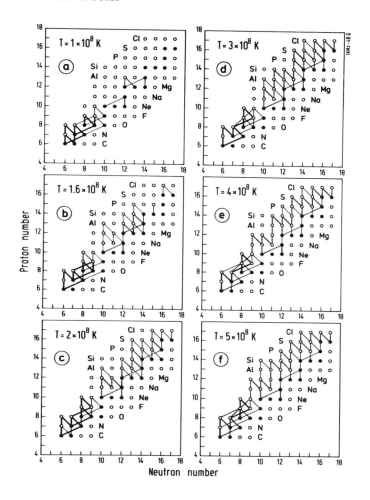

Figure 3. Reaction flow for different temperatures at constant density $\rho = 5 \cdot 10^3 \, g/cm^3$.

C-O-white dwarf matter (Starrfield et al. 1978, Hillebrandt, Thielemann, 1982). Table I shows the resulting isotopic abundances by number of nuclei produced at the base of the envelope where the highest temperatures occur. This neglects possible contributions from nucleosynthesis in the outer zones of the envelope at lower temperatures and from mixing between the burning zones. However, comparision with Figure 1 indicates clearly that for these model conditions, even with the modified, mostly increased reaction rates, no enrichment for Ne, Mg, Al, Si was found. This is because only above $T = 3.5 \cdot 10^8$ K CNO-material will be processed by the reaction sequence $^{15}O(\alpha,\gamma)^{19}Ne(p,\gamma)^{20}Na$ into NeNa, while the nova models used only predict peak temperatures far below $T = 3 \cdot 10^8$ K.

TABLE I				
Predicted Ejecta from Nuclesynthesis Calculations Model*				
	1	2	4	8
^{1}H	0.690	0.693	0.370	0.639
^{4}He	5.63(-2)	5.74(-2)	6.67(-2)	5.68(-2)
^{11}B	1.88(-10)	4.09(-11)	1.25(-11)	3.94(-10)
^{12}C	1.55(-4)	1.70(-4)	1.20(-3)	4.63(4)
^{13}C	8.325(-4)	2.03(-3)	1.08(-2)	3.61(-3)
^{14}N	1.891(-3)	1.60(-3)	1.30(-2)	3.25(-3)
^{15}N	2.69(-3)	1.44(-3)	1.10(-3)	2.07(-3)
^{16}O	8.75(-6)	2.39(-6)	1.13(-4)	4.83(-6)
^{17}O	5.29(-7)	4.95(-7)	4.42(-5)	2.26(-5)
^{18}O	1.09(-7)	1.56(-7)	7.81(-5)	2.20(-5)
^{19}F	6.42(-10)	7.64(-11)	8.63(-8)	6.39(-9)
^{20}Ne	4.60(-6)	2.34(-6)	5.21(-5)	5.37(-5)
^{21}Ne	4.96(-9)	3.03(-9)	4.21(-8)	5.32(-8)
^{22}Ne	0.00	0.00	0.00	0.00
^{22}Na	8.01(-10)	1.29(-9)	2.38(-8)	2.45(-8)
^{23}Na	1.42(-7)	2.79(-8)	1.37(-7)	2.46(-7)
^{24}Mg	2.40(-9)	5.22(-10)	8.97(-9)	5.92(-9)
^{25}Mg	5.23(-7)	2.34(-7)	2.04(-5)	4.68(-6)
^{26}Mg	2.74(-10)	9.11(-12)	3.68(-10)	1.16(-10)
^{27}Al	2.11(-8)	4.28(-9)	8.87(-8)	4.86(-5)
^{28}Si	4.42(-6)	2.52(-6)	1.94(-5)	5.38(-5)
^{29}Si	4.05(-8)	2.00(-8)	9.00(-7)	4.08(-7)
^{30}Si	5.01(-7)	3.44(-7)	6.71(-7)	2.95(-7)
^{31}P	2.42(-7)	1.13(-7)	1.25(-7)	2.29(-6)
^{32}S	1.55(-6)	8.26(-7)	9.52(-6)	1.05(-5)
^{33}S	1.31(-6)	6.89(-7)	1.16(-7)	4.82(-6)
^{34}S	1.38(-6)	7.35(-7)	4.2(-7)	2.20(-6)
^{36}S	0.00	0.00	0.00	0.00
^{35}Cl	5.93(-6)	3.36(-6)	7.60(-8)	1.12(-6)
^{36}Ar	1.26(-4)	1.38(-4)	1.80(-6)	2.43(-6)

*Following temperature and densities of models given in Starrfield, Truran, and Sparks (1978).

This indicates, assuming the modified reaction rates are sufficiently correct, that nova events may operate at slightly higher temperatures up to $T = 5 \cdot 10^8$ K as already suggested by Sugimoto et al. (1980). Hydrogen burning at these high temperatures certainly will allow to produce enrichment in Ne, Na, Mg, etc. This is shown in Figure 4a which displays the resulting elemental abundances relative to solar abundances after hydrogen burning at $T = 5 \cdot 10^8$ K. Similar results may be obtained by nova explosions on a Ne-O-Mg-white dwarf, which may exist in nature as well (Nomoto 1985, Law, Ritter 1983). In such a case, the enrichment in Ne, Mg does not require high temperature burning, but results from mixing of white dwarf matter into the envelope, being preprocessed by carbon burning. Figure 4b shows the calculated elemental abundances for hot hydrogen burning at a typical nova temperature $T = 1.6 \cdot 10^8$ K, assuming for the envelope solar abundances plus 10% enrichment from material which underwent prior carbon burning (Arnett, Thielemann, 1985).

To prove either suggestion for the observed nova events to be correct requires more detailed hydrodynamic nova calculations, considering not only the burning conditions at the base of the envelope but also the effects of shell burning at lower temperatures and densities in the outer zones of the envelope.

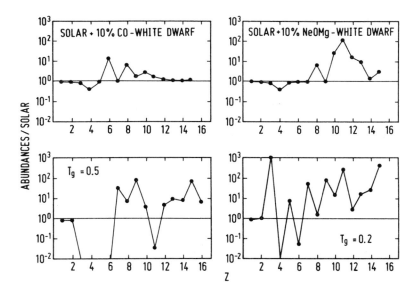

Figure 4. Start abundances and final elemental abundances for hot hydrogen burning ($5 \cdot 10^8$K) in solar matter enriched in C and O and for hydrogen burning at lower temperatures ($1.6 \cdot 10^8$K) in solar matter enriched in O, Ne Na Mg Al.

REFERENCES

1. Ajzenberg-Selove, F., 1983, *Nucl. Phys.* **A392**, 1
2. Arnett, W. D., and Thielemann, F.-K., 1985, *Ap. J.*, in press.
3. Arnould, M., Nørgaard, H., Thielemann, F.-K., and Hillebrandt, W., 1980, *Ap. J.*, 237, 931
4. Cole, B. J., Watt, A., and Whitehead, R. R., 1976, *J. Phys.* G**2**, 501
5. Endt, P. M., and Van der Leun, C., 1978 *Nucl. Phys.* **A310**, 1
6. Endt, P. M., 1979, *At. Data Nucl. Data Tables* **23**, 3
7. Fowler, W. A., Caughlan, G. R., and Zimmerman, B. A., 1967, *Ann. Rev. Astron. Astrophys.* **5**, 525
8. Fowler, W. A., Caughlan, G. R., and Zimmerman, B. A., 1975, *Ann. Rev. Astron. Astrophys.*, **13**, 69
9. Harris, M., Fowler, W. A., Caughlan, G. R., and Zimmerman, B. A., *Ann. Rev. Astron. Astrophys.* **21**, 165
10. Hillebrandt, W., and Thielemann, F.-K., 1982, *Ap. J.* **255**, 612
11. Law, W. Y., and Ritter, M., 1983, *Astron. Astrophys.* **123**, 33
12. Nomoto, K. 1985, private communication
13. Rolfs, C., 1973, *Nucl. Phys.* **A217**, 29
14. Snijders, M. A. J., Batt, T. J., Seaton, M. J., Blades, F. C., and Morton, D. C., 1984, preprint
15. Starrfield, S. Truran, D. W., and Sparks, W. M., 1978, *Ap. J.*, **226**, 186
16. Sugimato, D., Fujimoto, M. Y., Nariai, K., and Nomoto, K., 1980, in *Proc. Inst. Astr. Union Coll. No.* **53**
17. Thomas, R. G, 1952, *Phys. Rev.* **88**, 1109
18. Wallace, R. K., and Woosley, S. E., 1980 *Ap. J. Suppl.* **45**, 389
19. Williams, R. E., Ney, E. P. Sparks, W. M., Starrfield, S. G. Truran, J. W., and Wyckoff, S., 1984, *Mon. Not. Roy. Astr. Soc* submitted.
20. Woosley, S. E., Fowler, W. A., Holmes, J. A., and Zimmerman, B. A, 1978, *At. Data Nucl. Data Tables* **22**, 371

TYPE I SUPERNOVAE

J. Craig Wheeler
Department of Astronomy
University of Texas
Austin Texas
USA

ABSTRACT. Several projects involving Type I supernovae and their impact on nucleosynthesis are described. 1) A major effort is underway to compute the spectrum at epochs near maximum light. 2) Dynamical calculations have been done exploring the range of possible outcomes of the evolution of binary white dwarfs. 3) Consideration is given to the need to eject iron from the Galaxy to avoid overproduction by Type I events. 4) The nature of the class of peculiar Type I supernovae is explored.

1. INTRODUCTION

A number of other papers in these proceedings discuss various aspects of Type I supernovae (see contributions by Canal, Matteucci, Thielemann, and Woosley). To avoid duplication, this paper will present a potpourri of four topics which all bear on the subject of nucleosynthesis by Type I supernovae.

A crucial problem is the direct determination of abundances of the ejecta by analysis of the spectrum. Three components are necessary, a thorough set of modern observational data, detailed hydrodynamic and nucleosynthetic models, and theoretical spectra to compare the two. Great progress has been made in the first two areas. McDonald Observatory has been particularly fruitful in providing a detailed spectral record of many supernovae (Branch et al 1983; Wheeler 1985). Elaborate numerical calculations give a very detailed picture of the expected conditions and abundances in specific models (Nomoto, Thielemann, and Yokoi 1984; Woosley, Axelrod, and Weaver 1984). Until recently the third leg of this analytical triad has been the weakest. The pioneering work of Branch (Branch et al 1982) and Axelrod (1980 a,b) has ushered in a new era of spectral modeling. Section 2 describes the effort underway at Texas to calculate detailed model atmospheres of Type I supernovae.

The contribution of Type I supernovae to nucleosynthesis will not be completely understood without an understanding of their astrophysical origin. The carbon deflagration model has proved capable of reproducing many observational aspects of Type I supernovae: light curves, spectra, kinematics, and abundances. There are manifest problems, however, in determining the origin of Type I supernovae. In particular, while binary transfer of mass onto a white dwarf has long been discussed, no self-consistent way has been discovered to produce a Type I supernova by means of the transfer of hydrogen onto a carbon/oxygen white dwarf. Sutherland and Wheeler (1984) summarize the problems in the following fashion. At low mass accretion rates, nova explosions drive off a majority of the accreted hydrogen, and perhaps some of the dwarf core material as well, so the Chandrasekhar limit is not attained quickly enough. At somewhat higher rates, "double detonation" driven by the ignition of a thick degenerate helium shell consumes the star leaving insufficient elements of intermediate mass. At higher mass accretion rates, hydrogen and helium will burn quiescently on the surface of the dwarf, but produce a luminosity exceeding observed limits. At very high accretion rates, the matter can not be assimilated by the dwarf and so will linger around the binary, presumably to contaminate any explosion with hydrogen, which is not observed in the spectrum. Thus, according to constraints as they are currently perceived, no hydrogen accretion rate from zero to infinity can yield a supernova of the observed properties to be a Type I. Iben and Tutukov (1984) and Webbink (1984) have proposed that these problems can be avoided if the accreted matter is not hydrogen. They outline an evolutionary sequence that will lead to the formation and coalescence of two white dwarfs composed of helium or carbon and oxygen. The process by which the two dwarfs coalesce is not well understood. In Section 3 a set of initial models is described in which the outcome of the complex evolution is parametrized. Dynamic models then set constraints on the possible evolutionary outcomes which could be consistent with the observations of Type I supernovae.

A well known problem with the carbon deflagration model is that it predicts the ejection of order a solar mass of iron per event There is a danger of overcontaminating the Galaxy with iron in this model, although the severity of the problem is controversial. In Section 4 the question of the overproduction of iron is addressed, and the possibility that the ejecta of Type I supernovae are expelled from the Galaxy is considered.

Recent observations confirm the existence of a spectrally homogeneous class of peculiar Type I supernovae. These observations and the nature of this class of supernovae are described in Section 5.

2. ATMOSPHERE MODELS

The attempt to better understand the spectra of Type I (and other) supernovae is being undertaken with the aid of a code developed by Robert Harkness (1985) expressly to handle this problem. The code solves the special relativistic radiative transfer equations in a spherical geometry using the co-moving form and numerical methods outlined by Mihalas, Kunasz, and Hummer (1976) and Mihalas (1980). The code can handle the full non-LTE problem, and is efficiently vectorized and optimized to run on a Cray 1s or multiprocessor XMP. Model W7 of Nomoto, Thielemann, and Yokoi (1984) which represents a deflagrating carbon/oxygen white dwarf is used as input data. For the Type I problem LTE is assumed because of the large number of heavy elements which must be handled.

One of the outstanding problems of the nature of the spectrum of Type I supernovae near maximum light is the origin of the severe UV deficiency which sets in at about 3000 Å (see, eg, Branch et al (1983). This deficiency and the attendant flux redistribution must be understood before Type I supernovae can be used with complete confidence as distance indicators, and before the composition and structure of the ejecta can be accurately deduced from the spectral lines. Close attention has been paid to the origin of the continuous opacity in order to understand the shape of the continuum in Type I events. At low frequencies the continuous opacity is principally free-free. At high frequencies, it is due in great part to bound-free absorption from the ground states of cobalt and iron in the core of the model. The ground state bound-free opacities are very large, of order 10^4 cm^2/gm, and will maintain a large optical depth until well after maximum light, but they contribute little to the opacity at wavelengths < 300 Å. Excited state bound-free opacites due to Co and Fe are expected to contribute between 3000 and 300 Å but atomic data on these transitions is lacking. As a stop-gap, the appropriate levels have been included in the code with LTE populations and a constant reasonable but arbitrary cross section of 10 megabarns has been assigned. The resulting contributions cause the bound-free opacity to rise from the minimum at about 3000 Å resulting from free-free to the ground state contribution at 300 Å This gives a maximum of the transmitted flux at about 3000 Å qualitatively in agreement with the observations. Quantitatively, however, the decline in flux at short wavelengths is not as severe as observed. The excess flux in the UV severely affects the formation of the spectral lines as well as the continuum. Without the excited state opacity, the UV radiation from the hot Co/Fe core blasts directly into the mantle and changes the ionization and excitation structure, thus degrading the fit to the lines.

Whether the disagreement is due to some source of opacity which has not yet been included, to the simplicity of the current method of treating excited states or to the lack of some physical process in the model remains to be determined. Thanks to communications with D. Dearborn and M. Howard at this meeting, we may be able to obtain relevant excited-state bound free transition probabilities from Livermore.

3. BINARY WHITE DWARF MODELS

To explore the importance of binary white dwarf evolution to the origin of Type I supernovae, a number of simple parametrized models have been been computed (Wheeler, Li, Sutherland, and Swartz 1985). We assume that gravitational radiation causes two He or C/O dwarfs to spiral together until the smaller mass, larger volume star fills its Roche lobe as shown in the first part of Figure 1. The process of mass loss from a degenerate star filling its Roche lobe is not well understood. Departures from spherical symmetry of the Roche lobe may affect the systematics drastically, and the effect of the angular momentum of the disrupted matter is a severe complication. If the two dwarfs have the same composition the likely outcome is for the smaller dwarf to be disrupted on nearly a dynamical timescale and its mass deposited suddenly as an envelope on the larger mass dwarf gas illustrated in the second part of Figure 1. (Iben and Tutukov 1984; Webbink 1984). If a He dwarf is paired with a C/O dwarf, the helium dwarf is likely to be the less massive, and to be disrupted on a longer timescale such that it gets engulfed in a common envelope of its own making. This extra complication and the fact that accretion of He onto a C/O core is relatively mundane compared to the other possibilities has led us to set this possibility aside and consider the cases of two white dwarfs of similar composition.

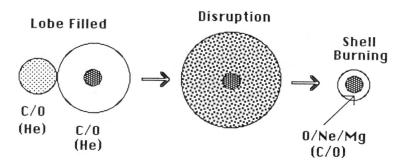

Figure 1 - The disruption of a white dwarf which fills its Roche lobe produces an envelope around its companion. A burning shell is assumed to consume the envelope and a portion of the inner core.

Studies of the possible outcome of the accretion of C/O onto C/O are just beginning (Nomoto and Saio 1985; Woosley 1985). The first calculations assume spherically symmetric accretion at a constant rate near the Eddington limit. These studies show that a burning shell ignites which consumes the accreted matter, and which eats into the inner core as shown in the third part of Figure 1. Even with the simplifying assumption of spherical symmetry, the validity of the assumption of a constant accretion rate can be questioned. A more representative situation might be to place an envelope of fixed mass (equal to that of the disrupted dwarf) on top of the core, and allow the shell burning to proceed in quasi-thermal equilibrium. Calculations corresponding to the accretion of He onto He have apparently not been done.

Because of the uncertainties in the evolution of binary dwarfs, we have chosen to adopt a different approach. We have constructed a series of parametrized models which correspond to possible outcomes of the phase of disruption and shell burning. We have then initiated thermonuclear explosions in these models and studied the dynamics to determine which are viable candidates for Type I supernovae. This allows us to set constraints on the evolution which must be satisfied if it is to lead to Type I supernovae.

We envisage the general result of the evolution of two white dwarfs of identical composition to yield a structure with an inner core of the original composition, and an outer mantle composed of the burning products of that compositon (see Figure 1). That is, the result of accretion and shell burning of He is a C/O mantle on a He core, and the result for C/O is an O/Ne/Mg mantle on a C/O core. In our calculations the mass of the inner core is the principle free parameter.

For the case of two C/O dwarfs, we assume a total mass very near the Chandrasekhar limit and a central density of 10^9 g/cm^3. The latter is a bit lower than normal on the assumption that the rapid accretion will promote early ignition. Ignition is initiated in the center and followed by a deflagration scheme (Sutherland and Wheeler 1984). The burning in the outer O/Ne/Mg mantle is followed by a simple alpha chain plus ^{16}O-^{16}O. We find that the mantle is very difficult to deflagrate, and provides virtually no energy to power the explosion. This means that the C/O core must itself be massive enough to drive the thermonuclear explosion, and hence in excess of about 0.8 M_o (Sutherland and Wheeler 1984). We find, for instance that a model with a C/O core mass of only 0.5 M_o produces expansion velocities of less than 7000 km/s in the bulk of the matter, insufficient to correspond to a Type I explosion.

The exploratory calculations of the accretion of C/O onto C/O have indicated that the burning shell penetrates to the center of the star (Nomoto and Saio 1985; Woosley 1985). If this result continues to hold even approximately in future more realistic calculations, then double C/O dwarf evolution can not lead to Type I supernovae.

The case of two He dwarfs has another degree of freedom in that the mass needed to trigger an explosion can be less than the Chandrasekhar mass. We have considered models with a total mass of 0.84 M_o guided by the calculation of accretion of H onto He dwarfs by Nomoto and Sugimoto (1977). In these models the outer mantle of C/O overlying the He core is at low densities, $<10^{+7}$ gm/cm^3. Such densities normally cannot support a self-propagating detonation in C/O (Buchler,

Wheeler, and Barkat 1971). We anticipated that the envelope would be only slightly processed to provide the observed intermediate mass elements, and the high specific energy of the helium would provide the requisite kinematics in a model which was not locked into the Chandrasekhar mass. This feature might be useful to explain the Pskovskii/Branch correlation between light curve shape and velocity (Branch 1982).

We found, however, that the helium is so volitile that it acts as an inner piston and gives rise to an *overdriven* detonation in the mantle. The result is to consume the whole model to iron peak elements, leaving none of the observed intermediate mass elements. Only if we reduced the mass fraction of C to be well below 0.1 could we avoid this consequence. As a result, we must conclude that the evolution of double He white dwarfs also fails to produce reasonable models for Type I explosions. The evolutionary people must be yet even more clever!

4. IRON PRODUCTION AND EJECTION FROM THE GALAXY

The carbon deflagration model for Type I supernovae raises questions about a possible overproduction of iron in the Galaxy. The model requires > 0.4 M_o of ^{56}Fe per event (Arnett, Branch, and Wheeler 1985). The rate of production of Fe in the Galaxy is about 0.005 M_o pc^{-2}Gyr^{-1} (Twarog 1980). Twarog and Wheeler (1982) used a chemical enrichment model with constant infall to deduce that the allowed production of Fe per Type I event was 0.3 M_o r_{-11} where r_{-11} is the rate of Type I explosions in units of 10^{-11} pc^{-2} yr^{-1} which in turn corresponds approximately to one event in 100 years in the whole Galaxy. At the lower limit of the model production rate, and considering the various uncertainties, the problem seems to be nearly explained away.

Ostriker and Wheeler (1985) have re-examined this problem to reconcile their differing estimates of the severity of the iron production problem. An examination of the contribution of the individual components contributing to the enrichment at the current epoch revealed that of the 0.3 M_o per event estimated by Twarog and Wheeler, 0.2 M_o was allowed by dilution due to infall and only 0.1 M_o was allowed by removal of Fe from the ISM by ongoing star formation. Thus if there is no current infall, Type I supernovae can only eject of order 0.1 M_o per event, and the discrepancy with the models is of order a factor of four.

A simple estimate of the infall power due to matter falling into the Galactic gravitational potential (a solar mass per year at 1000 km/s) is of order 10^{41} erg/s. The power generated by Type I supernovae (10^{-2} per year with 10^{51} ergs) is of the same order. The observed soft X-ray flux, however, is only of order 10^{39} erg/s. These arguments suggest that there can not be infall currently, and that the thermal energy as well as the iron ejecta must be expelled from the Galaxy. Such a selective expulsion process for Type I events which are naturally objects of large scale height might prevent Type I remnants from contributing to the population of old radio remnants (Li, Wheeler, and Bash 1984) and account for the accumulation of Fe in the intracluster gas.

5. PECULIAR TYPE I SUPERNOVAE

Two recent supernovae SN 1983n in M83, and SN19841 in NGC 991 have confirmed the existence of a spectrally distinct class of peculiar Type I supernovae (Wheeler and Levreault 1985). The spectra of these two events are virtually identical. They do not resemble the spectra of Type II supernovae at maximum light, and show no apparent evidence for hydrogen. On the other hand, they do not have the major absorption feature at 6100 Å that characterizes classical Type I events and otherwise the maximum light spectra resemble classical Type I events a month or two past maximum.

The peculiar Type I events are dimmer at maximum light than classical Type I supernovae by about a factor of 4, although the shape of the light curve is not much different from that of classical Type I events. The peculiar Type I supernovae seem to always be associated with Population I environments, H II regions, spiral arms, bars, etc. in strong contrast to the classical Type I events. The similarity of the velocity from the Doppler width of the spectral features, and of the ejected mass from the width of the light curve peak, suggests that the kinetic energy of the ejecta is nearly the same for the classical and peculiar events. If these peculiar events are powered by radioactive decay of nickel, however, then only a small amount can be ejected, of order 0.2 M_o. This suggests that the kinetic energy of the peculiar events can not arise from thermonuclear energy alone, the popular hypothesis for the classical events. The only other obvious source of energy is core collapse.

If the peculiar events eject about 1.5 M_o, the same as the classical events, and leave behind a neutron star, then the mass of the immediate progenitor must be of order 2 - 3 M_o. This mass represents the core of a star of original main sequence mass of 10 - 20 M_o. A progenitor of this mass is consistent with a population I progenitor, but not necessarily with a Wolf-Rayet star, which would be expected to have considerably more mass and hence to produce a light curve of excessive width (but see the contribution here of Cahen).

These arguments suggest that although the spectra of the peculiar events qualitatively resembles those of Type I events, the physics of the explosion may be more closely related to Type II events. The estimates of the mass range suggest that the peculiar Type I events may arise from the same stars that normally produce Type II supernovae, but under circumstances that produce a shedding of the envelope by mass ejection or transfer. Here is another fascinating challenge to the theory of stellar evolution.

References.

Arnett, W. D., Branch, D., and Wheeler, J. C. 1985, Nature, in press.
Axelrod, T. S. 1980a, Thesis, University of California, Santa Cruz.
Axelrod, T. S. 1980b, in Type I Supernovae, ed. J. C. Wheeler (Austin: University of Texas) p. 80.
Branch, D. 1982, Ap. J., **258**, 35.
Branch, D., Buta, R., Falk, S. W., McCall, M. L., Sutherland, P. G., Uomoto, A., Wheeler, J. C., and Wills, B. J. 1982, Ap. J. Letters, **252**, L61.
Branch, D., Lacy, C. H., McCall, M. L., Sutherland, P. G., Uomoto, A., Wheeler, J. C., Wills, B. J. 1983, Ap. J., **270**, 123.
Buchler, J. -R., J. C. Wheeler, and Barkat, Z., 1971, Ap. J., **167**, 465.
Harkness, R. P. 1985, Supernovae as Distance Indicators, ed. N. Bartel, (Berlin, Springer- Verlag) p. 183.
Iben, I., Jr., and Tutukov, A. V. 1984, Ap. J. Supplement, **54**, 335.
Li, Z. W., Wheeler, J. C., and Bash, F. N. 1984, Stellar Nucleosynthesis, ed. C. Chiosi and A. Renzini, (Dordrecht: Reidel) p. 49.
Mihalas, D. 1980, Ap. J., **237**, 574.
Mihalas, D., Kunasz, P. B., and Hummer, D. G. 1976, Ap. J., **206**, 515.
Nomoto, K., and Saio, H. 1985, private communication.
Nomoto, K., Sugimoto, D., 1977, Publ. Astr. Soc. Japan, **29**, 765.
Nomoto, K., Thielemann, F. -K, and Yokoi, K. 1984, Ap. J., **286**, 644.
Ostriker, J. P. and Wheeler, J. C. 1985, in preparation.
Sutherland, P. G. and Wheeler, J. C. 1984, Ap. J., **280**, 282.
Twarog, B. A. 1980, Ap. J., **242**, 242.
Twarog, B. A. and Wheeler, J. C. 1982, Ap. J., **261**, 636.
Webbink, R. F. 1984, Ap. J. Letters, **277**, 355.
Wheeler, J. C. 1985, in Supernovae as Distance Indicators, ed. N. Bartel (New York: Springer-Verlag) p. 34.
Wheeler, J. C. and Levreault, R. 1985, Ap. J. Letter, in press.
Wheeler, J. C., Li, Z. W., Sutherland, P. G., and Swartz, D. 1985, in preparation.
Woosley, S. E. 1985, private communication.
Woosley, S. E., Axelrod, T. S., and Weaver, T. A. 1984, Nucleosynthesis, ed. C. Chiosi and A. Renzini (Dordrecht, Reidel) p. 263.

NUCLEOSYNTHESIS AND TYPE I SUPERNOVAE

R. Canal
Departamento de Física de la Tierra y del Cosmos
University of Granada
and
Instituto de Astrofísica de Andalucía (C.S.I.C.)
and
Grup d'Astrofísica del Institut d'Estudis Catalans
18001 Granada, Spain

J. Isern, J. Labay, and R. López
Departamento de Física de la Tierra y del Cosmos
University of Barcelona
and
Grup d'Astrofísica del Institut d'Estudis Catalans
08028 Barcelona, Spain

ABSTRACT. The white dwarf model of Type I supernovae is discussed with particular emphasis on the often neglected effects of crystallization upon thermonuclear ignition and burning propagation. Preliminary hydrodynamic results are presented for different assumptions concerning the outcome of phase transition in cooling white dwarfs. Light curves associated with those models are shown to reproduce the "Pskovskii-Branch effect". The possible relevance of this to the determination of Hubble's constant and ways to avoid overproduction of ^{58}Fe are finally discussed.

1. INTRODUCTION

1.1 The white dwarf model.

Exploding white dwarfs are now the most widely accepted model for Type I supernova outbursts (SNI). In this model, hydrodynamic thermonuclear ignition propagating through an electron-degenerate core produces variable amounts of ^{56}Ni from incineration of a fraction of the core, plus intermediate-mass elements (O, Mg, S, Si, Ca) from partial burning of the rest of the material (Müller and Arnett 1982, 1985; Sutherland and Wheeler 1984; Nomoto, Thielemann, and Yokoi 1984). The decay of ^{56}Ni through ^{56}Co to ^{56}Fe would account for the shape of the light curves and the late-time spectra (Arnett 1982; Axelrod 1980), while the yield of intermediate mass elements would account for the spectra at maximum light (Branch et al. 1982, 1983). The corresponding scenarios involve white dwarfs in close binary systems, accreting mass up to the point of ignition (Schatzman 1963; Iben and Tutukov 1984). This includes both He and C-O white dwarfs, with either main-sequence stars, red giants or white dwarfs (again He or C-O) as possible companions.

1.2 Modes of burning propagation and their implications

Detonation has been the first suggested mechanism for propagating the thermonuclear burning throughout the star, in the context of degenerate carbon ignition in the cores of red giants of intermediate mass: $4 M_\odot \lesssim M \lesssim 8 M_\odot$ (Arnett 1969). The burning would be supersonically propagated by the shock wave generated by the initial flash. In the case of a central ignition, this would lead to complete incineration and disruption of the white dwarf. The reason is that

$$\tau_{prop} < \tau_{exp} < \tau_{EC} \qquad (1)$$

those being the time scales for burning propagation across the core, for core expansion, and for electron captures on the incinerated material. A typical value of τ_{prop} for a detonation wave is ~ 0.1 sec. Core expansion during this time is negligible and densities $\sim 10^{10}$ g cm^{-3} are required for the electron captures to compete with hydrodynamic expansion. Deflagration has been later advocated as the relevant mechanism of burning propagation, for carbon ignition at least (Mazurek, Meier, and Wheeler 1977). It consists in the subsonic propagation by Rayleigh-Taylor instability and it results in the partial incineration and total disruption of the core (Sutherland and Wheeeler 1984; Nomoto, Thielemann, and Yokoi 1984). We have, in this case:

$$\tau_{prop} \sim \tau_{exp} < \tau_{EC} \qquad (2)$$

τ_{prop} being ~ 1 sec. The core has enough time to expand before burning is complete. Intermediate-mass elements, as well as some unburnt oxygen and carbon can be ejected.
Hydrodynamic burning. Detailed hydrodynamic calculations (Müller and Arnett 1982, 1985) show the propagation of a turbulent, inhomogeneous combustion front. Only a fraction of the core is incinerated but the outcome is also total disruption. The two-dimensional numerical experiments so far performed show the unreliability of one-dimensional schemes for predicting detailed nucleosynthesis (see Arnett, this volume).

All of the aforementioned calculations do assume a fluid core at the start of ignition. Burning propagation thus happens on hydrodynamical time scales. This determines the final disruption. Fluid cores, however, are not the most general case when dealing with mass-accreting white dwarfs. We will see that, by also considering partially solid cores, the range of possible outcomes is appreciably broadened. In particular, bound remnants can be left after a SNI outburst and variable amounts of material can be ejected.

2. WHITE DWARF COOLING

Cooling C-O white dwarf models predict the growth of a solid core when the internal temperatures fall below the freezing point of the C-O mixture (which is function of the density). Specifically, the time for crystallization of the inner half of a C-O white dwarf of 1 $M_\odot$ is $\tau_{crys} = 7.3 \times 10^8$ yr (Lamb and Van Horn 1975). Times of this order are short enough to be accommodated in the interval following white dwarf formation and preceding mass accretion, for most scenarios of SNI production in close binary systems (see, for example, Iben and Tutukov 1984). Thus, partially solid cores should be expected at ignition, in a significant fraction of cases at least. Their size will depend on the cooling times, initial masses, and accretion rates (Hernanz et al. 1985, in preparation).

3. PARTIALLY SOLID CORES AND THEIR PHYSICS

According to recent Monte Carlo simulations (Slattery, Doolen, and De Witt 1980), solidification in a cooling, high-density, one-component plasma happens for $\Gamma_{crit} = 177 \pm 3$, this parameter being the ratio of Coulomb to thermal energy. What happens in a two-component plasma such a $^{12}C-^{16}O$ mixture is still an open question. Since Γ is proportional to Z^2 for fixed temperature and density, it is not obvious that carbon and oxygen will crystallize together. We have three different possibilities:
a) <u>Formation of a random C-O alloy</u>. In this case, the main effect is that thermonuclear burning can only propagate by conduction. The electron captures can then compete with expansion, since pycnonuclear reactions start at high densities ($\sim 10^{10}$ g cm^{-3}) and the velocities of the conductive burning front are small, compared to the hydrodynamical time scale (Canal and Isern 1979; Isern, Labay, and Canal 1984):

$$\tau_{EC} < \tau_{exp} < \tau_{prop} \qquad (3)$$

τ_{prop} being ~ 500 sec. This value is obtained by using the diffusion approximation for calculating the speed of the heat wave. A more accurate treatment, dealing with the mi-

croscopic structure of the front is currently under way (Isern and Schatzman 1985, in preparation). The value for the pycnonuclear ignition density ($\rho \approx 1 \times 10^{10}$ g cm^{-3}) must be stressed, since spuriously low values have recently been quoted in the literature (see Mochkovich, this volume, for a complete discussion).
b) <u>Non-random C-O alloy</u>. Schatzman (1983) has shown that, in the solid phase, it is more energetically favourable to have the carbon ions surrounded by oxygen ions, rather than having other carbon ions as neighbours. Thus, were the C/O ratio only slightly below unity at the end of helium burning, the carbon ions would become isolated from each other upon freezing. Recent upward revisions of the $^{12}C(\alpha,\gamma)^{16}O$ cross-section would favour this hypothesis. Its main consequence would be that $^{12}C + ^{16}O$ and $^{16}O + ^{16}O$ should become the relevant reactions for ignition in the pycnonuclear regime. Thus, besides the relative slowness of burning propagation, the ignition would be delayed to higher densities, up to $\rho_c = 2 \times 10^{10}$ g cm^{-3} and it would be induced by the electron captures on ^{16}O.
c) <u>C/O separation in the solid phase</u>. This is suggested by Stevenson's (1980) phase diagram for a C-O mixture. The predicted oxygen-poor eutectic means that, for $X_C = X_O = 0.50$, oxygen would crystallize first and accumulate at the center of the star. Mochkovich (1983) has shown in detail how this could even lead to a complete C/O separation. The effect is to reduce the reaction rates (the central layers would be made up of ^{16}O). Ignition is delayed to $\rho_c \approx 2 \times 10^{10}$ g cm^{-3} and it is due to the electron captures. Depending on the size of the solid core, we can have either central ^{16}C ignition or off-center ^{12}C ignition, the last taking place at the bottom of the still fluid layers. (Canal, Isern, and Labay 1980; Isern <u>et al</u>. 1983).

4. HYDRODYNAMICS

Several cases have been studied thus far, by one-dimensional numerical simulations only:
a) Models assuming complete C/O separation, initial white dwarf masses $M_{WD} \gtrsim 1.2 M_\odot$, and accretion rates 4×10^{-8} $M_\odot$ yr$^{-1} \lesssim \dot{M} \lesssim 10^{-7}$ $M_\odot$ yr^{-1}, evolve towards homologous collapse of the entire core. The study of the later phases of the collapse, up to and after bounce, is currently in progress (Arnett and Canal 1985, in preparation).
b) Models assuming complete mixing in both the fluid and the solid phases give central carbon ignition, collapse of the solid layers to a neutron star, and explosion with ejection of the fluid layers (Isern, Labay, and Canal 1984).
c) Models with partial C/O separation give off-center igni-

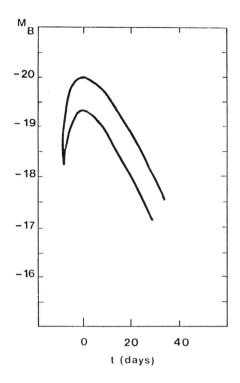

Figure 1. Light curves for models a (upper curve) and e (lower curve) of Table 1, the two extreme "slow"(a) and "fast" (e) outbursts considered. Model a corresponds to total disruption.

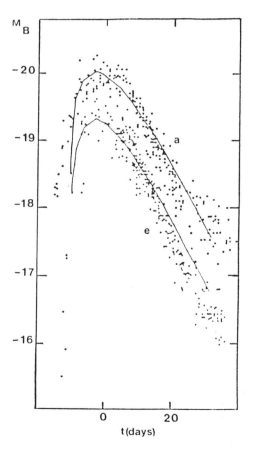

Figure 2. The light curves of models a and e, superimposed to the composite B-band light curves (dots) for "slow" and "fast" SNI obtained by Barbon, Ciatti, and Rosino (1973). Only the slopes are significant, since the composite light curves were not calibrated in absolute magnitudes.

tions and partially explode, leaving a white dwarf remnant (López et al. 1985). Most of the aforementioned results are still preliminary and to be confirmed by more complete calculations.

5. LIGHT CURVES FROM PARTIALLY SOLID MODELS

We will now concentrate upon the third class of models just enumerated: those with partial C/O separation. Models in this class eject variable amounts of partially incinerated matter. This last characteristic might give the key to the "Pskovskii-Branch effect" in the SNI light curves (Pskovskii 1977; Branch 1982). This effect consists in a correlation between peak absolute magnitude, the rate of decline of the initial post-peak light curve, and the photospheric velocities at maximum light. "Slow" SNI (those with broader maxima) are brighter and expand faster than those with narrower maxima (the "fast" ones). Arnett's (1982) analysis already suggests that the ejection of different amounts of material might explain this effect. In the context of models based upon degenerate carbon ignition, this would imply leaving a bound remnant.

We have calculated the light curves for several models defined by their total ejected mass and the mass of ^{56}Ni synthesized. Different characteristics of the models are given in Table 1. Incineration of all but the outermost 0.2 $M_\odot$ of the ejected material has been prescribed, in accordance with our preliminary results for the explosion hydrodynamics. A detailed account of the input physics and the numerical treatment can be found in López et al.(1985).

In Figure 1 we show the light curves for models a and e from Table 1. Model a would correspond to the "slow" and model e to the "fast" end of the range. The same models are displayed in Figure 2, superimposed to the composite light curves obtained by Barbon, Ciatti, and Rosino (1973) for "slow" and "fast" SNI. We see that the observed and the predicted slopes do agree quite well.

6. LIGHT CURVES AND THE HUBBLE CONSTANT

SNI are in principle good standard candles and can be used to calibrate Hubble's constant, H_0 (Arnett, Branch, and Wheeler 1985). As indicated in the same paper, the calibration based on the hypothesis of total disruption of the white dwarf and almost identical peak luminosities for the different outbursts would fail if only a portion of the matter is ejected and/or the peak luminosities do cover a finite range.

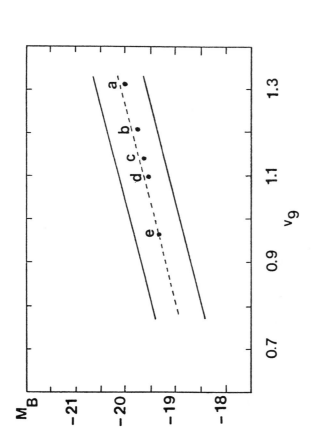

Figure 3. Comparison of the B-magnitude versus ejection velocity (both at maximum light) relationship between our models and Pskovskii's observational data. The strip is the loci of the observations and the dashed line is their best fit for $H_0 = 60$ km sec^{-1} Mpc^{-1}.

TABLE 1

Model Characteristics

Model	M_{ej} ($M_\odot$)	M_{Ni} ($M_\odot$)	E_{kin} (10^{51} erg)	R_p (10^{15} cm)	V_p (km/sec)	L_{43} (erg/sec)	M_B
a	1.435	1.2	1.12	1.60	13118	2.26	-20.01
b	1.2	1.0	0.81	1.52	12075	1.83	-19.76
c	1.0	0.8	0.60	1.38	11408	1.61	-19.64
d	0.9	0.7	0.50	1.30	10975	1.49	-19.55
e	0.8	0.6	0.34	1.18	9645	1.22	-19.34

The velocity-magnitude relation given by the Pskovskii-Branch effect provides a new method of calibration, however. In Figure 3 we compare the models from Table 1 with the velocity-magnitude relation obtained from the observational data for H_0 = 60 km sec^{-1} Mpc^{-1}. We see that our models do follow reasonably well the observed relation. One should keep nonetheless in mind the error bars in the photospheric velocities ($\pm$ 800 km sec^{-1}) and in the magnitudes ($\pm$ 0.5^m), as well as the character of the models, still not self-consistent.

7. NUCLEOSYNTHESIS PROBLEMS

In the models we have just considered for the light curves, the off-center carbon ignition happens at $\rho \sim 10^{10}$ g cm^{-3}. This would imply overproduction of ^{58}Fe (see Woosley, this volume). Higher accretion rates than those thus far considered may give (in combination with big central oxygen cores) thermonuclear ignition at lower densities, but it is still unclear whether they can be lowered down to $\rho \simeq 2 \times 10^9$ g cm^{-3} and and how the allowed range of ejected masses would then change. The models assuming complete C-O mixing in the solid phase (random alloys) are more promising on this respect, since the matter initially at high densities goes into the collapsed remnant. Light curves for those models can only be calculated, however, when the dynamics of the collapsing cores has been followed up to their bounce at supranuclear densities, since some of the kinetic energy of the ejecta will come from the binding energy of the neutron star.

8. CONCLUSION

Partially solid C-O cores are to be expected on the basis of the white dwarf model of SNI. The dynamics of their thermonuclear ignition does substantially differ from that of the completely fluid cores usually considered, and several assumptions are possible as to the distribution of their chemical composition. The models predict the ejection of variable amounts of partially incinerated material, simultaneously with the formation of a bound remnant (either neutron star or white dwarf). Their light curves might explain the Pskovskii-Branch effect and would provide a calibration of Hubble's constant. Some of them would avoid the nucleosynthetic difficulties associated with carbon ignition at high densities. The present results are, however, still fragmentary and they suffer from the common drawbacks of all one-dimensional approaches to stellar hydrodynamics.

REFERENCES

Arnett, W.D. 1969, Ap. and Space Sci., 5, 180
Arnett, W.D. 1982, Ap. J., 253, 785
Arnett, W.D., Branch, D., and Wheeler, J.C. 1985, Nature, 314, 337
Axelrod, T.S. 1980, in Proceedings of the Texas Workshop on Type I Supernovae, ed J.C. Wheeler (University of Texas Press, Austin), p. 80
Barbon, R., Ciatti, F., and Rosino, L. 1973, Astr. Ap.,25,24
Branch, D. 1982, Ap. J., 258, 35
Branch, D., Buta, R., Falk, S.W., McCall,M.L., Sutherland, P.G., Uomoto, A., Wheeler, J.C., and Wills, B.J. 1982, Ap. J. (Letters), 252, L61
Branch, D., Lacy, C.H., McCall, M.L., Sutherland, P.G., Uomoto, A., Wheeler, J.C., and Wills, B.J. 1983, Ap. J., 270, 123
Canal, R., and Isern, J. 1979, in Proceedings of IAU Colloquium 53, White Dwarfs and Variable Degenerate Stars, ed H.M. Van Horn and V. Weidemann (University of Rochester Press, Rochester), p. 52
Canal, R., Isern, J,. and Labay, J. 1980, Ap. J. (Letters), 241, L33

Iben, I., and Tutukov, A.V. 1984, Ap. J. Supplement, 54, 335
Isern, J., Labay, J., Hernanz, M., and Canal, R. 1983, Ap. J., 273, 320
Isern, J., Labay, J., and Canal, R. 1984, Nature, 309, 431
Labay, J., Canal, R., and Isern, J. 1983, Astr. Ap., 117, L1
Lamb, D.Q., and Van Horn, H.M. 1975, Ap. J., 200, 306
López, R., Isern, J., Canal, R., and Labay, J. 1985, Astr. Ap., in press
Mazurek, T.:., Meier, D.C., and Wheeler, J.C. 1977, Ap. J., 213, 518
Mochkovich, R. 1983, Astr. Ap., 122, 212
Müller, E., and Arnett, W.D. 1982, Ap. J. (Letters), 261, L109
Müller, E., and Arnett, W.D. 1985, Ap. J., in press
Nomoto, K., Thielemann, F.K., and Yokoi, K. 1984, Ap. J., 268, 664
Pskovskii, Yu. P. 1977, Soviet Astr., 21, 675
Schatzman, E. 1963, in Star Evolution, Proceedings of the International School of Physics Enrico Fermi Nº 28, ed L. Gratton (Academic Press, New York), p. 389
Schatzman, E., 1983, in Proceedings of IAU Colloquium 72, Cataclysmic Variables and Related Objects, p. 149
Slattery, W.L., Doolen, G.D., and DeWitt, H.E. 1980, Phys. Rev. A, 21, 2087
Stevenson, D.J. 1980, J. de Phys. Suppl., Nº 3, Vol. 41, p. C2-53
Sutherland, P.G., and Wheeler, J.C. 1984, Ap. J., 280, 282

EXPLOSIVE NUCLEOSYNTHESIS IN CARBON DEFLAGRATION MODELS
OF TYPE I SUPERNOVAE

Friedrich-Karl Thielemann[1,2]
Ken'ichi Nomoto[3]
Koichi Yokoi[4]

[1] University of Illinois, Department of Astronomy, Urbana, Illinois 61801, USA

[2] on leave from Max-Planck-Institut für Physik und Astrophysik, Garching b. München, FRG

[3] Department of Earth Science and Astronomy, College of Arts and Science, University of Tokyo, Tokyo, Japan

[4] Kernforschungszentrum Karlsruhe, Institut für Kernphysik III, Karlsruhe, FRG

ABSTRACT There is increasing evidence that Type I supernovae (SN I) are the main producers of iron-peak elements in the Galaxy. In addition observations of SN I also indicate the existence of appreciable amounts of intermediate elements like O, Mg, Si, S, and Ca in the outer layers of the exploding star. Such an abundance pattern can be produced by carbon deflagration models of accreting carbon-oxygen white dwarfs in binary systems or stars on the asymptotic giant branch (AGB) which ignite central carbon burning explosively. The present study discusses the nucleosynthesis results of those carbon deflagration supernovae in detail. Special emphasis is given to the discussion of burning conditions and corresponding nucleosynthesis products. The overproduction of ^{54}Fe+^{58}Ni, mentioned in earlier publications, is still existing. Assuming that SN I which contributed to the abundances in the solar system, originated from white dwarfs with a metallicity range $0.1 < Z/Z_\odot < 1$, might remove this overproduction. This would allow for SN I to be the major contributors of Fe-group nuclei in galactic nucleosynthesis.

1. INTRODUCTION

SN I have been suggested to produce a significant amount of iron peak elements and intermediate mass elements. From the observational point of view, there is some evidence for the existence of a large amount of iron in SN I; i.e. late time spectra and light curves of SN I are well explained by the radioactive decay model of ^{56}Ni. Freshly produced ^{56}Ni decays via ^{56}Co to ^{56}Fe (Arnett 1979; Colgate, Petsheck, and Kriese 1980; Meyerott 1980; Chevalier 1981; Axelrod 1980; Weaver, Axelrod, and Woosley 1980; Schurmann 1983; Sutherland and Wheeler 1984). These models require a total amount of $0.2-1.0 M_\odot$ of ^{56}Ni ejected per SN I event.

The evidence for the presence of intermediate mass elements in SN I was found in early time spectra of SN I; in particular the features of SN I 1981b at maximum light are well identified with lines of Ca, Si, S, Mg, and O (Branch et al. 1981). These observations suggest the following composition structure of SN I; i.e. the outer layers, observed at maximum light, are composed of O-Si-Ca and the inner layers, exposed at late times, contain $0.2-1 M_\odot$ ^{56}Ni.

From a theoretical point of view, the most plausible model for SN I is the thermonuclear explosion of accreting white dwarfs in close binaries (see Wheeler 1982 for a review). In particular the carbon deflagration model has been suggested to be consistent with many of the observed features of SN I (Nomoto, Sugimoto, and Neo 1976; Nomoto 1980, 1981; Chevalier 1981; Müller and Arnett 1982, 1985).

Recently Nomoto, Thielemann and Yokoi (1984) have calculated hydrodynamical models of carbon deflagration supernovae and the associated nucleosynthesis in detail (hereafter NTY models). The NTY models produce $0.5-0.6 M_\odot$ ^{56}Ni in the inner layers and substantial amounts of O-Si-Ca in the outer layers. These models, therefore, can naturally account for the suggested composition structure of SN I. Moreover, Branch (1984) and Branch et al. (1985) have calculated synthetic spectra of the NTY model and obtained a rather good fit to the observed spectra near maximum light, provided that the outer layers of the white dwarf are mixed.

These studies indicate the importance to investigate the nucleosynthesis process in carbon deflagration models in detail. This was performed by Thielemann, Nomoto, and Yokoi (1985). In the present contribution, we want to review these results briefly.

2. ACCRETING WHITE DWARFS AND CARBON DEFLAGRATION SUPERNOVAE

The fate of accreting white dwarfs in binary systems depends mainly on their initial mass of the C+O core and the mass accretion rate (see e.g. Nomoto 1980, 1981, 1982a). Accretion leads to the build-up of a He-layer on top of the C+O white dwarf. If He does not burn steadily or with weak flashes at the base of the He-layer, it will ignite in a violent flash causing either only an outward detonation wave through the He-layer (single detonation: Nomoto 1980, 1982b; Woosley, Axelrod, and Weaver 1984) or also an inward detonation wave through the C+O core (double detonation: Nomoto 1980, 1982b; Woosley, Weaver, and Taam 1980). In the case of steady He-burning at the base of the

He-layer the C+O core grows and carbon will ignite at the center and drive an outward deflagration wave.

The main characteristics of detonation and deflagration are the following: (a) in a detonation a strong shock is formed and peak temperatures in the shock are sufficient for explosive ignition of the fuel, (b) in the deflagration case the peak temperature due to compression is not sufficient to ignite the fuel and the burning front propagates rather via convective mixing of burning and unburnt material behind the shock wave. The difference between these two models is due to the larger energy release and lower ignition temperature in degenerate He-burning in comparison to C-burning (see Figs. 1 and 2 in Nomoto, 1982b). In the detonation the burning front propagates supersonicly with respect to the unburnt matter, resulting in complete burning to nuclear statistical equilibrium and an iron-peak abundance composition. In the deflagration the front propagates subsonically and matter can adjust before the burning front arrives; i.e. the star expands to weaken the deflagration. Depending on the initial temperatures and densities only partial burning to intermediate mass nuclei occurs and such a model predicts certain amounts of nuclei like Mg, Si, S, and Ca in the outer layers where the deflagration front is weakened already, while iron-peak nuclei dominate in the central parts of the exploding C+O core.

The probability of occurrence for this kind of scenario in binary systems and the reason why detonation models are apparently not present in SN I events have been discussed by Iben and Tutokov (1984), Webbink (1984), and Nomoto, Thielemann and Yokoi (1984). It has to be noticed, however, that those considerations still contain a large degree of uncertainty. Despite these uncertainties, the carbon deflagration model is the most plausible one in the sense that it can account for many of the observed features of SN I as discussed in the introduction. Thus we present here nucleosynthesis products from carbon deflagration supernovae based on the model calculations of Nomoto, Thielemann, and Yokoi (1984). The details about the reaction network, nuclear reaction rates and computational techniques have been discussed in Thielemann, Nomoto, and Yokoi (1985).

3. EXPLOSIVE NUCLEOSYNTHESIS IN THE CARBON DEFLAGRATION FRONT

3.1. General Features of Explosive Nuclear Burning in the Deflagration Wave

From a nucleosynthesis point of view the peak temperatures T_p of each mass zone, achieved during the propagation of the deflagration front are the most important quantities and displayed in Fig. 1 as a function of the Lagrangean mass coordinate. The values range from $\approx 9 \times 10^9$K at the center down to $\approx 2 \times 10^9$K at the position where the deflagration wave dies out.

Thus the material undergoes various modes of explosive Si-burning for peak temperatures in excess of $T_{9,p} \equiv T_p/(10^9 K) \approx 4.5-5.0$ out to $\approx 1.0 M_\odot$, explosive O-burning for $T_{9,p} \gtrsim 3.2$ and $1.0 \leq M_r/M_\odot \leq 1.15$,

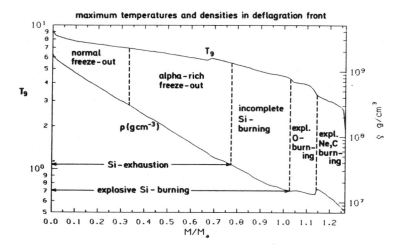

Fig. 1: Maximum temperatures and densities obtained during the outward propagation of the deflagration front, as a function of Lagrangean mass coordinate $M/M_\odot$. Zones of different explosive burning conditions are indicated.

Thus, the material undergoes various modes of explosive Si-burning for peak temperatures in excess of $T_{9,p} \equiv T_p/(10^9 K) \simeq 4.5$–$5.0$ out to $\approx 1.0 M_\odot$, explosive O-burning for $T_{9,p} \geq 3.2$ and $1.0 \leq M_r/M_\odot \leq 1.15$, explosive Ne-burning for $T_{9,p} \geq 2.2$ and $1.15 \leq M_r/M_\odot \leq 1.28$ and explosive C-burning for $T_{9,p} \leq 2.2$ in the outmost parts of the C+O core which is processed explosively before the deflagration front dies out. The total overproduction of material, compared to solar abundances and normalized to ^{56}Fe is displayed in Fig. 3. The results of the individual explosive burning processes can be seen in Fig. 2 and clearly identified by their burning products:

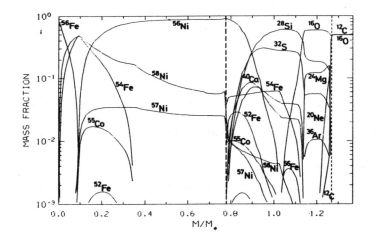

Fig. 2: Resulting nucleosynthesis after explosive processing. Shown are the major abundances. The inner dashed line indicates the transition from an alpha-rich freeze-out to incomplete silicon burning. The outer dashed line marks the quenching of the deflagration front beyond which unburned carbon and oxygen are left.

(1) Fe-peak nuclei (e.g. ^{56}Ni) for complete Si-burning, being mixed with ^{40}Ca, ^{32}S, ^{28}Si and ^{36}Ar in the partially burnt outer zones
(2) ^{28}Si and ^{32}S in the O-burning zones
(3) ^{16}O, ^{28}Si, ^{24}Mg in Ne-burning and
(4) ^{20}Ne and ^{24}Mg coming up in the outermost zones of C-burning.

In the following subsections we want to discuss and analyze our results from different zones of a carbon deflagration supernova with regard to those parametrized explosive nucleosynthesis calculations.

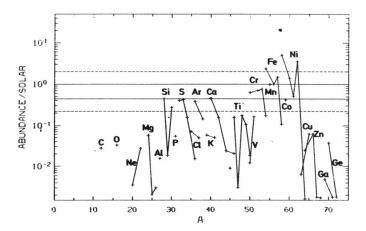

Fig. 3: Nucleosynthesis of a carbon deflagration supernova. The abundances of stable isotopes relative to their solar values are shown. The ratio is normalized to ^{56}Fe. Note the strong overabundances of ^{58}Ni, ^{62}Ni, and ^{54}Fe.

3.2. Explosive Silicon Burning

All mass zones in the inner core with $M \leq 1.0 M_\odot$ experience maximum burning temperatures in excess of 4.5×10^9K during the passage of the deflagration front. This allows (depending on the time scale) the attainment of complete nuclear statistical equilibrium (NSE) or at least quasi-equilibirum (QSE) for nuclei in the mass range $28 \leq A \leq 45$. Peak temperatures larger than 5×10^9K lead to silicon exhaustion; otherwise only incomplete Si-burning takes place in agreement with Woosley, Arnett, Clayton (1973). This is the case in the range $0.8 \leq M/M_\odot \leq 1.0$ as can be deduced from Figs. 1 and 2.

As known from parametrized explosive nucleosynthesis calculations, three different kinds of freeze-out conditions can occur. When assuming an adiabatic expansion on a hydrodynamic time scale $\tau_{HD}=446 \kappa \rho_p^{-\frac{1}{2}}$ ($\kappa \sim 1$) with $\rho(t)=\rho_p \exp(-t/\tau_{HD})$ and $T(t)=T_p(\rho/\rho_p)^{1/3}$. Fig. 20 in Woosley et al. (1973) divides the (ρ_p, T_p)-plane into regions of (1) incomplete silicon burning, (2) normal freeze-out and (3) α-rich freeze-out. Peak temperatures below 5×10^9K lead to incomplete Si burning. At higher temperatures, low densities cause an α-rich (particle-rich) freeze out and high densities lead to a normal freeze-out. Those boundaries are displayed in Fig. 4. Also shown are the maximum temperatures and densities of each mass zone of the carbon deflagration supernova during the passage of the deflagration front. The parametrized calculations with an adiabatic expansion are not completely comparable with the present investigation but can give a rough estimate on the expected behavior, as the boundaries are not very dependent on the parameter κ in the hydrodynamic timescale τ_{HD}.

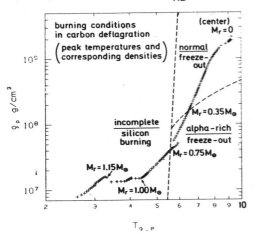

Fig. 4: The maximum temperatures and densities ($T_{9,p}$, ρ_p) of each mass zone, experienced during the propagation of the deflagration front, are indicated as crosses (see also Fig. 1). The dahshed lines show the boundaries of normal freeze-out, alpha-rich freeze-out and incomplete silicon burning, extensively investigated by Woosley, Arnett, and Clayton (1973) in parametrized explosive nucleosynthesis calculations.

The freeze-out abundances, however, depend not only on the freeze-out temperatures and densities but also on the neutron excess $\eta = \Sigma(N_i - Z_i)Y_i$ (N_i and Z_i being neutron and proton number of nucleus i, respectively), as equilibirum abundances depend on ρ, T, and η. η is greatly changed via electron captures on protons and nuclei during the explosive processing at high densities and temperatures. This leads to a neutron excess of up to 8×10^{-2} at the center as displayed in Fig. 5 (after freeze-out, before long-term β-decays). There are uncertainties in η which is determined by electron captures in the inner core and the initial abundance of ^{22}Ne in the outer zones. η has a strong influence on the products of explosive nucleosynthesis. The detailed discussion about explosive Si-burning in terms of parametrized explosive nucleosynthesis calculations by Woosley et al. (1973) has only been carried out for neutron excesses $\eta < 4.6 \times 10^{-3}$. Thus, with regard to Fig. 5, an exact comparison can only be made for zones at $M_r > 0.4 M_\odot$.

3.2.1. Incomplete Si-burning

Explosive conditions which leave about 10-20% ^{28}Si by mass after freeze-out result in roughly the same amount of ^{56}Ni (decaying to ^{56}Fe). Woosley et al. (1973) showed such a result in Fig. 19 where the parametrized calculations were performed with a peak temperature $T_{9,p} = 4.7$ and a peak density $\rho_p = 2 \times 10^7$ g cm^{-3}. The main abundances (in decreasing order) are ^{28}Si, ^{32}S, ^{56}Fe, ^{40}Ca, ^{36}Ar, ^{54}Fe, ^{52}Cr, ^{55}Mn, ^{58}Ni, and ^{57}Fe. Similar conditions can be found in the carbon deflagration supernova for mass zones at $0.75 \leqslant M_r/M_\odot \leqslant 1.0$ (see Fig. 2). The peak temperatures and densities for those zones vary from 4.6 to

EXPLOSIVE NUCLEOSYNTHESIS

5.3×10^9K and 1.5 to 4×10^7g cm^{-3}, respectively. The nuclei ^{56}Fe, ^{52}Cr, ^{55}Mn and ^{57}Fe are decay products of the parent nuclei ^{56}Ni, ^{52}Fe, ^{55}Co, and ^{57}Ni.

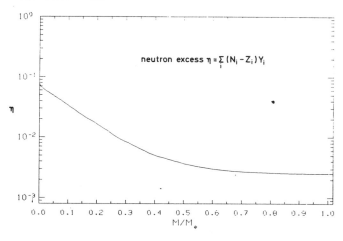

Fig. 5: Neutron excess $\eta = 1 - 2Y_e$, due to electron captures on protons and heavy nuclei during the high temperature and density phase of the explosion. η dominates the abundance composition in nuclear statistical equilibrium (NSE).

3.2.2. alpha-rich freeze-out

Peak temperatures in excess of 5×10^9K lead to a complete NSE and ^{28}Si exhaustion. When peak densities larger than 2×10^8g cm^{-3} are achieved, a so-called normal freeze-out occurs. This means that freezing occurs because of deficient α-particles. This is in contrast to the alpha-rich freeze-out where α-capture reaction rates would still allow the build-up of heavier nuclei. Woosley et al. (1973) present such a case in Fig. 23 which gives the results of a nucleosynthesis run for $T_{9,p} = 5.5$ and $\rho_p = 2 \times 10^7$g cm^{-3}. The main constituents are ^{56}Fe, ^{58}Ni, ^{57}Fe, and ^{60}Ni followed by ^{62}Ni and ^{59}Co. Similar conditions are obtained in the carbon deflagration supernova for mass zones at $0.35 \leq M_r/M_\odot \leq 0.75$ with $5.3 \leq T_{9,p} \leq 7.0$ and $4 \times 10^7 \leq \rho_p$(g cm^{-3})$\leq 3 \times 10^8$. The comparison can again be made with Fig. 2. Note, however, that ^{60}Ni, ^{62}Ni, and ^{59}Co are not plotted. Again ^{56}Fe, ^{57}Fe, ^{60}Ni, ^{62}Ni and ^{59}Co are produced via the parent nuclei ^{56}Ni, ^{57}Ni, ^{60}Zn, ^{62}Zn and ^{59}Cu, respectively. Slight deviations occur as not exactly the same burning conditions are met and in addition the inner zones have a larger neutron excess η than considered in the calculations by Woosley et al. (1973).

3.2.3. the neutron-rich inner core

The inner core with $0 \leq M_r/M_\odot < 0.35$ experiences the highest densities and temperatures in explosive burning and thus the largest amount of electron capture, resulting in a neutron excess roughly between 10^{-2} and 10^{-1}. This leads to the build-up of neutron-rich nuclei, so that after freeze-out a nucleus like ^{56}Fe has the dominant abundance at the very center. With decreasing η (going outwards) ^{56}Fe is replaced by ^{54}Fe and ^{58}Ni until finally ^{56}Ni becomes the dominant nucleus. This can be seen in Fig. 2.

Other important contributions are given for the nuclei ^{57}Fe, ^{55}Mn, and ^{52}Cr (decay products of ^{57}Ni, ^{55}Co, and ^{52}Fe). This neutron-rich inner core has a strong influence on the isotopic composition of iron group nuclei in the supernova ejecta. It is also dependent on the amount of electron captures occuring during explosive burning. For the values of the neutron-excess η, present under those conditions, no parametrized nucleosynthesis calculations are available. For a detailed discussion see Thielemann, Nomoto and Yokoi (1985).

3.3 Explosive Oxygen Burning

Temperatures in excess of roughly 3.5×10^9K lead to a quasi-equilibrium in the equilibrium cluster $28 \leqslant A < 45$ (Woosley et al., 1973). Those conditions are accomplished in explosive oxygen burning and the main burning products are ^{28}Si, ^{32}S, ^{36}Ar, ^{40}Ca, ^{38}Ar, and ^{34}S. This can be seen in Fig.2, for $1.03 \leqslant M_r/M_\odot \leqslant 1.15$.

The abundances in the quasi-equilibrium cluster are determined by the alpha, neutron, and proton abundances. As we have material which does not experience high densities and temperatures during explosive burning, electron captures on nuclei are negligible. Thus η stays constant at values ≈ 0.002 and the neutron to proton ratio is automatically determined. Under those conditions the resulting composition is only dependent only on the alpha and neutron abundances at freeze-out. Woosley et al. (1973) used the isotopic pair ^{32}S/^{28}Si to define the alpha abundance at freeze-out and ^{38}Ar/^{36}Ar to define the neutron abundance. The necessary conditions for solar abundance ratios are then given by the quantities ρX_n and ρX_α as functions of the freeze-out temperature. They also found out that under the same conditions the pairs ^{39}K/^{35}Cl, ^{40}Ca/^{36}Ar, ^{36}Ar/^{32}S, ^{37}Cl/^{35}Cl, ^{38}Ar/^{34}S, ^{42}Ca/^{38}Ar, ^{41}K/^{39}K, and ^{37}Cl/^{33}S have solar ratios within a factor of 2, as long as the freeze-out temperatures, T_f fall within the range $3.1 < T_{9,f} < 3.9$. Under those conditions, ^{28}Si, ^{32}S, ^{34}S, ^{36}Ar, ^{38}Ar, ^{40}Ca, and ^{42}Ca and maybe ^{46}Ti are reproduced in solar proportions while ^{33}S, ^{35}Cl, ^{37}Cl, ^{39}K, and ^{41}K are underproduced by a factor of roughly 3. The latter was not clearly stated in Woosley et al. (1973), but changes in solar abundances from Cameron (1968) to Cameron (1982) and Anders and Ebihara (1982) (especially for Cl) yield this change.

The values of ρX_n, ρX_α, and $T_{9,f}$ found in the oxygen burning zones of our model fit quite well into the scheme outlined above and so does the nucleosynthesis of these zones (for details see Thielemann, Nomoto, Yokoi, 1985).

The importance of explosive oxygen burning to the carbon deflagration supernova can be evaluated from Fig.3. ^{34}S, ^{38}Ar, and ^{42}Ca are roughly on a constant line. ^{35}Cl, ^{37}Cl, ^{39}K, and ^{41}K are down by a factor of 3 compared to the previous nuclei, in agreement with the discussion outlined above. The values for ^{28}Si, ^{32}Si, ^{36}Ar, and ^{40}Ca, however, are up by a factor of 2 to 5. This means that the dominant production site for those nuclei is not explosive oxygen burning. It was discussed in the previous subsection that their major source is incomplete silicon burning, as can also be concluded from Fig. 2.

3.4. Explosive Neon and Carbon Burning

Peak temperatures for explosive carbon and neon burning are close to 2×10^9K and up to 3×10^9K, respectively. These conditons are found for the outer zones of the exploding C+O white dwarf with mass coordinates $M_r > 1.15 M_\odot$. Here the deflagration front is already weakened considerably and the peak temperatures drop rapidly until the burning front finally quenches.

In Ne-burning we find that Fig. 11a shows, after the build-up of ^{20}Ne via ^{12}C(^{12}C,α)^{20}Ne and ^{12}C(^{12}C,p)^{23}Na(p,α)^{20}Ne, how the photodisintegration of ^{20}Ne leads to an additional increase in the pre-existing ^{16}O and the build-up of ^{24}Mg and ^{28}Si. Besides these major abundances explosive neon burning also provides substantial amounts of ^{27}Al, ^{29}Si, ^{30}Si, and ^{31}P (see also Morgan 1980).

In explosive carbon burning ^{12}C is mainly burned to ^{20}Ne. Besides the pre-existing ^{16}O; ^{20}Ne, ^{24}Mg and ^{28}Si are the dominant abundances. Explosive carbon burning is suited to produce ^{20}Ne, ^{23}Na, ^{24}Mg, ^{25}Mg, ^{26}Mg, ^{28}Al, ^{29}Si and ^{30}Si in approximately solar proportions (Arnett 1969, Howard et al. 1972, Truran and Cameron 1978, Arnett and Wefel 1978, Morgan 1980). While in "normal" explosive carbon burning conditions around 2×10^9K considerable quantities of ^{23}Na are produced, higher temperatures lead to the destruction of ^{23}Na via ^{23}Na(p,α)^{20}Ne (see also Morgan 1980).

In the present calculation we find minimum peak temperatures around 2.2×10^9K before the deflagration front quenches and therefore the amount of ^{23}Na is negligible. It has been stated already in Nomoto, Thielemann, and Yokoi (1984) that the numerical treatment of the deflagration front is a difficult task in the quenching region and that the present method is only approximate. Thus we could expect ^{23}Na and also more ^{20}Ne and ^{24}Mg to be existent in outer layers.

Generally it can be said that in SN I events the major burning products of explosive neon and carbon (and partially oxygen) burning are not produced with sufficient amounts to account for galactic nucleosynthesis (see Fig. 3). The nucleosynthesis products of SN I are dominated by incomplete and complete explosive silicon burning (see Fig. 1 for the mass coordinates of the different burning zones). The produced amount of Mg and O is however sufficient to explain the observed features in SN I spectra (Branch et al. 1985).

4. CONCLUSIONS

The main nucleosynthesis results of a carbon deflagration supernova can be summarized as follows (see Fig. 1):

1. In the central part of $M_r \leq 0.75 M_\odot$, peak temperatures in excess of 5×10^9K is attained so that the material is incinerated into NSE. For $M_r \leq 0.35 M_\odot$, the density at the peak temperature is higher than 2×10^8 g cm^{-3} and thus a normal freeze-out (i.e., freezing due to α-particle

deficiency) takes place. At $0.35 < M_r/M_\odot \leq 0.75$, the density is lower and an α-rich freeze-out resulted. In this region, production of ^{54}Fe and a small amount of ^{40}Ca and ^{36}Ar, found in the NTY model, does not take place because these elements are processed by (α,γ) reactions. Instead ^{58}Ni is produced via ^{54}Fe (α,γ) ^{58}Ni.

  2. At $0.75 < M/M_\odot \leq 1.0$, incomplete Si burning takes place and ^{56}Ni, ^{40}Ca, ^{36}Ar, ^{32}S and ^{28}Si are the major products.

  3. Still outer layers undergo explosive O, Ne, and C-burning, producing mainly (i) ^{28}Si, ^{32}S, ^{36}Ar, and ^{40}Ca, (ii) ^{16}O, ^{24}Mg, and ^{28}Si, and (iii) ^{20}Ne and ^{24}Mg. Nuclear abundances with minor contributions to the solar mix are coproduced and a solar abundance distribution is obtained within the atomic mass range, characterizing each of the mentioned explosive burning processes. The mass contained in the explosive O-, Ne-, and C-burning zones is, however, too small in comparision to matter in complete and incomplete Si-burning zones, to attain a solar abundance pattern for the total SN I ejecta. Therefore SN I ejecta are dominated by products of complete and incomplete Si-burning, being thus also main contributors to Fe-group nucleosynthesis in Galactic evolution. This can be seen in Fig. 3.

The general abundance pattern ($0.5-0.6 M_\odot$ ^{56}Ni in the inner core and some intermediate mass elements, Ca-Si-O, in the outer layers) are quite consistent with all presently observable features of SN I, i.e., light curves, late time spectra (Axelrod 1980), and early time spectra (Branch 1985; Branch et al. 1985).

The remaining problem in the present carbon deflagration model is the overproduction of ^{54}Fe (slightly), ^{58}Ni, and ^{62}Ni relative to the solar abundances. We have analyzed the uncertainties of nuclear partition functions and electron capture rates in Thielemann, Nomoto, and Yokoi (1985). These uncertainties do not seem to be large enough to remove the overproduction difficulty although complete investigation with a full network calculation for different sets of partition functions and electron captures remain to be done. Other possibilities to improve this difficulty are as follows:

  1. The central density at the carbon ignition is in most cases appreciably lower than the present model of W7; in other words the typical accretion rate leading to the carbon deflagration is significantly higher than $4 \times 10^{-8} M_\odot$ yr^{-1}.

  2. The mixing length parameter $\alpha = \ell/H_p$ is not constant but changing during the propagation of the deflagration wave.

  3. The ^{12}C/^{16}O ratio in the initial fuel can vary from one white dwarf progenitor to another, dependent on the initial main sequence mass and He-burning temperatures. There might also be a radial discontinuity due to different temperatures in core and shell He-burning, which can influence the propagation of the burning front.

  4. The carbon deflagration supernova is not the only major source of iron peak elements in the Galaxy. Possible other sources include SN II and the subclass of SN I which is characterized by the absence of the Si feature at about 6100 Å in the early time spectra (e.g., SN 1983n in M83 and SN 1984l in NGC0991; see Wheeler 1985). The amount of iron peak elements and the ^{58}Ni/^{56}Fe ratio in the ejecta of SN II depend on the mass cut and the explosion energy, which awaits

for further investigation. The subclass of SN I might be due to the detonation type explosion because the detonation models produce negligible amount of intermediate mass elements. If so, the $^{58}Ni/^{56}Fe$ ratio must be small in these SN I because the detonation models experience negligible neutronization owing to the low ignition density and synthesize mostly ^{56}Ni with an underproduction of ^{58}Ni, ^{62}Ni, and ^{54}Fe (Nomoto and Sugimoto 1977; Nomoto 1982b; Woosley et al. 1984).

5. Another, quite promising possibility is the following:
White dwarfs in binary systems can have ages which cover the whole range of the galactic disk population. This leads to limits on the metallicity of $0.1 < Z/Z_\odot < 1$ for systems having contributed nucleosynthesis products to the solar system. The metallicity determines original CNO abundances, the ^{14}N content after hydrogen burning and the ^{22}Ne abundance after He-burning which is contained in the C+O white dwarf. The neutron excess η is only strongly affected by electron captures within the inner core. The neutron excess which determines the abundance of ^{58}Ni and (or) ^{54}Fe, out to $M_r \approx 1.0 M_\odot$ is given by the initial ^{22}Ne content or in other words the initial metallicity. A reduction in η by a factor of 10 reduces ^{58}Ni and (or) ^{54}Fe by more than a factor of 100 (Truran and Arnett, 1971; see tables III and IV). Thus a moderate metallicity of ≈ 0.5 for SN I averaged over the history of nucleosynthesis contributions to the solar system might still result in negligible amounts of ^{54}Fe and ^{58}Ni for $M_r > 0.35 M_\odot$. This would lead to a total reduction of ^{58}Ni by 35-40% and ^{54}Fe by 25%.

^{54}Fe will then be within the uncertainty range of a factor of 2 in Fig. 3, and ^{58}Ni comes very close to it. ^{62}Ni being produced in form of ^{62}Zn via $^{54}Fe(\alpha,\gamma)^{58}Ni(\alpha,g)^{62}Zn$ will change accordingly.

We might conclude that this effect, when also taking into account the uncertainties mentioned in 1.-3., could actually lead to a solar abundance pattern of Fe-group nuclei, after averaging all SN I contributions in the past. However, an additional calculation would have to prove this conclusion and also show that no problematic affects on other abundances are introduced.

REFERENCES

Anders, E., Ebihara, M., 1982, Geochim. Cosmochim. Acta **46**, 2363
Arnett, W. D. 1969, Astrophys. J. **157**, 1369
Arnett, W. D., 1979, Astrophys. J. (Letters) **230**, L37
Arnett, W. D., Wefel, J. P., 1978, Astrophys. J. (Letters) **224**, L139
Axelrod, T. S., 1980, in Type I Supernovae, ed. J. C. Wheeler (University of Texas, Austin) p. 80
Bodansky, D., Clayton, D. D., Fowler, W. A., 1968, Astrophys. J. Suppl. **16**, 299
Branch, D., 1985, in Challenges and New Developments in Nucleosynthesis, ed. W. D. Arnett and J. W. Truran (University of Chicago Press)
Branch, D., Dogget, J. B., Nomoto, K., Thielemann, F.-K., 1985, submitted to Astrophys. J., preprint MPA 169

Branch, D., Falk, S.W., McCall, M.L., Rybski, P., Uomoto, A.K., Wills, B.J., 1981, Astrophys. J. **244**, 780
Cameron, A.G.W., 1968, in Origins and Distribution of the Elements, ed. L.H. Ahrens (Pergamon Press) p. 125
Cameron, A.G.W., 1982, in Essays in Nuclear Astrophysics, eds. C.A. Barnes, D.D. Clayton, D.N. Schramm (Cambridge University Press, Cambridge) p. 23
Chevalier, R.A., 1981, Astrophys. J. **246**, 267
Colgate, S.A., Petsheck, A.G., and Kriese, J.T., 1980, Astrophys. J. (Letters) **237**, L81
Howard, W.M., Arnett, W.D., Clayton, D.D., Woosley, S.E., 1972, Astrophys. J. **175**, 201
Iben, I., Jr., Tutokov, S., 1984, Astrophys. J. Suppl. **54**, 335
Meyerott, R.E., 1980, Astrophys. J. **239**, 257
Morgan, J.A., 1980, Astrophys. J. **238**, 674
Müller, E., Arnett, W.D., 1982, Astrophys. J. (Letters) **261**, L107
Müller, E., Arnett, W.D., 1985, submitted to Astrophys. J.
Nomoto, K., 1980, in Type I Supernovae, ed. J.C. Wheeler (University of Texas, Austin) p. 164
Nomoto, K., 1981, in IAU Symposium 93, Fundamental Problems in the Theory of Stellar Evolution, eds. D. Sugimoto, D.Q. Lamb, D.N. Schramm (Reidel, Dordrecht), p. 295
Nomoto, K., 1982a, Astrophys. J. **253**, 798
Nomoto, K., 1982b, Astrophys. J. **257**, 780
Nomoto, K., Sugimoto, D., Neo, S., 1976, Astrophys. Space Sci. **39**, L37
Nomoto, K. and Sugimoto, D., 1977, Publ. Astron. Soc. Japan **29**, 165
Nomoto, K., Thielemann, F.-K., Wheeler, J.C., 1984, Astrophys. J. (Letters) **279**, L23; **283**, L25
Nomoto, K., Thielemann, F.-K., Yokoi, K., 1984, Astrophys. J. **286**, 644
Schurmann, S.R., 1983, Astrophys. J. **267**, 779
Sutherland, P.G., Wheeler, J.C., 1984, Astrophys. J. **280**, 282
Thielemann, F.-K., Nomoto, K., Yokoi, K., 1985 submitted to Astron. Astrophys.
Truran, J.W., Arnett, W.d., 1971, Astrophys. Space Sci. **11**, 430
Truran, J.W., Cameron, A.G.W., 1978, Astrophys. J. **219**, 226
Weaver, T.A., Axelrod, T.S., Woosley, S.E., 1980, in Type I Supernovae, ed. J.C. Wheeler (University of Texas, Austin), p. 113
Webbink, R.E., 1984, Astrophys. J. **277**, 355
Wheeler, J.C., 1982, in Supernovae: A Survey of Current Research, eds. M.J. Rees, R.J. Stoneham (D. Reidel: Dordrecht) p. 79
Wheeler, J.C., 1985, Preprint to appear in Proceedings of Harvard-Smithonian Workshop on Supernovae as Distance Indications.
Woosley, S.E., Arnett, W.D., Clayton, D.D., 1973, Astrophys. J. Suppl. **26**, 231
Woosley, S.E., Axelrod, T.S., Weaver, T.A., 1984, in Stellar Nucleosynthesis, eds. C. Chiosi, A. Renzini (Reidel, Dordrecht), p. 263
Woosley, S.E., Weaver, T.A., Taam, R.E., 1980, in Type I Supernovae, ed. J.C. Wheeler (University of Texas, Austin), p. 96

DIFFERENTIALLY ROTATING EQUILIBRIUM MODELS AND THE COLLAPSE OF ROTATING DEGENERATE CONFIGURATIONS

Ewald Müller
Max-Planck-Institut für Physik und Astrophysik
Institut für Astrophysik
Karl-Schwarzschild-Str. 1
D-8046 Garching b. München, Fed. Rep. of Germany

and

Yoshiharu Eriguchi
Department of Earth Science and Astronomy
College of Arts and Sciences
University of Tokyo
Komaba, Meguro-Ku, Tokyo, Japan

ABSTRACT

Using a recently developed method (Eriguchi and Müller, 1985a) two kind of problems involving axisymmetric, differentially rotating degenerate configurations are investigated:

(i) Equilibrium models of rotating polytropes are used to estimate the properties of the degenerate core of a massive star $M \geqslant 8M_\odot$ at the endpoint of its collapse without performing a detailed collapse calculation. Our analysis shows (Eriguchi and Müller, 1985b) that a rotating stellar core will not collapse all the way to neutron star densities on a dynamical time-scale, if its initial ratio of rotational to gravitational energy β_i is larger than some minimum value: $\beta_{min}=0.01$, 0.03, and 0.08 for $\gamma=1.30$, 1.25, and 1.20, respectively. Instead the collapse is stopped due to rotation at an intermediate, dynamically stable, axisymmetric equilibrium state. The further evolution will proceed on a secular time-scale.

(ii) Rotating, completely catalyzed, zero-temperature Newtonian configurations with central densities in the range $10^7 \text{gcm}^{-3} \leqslant \rho_c \leqslant 5 \cdot 10^{14} \text{gcm}^{-3}$ are calculated. Based on these models we have then studied the question, if there exist evolutionary scenarios of rotating white dwarfs, where due to angular momentum losses a white dwarf with a mass larger than the Chandrasekhar mass will evolve towards a neutron star on a secular time-scale, i.e. without a sudden release of gravitational potential energy in the form of an optical supernova outburst.

We find (Müller and Eriguchi, 1985) that dynamically stable rotating equilibrium models exist up to densities of $\approx 10^{11} \text{gcm}^{-3}$ (without rotation $\approx 10^9 \text{gcm}^{-3}$) and with masses up to $1.7 M_\odot$ (without rotation $1.0 M_\odot$). Configurations with masses in the range $1.7 \leq M/M_\odot \leq 2.2$ are also dynamically stable, but secularly unstable against non-axisymmetric perturbations. We find that for all studied combinations of mass, angular momentum and angular momentum distribution the evolution of a rotating (cold) white dwarf must become dynamic at densities around 10^{12}gcm^{-3}, i.e. roughly 80% of the neutron star's binding energy will be released on a dynamical time-scale.

REFERENCES

Eriguchi, Y. and Müller, E.: 1985a, Astron. Astrophys. (in press).
Eriguchi, Y. and Müller, E.: 1985b, Astron. Astrophys. (in press).
Müller, E. and Eriguchi, Y.: 1985, submitted to Astron. Astrophys.

Theoretical Models for Type I and Type II Supernova

S. E. Woosley
Board of Studies in Astronomy and Astrophysics
Lick Observatory, University of California at Santa Cruz
Santa Cruz CA 95064
and
Special Studies Group, Lawrence Livermore National Laboratory
Livermore CA 94550
and
Thomas A. Weaver
Special Studies Group, Lawrence Livermore National Laboratory
University of California, Livermore CA 94550

Abstract

Recent theoretical progress in understanding the origin and nature of Type I and Type II supernovae is discussed. New Type II presupernova models characterized by a variety of iron core masses at the time of collapse are presented and the sensitivity to the reaction rate $^{12}C(\alpha,\gamma)^{16}O$ explained. Stars heavier than about 20 $M_\odot$ must explode by a "delayed" mechanism not directly related to the hydrodynamical core bounce and a subset is likely to leave black hole remnants. The isotopic nucleosynthesis expected from these massive stellar explosions is in striking agreement with the sun. Type I supernovae result when an accreting white dwarf undergoes a thermonuclear explosion. The critical role of the velocity of the deflagration front in determining the light curve, spectrum, and, especially, isotopic nucleosynthesis in these models is explored.

1. New Presupernova Models of Massive Stars

As a part of an ongoing project to calculate and understand the final evolution of massive stars and the consequent Type II supernova display, we have recently simulated the evolution of stars of a variety of masses: 11, 12, 15, 20, 25, 50, and 100 $M_\odot$ (Table 1). Some aspects of these models have been published previously (Wilson et al. 1985; Weaver and Woosley 1985) and the detailed evolution of the remainder is the subject of papers now in preparation. Here we briefly review the properties of the presupernova models and, especially the sensitivity of structure and nucleosynthesis to the new reaction rate for $^{12}C(\alpha,\gamma)^{16}O$. Stars have been studied using a revised version of the stellar evolution program KEPLER (Weaver, Zimmerman, and Woosley 1978; Weaver, Woosley, and Fuller 1984) which incorporates the new weak interaction rates of Fuller, Fowler, and Newman (1982ab, 1985), an improved implementation of weak processes at an earlier stage of the star's life (near oxygen core depletion), revisions to nuclear screening corrections (although see paper by Mochkovitch, this volume; photodisintegration reactions should *not* be screened), finer time step and zoning criteria, and Cameron (1982) initial abundances. Because of the smaller helium abundance employed in the new studies, helium core masses are smaller than in some of our earlier studies. For example, the new 11 $M_\odot$ star has a helium core at the end of its life of 2.4 $M_\odot$, slightly *smaller* than the helium core in our old 10 $M_\odot$ star (Woosley, Weaver, and Taam 1980). Helium core masses in the larger stars are also reduced by about 10%. More importantly, the revised weak rates and prescription for their implementation leads to *iron* core masses that are considerably smaller in 12 and 15 $M_\odot$ stars than in Weaver, Zimmerman, and Woosley (1978). Our present 15 $M_\odot$ model (Table 1) is an example with an iron core mass reduced from its 1978 value, 1.56 $M_\odot$, to 1.33 $M_\odot$. On the other hand, revisions to the reaction rate for $^{12}C(\alpha,\gamma)^{16}O$ lead to *larger* iron core masses in the bigger stars. The new 25 $M_\odot$ model, for example, which had a mass in 1978 of 1.61 $M_\odot$ and in Weaver, Woosley, and Fuller (1985) of 1.35 $M_\odot$, now has a mass of 2.1 $M_\odot$. These differences will be explored later. First we deal with the evolution of the lighter stars.

TABLE 1

PRESUPERNOVA MODELS AND EXPLOSIONS

Main Sequence Mass ($M_\odot$)	Iron Core Remnant ($M_\odot$)	Explosion Energy^a(10^{50}erg)	Baryon Mass^a($M_\odot$)	Neutron Star^a ($M_\odot$)
11	—^{b}	3.0	1.42	1.31
12	1.31	3.8	1.35	1.26
15	1.33	2.0	1.42	1.31
20	1.80	—	—	—
25	2.10	4.0	2.44	1.96
50	2.45	—	—	—
100	—^{b}	~4	39	Black Hole

^{a} All except for 100 $M_\odot$ determined by Wilson et al. (1985).
^{b} Never developed iron cores in hydrostatic equilibrium.

Our new 11 $M_\odot$ star had an evolution that differed markedly from all the other stars examined here. It was also the only model reported here which employed an old (Fowler, Caughlan, and Zimmerman 1975) rate for the $^{12}C(\alpha,\gamma)^{16}O$ reaction rate, although, in this particular case, we do not view that as critical. Qualitatively, the life history was similar to that which has been reported previously for a 9.6 $M_\odot$ star by Nomoto (1984ab; see also Miyaji et al. 1980). The helium core mass in the Nomoto study was 2.4 $M_\odot$, roughly the same as here *at the end of the evolution*. However, just after depletion of carbon in the center of our 11 $M_\odot$ star, the helium core mass was 2.6 $M_\odot$, hydrogen dredge-up of the helium core during off-center carbon burning accounts for the difference. The later stages of evolution are characterized by temperature inversion in the core (owing especially to plasma neutrino losses) and off-center burning. A neon core develops (60% neon and 20% each of magnesium and oxygen) containing very nearly a Chandrasekhar mass (1.45 $M_\odot$). That this mass exceeds 1.37 $M_\odot$ marks a critical departure between the present work and that of Nomoto (1984ab). For the smaller mass, neon and subsequent burning stages never ignite (under stable conditions) whereas in our 11 $M_\odot$ model, both neon and oxgen burning do occur in a shell from about 0.6 $M_\odot$ to 1.3 $M_\odot$ (Fig. 1) and, in a second convective burning stage from 1.3 $M_\odot$ to 1.4 $M_\odot$. Thus at the end of its life this star consists of a cold ($T \sim 5 \times 10^8$ K) central core of about 0.5 $M_\odot$ of neon surrounded by silicon and sulfur with a total mass slightly exceeding the (zero entropy) Chandrasekhar mass. Further cooling then leads to the collapse of the core containing combustible nuclear fuel.

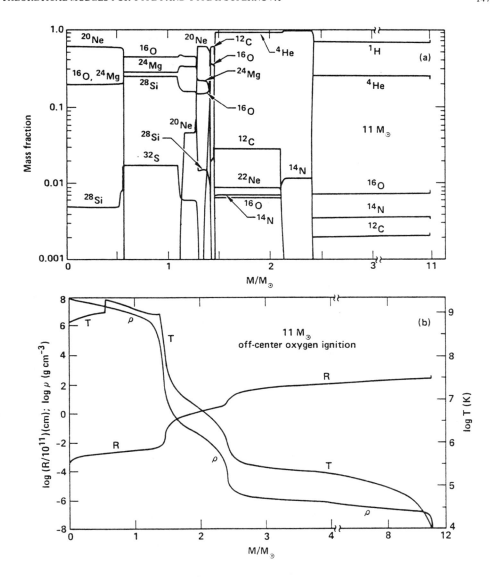

Fig. 1 - Composition (a) and structure (b) of an 11 $M_\odot$ star just after the off-center ignition of oxygen burning. The temperature at the base of the convective oxygen shell at this point is 1.87×10^9 K. The central density and temperature are 8.19×10^7 g cm^{-3} and 5.63×10^8 K respectively. The surface luminosity, effective temperature, and radius, all of which should not change prior to the stellar explosion, are 1.47×10^{38} erg s^{-1}, 4290 K, and 2.47×10^{13} cm. The neutrino luminosity is 1.21×10^{42} erg s^{-1}. Note scale breaks at 3.0 $M_\odot$ (a) and 4.0 $M_\odot$ (b).

It must be acknowledged at this point that the onset of collapse in this star has not been calculated in an entirely self-consistent fashion. Electron capture reactions were not included in the calculation of neon and oxygen burning although oxygen burns at such a high density off-center here that such captures would certainly have been important, probably triggering core collapse at an earlier stage. At a density of $\sim 10^8$ g cm^{-3}, for example, capture on the products of oxygen burning (e.g. ^{33}S$(e^-,\nu)^{33}$P; see Woosley, Arnett, and Clayton 1972) would lead to $\langle Y_e \rangle \sim 0.49$ reducing the zero entropy Chandrasekhar mass to 1.38 M$_\odot$ and causing the core to collapse. We do not think that the continued evolution is especially sensitive to this detail, but a self-consistent calculation would obviously be desirable. It is interesting to note that the contraction of the core is not without nucleosynthetic import in the overlying layers. We find that the heating of the helium burning shell leads to extension of the helium convective zone *into the hydrogen envelope*. The mixing of protons into a super-heated helium shell has obvious and far reaching implications for r and s-process nucleosynthesis which will be discussed elsewhere.

As the core contracts to 2.5×10^{10} g cm^{-3}, oxygen burning, greatly enhanced by electron screening, ignites at a temperature of $\sim 5 \times 10^8$ K. We note that at this point the core is also hovering on the verge of *general relativistic* instability (Shapiro and Teukolsky 1983) because Γ is so nearly equal to 4/3. Future calculations should include post-Newtonian gravity. Owing to the extremely degenerate nature of the core the nuclear runaway that ensues leads to complete combustion to iron group nuclei. Thereafter the evolution becomes quite similar to that described by Nomoto (1984ab) for his 2.4 M$_\odot$ helium core. Iron group products capture electrons efficiently leading to a sudden drop in Y_e at the center of the star to about 0.40. The thermal increment to the pressure from burning is inconsequential compared to the great loss in electron degeneracy pressure and the core begins to implode dynamically. As neon and magnesium and, later, silicon and sulfur, fall down they are heated by compression and burn explosively so that a standing combustion front exists at $\sim$110 to 170 km. Once the collapse is well under way, energy transport by convection is negligible.

The explosion of this model was calculated by Wilson et al. (1985) and the reader should see that reference for details. We note, however, that the explosion was of the "delayed" variety, occurring only after about 0.07 s. This differs qualitatively from the results of Hillebrandt, Nomoto, and Wolff (1984) who obtained a prompt hydrodynamical explosion for Nomoto's 9.6 M$_\odot$ star. We suspect that the difference may lie in different nuclear equations of state employed in the two codes or perhaps slight differences in the initial models. This point is under investigation by Wilson and Mayle.

Other models were also calculated with the improved version of KEPLER but using the recently revised rate for ^{12}C$(\alpha,\gamma)^{16}$O. Measurements by Kettner et al. (1982) and a reanalysis of new and old data by Langanke and Koonin (1984) have led to a revised rate, as tabulated by Caughlan et al. (1985), that under typical conditions in massive stars, is about 3 times the old value used in our previous calculations (Fowler, Caughlan, and Zimmerman 1975). Such a large revision has, as we shall see, major implications not only for nucleosynthesis, but also for the *structure* of the presupernova star. The iron core mass in the 12 and 15 M$_\odot$ models are not greatly altered from those obtained (for 20 and 25 M$_\odot$ stars) by Weaver, Woosley, and Fuller (1984), although, as it turns out, the extent of the oxygen burning *shell* is much larger with the new rate), but the new 25 M$_\odot$ model has a *much* larger iron core (Figs. 2 and 3).

Why should there be such great variation in the iron core mass for stars of 15 and 25 M$_\odot$ or for two stars, both of 25 M$_\odot$, differing only in the magnitude of the reaction rate for ^{12}C$(\alpha,\gamma)^{16}$O? The answer resides in the nature of carbon and neon burning and how they affect the entropy structure of the stellar core (Fig. 4). In the 15 M$_\odot$ star and in the older 25 M$_\odot$ star with low α-rate, carbon burning ignites as a well developed, exoergic, convective burning

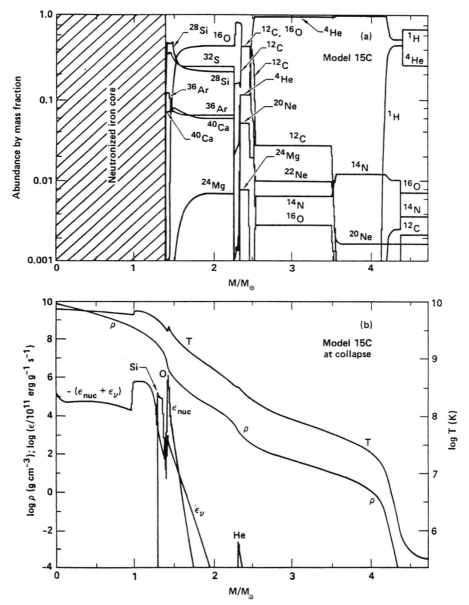

Fig. 2 - Composition (a) and structure (b) of a 15 $M_\odot$ star at the onset of core collapse (as defined by collapse velocity equals 1000 km s^{-1}). Interior to the iron core (1.33 $M_\odot$) energy is being lost due to a combination of neutrino emission from electron capture and photodisintegration. Peaks in the nuclear energy generation, ϵ_{nuc}, are apparent at the silicon, oxygen, and helium burning shells. Also plotted are the neutrino losses due to plasma processes (ϵ_ν in the region outside the iron core). Note the rapid fall off in density just outside the iron core and the substantial abundances of the elements silicon thru calcium in the oxygen convective shell.

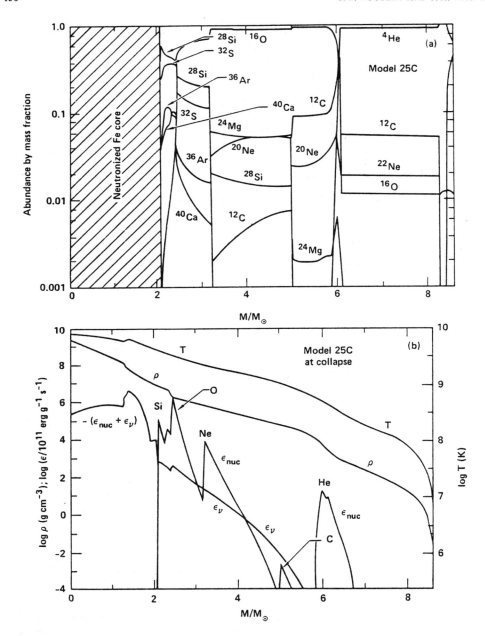

Fig. 3 - Composition (a) and structure (b) of a 25 $M_\odot$ star at the onset of core collapse. Note the large iron and the less rapid fall off in density (compared to Fig. 2) outside that iron core. Notation and sampling time are as in Fig. 3.

stage. In fact, the 15 $M_\odot$ star goes though *three* distinct stages of carbon convective burning before igniting neon burning at its center. The first stage depletes carbon in a region out to about 0.4 $M_\odot$; the second convective shell goes from $\sim$ 0.4 out to 1.0 $M_\odot$; and finally a third stage burns carbon out to about 1.5 $M_\odot$. The initial carbon abundance following helium burning in this star is 0.14. While carbon burning goes on, neutrino losses cool the core and its entropy decreases. During the first carbon convective stage, the central conditions are $T_c = 8.1 \times 10^8$ K, $\rho_c = 3.8 \times 10^5$ g cm^{-3} while at the end of the third stage, (at neon ignition) $T_c = 1.6 \times 10^9$ K, $\rho_c = 8.9 \times 10^6$ g cm^{-3}. This loss of entropy from the core allows the evolution to "keep in touch" with the Chandrasekhar mass (as adjusted for the finite entropy of the electrons). Thus a strong carbon burning shell is established at the edge of a (semi-)degenerate core. From this point onwards there always exists a sharp increase in the entropy at $M \sim$ 1.4 to 1.5 $M_\odot$ that restrains the outward extension of convective shells during oxygen and silicon burning so that when the star finally collapses it does so with an iron core of 1.33 $M_\odot$, close to the traditional Chandrasekhar mass for material with equal numbers of neutrons and protons. At least a portion of the decrease from $\sim$ 1.44 to 1.33 $M_\odot$ is a consequence of electron capture during and shortly after oxygen burning.

The 25 $M_\odot$ star with the most recent α-capture rate (Table 1) is quite another story. Because the carbon abundance following helium depletion is so low, $\sim$9% by mass, *carbon burning and neon burning never ignite in the center of the star as exoergic, convective burning stages.* The trace abundances of carbon and neon burn away radiatively, without the nuclear energy generation ever exceeding neutrino losses. Because there is no cooling stage, and because the 25 $M_\odot$ star had a larger entropy in its core to begin with, the core does not become especially degenerate (Fig. 4) and is not sensitive to the Chandrasekhar mass. Carbon is depleted radiatively out to a mass of about 2.5 $M_\odot$. Neon burns radiatively out to about 1.5 $M_\odot$. The next fully developed burning stage after helium burning is oxygen burning. Prior to silicon core ignition, oxygen burns in a convective core first out to about 1.35 $M_\odot$ and then in a convective shell out to 2.4 $M_\odot$. When silicon does ignite, it is within a 2.4 $M_\odot$ core comprised of almost pure silicon and sulfur. The large entropy increase associated with the oxygen shell is then at 2.4, not 1.4 $M_\odot$ as it was in the 15 $M_\odot$ model. Silicon burns out to 1.3 $M_\odot$ but, even with $Y_e \sim$ 0.46, the core is too small to collapse given its thermal content. Thus a silicon convective shell burns out to 2.1 $M_\odot$ before the core collapses. It is interesting that the property of a massive star that most sensitively determines its final evolutionary state, neutron star or black hole, namely the iron core mass at collapse, is so sensitive to occurrences during its relatively early life.

The 50 $M_\odot$ stellar evolution was qualitatively similar to that of the 25 $M_\odot$ model although, owing to a larger entropy, less centrally condensed core, and smaller carbon abundance, the iron core at the end was even larger, 2.45 $M_\odot$.

The 100 $M_\odot$ star, on the other hand, met a death unlike any reported thus far. Near the end of helium burning, the hydrogen envelope of the star was removed from the calculation when it became apparent that no reasonable choice of surface boundary conditions would allow it to stay on with a luminosity that was very nearly super-Eddington. Thus in its final stages the star was a 42 $M_\odot$ helium core, *i.e.*, a massive Wolf-Rayet star. Removal of the envelope has no important effect on the subsequent evolution of the helium core although it obviously affects the observational properties of the star. Following the central (radiative) exhaustion of carbon and neon, this core encountered the electron-positron pair instability upon attempting to ignite oxygen. The first collapse resulted in a peak central temperature of 3.0×10^9 K at a density of 1.5×10^6 g cm^{-3} which led to explosive oxygen burning and expansion. The initial explosion was far too weak to unbind the entire star, but a portion, about 1/4 $M_\odot$, was ejected from the

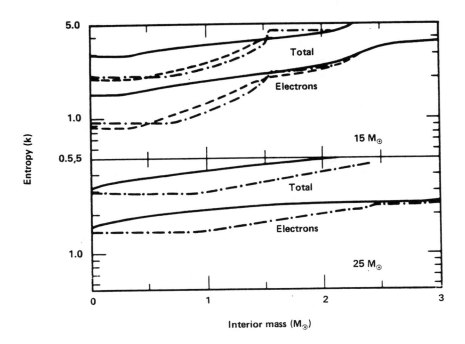

Fig. 4 - Entropy distributions in the inner regions of 15 and 25 $M_\odot$ model star just prior to carbon ignition (solid line), neon ignition (dashed line), and oxygen ignition (dot-dashed line). The total entropy and partial entropy in the electrons is given for both stars in dimensionless units of Boltzmann's constant. Stellar structure is most sensitive to the electronic entropy since the electrons are the major source of pressure. The locations of convective shells are sensitive to the distribution of total entropy. Entropy decreases substantially in the 15 $M_\odot$ model during carbon burning.

surface at velocities of several thousand kilometers per second (2.2×10^{49} erg). The central density of the star declined to 7.1×10^4 g cm^{-3}, and roughly one month long Kelvin-Helmholtz stage ensued as the star contracted and encountered the pair instability again. This time a higher peak temperature, 3.3×10^9 K is reached and a more violent explosion ensues, 6.5×10^{49} erg carried by an additional quarter solar mass of ejecta. For a time, again of about a month, the star oscillates violently but eventually settles down to encounter the pair instability a third time, the peak temperature this time being 3.7×10^9 K. This leads to a strong explosion and the ejection of 2.7 $M_\odot$ of surface material (helium, carbon, and oxygen) with energy 3.4×10^{50} erg. Oscillations and relaxation then lead to a fourth instability with $T = 4.6 \times 10^9$ K and a weak explosion. On the fifth time down, 2 years after the onset of the first explosion, there is nothing left at the center of the star to burn, hence no explosion. The remainder of the star, neglecting rotation, becomes a black hole. This model and its observational consequences should amuse and employ us theoreticians for years. Stay tuned!

As a general result of our study of stars in the 11 to 100 $M_\odot$ range, it appears that the compact remnants of stars having main sequence mass $\gtrsim$ 20 $M_\odot$ [with exact value sensitive to ^{12}C$(\alpha,\gamma)^{16}$O], will have quite different properties from those of lower mass. In particular, since nuclear equations of state suggest an upper bound to the mass of a stable neutron star of $\sim$2.0 $M_\odot$, stars heavier than $\sim$20 $M_\odot$ may leave black holes and lighter stars (but heavier than $\sim$10 $M_\odot$) will leave neutron stars. It is important also to note that an object which eventually, after cooling and deleptonization, becomes a black hole may be the residual of an explosion which ejects matter, produces explosive nucleosynthesis, and exhibits a light curve not markedly discrepant with observations of Type II supernovae (Wilson et al. 1985).

2. Delayed Explosions of Type II Supernovae

For many years the outward propagation of the shock wave produced by the bounce of these iron cores has been studied as a possible mechanism for the explosion of the star (cf. Bowers and Wilson 1982; Arnett 1980, 1983; Brown, Bethe, and Baym 1982; Hillebrandt 1984; Bruenn 1984; Cooperstein 1982; Kahana, Baron, and Cooperstein 1984). For the most part, the results of these studies have not been particularly encouraging, except, perhaps, in the case of very low mass iron cores (Hillebrandt 1982; Hillebrandt, Nomoto, and Wolff 1984; Cooperstein 1982). The shock stalls, overwhelmed by photodisintegration and neutrino losses, and the star does not explode. More recently, slow late time heating of the envelope of the incipient neutron star has been found to be capable of rejuvenating the stalled shock and producing an explosion after all (Wilson 1984; Bethe and Wilson 1985; Wilson et al. 1985). The basic mechanism is energy transport by electron neutrinos and anti-neutrinos which capture on nucleons and scatter on electron-positron pairs just behind the stalled (accretion) shock. The source of the neutrinos is the cooling proto-neutron-star beneath the shock and, to a lesser but important extent, the neutrinos liberated by the accretion flux itself. A complete discussion of the complexities (and attendant uncertainties) associated with this particular explosion mechanism are best left to the referenced papers by Wilson and co-workers and to future papers. For now we merely note that each of the explosions listed in Table 1 were calculated by Wilson and Mayle and were of this "delayed" variety (Wilson et al. 1985). Credible variations in the nuclear equation of state may allow the prompt explosion of the 11, 12, and 15 $M_\odot$ models (E. Baron and H. Bethe; private communication; Bethe and Brown 1985), but it appears highly unlikely that explosion of iron cores as large as 2.1 $M_\odot$ (i.e. the 25 $M_\odot$ model) will ever be achieved solely by a hydrodynamical bounce. Thus the nucleosynthesis we shall now discuss is quite contingent upon the successful occurrence of a delayed explosion.

Otherwise the complexity of rotation will have to be considered (cf., Bodenheimer and Woosley 1983).

3. Nucleosynthesis in Type II Supernovae

The passage of the shock wave through the overlying mantle of the star, in addition to providing the impulse for its ejection, leads to high temperatures and nuclear reactions that have been followed in detail for the 15 and 25 $M_\odot$ models (Weaver and Woosley 1985; Wilson et al. 1985). Figure 5 shows the explosive nucleosynthesis in the 25 $M_\odot$ model and nucleosynthetic results of 15 and 25 $M_\odot$ models are both shown compared to solar abundances in Figure 6. In the case of the 15 $M_\odot$ model, the abundances ejected are almost entirely, with the obvious exception of the iron group, produced in the pre-explosive stages of evolution and merely shoved off the star by the explosion. The diminished importance of explosive nucleosynthesis, as compared, for example, to past studies (Weaver and Woosley 1980) is a consequence both of the low explosion energy and the rapid fall off in density around the core of the preexplosive star. The large abundances of silicon through calcium are especially a result of an extensive oxygen burning convective shell just outside the core (Fig. 2). It is interesting to note that the energy liberated by nuclear reactions was not an inconsequential fraction of the final explosion energy in the 15 $M_\odot$ model. The final kinetic energy at infinity was calculated to be 2×10^{50} erg, 30% of which was generated by nuclear burning during shock wave passage.

Because of the larger explosion energy in the 25 $M_\odot$ model, there was a considerable amount of *explosive* nucleosynthesis and nuclear energy generation formed an even greater fraction of the final kinetic energy at infinity. The initial shock contained roughly 10^{51} erg as calculated in the core bounce program, but the binding energy of the mantle was also about 10^{51} erg. Nuclear burning gave 4×10^{50} erg and the final kinetic energy was also 4×10^{50} erg. *Thus without the energy from nuclear burning following shock wave passage a portion of this star may well have reimploded.* In the 15 $M_\odot$ model calculated by Wilson et al. (1985) some mass reimplosion was actually observed.

Figure 7 shows the isotopic nucleosynthesis resulting from the explosion, by Wilson's delayed mechanism (Wilson et al 1985), of our 25 $M_\odot$ presupernova model. The comparison with solar abundances is very good, much better, for example, than that published for a previous 25 $M_\odot$ model by Woosley and Weaver (1982). The changes reflect principally the altered structure of the presupernova mantle and core brought about by alterations in weak interaction rates (Weaver, Woosley, and Fuller 1984) and in the reaction rate for $^{12}C(\alpha,\gamma)^{16}O$. It is worth noting that 34 species out of a total of 61 in this mass range are produced within a factor of two of their relative solar abundances and 52 are produced within a factor of 4. In the sun these abundances span a range in mass fraction of 7 orders of magnitude. Of the remaining 9, ^{13}C and ^{15}N are probably the products of hydrogen burning in lower mass red giant stars and novae respectively (cf. Wallerstein 1973; Lambert and Tomkin 1974; Sneeden and Lambert 1975; Wallace and Woosley 1981); the origin of fluorine is unknown (although see Howard, Arnett, and Clayton 1971; Scalo and Despain 1976); ^{18}O is produced in a 15 $M_\odot$ star to be reported elsewhere; ^{46}Ca and ^{47}Ti are very rare species whose production might be quite sensitive to poorly determined nuclear reaction rates (or a change in stellar mass); ^{48}Ca and ^{54}Cr can be made in a neutron-rich nuclear statistical equilibrium (Hainebach et al. 1974; Hartmann, Woosley, and El Eid 1985); and ^{63}Ni was near the end of our nuclear reaction network and may not have been tracked accurately.

Although the agreement is really very good overall, some annoying discrepancies can be noted. Carbon is down relative to oxygen. That's not so bad. Perhaps carbon is produced in low

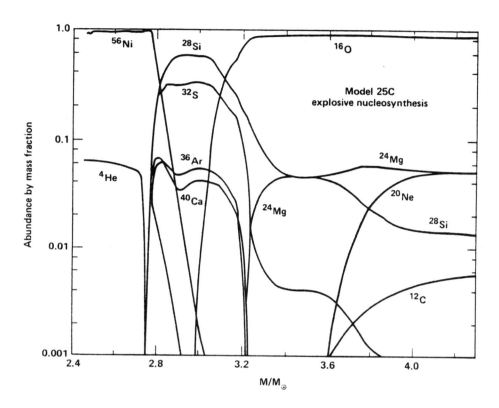

Fig. 5 - Composition of the inner regions of a 25 $M_\odot$ supernova following shock passage and ejection. The pre-explosive composition (see Fig. 3) is substantially altered out to about 3.6 $M_\odot$. Composition interior to the mass cut, 2.44 $M_\odot$, is not plotted. The explosion was calculated by Wilson et al. (1985).

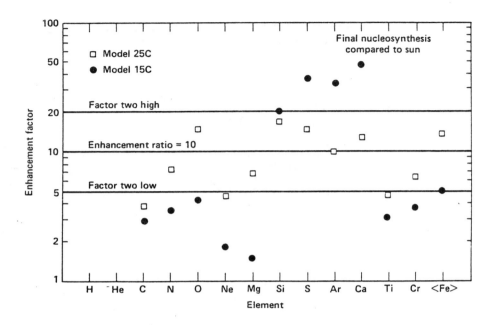

Fig. 6 - Comparison of bulk elemental nucleosynthesis in 15 and 25 $M_\odot$ supernovae to solar abundances (Cameron 1982). The enhancement factor is the mass fraction of a given element in the ejecta compared to its mass fraction in the sun (which was also its initial abundance in the calculation). The deficient production of carbon, neon, and magnesium, especially in the 15 $M_\odot$ study, is a consequence of the revised rate for $^{12}C(\alpha,\gamma)^{16}O$. The large productions of silicon thru calcium in the 15 $M_\odot$ model do not exist if this material falls back onto the neutron star following hydrodynamical interaction with the stellar envelope. Such was the result in the present study, but this occurrence is quite model dependent.

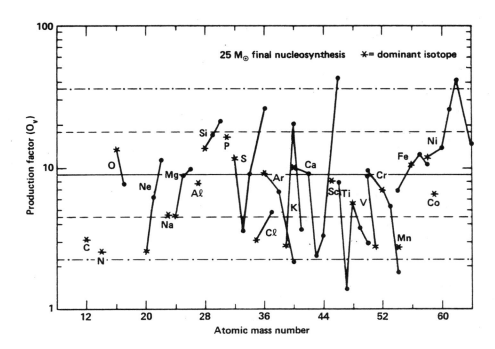

Fig. 7 - Isotopic nucleosynthesis in a 25 $M_\odot$ explosion. Final abundances in the ejecta are plotted for isotopes from ^{12}C to ^{64}Ni compared to their abundances in the sun (Cameron 1982). An average *overproduction factor* of 9 characterizes the distribution. If one gram in 9 of the matter in the Galaxy has experienced conditions like those in a 25 $M_\odot$ star, its metallicity will resemble the sun with an abundance pattern as shown.

mass stars (Dufour 1984). Of greater concern are the products of carbon burning, eg. ^{20}Ne, ^{23}Na, and ^{24}Mg. It appears unlikely that all these abundances could systematically be underestimated by a factor of two in the sun (relative to oxygen, say), nor does it appear probable that other masses of supernovae, eg. 15 $M_\odot$ are going to fill in the gap. All things considered, it would have been better if the rate for ^{12}C$(\alpha,\gamma)^{16}$O had not increased quite so much. Further increases (beyond the roughly factor of 3 times FCZ75 used here) would make matters worse still. We shall see. The rates will, of course, be whatever the nuclear physicists determine. It is interesting however, that in our previous model (Weaver, Woosley, and Fuller 1984), these same species were major *overproductions* while the silicon through calcium group was underproduced; the only difference, a smaller rate for ^{12}C$(\alpha,\gamma)^{16}$O.

Following the observations of Mahoney et al. (1982), increased attention has been focused upon the production of ^{26}Al in Wolf Rayet stellar winds (see papers by Casse and by Dearborn this volume), novae (see paper by Truran this volume and Wallace and Woosley 1981) and supernovae. Clayton (1984), in particular, has emphasized the difficultiy associated with producing the (strong) observed signal using ^{26}Al produced solely during supernova explosions. While Clayton's conclusion is somewhat sensitive to both an assumed model for Galactic chemical evolution and the unknown distribution of ^{26}Al interior to the solar orbit, we accept it for now and merely point out that revisions of reaction rates and stellar model have lead to a substantial increase in the supernova production of ^{26}Al over that estimated by Woosley and Weaver (1980), ^{26}Al/^{27}Al $\approx 10^{-3}$. Here, for the new 25 $M_\odot$ model, ^{26}Al/^{27}Al $= 6 \times 10^{-3}$ and 6.9×10^{-5} $M_\odot$ of ^{26}Al are ejected. Of this ^{26}Al, a lesser fraction, 1.9×10^{-5} $M_\odot$ comes from proton capture on magnesium in the hydrogenic envelope (see also Dearborn and Blake 1985). Similarly, in the 15 $M_\odot$ model, 7.1×10^{-6} $M_\odot$ was ejected in the hydrogenic envelope. These values would have been larger if the initial metallicity had been greater or if mass loss had been included in the calculation. On the other hand, this is an upper bound to the ^{26}Al that could be produced in 15 or 25 $M_\odot$ stars of constant mass and solar metallicity since we considered *all* reactions on ^{25}Mg as going to the long lived ground state of ^{26}Al. In fact, some would go to the isomeric state (Harris et al. 1983) and be lost. Also recent re-examination of the ^{25}Mg$(p,\gamma)^{26}$Al$_g$ reaction rate by Peter Parker at Yale (private communication) substantially reduces the value over that reported by Champagne, Howard, and Parker (1983) and included in the tables of Caughlan et al. (1985) for temperatures less than 40 million degrees (and increases it for higher temperatures). Typically hydrogen burns at temperatures at the cooler end of this range in 15 and 25 $M_\odot$ stars and the synthesis of ^{26}Al in such stars will be correspondingly reduced. On the other hand, losing the ^{26}Al in a wind can increase its effective production in the hydrogen envelope since getting the material off of the star reveals the radioactivity to possible observation before a portion has had time to decay while waiting for the supernova (Dearborn and Blake 1985). More calculations are needed, especially using the modified ^{25}Mg$(p,\gamma)^{26}$Al reaction rate. Calculations are also needed of nucleosynthesis in nova ejecta enriched in neon and magnesium (see Truran, this volume). It is interesting that supernovae like the 25 $M_\odot$ star described here, though perhaps still unable to account for all of the signal observed by Mahoney et al. should still give a substantial flux. We crudely estimate this to be about 10% of the HEAO-C signal (ignoring production in the hydrogen envelope). Since the Gamma-Ray Observatory will have at least 10 times the sensitivity of HEAO-C to these γ-lines (Kurfess et al. 1983), the signal from supernovae may still be discernable (depending upon the poorly determined Galactic distribution of the HEAO-C signal).

4. Type I Supernova Models - General Comments

There are many reasons for believing that Type I supernovae might be the result of accreting white dwarfs provoked into thermonuclear explosion. Observational evidence favors the association of such supernovae with a low mass population. They are not preferentially situated in the spiral arms of spiral galaxies (cf. Maza and Van den Bergh 1976) and they do occur in elliptical galaxies where no Type II supernovae are seen (Tammann 1977) and no young stars are expected (although see Oemler and Tinsley 1979). Type I supernovae, by definition, lack hydrogenic lines in their spectra as would be the case if a white dwarf exploded. The velocities inferred from spectral measurements of Type Is and the energies of the explosions agree with what one would obtain by converting a fraction of a white dwarf mass to iron (or ^{56}Ni). Further observational evidence supporting this inference is provided by the fact that iron is seen in the explosions (Kirshner and Oke 1975; Wu et al. 1983) as well as the radioactive decay product of ^{56}Ni, ^{56}Co (Axelrod 1980a,b; Branch 1984a,b). Furthermore the degenerate nature of a white dwarf guarantees that a nuclear runaway will convert a substantial fraction of its mass to iron on a short time scale (Hoyle and Fowler 1960) with the resulting light curve generated by the decay of these same radioactive species (Pankey 1962; Truran, Arnett, and Cameron 1967; Colgate and McKee 1969; Arnett 1979; Chevalier 1981; Weaver, Axelrod, and Woosley 1980). Finally, Type I supernovae are a very uniform class of events which might be understood if they all had a very similar origin, a compact object that creates 0.3 to 1.1 $M_\odot$ of ^{56}Ni (Axelrod 1980a,b). For all these reasons it is generally presumed that Type I supernovae are the consequence of the thermonuclear disruption of an accreting white dwarf star.

Generally speaking, there are two mechanisms for the explosion: *detonations* and *deflagrations*. Detonations have been recently reviewed by Nomoto (1982ab) and by Woosley, Taam, and Weaver (1985); deflagrations have been recently examined by Nomoto, Thielemann, and Yokoi (1984) and by Woosley, Axelrod, and Weaver (1984). See these reviews for extensive discussions of specific models and references to the published literature. Burning occurs by detonation when the runaway ignites in *helium*-rich material. This may occur either at the center of an accreting helium white dwarf or, off-center, at the base of an accreted helium layer on a carbon-oxygen or an oxygen-neon white dwarf. Ignition in helium occurs at sufficiently low density that burning to nuclear statistical equilibrium at $T \sim 8 \times 10^9$ generates a large local increase in the pressure, a factor of 5 being broadly representative. A higher temperature cannot be attained by the combustion to iron since photodisintegration truncates the energy deposition. This large overpressure leads to a shock wave. As material passes through the shock wave, typically at speeds of 2 to 3 times the speed of sound, the temperature and density rise, virtually instantaneously, and nuclear burning gives enough energy to keep the shock wave going. Because the expansion is supersonic, material does not have time to "get out of the way" ahead of the burning front and essentially the entire layer of helium is converted to iron group elements, especially ^{56}Ni. Indeed, if the helium layer is thick enough and the density in the carbon-oxygen core not too high, the shock wave driven into the carbon oxygen core may propagate as a successful detonation wave as well in which case the star is totally disrupted and there is no bound remnant. More typically, a portion of the carbon oxygen core stays behind as a white dwarf remnant of the supernova.

The detailed computer models of this variety (referenced above) have been quite successful in reproducing the *light curves* observed in Type I explosions and in synthesizing, in proper solar proportions, some of the major isotopes of the iron group elements. Unfortunately this class of model is not consistent with spectroscopic restrictions. Ignition at a lower density *and* conversion of all the ejected material into iron gives an explosion that is both too energetic, i.e., the velocities are too high compared to observed line widths and Doppler shifts, and lacking in substantial quantities of intermediate mass elements, especially silicon and calcium, that are quite prominent in the early time spectrum. For this reason, greater attention is given, at least presently, to deflagrating models.

Deflagrations generally occur when an accreting carbon-oxygen white dwarf approaches the Chandrasekhar mass and ignites in its center. Typical ignition conditions require a balance of neutrino losses by the plasma process to balance the nuclear energy generation by a highly screened carbon fusion reaction and imply $\rho \sim 2 \times 10^9$ g cm^{-3}. At this high density a temperature of 8×10^9 K gives only a small increment in the large Fermi pressure, $\lesssim 10\%$, which is *not* enough to generate a self-consistent detonation wave. Instead, once the burning front begins to travel, unburned fuel expands ahead of the front, having been notified by a sonic precursor of the events transpiring in the core. Density decreases as material crosses the burning front, generates heat, and expands. Perhaps most importantly, the expansion of a portion of the white dwarf is rapid enough that the (subsonic) burning front never overtakes it and thus unburned fuel is ejected, as well as a portion of the star that experiences intermediate burning temperatures and produces intermediate mass elements.

5. Problems with Deflagrations and Some Possible Solutions

The problems with deflagrating models center upon the poorly defined nature of the burning front and gross uncertainty in the velocity with which it propogates. The only clear limits are that it be subsonic yet lead, ultimately, to the rapid expansion of the entire star. If the deflagration moves slowly compared to sound, then the expansion of the star is well underway before the burning front passes a fiducial point, say half-way out in mass. Thus less matter is burned to iron and, for starting points of similar gravitational binding, the supernova energy and velocity are smaller. Less ^{56}Ni produced also means a dimmer, briefer light curve, eventually, for $M_{56} \lesssim 0.3$ $M_\odot$, too dim and too brief to be in accord with observations. A rapid deflagration velocity, on the other hand, gives the converse of these properties.

A generic problem with deflagrations is that, so far, they give unacceptable nucleosynthesis for the isotopes of the iron group. In the work of Nomoto, Theilemann, and Yokoi (1984), for example, ^{54}Fe/^{56}Fe is overproduced by a factor of 3.9 (Model W7) compared to the sun. Woosley, Axelrod, and Weaver (1984) find an even larger overproduction in a similar model. In this volume, F. Thielemann presents a recalculation of Model W7 that properly includes the "α-rich freeze-out" of Woosley, Arnett, and Clayton (1973). In the new study, he finds much of the material that previously was in the form of ^{54}Fe is transformed into ^{58}Ni. This helps things a bit since elemental ratios in the sun are more poorly known than isotopic ones, but the abundance of ^{58}Ni in the sun is comparable to that of ^{54}Fe and the overproduction is still quite large, about a factor of 5 in the new calculation for any reasonable value of electron capture rates.

The difficulty stems from the large ignition density of the deflagrating models which leads to a great deal of electron capture during the explosion. One would like to have almost no capture since adequate ^{54}Fe (or ^{58}Ni) can be created by just those neutrons available from conversion of the initial metallicity of the star to ^{22}Ne. In fact, ^{54}Fe/^{56}Fe resulting from nuclear statistical equilibrium *with no electron capture* is about 0.05 (Z/0.015), i.e., about solar for solar metallicity. Perhaps those white dwarfs that make Type I supernovae are metal deficient, but obviously the amount of electron capture must be kept to a minimum.

Two resolutions have been suggested for this dilemma (the second being reported here in published form for the first time). The first attempts to decrease the central ignition density by raising the accretion rate to a very high value. Iben and Tutukov (1984) and Webbink (1984) have discussed the possibility that two carbon oxygen white dwarfs might merge as a consequence of gravitational radiation in a binary pair. Such probabilities as one can legitimately assign to such uncertain events are consistant with observed SNI rates. In the final merger, the effective accretion rate must be very large; for lack of any calculation, an Eddington value is usually assumed. Unfortunately studies (thus far unpublished) by Nomoto and by us show that such high accretion rates do not, in the general case, lead to central carbon ignition. Compression of the outer layers by the rapid mass addition leads to off-center burning that gradually, on a thermal diffusion time, propagates inwards until the entire white dwarf has been converted to a oxygen-magnesium-neon composition. In the specific calculation leading to Figure 8, a carbon-oxygen white dwarf of 0.7 $M_\odot$ and central temperature 5×10^7 K was allowed to accrete carbon and oxygen at a rate 2×10^{-5} $M_\odot$ y^{-1}. Once the mass had increased to 1.06 $M_\odot$ carbon burning ignited in a shell at 0.77 $M_\odot$. Over the next 19,000 years the carbon shell burned into the star at the same time that the star's mass continued to increase eventually reaching the center about the time the Chandrasekhar mass was achieved. No explosive burning or dynamic events were observed. This star presumably does not ignite neon or oxygen burning until such high central densities are reached that electron capture causes a collapse to a neutron star (Miyaji et al. 1980). Thus while such objects are very interesting candidates for producing certain x-ray variables, they are hardly Type I supernovae. No radioactive ^{56}Ni will be produced and very little matter will be expelled.

More recently we have examined this same sort of model for larger initial white dwarf masses and larger accretion rates. So long as the time scale for reaching the Chandrasekhar mass, $(1.44 - M_i)/\dot{M}$, is less than the thermal difusion time to the center of the star, the burning front will not have time to reach the center before a carbon deflagration ignites. Also, larger white dwarf masses diminish the effects of off-center compression relative to central heating. For a white dwarf of 1.37 $M_\odot$ accreting at 10^{-5} $M_\odot$ yr^{-1}, for example, we find that carbon ignites in the center. Accretion onto a 1.0 $M_\odot$ white dwarf is under study. However, since these larger masses and accretion rates imply a central ignition density comparable to what one generally uses with a Chandrasekhar mass white dwarf anyway, part of the motivation for going to the high accretion rate is lost (although one does retain the possibility of making a Type I supernova in the fashion suggested by Iben and Tutukov). We prefer to solve the problem of abnormal nucleosynthesis in deflagrating models a second way, by a more careful consideration of the properties of the propagating burning front.

The velocity of a *conductive* deflagration is $v_b \sim \delta/\tau \sim (l_{mfp} c/\tau)^{1/2}$ where v_b is the velocity of the deflagration (flame) front, δ is its thickness, l_{mfp} is the mean free path of the electrons ($= 1/\kappa\rho$), τ is a typical burning time in the front, and c is the speed of the electrons which here are relativistic. For a density $\rho \sim 2 \times 10^9$ g cm^{-3}, the opacity is quite low owing to the long path length of the degenerate electrons. At $T_9 = 1$, $\kappa = 2 \times 10^{-5}$ cm^2 g^{-1}, and at $T_9 = 8$, $\kappa = 8 \times 10^{-3}$ cm^2 g^{-1}. The nuclear burning time is of course very sensitive to the temperature which rapidly changes in the burning front. Typically half of the carbon burns, releasing about 2×10^{17} erg g^{-1}, by the time the temperature has risen to 4 billion degrees. At this temperature, $\tau \sim 10^{-10}$ s and thus, with obvious great uncertainty, $\delta \sim 10^{-3}$ cm and $v_b \sim 10^7$ cm s^{-1}. These values are interesting, but largely irrelevant since the flame will not propagate as a stable conductive front (cf. Müller and Arnett 1982).

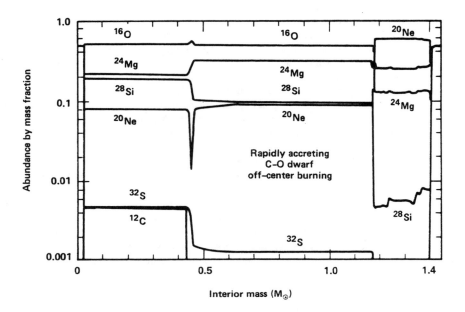

Fig. 8 - Composition of a white dwarf that has experienced a very high accretion rate and off-center carbon ignition. The composition of an initially 0.7 $M_\odot$ white dwarf of carbon and oxygen is sampled 37,000 years after the onset of accretion at 2×10^{-5} $M_\odot$ yr^{-1}. Very small regions of unburned carbon and oxygen still exist in the inner 0.02 $M_\odot$ and outer 0.03 $M_\odot$ of the star. The central density at this point is 2.57×10^8 g cm^{-3} as the white dwarf is nearing the Chandrasekhar mass. Continued evolution will lead to a neutron star.

For any reasonable conditions in the white dwarf star, the kinematic viscosity (Spitzer 1962; Landau and Lifshitz 1959) will be sufficiently small and the characteristic velocities and dimensions suffiently large that the Reynolds number is very great. Thus the flow will be turbulent. Specifically, for $R > 10^5$, as noted by Zeldovich and Rozlovski (1947 as referenced on p. 477 of Landau and Lifshitz 1959), a spherically propagating combustion front in free space will have self-turbulence with characteristic dimension equal to the radius of the spherical flame. Thus we envision the runaway initiating as a point in the center of the star, stabilized perhaps for some brief period by convection, but eventually localized to a small region. The pressure scale height as the runaway commences is 5.3×10^7 cm and contains 0.35 $M_\odot$, but carbon burns at a rate less than the sound crossing time for this region once the temperature exceeds 1.5×10^9 K. As the temperature in the center continues to rise, ultimately reaching 8×10^9 K, the runaway becomes even more localized. Thus the flame front initially propagates in a very small region and, as we have just discussed, this means that the initial scale of the turbulence will also be quite small and the velocity of the flame relatively slow. As the burning front encompasses more and more material, the turbulent scale grows and so does the velocity. Thus one may approximate $v_b \sim k r_b$ where k is a constant of proportionality and r_b is the radius of the burning front. The solution to this equation is $r_b = r_o\ exp(kt)$ and hence $v_b = v_o\ exp(kt)$. Of course, one must impose the restriction that the turbulent motion remain subsonic, so as a maximum upper bound (which may, in fact, be approached), $v_b^{max} \lesssim c_s$, the sound speed. Thus we adopt

$$v_b(r) = f c_s \left(1 - e^{r/R_o}\right)$$

where f and R_o are adjustable constants, $f < 1$, R_o less than the radius of the (expanding) white dwarf, which, in practice, will be determined by requirements on explosion energy, light curve, and nucleosynthesis.

This prescription is obviously arbitrary, but we feel that it relates more physically to the actual circumstances than any formulation based upon a convective mixing length, i.e., pressure scale height as used in all previous studies (cf. Nomoto, Thielemann, and Yokoi 1984; Woosley, Axelrod, and Weaver 1984). Certainly in its initial stages, the burning front has little knowledge of the pressure scale height. It propagates in a region of nearly constant density and pressure. The relevant scale is the scale of the turbulence.

We are quite concerned about the very early propagation of the front because the isotopic nucleosynthesis is so sensitive to it. A white dwarf near the Chandrasekhar mass obviously has $\Gamma \approx 4/3$ and is only marginally stable against large excursions in radius either inwards or outwards. Thus burning only a tiny fraction (we estimate $\lesssim 0.03$ $M_\odot$) of carbon and oxygen in the center of the star releases enough energy that it would expand to the point where its central density has dropped by a factor of 10. What is needed then, in order to avoid the excess electron capture that has plagued the deflagration model, is only that the star have time to respond to this burning and expand before the burning front encompasses a substantial fraction of its mass. Later, when the expansion has occurred, one then wants the burning front to go very fast in order to burn enough of the star that an energetic explosion results. This is the behavior that equation (1) will give and seems, to us at least, to be what is physically expected. We are carrying out a calculation to see if it will all work quantitatively.

This work has been supported by the National Science Foundation (AST-84-18185) and, at LLNL, by the Department of Energy through contract number W-7405-ENG-48.

References

Arnett, W. D. 1979, *Astrophys. J. Lett.*, **230**, L37.

_____. 1980, *Ann. N. Y. Acad. Sci.*, **336**, 366.

Axelrod, T. S. 1980a, Ph.D. thesis UCSC, available as UCRL preprint 52994 from Lawrence Livermore National Laboratory.

_____, 1980b, in *Type I Supernovae*, ed. J. C. Wheeler, (Austin: University of Texas), p. 80.

Bethe, H. A., and Brown, G. 1985, *Sci. Am.*, **252**, 60.

Bethe, H. A., and Wilson, J. R. 1985, *Astrophys. J.*, in press.

Bodenheimer, P., and Woosley, S. E. 1983, *Astrophys. J.*, **269**, 281.

Bowers, R. L., and Wilson J. R. 1982, *Astrophys. J.*, **263**, 366.

Branch, D. 1984a, *Ann. N. Y. Acad. Sci.*, **422**, 186.

_____. 1984b, in *Stellar Nucleosynthesis*, eds. C. Chiosi and A. Renzini, (D. Reidel: Dordrecht), p. 19.

Brown, G. E., Bethe, H. A., and Baym, G. 1982, *Nucl. Phys. A*, **A375**, 481

Bruenn, S. 1984, preprint submitted to *Astrophys. J. Suppl. Ser*

Cameron, A. G. W. 1982, in *Essays in Nuclear Astrophysics*, ed. C. A. Barnes, D. D. Clayton, and D. N. Schramm, (Cambridge Univ. Press: Cambridge), p. 23.

Caughlan, G. R., Fowler, W. A., Harris, M. J., and Zimmerman, B. A. 1985, *Atomic Data and Nuclear Data Tables*, **32**, 197.

Champagne, A. E., Howard, A. J., and Parker, P. D. 1983, *Astrophys. J.*, **269**, 686.

Chevalier, R. A. 1981, *Astrophys. J.*, **246**, 267.

Clayton, D. D. 1984, *Astrophys. J.*, **280**, 144.

Colgate, S. A., and McKee, C. 1969, *Astrophys. J.*, **157**, 623.

Cooperstein, J. 1982, Ph.D thesis, SUNY-Stony Brook.

Dearborn, D. S. P., and Blake, J. B. 1985, *Astrophys. J. Lett.*, **288**, L21.

Dufour, R. J. 1984, *Bull. Am. Astron. Soc.*, **16**, No. 4, 888.

Fowler, W. A., Cauglan, G. R., and Zimmerman, B. A. 1975, *Ann. Rev. Astron. Astrophys.*, **13**, 69.

Fuller, G. M., Fowler, W. A., and Newman, M. J. 1982a, *Astrophys. J. Suppl. Ser.* **48**, 279.

_____. 1982b, *Astrophys. J.*, **252**, 715.

_____. 1985, submitted to *Astrophys. J.*

Hainebach, K. L., Clayton, D. D., Arnett, W. D., and Woosley, S. E. 1974, *Astrophys. J.*, **193**, 157.

Harris, M. J., Fowler, W. A., Caughlan, G. R., and Zimmerman, B. A. 1983, *Ann. Rev. Astron. Astrophys.*, **21**, 165.

Hartmann, D., Woosley, S. E., and El Eid, M. F. 1985, *Astrophys. J.*, in press.

Hillebrandt, W. 1982, *Astron. Astrophys. Lett.*, **110**, L3.

Hillebrandt, W. 1984, *Ann. N. Y. Acad. Sci*, **422**, 197.

Hillebrandt, W., Nomoto, K., and Wolff, R. G. 1984, *Astron. Astrophys.*, **133**, 175.

Howard, W. M., Arnett, W. D., and Clayton, D. D. 1971, *Astrophys. J.*, **165**, 495.

Hoyle, F., and Fowler, W. A. 1960, *Astrophys. J.*, **132**, 565.

Iben, I., Jr. and Tutukov, A. V. 1984, *Astrophys. J. Suppl. Ser.* **54**, 335.

Kahana, S., Baron, E., and Cooperstein, J. 1984, in *Problems of Collapse and Numerical Relativity*, ed. D. Bancel and M. Signore, (D. Reidel: Dordrecht), p. 163.

Kettner, K. U., Becker, H. W., Buchmann, L., Görres, J., Kräwinkel, H., Rolfs, C., Schmalbrock, P., Trautvetter, H. P., and Vlieks, A. 1982, *Z. Phys.*, **A308**, 73.

Kirshner, R. P., and Oke, J. B. 1975, *Astrophys. J.*, **200**, 574.

Kurfess, J. D., Johnson, W. N., Kinzer, R. L., Share, G. H., Strickman, M. S., Ulmer, M. P., Clayton, D. D., and Dyer, C. S. 1983, *Adv. Space Res.*, **3**, 109.

Lambert, D. L., and Tomkin, J. 1974, *Astrophys. J. Lett.*, **194**, L89.

Langangke, K. and Koonin, S. 1984, preprint MAP-56, Kellogg Radiation Lab., Cal Tech, submitted to *Nucl. Phys. A*.

Landau, L. D., and Lifshitz, E. M. 1959, *Fluid Mechanics*, (Pergamon Press: Oxford), p. 62.

Mahoney, W. A., Ling, J. C., Jacobson, A. S., and Lingenfelter, R. E. 1982, *Astrophys. J.*, **262**, 742.

Maza, J., and van den Bergh, S. 1976, *Astrophys. J*, **204**, 519.

Miyaji, S., Nomoto, K., Yokoi, K., and Sugimoto, D. 1980, *Publ. Astron. Soc. Japan*, **32**, 303.

Muller, E., and Arnett, W. D. 1982, *Astrophys. J. Lett.*, **261**, L109.

Nomoto, K. 1982a, *Astrophys. J.*, **253**, 798.

──────. 1982b, *Astrophys. J.*, **257**, 780.

──────. 1984a, in *Stellar Nucleosynthesis*, eds. C. Chiosi and A. Renzini, (D. Reidel: Dordrecht), p. 205.

──────. 1984b, in *Problems of Collapse and Numerical Relativity*, ed. D. Bancel and M. Signore, (D. Reidel: Dordrecht), p. 89.

Nomoto, K., Thielemann, F.-K., and Yokoi, K. 1984, *Astrophys. J.*, **286**, 644.

Oemler, A., and Tinsley, B. M. 1979, *Astron. J.*, **84**, 985.

Pankey, T. 1962, Ph.D. Thesis Howard University, *Diss. abstr.*, **23**, 4.

Scalo, J. M. and Despain, K. H. 1976, *Astrophys. J.*, **203**, 667.

Shapiro, S. L., and Teukolsky, S. A. 1983, *Black Holes, White Dwarfs, and Neutron Stars*, (Wiley Interscience: New York), p. 160.

Sneeden, C., and Lambert, D. L. 1975, *Mon. Not. R. Astron. Soc.*, **170**, 533.

Spitzer, L. Jr. 1962, *Physics of Fully Ionized Gases*, Second Revised Edition, Interscience Tracts on Physics and Astronomy, ed. R. E. Marshak, (John Wiley: New York), p. 146.

Tammann, G. A. 1977, in *Supernovae*, Vol. 66 Proc. IAU Symp., ed. D. N. Schramm, (D. Reidel: Dordrecht), p. 95.

Truran, J. W., Arnett, D., and Cameron, A. G. W. 1967, *Canadian J. Phys.*, **45**, 2315.

Wallerstein, G. 1973, *Ann. Rev. Astron. Astrophys.*, **11**, 115.

Wallace, R. K., and Woosley, S. E. 1981, *Astrophys. J. Suppl. Ser.*, **45**, 389.

Weaver, T. A., Axelrod, T. S., and Woosley, S. E. 1980, in *Type I Supernovae*, ed. J. C. Wheeler (Austin: University of Texas), p. 113.

Weaver, T. A., and Woosley, S. E. 1980, *Ann. N.Y. Acad. Sci.*, **336**, 335.

———. 1985, *Bull. Am. Astron. Soc.*, **16**, 971.

Weaver, T. A., Woosley, S. E., and Fuller, G. M. 1985, in *Numerical Astrophysics*, ed. J. Centrella, J. LeBlanc, and R. Bowers, (Jones and Bartlett: Boston), p. 374.

Weaver, T. A., Zimmerman, G. B., and Woosley, S. E., 1978, *Astrophys. J.*, **225**, 1021.

Webbink, R. F. 1984, *Astrophys. J.*, **277**, 355.

Wilson, J. R. 1984, in *Numerical Astrophysics*, ed. J. Centrella, J. LeBlanc, and R. Bowers, (Jones and Bartlett: Boston), p. 422.

Wilson, J. R., Mayle, R., Woosley, S. E., and Weaver, T. A. 1985, *Proc. of Twelfth Texas Symp. on Rel. Ap., Ann. N. Y. Acad. Sci.*, in press.

Woosley, S. E., Arnett, W. D., and Clayton, D. D. 1972, *Astrophys. J.*, **175**, 731.

———. 1973, *Astrophys. J. Suppl. Ser.*, **26**, 231.

Woosley, S. E., and Weaver, T. A. 1980, *Astrophys. J.*, **238**, 1017.

Woosley, S. E., Axelrod, T. S., and Weaver, T. A. 1984, in *Stellar Nucleosynthesis*, ed. C. Chiosi and A. Renzini, (D. Reidel: Dordrecht), p. 263.

Woosley, S. E., and Weaver, T. A. 1982, in *Essays in Nuclear Astrophysics*, eds. C. A. Barnes, D. D. Clayton, and D. N. Schramm, (Cambridge Univ. Press: Cambridge), p. 377.

Woosley, S. E., Weaver, T. A., and Taam, R. E. 1980, in *Type I Supernovae*, ed. J. C. Wheeler (Austin: University of Texas), p. 96.

Woosley, S. E., Taam, R. E., and Weaver, T. A. 1985, submitted to *Astrophys. J.*

Wu, C.-C., Leventhal, M., Sarazin, C. L., and Gull, T. R. 1983, *Astrophys. J. Lett.*, **269**, L5.

PAIR CREATION SUPERNOVAE WITH ROTATION

W. Glatzel[1], M.F. El Eid[2], K.J. Fricke[3]
[1] Institute of Astronomy, Cambridge University, England
[2] Department of Physics, American University of Beirut (AUB), Lebanon
[3] Universitäts-Sternwarte, University of Göttingen, West Germany

I. INTRODUCTION

The study of very massive stars with masses $M \gtrsim 100\ M_\odot$ is motivated by the possibility that Pregalactic very massive stars (POP III stars) might have formed and evolved with important cosmological consequences (cf. Ober et al. 1983; Carr et al. 1984). There is also observational evidence (Humphreys 1981) for the existence of very massive stars in both the Galaxy and the LMC, and masses up to 200 $M_\odot$ may well be expected also from theoretical point of view (Maeder 1980). Evolutionary calculations (Woosley and Weaver 1982; El Eid et al. 1983; Carr et al. 1984) indicate that very massive stars ($M \gtrsim 100\ M_\odot$) encounter the electron-positron pair instability and collapse during Carbon burning or at the onset of oxygen burning. The ensuing incomplete explosive oxygen burning leads in certain cases to the so called pair creation supernovae (PCSN).
Previous spherically symmetric computations of PCSN (Woosley and Weaver 1982; Ober et al. 1983; Carr et al. 1984) explored the mass range for which PCSN occur. A general agreement has been achieved that carbon-oxygen cores explode within the mass range $40 \lesssim M_{CO}/M_\odot \lesssim 100$. It is, however, a difficult task to relate these core masses to the main sequence ones, since mass loss and mixing are still not well understood (cf. Langer et al., these proceedings). The present best estimate is: $100 \lesssim M/M_\odot \lesssim 300$.

In this contribution we present a brief summary of the results we have obtained for the mass range of the PCSN including the effect of rotation in a simple but illustrative scheme. Detailed discussion will be published elsewhere (cf. Glatzel et al. 1985). We constrain the mechanical structure of the rotating C-O cores to MacLaurin spheroids (axisymmetric homogeneous ellipsoids) to describe their stability and dynamical evolution. This method which has been previously applied to rotating iron-nickel-cores (Glatzel et al. 1981; hereafter GFE) allows consistent study of marginally unstable configurations from which the nonspherical collapse proceeds.

Our treatment does not intend to replace detailed 2D calculations (Stringfellow and Woosley 1983). Rather it will illustrate the main effects of rotation on PCSN, and will yield guide lines for such simulations.

II. BASIC EQUATIONS AND ASSUMPTIONS

For Maclaurin spheroids, the gravitational potential is a quadratic function of the coordinates r_i (Chandrasekhar 1969), and the homogeneity condition requires that the velocity field is linear in the coordinates (rigid rotation). Hence the pressure is also a quadratic function of r_i. The boundary condition of zero surface pressure requires the isobaric surfaces to be similar to the actual boundary of the spheroid. In the notation of GFE, the pressure distribution is given by

$$P = P_c (1 - \sum_{i=1}^{3} r_i^2/a_i^2) \qquad (1)$$

where P_c is the central pressure, and a_i are the axes of the spheroid.

For a velocity field describing nonspherical collapse, the momentum equations for the semi-major axis a_1 and the semi-minor axis a_3 are:

$$\frac{d^2 a_1}{dt^2} = a_1 \omega^2 - 2\pi G \rho A_1 a_1 + 2P_c/\rho a_1 \qquad (2)$$

$$\frac{d^2 a_3}{dt^2} = -2\pi G \rho A_3 a_3 + 2P_c/\rho a_3 \qquad (3)$$

where ω is the angular frequency, $2A_1 + A_3 = 2$ and A_3 is a function of the eccentricity $e = \{1 - (a_3/a_1)^2\}^{1/2}$ only (cf. GFE). The momentum equation for the axis a_2 is the same as Eq.(2), since $a_1 = a_2$ for Maclaurin spheroids.

The mass M and angular momentum L are given by:

$$M = \rho \frac{4\pi}{3} a_1^2 a_3$$
$$L = \frac{2}{5} M a_1^2 \omega \qquad (4)$$

In addition we assume $\frac{dM}{dt} = \frac{dL}{dt} = 0$.

The energy equation is used to calculate the temperature T implicity. It has the form:

$$\frac{d\varepsilon_c}{dt} = \frac{1}{\rho} (P_c + \varepsilon_c) \frac{d\rho}{dt} - Q \qquad (5)$$

where $\varepsilon_c = \varepsilon_c (\rho, T, [x_i])$ is the central energy density, and Q is the energy loss rate (erg $cm^{-3} s^{-1}$) is mainly due to neutrinos in our computations.

The equation of state has been computed for a mixture of e^-, e^+ and photons in thermodynamic equilibrium. The appropriate complete expressions (cf. Cox and Giuli 1968) for pressure, internal energy and entropy have been numerically evaluated assuming charge neutrality to fix the chemical potentials of e^- and e^+. The ions are treated as Maxwell-Boltzmann gas; their contribution is simply added.

The net-work of the nuclear reactions is essentially the same as that used by Ober et al. (1983). It is a simplified one consisting of 13 alpha-nuclei starting at ^{4}He and also including $^{12}C(^{12}C,\alpha)^{20}Ne$, $^{12}C(^{16}O,\alpha)^{24}Mg$ and $^{16}O(^{16}O,\alpha)^{28}Si$.

At high temperatures and densities we have included photodissociation of α-particles into nucleons via the equilibrium process $^4He \rightleftarrows 2n + 2p$. The abundances have been calculated using the Saha equation in this case.

The energy loss term Q in Eq.(5) is due to photo-, pair-annihilation-, and plasma-neutrino processes (Beaudet et al. 1967). The correction factors of Ramadurai (1976) have been also included.

Using standard techniques the ordinary differential equations (2), (3) and (5) have been integrated simultaneously with the reaction network. The mass M and the angular momentum L and the initial conditions for the integration are chosen as described in the following section.

III. INITIAL MODELS

The initial models are assumed to consist of oxygen and carbon with $X_O = 0.95$ and $X_C = 0.05$ in accordance with evolutionary calculations (cf. El Eid et al. 1983). The initial quantities (T, ρ, P, M, L) of the rotating cores at the onset of dynamical instability are determined by the requirement of equilibrium and marginal stability.

As described in detail by GFE the condition of marginal stability is equivalent to $\gamma(\rho) = \gamma_{cr}(e)$ where γ_{cr} is a critical adiabatic index which is a function of eccentricity e only (cf. Eqs. (12) and (13) of GFE), and $\gamma \equiv (\partial \ln P_c/\partial \ln \rho)_S$ is the adiabatic index. For given dimensionless entropy S the isentropes $T(\rho)$, $P_c(\rho)$, and $\gamma(\rho)$ are calculated using the equation of state. The parameter e determines $\gamma_{cr}(e)$ which is equated to $\gamma(\rho)$ to obtain ρ from which T and P follow.

The initial mass and angular momentum of the rotating cores are determined from the equilibrium conditions. Putting the l.h.s. of Eqs. (2) and (3) equal to zero, and inserting the resulting equations into Eq.(4), M and L can be expressed in terms of e, ρ, and P_c. The results are given in GFE (their Eqs. 8 and 9).

Up to now we have treated homogeneous models which correspond to a polytropic index n = 0. It is however possible to correct M and L for polytropes of n = 0. For the realistic case n = 3, which allows comparisons with other investigations, GFE found correction factors 3.36 and 19.69 for M and L, respectively.

IV. RESULTS

a) Spherically symmetric calculations

The method described in sect. II has been applied to the case of zero eccentricity. The only parameter here is the initial entropy S_i which can be uniquely related to the core mass (cf. Eq. 8 of GFE). We found a relationship $M_{CO} = 59(S_i/10)^{2.50}$ over a range $8 < S_i < 15$, where M_{CO} denotes the mass of the C/O core corrected for n = 3 polytropes. This result is in good agreement with the work of Carr et al. (1981).

Our results for e = 0 are summarized in Table 1. A remarkable overall agreement is achieved with other more elaborated computations. In the range $7.3 < S_i < 11.6$ explosive oxygen burning reverses the collapse leading to explosion, but for $S_i > 11.6$ the collapse velocities are too large and the equation of state is too soft to stop the collapse before the nuclei photo-dissociate. These endothermic processes lead to total collapse.

b) Dynamical evolution with rotation

The initial rotating cores are characterized by the parameters (S_i, e_i) which are uniquely related to the specific angular momentum $j = L/M$ and the mass M via the equilibrium and marginal stability conditions. The advantage of using j and M is that these quantities may be assumed conserved during collapse.

The results of the 2D dynamical homogeneous calculations are contained in the j-M diagram in Fig. 1. Following the lines of constant S_i below the stability boundary we find with decreasing e_i regions of oscillation, explosion, and collapse. In the region of oscillation, e_i is large enough so that nuclear burning and rotation reverse the collapse at moderate values of T and ρ, but the energy release from explosive oxygen burning is insufficient to disrupt the core. There is, however, no sharp transition between this and the explosion region beneath.

The dynamical evolution in the regions of explosion and collapse is shown for selected cases in Figs. (2) and (3). For model 1 (see also Fig.1) with $(S_i, e_i) = (20, 0.40)$ the collapse is reversed just prior to the region of photo-dissociation indicated by the Ni$\rightarrow\alpha$ transition line in Fig.2. This is more pronounced for model 2 with $(S_i, e_i) = (40, 0.60)$, where the collapse is reversed even earlier. Fig.2 shows also that the pair creation instability regions become smaller with increasing eccentricity due to stabilization induced by rotation.

Table 1: The various domains regarding the evolutionary fate of nonrotating C/O-cores in terms of the initial entropy S_i or core mass M (in units of solar masses). The masses given in this Table for the present work have been corrected for a polytropic density stratification of index n = 3 and thus correspond to $\tilde{M}_{CO}$ as introduced in Sect. IIIa.

Authors	stability		oscillation		explosion		collapse	
	S_i	M	S_i	M	S_i	M	S_i	M
Woosley and Weaver (1982)	--	--	--	--	--	60-100	--	>100
Carr et al. (1981)	<8	<32	∼8	∼32	8-12.4	32-95	>12.4	>95
Ober et al. (1983)	--	<50	--	--	--	50-112	--	>112
present work	<6.8	<21.8	6.8-7.3	21.8-26.9	7.3-11.6	26.9-86	>11.6	>86

The models 3 and 4 both collapse beyond the Ni→α transition line. Consequently, the collapse proceeds to the stage where neutrino energy losses become essential on dynamical time scales. Although the energy release by oxygen burning would be sufficient to disrupt the core, the reinforcement of the collapse by photodissociation, and the energy dissipation by neutrinos lead to total collapse. We note that this type of evolution is accompanied by neutrino damped oscillations which are induced by rotation.

Summing up: all cases, in which rotation and nuclear burning succeed to stop the collapse prior to the region of photodissociation of nuclei constitute the region of explosions in Fig. 1, while the region of collapse contained all cases for which the endothermic photodissociation processes together with the neutrino energy losses ultimately lead to total collapse despite of rotation.

Meanwhile Stringfellow and Woosley (1985) have done inhomogeneous 2D hydrodynamical calculations for very massive He cores ($M/M_\odot$ = 200 - 500). Their results shown in Fig. 4 are qualitatively in good agreement with our results. Our curve (dashed line in Fig.4) separating the regions of explosion and collapse, is shifted upward by a factor 1.9. This is due to the assumption of rigid rotation in our computations.

V. CONCLUSION

We have followed the evolution of rotating massive C-O cores assuming Maclaurin spheroids for the description of their structure. This simplified treatment allows us to establish the marginally unstable initial configurations depending on the geometry and the thermodynamic state of the cores. These are then used as initial models for the nonspherical core collapse calculations.

For zero rotation our results are in reasonable agreement with those obtained from more elaborated treatments (cf. Table 1).

With rotation our results are contained in a "phase diagram" (cf. Fig. 1) in which the various regions of stability, oscillation, explosion and collapse are shown in terms of j and M. We arrive at the conclusion that rotation extends the region of pair creation supernovae, whenever the collapse is reversed prior to the stage, where photodissociation of nuclei occurs. Beyond this stage the endothermic photodissociation together with energy dissipation by neutrinos lead to total collapse despite of rotation.

Our results are also in good qualitative agreement with recent 2D inhomogeneous investigations (cf. Fig.4).

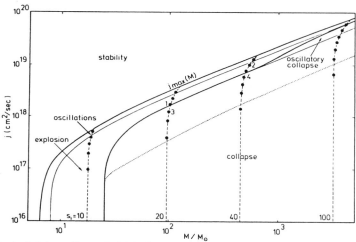

Fig.1: A "phase diagram" which shows the ultimate fate of a pair creation unstable C/O core in terms of specific angular momentum j, and mass M. If neutrino losses are neglected, the explosion domain extends down to the dotted line. The curve labeled $j_{max}(M)$ separates the regions of quasistationary evolution from that of dynamical evolution. Configurations of constant initial entropy S_i are connected by dashed lines. Dots on these curves mark different initial eccentricities e_i with a constant spacing of 0.1. For each initial entropy the lowest dot belongs to $e_i = 0.1$. The location of model 1-4, whose dynamical evolution is described with Figs. 2-3 is indicated.

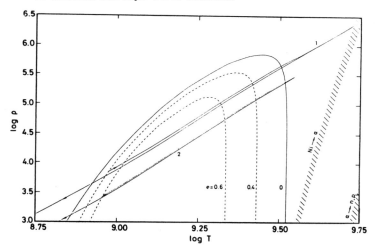

Fig.2: Central temperature T and density during the dynamical evolution of model 1 with $S_i,e_i) = (20, 0.4)$ and model 2 with $(S_i,e_i) = (40, 0.6)$. The transition regions Ni→α and α→ n,p, which are centered around $X_\alpha = X_{Ni}$ and $X_\alpha = X_p$, respectively, are indicated. The dashed lines enclose the regions of pair creation instability for the eccentricities they are labeled with; these are identical to the initial eccentricities of models 1 and 2. The curve labeled e = 0 encloses the region of pair creation instability for zero angular momentum. Models 1 and 2 explode.

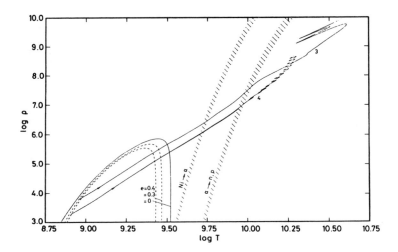

Fig.3: Same as Fig.2 for model 3 with $(S_i, e_i) = (20, 0.3)$ and model 4 with $(S_i, e_i) = (40, 0.4)$. Neutrino losses have been included in the dynamical calculations. Model 3 and 4 undergo rotationally induced oscillations which are damped by neutrino losses and finally collapse.

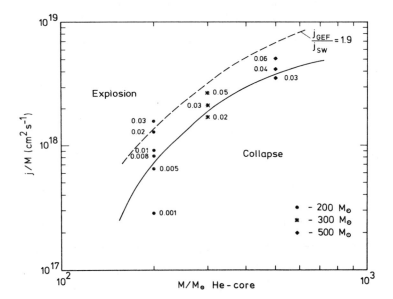

Fig.4: Specific angular momentum versus mass. The results of 2D inhomogeneous calculations of Stringfellow and Woosley (1985) are compared with the present 2D homogeneous calculations (dashed line).

ACKNOWLEDGEMENTS

This work has been supported in part by the Deutsche Forschungsgemeinschaft (DFG) through grants Fr 325/16-1 and 22-1. M.E. thanks the University of Göttingen and the Göttingen Akademie der Wissenschaften for a Gauss Visiting Professorship and W.G. the Institute of Astronomy, Cambridge, England, for hospitality. The numerical calculations were carried out with the Sperry 1100/83 of the GWDG, Göttingen and the PDP 11/44 of the Universitäts-Sternwarte Göttingen (grant Fr 325/15-2).

REFERENCES

Beaudet, G., Petrosian, V., Salpeter, E.E.: 1967, Astrophys. J. $\underline{150}$,979
Carr, B.J., Bond, J.R., Arnett, W.D.: 1984, Astrophys. J. $\underline{277}$, 445
Carr, B.J., Bond, J.R., Arnett, W.D.: 1981, in: The Most Massive Stars, S.d'Odorico, D. Baade, K. Kjär (Eds.), Garching, ESO
Chandrasekhar, S.: 1969, Ellipsoidal Figures of Equilibrium, Yale University Press, New Haven
Cox, J.P., Giuli, R.T.: 1968, Principles of Stellar Structure, Vol. II, Gordon and Breach
Glatzel, W., Fricke, K.J., El Eid, M.F.: 1981, Astron. Astrophys. $\underline{93}$, 395 (GFE)
Glatzel, W., El Eid, M.F., Fricke, K.J.: 1985, Astron. Astrophys., in press
Humphreys, R.M.: 1981, in: The Most Massive Stars, S.d'Odorico, D. Baade, K. Kjär (Eds.), Garching, ESO
Langer, N., El Eid, M.F., Fricke, K.J.: these proceedings
Maeder, A.: 1980, Astron. Astrophys. $\underline{92}$, 101
Ober, W., El Eid, M.F., Fricke, K.J.: 1983, Astron. Astrophys. $\underline{119}$, 61
Ramadurai, S.: 1976, Monthly Notices Roy. Astron. Soc. $\underline{176}$, 9
Stringfellow, G.S., Woosley, S.E.: 1983, Bulletin Americ. Astron. Soc. $\underline{15}$, 955
Stringfellow, G.S., Woosley, S.E.: 1985, (Woosley, private communication)
Woosley, S.E., Weaver, T.A.: 1982, in: Supernovae: A Survey of Current Research, M.J. Rees and R.J.Stoneham (Eds.), Reidel, Dordrecht

NUCLEOSYNTHESIS AND MASSIVE STAR EVOLUTION

N. Langer[1], M.F. El Eid[2], K.J. Fricke[1]
[1] Universitäts-Sternwarte Göttingen, Geismarlandstr. 11,
3400 Göttingen, F.R.G.
[2] American University of Beirut, Physics Department,
Beirut, Libanon

ABSTRACT. The uncertainty of the theoretical evolution of massive stars due to incomplete knowledge about internal mixing processes is examined, and consequences for the chemical evolution of galaxies are drawn. In contrast to earlier work, we found semiconvection to be very important in the post main sequence evolution of stars in the mass range $10 - 50 M_\odot$, when the mixing timescale of semiconvection is accurately taken into account. For more massive stars semiconvection is less important, but some mixing in addition to the usual convective mixing seems necessary to exlain details in the observed HR-diagram. An assumed enlargement by 1.9 pressure scale heights of the convective core in a $100 M_\odot$ star is not supported by the observations. Much smaller core enlargements are thereon proposed.

1. INTRODUCTION

Massive stars are main contributors to the galactic enrichment, both during their hydrostatic evolution and their explosive final states: They develop strong stellar winds in their Wolf-Rayet and Pre-Wolf-Rayet phases; their explosions as supernovae are caused in the lower mass range by the collapse of the iron nickel cores and in the upper mass range ($\geq 100 M_\odot$) by the production of e^+e^--pairs and subsequent explosive O-burning. Also abundance anomalies in the galactic cosmic rays are possibly produced by evolved massive stars, e.g. that of ^{22}Ne and ^{26}Al. At present, the evolution of massive stars is far from being fully understood; this even applies to the evolution of the main sequence. Models of galactic nucleosynthesis suffer mainly incertainties in the following processes: a) the mass loss rates and their changes during evolution, b) hydrodynamical instabilities and the resulting mxing of matter, heat, and angular momentum. As for a), the evolutionary phase of massive stars for which the mass loss rate is known best, is the main sequence. Here the uncertainty is about a factor of 2 (Lamers 1985). However, a change by a factor 2 in M implies already a change in the number of Wolf Rayet stars by a factor of 17 (Maeder 1984). Considering b), e.g. Lamers (1985) pointed out, that with the observed mass-loss rates for O-stars Conti's scenario for the production of WR-stars (cf. Maeder 1984) would not work even for very massive stars. That means that internal mixing on the main sequence may prevent very massive stars from becoming red supergiants (RSG). Mixing additional to the standard convective mixing also alleviates the problem with the observed very wide main sequence (cf. Maeder 1982, Doom 1985).

The present work deals mainly with point b). It should be pointed out that the effects of mass loss and hydrodynamical instabilities on nucleosynthesis are not independent of each other. For example, it is well known that a strong mass loss rate changes the internal structure of a star by increasing the temperature gradient outside the convective core and thereby reducing rotationally induced instabilities (Zahn, 1983), semiconvection (Langer et al. 1983), and convective overshooting. On the other hand, such instabilities may strongly affect the position of the star in the HR diagram, especially in the post main sequence evolution, and therefore modify the mass loss rate. A direct influence is possible when instabilities occur at or very close to the surface of the star.

In this contribution we present some ideas concerning the effects of 1) semiconvection and 2) enlarged convective cores (due to convective overshooting or other mechanisms) which could help to solve the puzzling problem of massive star evolution and nucleosynthesis (cf. Maeder 1984).

2. SEMICONVECTION

Since the work of Kato (1966) it is known that the Schwarzschild-criterion for convection $\nabla > \nabla_{ad}$ is valid only for the chemically homogeneous case (dlnµ / dlnP =: ∇_μ = 0), and the Ledoux-criterion
$\nabla > \nabla_{ad} + (\partial \ln T / \partial \ln \mu)_{P,\rho} \cdot \nabla_\mu =: \nabla_L$ should be use in the presence of a µ-gradient. For $\nabla_{ad} < \nabla < \nabla_L$ a vibrational instability occurs, which is denoted as semiconvection (SC) in the following. Nonradial g^+-mode-oscillations in µ-gradient zones have been found to be globally unstable due to enhanced heat diffusion in the perturbed state ("Kato mechanism") (e.g. Gabriel and Noels 1976). The timescale τ_{sc} on which SC acts was estimated by several authors (Kato 1966, Gabriel and Noels 1976, Stevenson 1979) to be of the order of the local Kelvin-Helmholtz timescale, eg. 10^3- 10^4 yrs for a 30 $M_\odot$ star. Since this timescale usually is very short compared to the evolutionary timescale τ_{ev} many authors concluded that semiconvection always establishes the neutrality condition according to the Schwarzschild criterion, i.e. $\nabla \equiv \nabla_{ad}$. One purpose of the present investigation is to point out that this is not true for several phases of the evolution of massive stars. Drastic consequences may appear when the actual timescale of semiconvection is properly taken into account.

Therefore evolutionary calculations have been carried out for a 15- and 30 $M_\odot$ star up to carbon ignition, using the semiconvection formatism of Langer, Sugimoto and Fricke (1983), who evaluated expressions for the semiconvective diffusion of matter and heat depending on the local physical conditions in the star. For comparison, sequences without semi-convection, i.e. ignoring μ-gradients, have been calculated (Langer, El Eid, Fricke 1985). In the following we give a brief outline of these results. For example, one stage in which $\tau_{ev} \simeq \tau_{sc}$ holds, is the phase between central H- and He-burning. Then the whole structure of the star changes on a Kelvin-Helmholtz timescale, and stars in the range $10\ M_\odot < M < 60\ M_\odot$ turn from the blue side of the HRD to the red supergiant stage. In this evolutionary phase the hydrogenprofile in the intermediate region of the star, i.e. above the H-burning shell, is established and is then retained during the entire He-burning phase. This H-profile determines the fraction of the He-burning phase which the star pends in the RSG-stage; a large part of the intermediate region of the star becomes superadiabatic in the previous core contraction phase. Ignoring μ-gradients , convective mixing homogenizes a large part of the intermediate zone up to layers very close to the H-burning shell, which therefore reaches quickly regions with a high H-content and burns very effectively, so that the star makes a loop to the blue part of the HR diagram. If μ-gradients are properly taken into account the convective mixing is strongly reduced, and the slope of the H-profile in the layers just above the H-burning shell stays relatively small. Therefore, the efficiency of the shell source is kept small and the star remains in the RSG region.

This mechanism decreases the fraction of blue/red-supergiants relative to standard evolutionary calculations (cf. Brunish, Gallagher and Truran 1983), as suggested by observations. Therefore, semiconvection influences indirectly the evolution by increasing the effective mass loss to the very high value which is characteristic for the RSG-stage as compared to that in the BSG-stage.

Furthermore, semiconvection affects the growth of the He-burning convective core, especially in the final phase of core-He-burning, and thus decreases the final C/O-ratio in the core (see Table 1):

	15	30	30 M	60 M	100 MO
with SC	0.90	0.47	0.60	0.55	0.02
without SC	0.97	0.61	0.85	0.58	0.05

Table 1: C/O ratios in the convective core at the end of core-He-burning for stars with initial masses of 15, 30, 60, and 100 $M_\odot$. The case with inclusion of SC is compared to that without SC. The symbol M indicates that mass loss is taken into account, O indicates that overshooting has been included (see below for the 100 $M_\odot$-star). Only the sequence at 100 $M_\odot$ has been calculated with the new $^{12}C\ (\alpha,\gamma)^{16}O$ rate (Caughlan et al. 1985), while in the others the rates of Fowler et al. (1975) have been used.

The penetration of the convective envelope into the stars in the RSG-phase is influenced by semiconvection, which is reflected in the CNO- and H, He-surface abundances. For details see however Langer, El Eid and Fricke (1985).

3. ENLARGED CONVECTIVE CORES (OVERSHOOTING)

The Schwarzschild-condition is strictly speaking a criterion for the onset of convection, but not necessarily for the convective regions in the steady state. Those may be somewhat larger than at the onset of convection due to the inertia of the moving convection cells. At present this enlargement by "overshooting" is unknown. Theoretical models predict overshooting distances from close to zero up to 1 or 2 pressure scale heights (for a summary see, e.g. Eggleton 1983, or Marcus et al. 1983). Many of these models predict the extent of overshooting in terms of the pressure scale height Hp to be constant during the evolution. In this situation we parametrize the convective core enlargement in the following way: the Schwarzschild-convective core is assumed to be increased by $\alpha_{over} H_p(r_s)$, where r_s is the boundary of the Schwarzschild core with α_{over} a free parameter taken as constant for a given star. In this picture a value $\alpha_{over} = 0$ corresponds to the standard case (no overshooting). The temperature gradient in the overshooting region is simply put equal to the adiabatic temperature gradient, in accordance with most overshooting theories.

It should be noted here, that an increased core could possibly also be produced by rotationally induced instabilities. In this case the mixing of matter resulting from linear theory is proportional to $(\nabla_{rad}-\nabla_{ad})^{-1}$ (cf. Zahn 1983) and may become very large near the convective core boundary. In our simplified procedure we account for such effects and proper overshooting in the same way by an parametrized enlargement of the convective core. All outer convective regions (not connected with the convective cores) have been treated here in the standard way for simplicity. We have studied the effects of overshooting on the evolution of a 100 $M_\odot$ star. The overshooting parameter α_{over} has been chosen as 1.9. This is the highest value found in the literature and corresponds to the prediction of the "Roxburgh-criterion" (Roxburgh 1978). The convective core at the ZAMS is then increased from the standard value of about 80 $M_\odot$ to 92 $M_\odot$ in this case. Such large overshooting may be taken as an upper limit to any realistic core enlargement. For comparison we have calculated a second sequence with $\alpha_{over} = 0$, i.e. without overshooting. In both sequences semiconvection has been taken into account as described in chapter 2. Mass loss has also been incorporated, adopting the formula proposed by Lamers (1981) for the main sequence phase. For He-burning, however, we have taken for the mass loss rate the maximum of the Lamers rate and $3 \cdot 10^{-5}$ $M_\odot$/yr, where the last value is typical for all kinds of WR-stars. In Fig. 1 we have plotted the total mass change and evolution of the interior structure with time for both sequences. The main differences between the two considered cases may be interpreted as follows. In the case $\alpha_{over} = 1.9$ the increased convective core transports the ashes of the central burning very closely towards the stellar surface. This is not only due to the enlargement of the core at the ZAMS by $\Delta M \sim 12$ $M_\odot$ but mainly due to the fact that the convective core shrinks much more slowly in the case $\alpha_{over} = 1.9$ as compared to the case $\alpha_{over} = 0$. The difference in the size of the convective cores at the end of H-burning is as large as almost 30 $M_\odot$ and has the following effects:

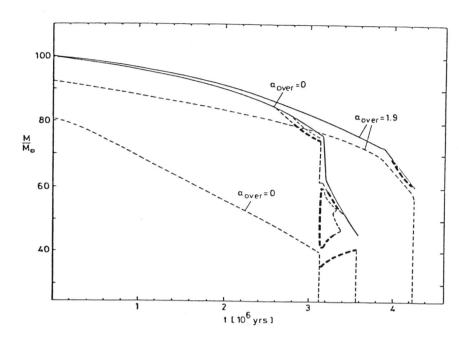

Fig. 1. The time evolution of the total mass (solid lines) and the interior unstable regions (broken lines) for two calculated sequences with an initial mass of 100 $M_\odot$ with (α_{over}= 1.9) and without (α_{over}= 0) overshooting up to central He exhaustion. Semiconvection occurs adjacent to several convective regions in rather thin layers and is marked by thick broken lines. The beginning of He-burning in the case α_{over}= 1.9 is coincident with the kink in the curve M(t), caused by the sudden increase of the mass loss rate to $3 \cdot 10^{-5}$ $M_\odot$/yr (typical for WR stars) at this stage.

1) H-burning: In the case α_{over} = 1.9 the surface abundance of hydrogen is reduced drastically very early. Therefore the star turns to the left of the ZAMS already during the main sequence phase (cf. Fig. 2). At the surface the star shows typical WN abundances after $2\ 10^6$ years, i.e. already after half of the hydrogen burning lifetime; on the other hand, if α_{over} = 0 an enhancement of ^{14}N and a reduction of H, ^{12}C and ^{16}O can only be seen at the surface after central H-exhaustion.

2) He-Burning: For α_{over} = 0 the star keeps a substantial part of its H-rich envelope until the end of core-H-burning. As a consequence it can develop an H-burning shell and result in an RSG-stage where it spends about 20% of its He-burning lifetime, until the strong mass loss in this part of the HR-diagram (several $10^{-4}\ M_\odot$/yr) has removed the main part of this envelope. Then the star turns to the blue and reaches effective temperatures between 30 000 and 80 000 K. In this stage it burns the rest of the central Helium like a typical WN-star (cf. Fig. 2).
In contrast, the star with α_{over} = 1.9 burns the central He at effective temperatures around 200 000 K, first as a WN-star, then turning into a WC star already at a central helium fraction Y_c = 0.8, and finally into a WO-star. It is important here to note, that (as in most published calculations for WC and WO-stars) effects of the ionisation at high C, O abundances are completely ignored concerning the opacity as well as the equation of state. Since C and O are the main constituents of matter in the WC and WO-phase, the resulting error in the effective temperature will probably be large. Taking the above effects into account, Sackmann and Boothroyd (1984) have analyzed carbon rich stellar envelopes. They found that it is a good approximation to simply replace the Carbon by Hydrogen. Doing so, we obtained the dotted track in the HR-diagram in Fig.2. We stopped the calculations when the surface abundance of carbon became greater than 10% by mass (this occurs at Y_c = 0.5), since the above approximation breaks down then. The effect of a large C abundance in surface layers is seen to be very large from Fig. 2 as the star again turns strongly to the right in the HR diagram. The nuclear enrichment due to the stellar wind is very different for the two sequences (cf. Table 2) due to the very different predicted durations of the WR-phases in both cases, which can easily be extrapolated from Fig. 1. The star with α_{over} = 0 never reaches the WC and WO phases. For this reason the enhancement factors X_{wind}/X_i for ^{12}C, ^{16}O, ^{18}O and ^{22}Ne are much smaller (in most cases even smaller than 1) than the corresponding values for the sequence with α_{over} = 1.9. Here X_{wind} is defined as

$$X_{wind} = \int X_{surface}\ \dot{M}\ dt\ /\ \int \dot{M}\ dt$$

and X_i is the initial abundance of isotope X.

Finally, some remarks will be made about the comparison of the calculated evolutionary tracks with the observed HR-diagram for massive stars (cf. Humphreys and Davidson 1979, Humphreys 1984). Concerning the main-sequence evolution the sequence with α_{over} = 0 seems to fit better, because there are several stars to the right of the ZAMS observed which are more luminous than a 100 $M_\odot$ ZAMS model. Also the post-RSG-evolution fits rather nicely in this case: (i) after the H-rich envelope has been removed from the star in the RSG-stage the star moves very rapidly

Tab. 2.
Nucleosynthesis output due to mass loss

$M_i = 100\ M_\odot$

	lost mass/$M_\odot$	X_{wind}	X_{wind}/X_i
H	12.3	.315	0.45
	27.6	.511	0.73
He	22.0	.564	2.09
	24.8	.460	1.70
^{12}C	2.10	.0538	12.88
	0.087	.00161	0.39
^{13}C	0.00196	5.03-5	1.12
	0.00347	6.43-5	1.43
^{14}N	0.342	.00877	6.96
	0.526	.00974	7.05
^{15}N	0.000020	5.13-7	1.14
	0.000029	5.29-7	1.18
^{16}O	1.57	.0404	3.21
	0.340	.0063	0.50
^{17}O	0.00368	0.94-4	20.98
	0.00817	1.51-4	33.62
^{18}O	0.0180	4.62-4	15.40
	0.0006	0.10-4	0.33
^{20}Ne	0.315	.00808	1.00
	0.435	.00800	1.00
^{22}Ne	0.235	.00603	6.70
	0.049	.00090	1.00

upper value: $\alpha_{over} = 1.9$
lower value: $\alpha_{over} = 0$

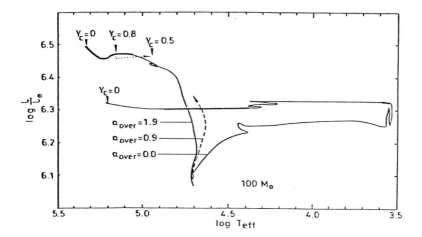

Fig. 2. Evolutionary tracks in the HR-diagram of two calculated sequences with initially 100 $M_\odot$ with (α_{over} = 1.9) and without (α_{over} = 0) overshooting. The main sequence phase of a star having α_{over}=0.9 is also shown (broken line). The thick fully drawn parts of the tracks indicate the stages of quiescent He-burning. The He-burning phase down to Y_C = 0.5 of the sequence with α_{over}=1.9 has also been calculated with the simplification, that carbon in the envelope mimics hydrogen in some respects (dotted line).

towards higher effective temperatures, performing some small loops on thermal timescales around log T_{eff} = 4.3. It is interesting to note that this is just the location of the Hubble-Sandage variables in the HR-diagram (e.g. P Cygni or η Carina), which are obviously stars in a somewhat unstable state. Also Lamers et al. (1983) argued from the observational point of view, that P Cygni is a star between the RSG and WR stage. (ii) the star settles down to burn most of its central helium content as a WN star just at the location in the HR-diagram where observer place the WN stars. With the sequence α_{over} = 1.9 both phenomena, the explanation of the Hubble-Sandage variables and the location of the WR stars in the HR-diagram are much more difficult to accomodate.

A strong discrepancy with observations in the α_{over} = 0 track is the observed absence of red supergiants with luminosities above M_{bol} = -10, since with this sequence a red supergiant with M_{bol} = -11.3 and a lifetime of ≈ 40000 yrs is produced. However, it should be kept in mind that the mass loss rate for very luminous RSGs is very large. For lack of better knowledge we have extrapolated the Lamers formula into this region and arrived at mass loss rates as large as $3 \cdot 10^{-4}$ $M_\odot$/yr. It may be quite possible that such high mass loss rates will obscure the star in optical wavelength, turning it into an IR-source. In fact in the HR-diagram of Humphreys and Davidson (1979) some IR-stars are indicated at luminosities of M_{bol} ≈ -9.5 and effective temperatures log T_{eff} ≈ 3.75. Now, the theoretical Hayashi-line at these luminosities is located at effective temperatures log T_{eff} ≈ 3.5. That the observed stars are found ∿ 2500 K left of the Hayashi-line is rather improbable on theoretical grounds. An explanation could be that the effective temperatures of these IR-stars might be overestimated and, at the same time, the bolometric corrections underestimated. Then, after correction, the stars could possibly be placed very close to the predicted location on the Hayashi-line.

As a further argument in favour of the α_{over} = 0 sequence might be considered the fact that in this case no artificial increase of the mass loss rate is required in the WR stage, since the Lamers formula predicts there Ṁ-values of the order of $3 \cdot 10^{-5}$ $M_\odot$/yr. For the case α_{over} = 1.9 during He-burning the mass loss rates according to Lamers have to be raised artificially by factors 5-10 in order to obtain a WR type mass loss.

To definitely solve the problem of the convective core enlargement, further work is required. Evolutionary calculations at different masses and with different overshooting parameters have to be carried out to address also the problem of the number statistics of WR stars. Also the effects of different mass loss rates have to be explored simultaneosly. However, from the above arguments we conclude, that a value of the overshooting parameter α_{over} of 1.9 is rather improbable. We also constructed a main sequence for α_{over} = 0.9. From Fig. 2 it can be seen that this value gives similar results as α_{over} = 1.9. We therefore propose that either much smaller values of α_{over} are appropriate or that the effects of internal mixing are much more complex than can be described by simply increasing the convective core.

4. CONCLUSION

No simple scenario of massive star evolution emerges from the present work. Many uncertainties involved in massive star evolution could not even be addressed in detail here. A specific mass loss rate had simply to be assumed in spite of the fact that it is particularly poorly known in the RSG phase (cf. Chiosi et al. 1978) and the WR stage (cf. Maeder 1981). Furthermore there remains a discrepancy concerning the opacities calculated by different groups (cf. Stothers and Chin 1977, and Bertelli et al. 1984). Also the effect of the initial metalicity on the evolution has not been investigated here.

Despite of these deficiencies the present investigation allows to draw the following conclusions: In contrast to widespread opinion, that semiconvection has only little influence on the evolution of massive stars, it has been found by properly taking into account the timescale of semiconvection that the post-main sequence evolution of stars in the massrange of 10 - 50 $M_\odot$ is strongly influenced by this phenomenon. The fraction of the He-burning lifetime which is spent in the RSG phase is determined by semiconvection. This has important consequences for the total mass lost during the He-burning phase and therefore for the number of WR stars and galactic enrichment. The central C/O-ratio at the end of He-burning has been shown to depend on the treatment of semiconvection. Although the analysis of Langer et al. (1985) was restricted to constant mass evolution without overshooting new calculations (Langer et al. 1985a) show, that the above mentioned effects are not weakened if mass loss and convective core enlargement is taken into account.

The evolution of a 100 $M_\odot$ star with two extreme assumptions on the core enlargement (by 0 and 1.9 pressure scale heights) has been analysed. Assuming the mass loss rate of Lamers (1981) the upper luminosity limit of M_{bol} = -10 as presently deduced from observations (Humphreys 1984) for red supergiants can only be reconciled with theory if some extra internal mixing ($\alpha_{over} > 0$) is postulated. Witout such mixing high mass stars would undergo a redward evolution with excessive luminosities after H-exhaustion. However, several points indicate, that this extra mixing (overshooting) cannot be very efficient, (i) the large width of the main sequence at log $L/L_\odot \approx 6.2$, (ii) the location of the Hubble-Sandage variables in the HR-diagram, and (iii) the observed position of the WR stars in the HRD. Therefore we propose that a value of α_{over} much smaller than 1 is appropriate.

Recently Audouze (1984) stressed that the constraints of the stellar evolution theory to the theory of the chemical evolution of galaxies are stronger than vice versa. Considering, however, our incomplete knowledge on massive star evolution we may suspect that the close interaction of both fields will be required to obtain answers to the numerous fascinating and yet unsolved problems in this area of research.

ACKNOWLEDGEMENT. We thank Dr. M. Arnould for valuable discussions. This work has been supported in part by the Deutsche Forschungsgemeinschaft (DFG) through grant Fr 325/22-1. The numerical calculations have been carried out at the Univac 1100 of the Göttinger university computer center (GWDG) and at the PDP 11/44 of the DFG (grant Fr 325/15-2).

REFERENCES

Audouze, J.: 1984, IAU-Symp. 105, 541 (Ed. A. Maeder & A. Renzini)
Bertelli, G., Bressan, A.G., Chiosi, C.: 1984, Astron. Astrophys. 130, 279
Brunish, W.M., Gallagher, J.S., Truran, J.W.: 1983, IAP 83-37
Caughlan, G.R., Fowler, W.A., Harris, M.J., Zimmerman, B.A.: 1985, Atomic Data and Nuclear Data Tables 32, 197
Chiosi, C., Nasi, E., Sreenivasan, S.R.: 1978, Astron. Astrophys. 63, 103
Doom, C.: 1985, Astron. Astrophys. 142,143
Eggleton, P.P: 1983, Monthly Notices Roy. Astron. Soc. 204, 449
Fowler, W.A., Caughlan, G.R., Zimmerman, B.A.: 1975, Ann. Rev. Astron. Astrophys. 13, 69
Gabriel, M., Noels, A.: 1976, Astron. Astrophys. 53, 149
Humphreys, R.M.: 1984, IAU-Symp. 105, 279 (Ed. A. Maeder & A. Renzini)
Humphreys, R.M., Davidson, K.: 1979, Astrophys. J. 232, 409
Kato, S.: 1966, Publ. Astron. Soc. Japan 18, 374
Lamers, H.J.G.L.M.: 1981, Astrophys. J. 245, 593
Lamers, H.J.G.L.M.: 1985, in Mass Loss of Stars, 3. Cycle in Astronomy and Astrophysics (Ed. P. Smeyers)
Lamers, H.J.G.L.M., de Groot, M., Cassatella, A.: 1983, Astron. Astrophys. 123, L8
Langer, N., Sugimoto, D., Fricke, K.J.: 1983, Astron. Astrophys. 126, 207
Langer, N., El Eid, M.F., Fricke, K.J.: 1985, Astron- Astrophys. 145, 179
Langer, N., El Eid, M.F., Fricke, K.J.: 1985a, in preparation
Maeder, A.: 1981, Astron. Astrophys. 99, 97
Maeder, A.: 1982, Astron. Astrophys. 105, 149
Maeder, A.: 1984, IAU-Symp. 105, 299 (Ed. A. Maeder & A. Renzini)
Marcus, P.S., Press, W.H., Ten Kolsky, S.A.: 1983, Astrophys. J. 267, 795
Roxburgh, I.W.: 1978, Astron. Astrophys. 65, 281
Sackmann, I.J., Boothroyd, A.I.: 1984, OAP 656
Stevenson, D.J.: 1979, Monthly Notices Roy. Astron. Soc. 187, 129
Stothers, R., Chin, C.W.: 1977, Astrophys. J. 211, 189
Zahn, J.P.: 1983, 13th Saas-Fee cource, p. 322 (Ed. B. Hauck & A. Maeder)

EVOLUTION AND NUCLEOSYNTHESIS OF MASSIVE STARS WITH EXTENDED MIXING

C. de Loore, N. Prantzos, M. Arnould and C. Doom
Brussels-Saclay-Brussels collaboration on massive
star evolution

ABSTRACT. We discuss the evolution of massive stars ($M_i \geq 50\ M_\odot$) with extended mixing. In the computations, we have taken into account detailed nucleosynthesis of 28 isotopes from ^1H to ^{30}Si, and all relevant reactions for hydrogen and helium burning, including the production of neutrons. The results show the same qualitative behaviour as in previous models computed with extended mixing but with no detailed nucleosynthesis. The principal difference with previous models is that we find a smaller hydrogen burning lifetime. This is due to the fact that the CNO cycle only reaches equilibrium on a timescale, comparable with the hydrogen burning timescale.

1. INTRODUCTION

During the last years it was realized that convective cores must be larger than given by the classical Schwarzschild criterion. The reason is that turbulent kinetic energy is transported beyond the clasical boundary of the convective core. The influence of this effect on the evolution of massive stars has been studied by several authors (Bressan et al. 1981, Doom 1982, Stothers and Chin 1985), showing that extended mixing is a key factor in the understanding of the evolution of massive stars. The first computations of lower mass stars with extended mixing (Maeder and Mermillod 1981) showed that it can be an important factor for lower mass stars as well. These results were strenghtened by recent computations (Doom 1985, Bertelli et al. 1985). This is a first reason to reinvestigate nucleosynthesis in massive stars during hydrogen and helium burning.
Since the extended nucleosynthesis computations of massive stars by Maeder (1983), many reaction rates have changed (Harris et al. 1983, Zimmerman et al. 1984). The most important change concerns the reaction $^{12}C(\alpha,\gamma)^{16}O$, the rate of which is a factor of three larger than previously assumed (see e.g. Trautvetter, this volume). This change is expected to alter the C/O ratio during core helium burning considerably. This is the second reason to reinvestigate the nucleosynthesis of

massive stars. Moreover, neutrons are explicitly included in our computations, in contrast with the Maeder (1983) computations.

2. INPUT PHYSICS

2.1 Mass loss rates

During core hydrogen burning, a massive star appears as an O or an Of star. During this phase, we have adopted the mass loss formalism of Lamers (1981):
$$-\dot{M}(M_\odot/yr) = 1.63 \; 10^{-13} \; L^{1.42} \; R^{0.61} \; M^{-0.99}$$
where all quantities are in solar units. This mass loss formalism represents well the observed increase of mass loss rate when a star evolves off the zero age main sequence (ZAMS). During the core helium burning phase, a massive stars which has lost its hydrogen rich envelope appears as a Wolf-Rayet (WR) star. In this case, we have adopted a constant mass loss rate of $3 \; 10^{-5} \; M_\odot/yr$, which is the average of the observed mass loss rates of 21 WR stars, presented by Barlow et al. (1981).

2.2 Mixing

In order to determine the boundary of the convective core, we have used the criterion derived by Roxburgh (1978). It has been shown by Doom (1985) that this criterion given an approximation of the integrated flux of turbulent kinetic energy $L_t(r)$ through a surface with radius r:

$$L_t(r) = T \int_0^r (L_r - L) \; T^{-2} \; \frac{dT}{dr'} \; dr'$$

where L is the total luminosity at level r and L_r is the integrated radiative flux. The boundary of the convective core is then given by the level where L_t vanishes:
$$L_t = 0$$
This criterion, derived by Roxburgh implies that convective cores are larger than given by the classical Schwarzschild criterion, since the flux of kinetic energy is maximal at this boundary, and then drops to zero in the zone of "extended mixing", just outside the classical core. In the whole convective zone, we use the adiabatic temperature gradient (Maeder 1975).

Computations of the evolution of massive stars with the Roxburgh criterion to determine the boundary of the convective core represent well the observed characteristics of the upper HR diagram in the mass range $1.2 \; M_\odot$ to $120 \; M_\odot$:

- The observed upper limit to the luminosity of late type supergiants at $M_{bol} = -9.8$ is reproduced: stars that are more luminous lose their hydrogen rich envelope during core hydrogen burning, and evolve towards the left of the main sequence, producing WR stars according to the Conti (1976) scenario;
- The observed extension of the main sequence band around $\log L/L_\odot = 5$ (down to $\log Te = 4.25$) is reproduced;

- The problem of the main sequence widening of intermediate mass stars (see e.g. Maeder and Mermillod 1981) is resolved, since the new computations produce a much wider main sequence band.
We will use the Roxburgh criterion in all models.

2.3 Radius correction

De Loore et al. (1982) have shown that the photosphere of an early type star, exhibiting an intense stellar wind, moves outwards into the wind. Since the evolutionary code only gives a "hydrostatic" radius R_*, it is necessary to apply this correction for all stars with a high mass loss rate. We have done this, using the wind velocity law
$$v(r) = 2800 \text{ km/sec } (1 - R_*/r)^2$$
which, according to de Loore et al. (1982) gives a reasonable agreement between theoretical and observed radii of WR stars.

2.4 Opacities and initial composition

We have used the opacity tables of Cox and Stewart, from which the tables for the appropriate composition were interpolated.
The intitial composition of all models is $X = 0.7$, $Y = 0.27$ and $Z = 0.03$., which is appropriate for Population I stars. The distribution of all elements, heavier than helium is taken from Cameron (1972).

2.5 The evolutionary code

The evolutionary code used is discribed by de Loore et al. (1978) and is based on the code HB8 of Paczynski. The Roxburgh criterion was introduced to determine the boundary of the convective core. The evolution of the chemical abundances is followed by solving a system of 30 coupled differential equations for each layer, with the linearization method of Wagoner (1969). In the convective regions, the average reaction rates are used, so only one system of equations has to be solved. An exception is made for neutrons, which reach equilibrium on a timescale shorter than the convection timescale. Hydrogen burning nucleosynthesis is not followed below 10^6 K, and helium burning is not followed below $7\ 10^7$ K. The decay of radioacitve species such as ^{26}Al is always followed.

3. RESULTS

We have computed the evolution of a 100, 80, 60 and 50 $M_\odot$ model from the ZAMS up to the end of core helium burning. The evolutionary tracks of these models are shown in Figure 1. This figure also shows the evolution of a 40 $M_\odot$ model up to core helium ignition for the purpose of comparison.

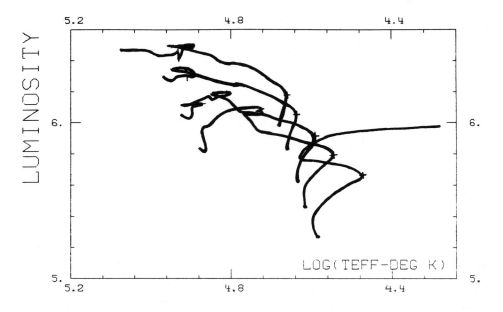

Figure 1: The evolutionary tracks of the 40, 50, 60, 80 and 100 $M_\odot$ models in the HR diagram. The 40 $M_\odot$ track is given up to the onset of core helium burning. The other tracks are shown up to core helium exhaustion.

3.1 Internal evolution

The evolution of the 60 $M_\odot$ model is typical for the evolutionary behaviour of these massive stars (see also Fig. 1 of Prantzos et al., this volume). During core hydrogen burning, the stellar mass decreases from 60 $M_\odot$ on the ZAMS to 45 $M_\odot$ at the end of core hydrogen burning, due to the stellar wind mass loss. The convective core decreases from 52 $M_\odot$ on the ZAMS to 42.5 $M_\odot$ when the central hydrogen abundance X_c is 10^{-4} and then drops rapidly to zero. After some $3.8 \; 10^6$ yr the stellar mass equals 52 $M_\odot$. At this moment, layers that have been in the initial convective core appear at the surface. The atmospheric hydrogen abundance X_{at} now begins to decrease, and the atmospheric nitrogen abundance abruptly rises by a factor of 5. The star has now turned from a normal O star into an Of star. Its mass loss rate has increased by a factor of 4 since the ZAMS, as is observed in Of stars. The deficiency of hydrogen in Of stars has been observed in one case (Kudritsky et al. 1983).

At the end of core hydrogen burning the Of star has evolved into a transition object between Of and WR (see also Doom et al. 1985).

These objects are very extreme Of supergiants, containing little hydrogen and with a spectrum resembling that of a WR star (see Andrillat 1982 for a review).

At the beginning of core helium burning the star has become a WN star and the mass loss rate rises to $3\ 10^{-5}\ M_\odot/yr$. In some 10^4 yr the convective core rises from zero until it reaches 98% of the total stellar mass (43 $M_\odot$). During this phase, some hydrogen is mixed into the convective core. We have assumed that this hydrogen is immediately burnt by the reaction $^{12}C(p,\gamma)^{13}N(\beta^+)^{13}C$. The zone of hydrogen shell burning is pushed outwards by the rapidlyly growing convective core and is soon extinguished.

Some $8\ 10^4$ yr after the beginning of core helium burning, the stellar mass equals 43 $M_\odot$. Layers that have been in the helium burning core now appear at the stellar surface. The atmospheric hydrogen abundance is now zero, and the stellar surface is carbon enhanced. The star has turned into a WC star. Later on, oxygen becomes also very abundant at the surface and the star could become a WO star.

The stellar mass now decreases very rapidly due to the large mass loss rate. At the end of core helium burning ($5.46\ 10^6$ yr after the beginning of core hydrogen burning) it is 32.3 $M_\odot$. During the whole phase of core helium burning the convective core represents some 98% of the stellar mass. When the central helium abundance Y_c is 0.01, the convective core drops to zero, and core helium burning is terminated.

The internal evolution of the other models is qualitatively equal to the 60 $M_\odot$ model. At the beginning of core helium burning in the 50 $M_\odot$ model, no hydrogen is mixed into the helium core. Stars more massive than 60 $M_\odot$ have relatively larger convective cores. They also evolve more rapidly into transition objects, and have only a very short WN phase. The most important timescales of these stars are summarized in Table I.

The hydrogen burning lifetimes of the actual models are shorter than found in previous computations including extended mixing (Doom 1985). This is a consequence of the inclusion of a detailed nuclear network in the computations. In the previous computations it was explicitly assumed that the CNO cycle is in equilibrium of the ZAMS. In the present computations this assumption is not made. The computations show that the CN cycle reaches equilibrium on a very short timescale

TABLE I: Relevant times of the evolution of massive stars with mass loss, extended mixing and including detailed nucleosynthesis. All times are in millions of years

	Initial Mass ($M_\odot$)			
	50	60	80	100
Beginning of Of phase	4.7	3.8	2.7	2.0
End core hydrogen burning	5.5	5.0	4.4	4.0
End core helium burning	6.1	5.5	4.8	4.4

(10^4 yr). However, the ON cycle only reaches equilibrium on a timescale, comparable to the hydrogen burning lifetime. This leads to a less efficient energy production during a large part of the hydrogen burning phase, since a part of the CNO cycle is "blocked". The star adjusts itself to this situation by rising slightly its central temperature, inducing a faster consumption of hydrogen. As a consequence, the star loses mass on a slightly longer timescale, which indirectly also shortens the hydrogen burning lifetime.

These results are in contrast with the results of Maeder (1983), who found that the CNO cycle reaches equilibrium relatively fast. However, in the computations of Maeder, extended mixing was not taken into account. In our models, the convective cores are larger, and therefore extend to lower temperatures. Since the evolution of the abundances in the core depends on the average reaction rates, these are lower in our models than in Maeders' models. Therefore, the CNO cycle reaches equilibrium on a longer timescale.

As in previous computations with extended mixing, we found no semiconvection in our models. This is clearly due to the combined effect of mass loss and overshooting (see also Bressan et al. 1981, Doom 1982, 1985).

3.2 The evolution in the HR diagram

The evolution of our models in the HR diagram is qualitatively equal to the results obtained with previous computations including extended mixing. Massive stars ($M_i > 42$ $M_\odot$ for the present models, $M_i > 33$ $M_\odot$ for previous models) lose their hydrogen rich envelope during core hydrogen burning. At the beginning of the core helium burning phase they do not produce red supergiants, but evolve to the left of the main sequence to form WN stars (see also Bressan et al. 1981, Doom 1982, 1985). Less massive stars produce red supergiants. The computations show that WR stars are formed in a natural way according to the Conti (1976) scenario if mass loss by stellar wind and extended mixing are combined. In extended mixing is not taken into account, WR stars are only formed if very large mass loss rates are adopted at some stage before core helium burning (Noels and Gabriel 1981, Maeder 1983).

The bifurcation mass we find between the formation of red supergiants and the formation of WR stars is somewhat larger than found in previous computations. Also, the maximum luminosity for late type supergiants we obtain is larger than in previous computations. This is due to the fact that our hydrogen burning lifetimes are shorter. According to the present computations, a 40 $M_\odot$ model loses less mass during core hydrogen burning than in previous computations and does not evolve to the left of the main sequence band after core hydrogen exhaustion. It produces a red supergiant with a luminosity of $\log L/L_\odot$ = 5.95. However, the exact bifurcation mass depends on the mass loss rates. It has been shown by Doom (1985) that the mass loss formalism, derived by Lamers (1981) contains possible systematic errors, due to the inconsistent derivation of the mass of the program stars. He showed that the real mass loss rate of very massive stars ($M_i > 40$ $M_\odot$) could be 20 - 50% smaller than given by Lamers and the mass loss rates for lower mass stars could be slightly larger. Although Doom (1985) corrected for this inconsistency, he did not obtain a more accurate fit to the observations. However, this shows that the mass loss rates, used in the computations are not certain. An increase in the mass loss rate of some 20 - 30% could lower the bifurcation mass and -luminosity noticably, removing the discrepancy.

Another indication that the mass loss rates for very massive stars during core hydrogen burning may be less than given by the Lamers formula is that the main sequence band for this stars is narrower than observed in the HR diagram (Humphreys and McElroy 1985). If smaller mass loss rates were adopted, the main sequence band would be broader (cf Stothers and Chin 1985).

Our computations do not admit to make an estimate of the ratio of WN to WC stars, since this ratio is determined by other factors such as the number of WR stars produced through the red supergiant scenario, the number of binary WR stars and the lower mass boundary of WR progenitors (see Doom et al. 1985).

The discussion of the detailled nucleosynthesis, included in the present computations is presented is a separate paper in this volume.

The Bussels-Saclay-Brussels collaboration on massive stellar evolution includes the following groups: Astrophysisch Instituut, Vrije Universiteit Brussel, Pleinlaan 2,B-1050 Brussel, Belgium (C. de Loore and C. Doom); Centre D'Etudes Nucléaires de Saclay, Service d'Astrophysique, DPh/EP/Ap, 91191 Gif sur Yvette Cedex, France (N. Prantzos), and Institut d'Astronomie et de Géophysique, CP 165, Université Libre de Bruxelles, Avenue F.D. Roosevelt 50, Bruxelles, Belgium (M. Arnould). The collaboration is supported by the Belgian Fund of Joint Fundamental Research (FKFO) under contract Nr 2.9002.82. M. Arnould is Chercheur Qualifié of the F.N.R.S. Belgium.

REFERENCES

Andrillat, Y. 1982, in "Reports on Astronomy", XVIIIth general assembly of the IAU, ed P. A. Wayman, 343
Barlow, M. J., Smith, L. J., Willis, A. J. 1981, Mon. Not. Roy. Astron. Soc. **196**, 101
Bertelli, G., Bressan, A. G., Chiosi, C. 1985, Astron. Astrophys. (in press)
Bressan, A. G., Bertelli, G., Chiosi, C. 1981, Astron. Astrophys. **102**, 25
de Loore, C., De Greve, J. P., Vanbeveren, D. 1978, Astron. Astrophys. Suppl. **34**, 363
de Loore, C., Hellings, P., Lamers, H. J. G. L. M. 1982, in "Wolf-Rayet stars, Observations, Physics, Evolution" Proc. IAU Symp. 99, eds C. de Loore and A. J. Willis (Dordrecht, Reidel), 53

Cameron, A. C. W. 1982, in "Essays in Nuclear Astrophysics", eds C.
   Barnes, D. Clayton and D. Schramm (Cambridge Univ. Press)
Conti, P. S. 1976, Mém. Soc. Roy. Liège **9**, 173
Doom, C. 1982, Astron. Astrophys. **116**, 303
Doom, C. 1985, Astron. Astrophys. **142**, 143
Doom, C., De Greve, J. P., de Loore, C. 1985, preprint
Harris, M. J., Fowler, W. A., Coughlan, G. R., Zimerman, B. A. 1983,
   Ann. Rev. Astron. Astrophys. **13**, 69
Humphreys, R. M., McElroy, E. L. 1984, Astrophys; J. **284**, 565
Kudritzki, R. P., Simon, K. P., Hamman, W. W. 1983, Astron. Astrophys.
   118, 245
Lamers, H. J. G. L. M. 1981, Astrophys. J. **245**, 593
Maeder, A. 1975, Astron. Astrophys. **40**, 303
Maeder, A. 1983, Astron. Astrophys. **120**, 113
Maeder, A., Mermillod, J. C. 1981, Astron. Astrophys. **93**, 136
Noels, A., Gabriel, M. 1981, Astron. Astrophys. **101**, 215
Roxburgh, I. W. 1978, Astron. Astrophys. **65**, 281
Stothers, R., Chin, W. 1985, preprint
Wagoner, R. V. 1969, Astrophys. J. Suppl. **18**, 247

NUCLEOSYNTHESIS IN MASSIVE, MASS LOSING, STARS

N. Prantzos
Service d'Astrophysique, CEN Saclay, France
C. De Loore, C. Doom
Astrophysics Institute, VUB, Brussells, Belgium
M. Arnould
Institut d'Astronomie et d'Astrophysique, ULB, Bruxelles, Belgique

ABSTRACT. We have made detailed nucleosynthesis computations for the H − and He burning phases in massive stars with mass loss and overshooting. We present here some results concerning the evolution of core and surface abundances, as well as the composition of the ejecta of these stars.

1. INTRODUCTION

The study of the evolution and nucleosynthesis of massive stars has regained interest among the theoreticians, as well as the observers in the past few years. On one hand, new light has been shed on some aspects of stellar physics, such as the mass loss (especially, the extremely high mass loss rates of WR stars) and the extent of the convective zones due to overshooting (e.g. Doom, 1982,a,b). On the other hand, due to the continuous efforts of the experimentalists, the rates of some key nuclear reactions have been revised recently. Consequently, a reassesment of the problem of nucleosynthesis in massive stars is indeed necessary.

In this paper, we present detailed results of our nucleosynthesis computations for the H − and He burning phases (§ 3), in massive ($M > 50\ M_\odot$) stars, the evolution of surface abundances during the O-Of-WN-WC phases and the yields of the stellar winds (§ 4). The physics of our stellar models as well as the main results of our stellar evolution computations are given in another paper in this volume (De Loore et al.).

It should be useful to recall here the interest of these massive, mass losing stars for nuclear astrophysics:
− the appearance of the core burning products at the stellar surface provides a unique opportunity for a direct test of our theories of stellar nucleosynthesis;
− the ejection of large quantities of material with non-standard composition through the intense stellar winds contributes to the enrichment of the interstellar medium in some particular species.

Among the most obvious implications of this enrichment we could mention:
. the (possible) explanation of the isotopic anomalies observed in the galactic cosmic rays;
. the (possible) contribution to the isotopic composition of the early solar system;
. the gamma-ray line astronomy, and particularly the origin of the 1.8 MeV line of ^{26}Al observed in the galactic center.

2. INPUT NUCLEAR PHYSICS

The nuclear reaction network used in our computations includes 28 isotopes, from H to ^{30}Si. The initial abundances are set to: X = 0.70, Y = 0.27, Z = 0.03, with a metal distribution corresponding to the one given by Cameron (1982). The nuclear reaction rates are from Harris et al.(1983), Fowler et al.(1975), Zimerman et al.(1984). The uncertainty factor f = (0. to 1.) is set to the value f = 0.1 whenever it appears in the formulae.

For H burning, the 3 p-p chains, the CNO tricycle as well as the Ne-Na and the Mg-Al cycles are considered. The nucleosynthesis of ^{26}Al has been carefully followed. In particular its isomeric state ^{26}Al^m is considered as a separate species, as it is not thermalized under the stellar conditions of interest in this work. For the ^{26}Al(p, γ) reaction the rate of Woosley et al.(1978) is adopted. Of considerable interest for He burning is the rate of the ^{12}C(α,γ)^{16}O reaction which has been revised upwards by a factor of 3 a few years ago, and could even be higher (e.g. Trautvetter, this volume). Neutrons are released during core He burning in massive stars, through the ^{22}Ne(α, n) reaction, the "fuel" of which (^{22}Ne) comes from the "ashes" of the previous CNO cycle (^{14}N) through the ^{14}N(α,γ)^{18}F(β^+)^{18}O(α,γ)^{22}Ne chain. The neutron capture cross-sections for all the nuclei in our network between ^{20}Ne and ^{30}Si are taken from Almeida and Kappeler (1982), whereas the ^{21}Ne(n,γ) rate is taken from Harris et al. (1983) and the ^{26}Al (n, p) and ^{26}Al (n,α) rates are given by Zimerman et al. (1984). In order to take into account the captures of neutrons by seeds heavier than ^{30}Si, a neutron "sink" with abundance (by number) $N_S = \sum_{A=31}^{210} N_{A\odot}$ and an effective neutron capture cross section $\sigma_S = \sum_{A=31}^{210} 6_A N_{A\odot} / N_S$ is introduced, 6_A being the n-capture cross-section of the nucleus with mass number A, and solar abundance by number $N_{A\odot}$. The results of the computations with a full s-process code (see Prantzos et al., this volume) confirm the validity of this approximation. More details about the techniques used for our treatment of nucleosynthesis are given elsewhere (Prantzos et al.,1985).

3. NUCLEOSYNTHESIS IN THE CORE: RESULTS AND UNCERTAINTIES

The main characteristics of the evolution of our 60 $M_\odot$ star are presented in fig.1, and are discussed by De Loore (this volume). The evolution of the chemical composition inside the large, homogeneous, convective core of the star is presented in fig.2 (for nuclides with

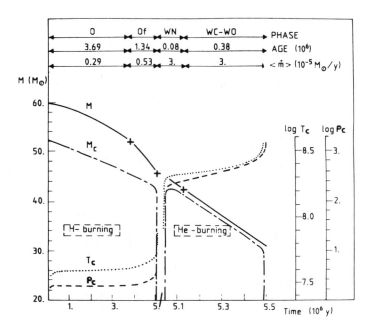

FIG. 1 : Evolution of a star with $M_{ZAMS} = 60\ M_\odot$ during H- and He burning.

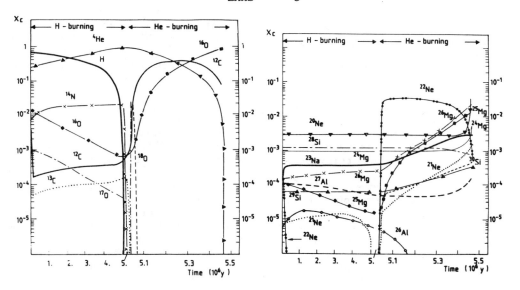

FIG. 2 : Evolution of the central composition for nuclides with mass number A < 20.

FIG. 3 : Evolution of the central composition for nuclides with mass number A > 20.

A < 20) and 3 (for nuclides with A > 20). The chemical evolution of the stars with M_{ZAMS} = 50, 80 and 100 $M_\odot$ has the same qualitative aspects (with some minor quantitative differences) and will not be presented here.

As a result of the CN cycle, ^{12}C, ^{13}C, ^{14}N and ^{15}N reach very rapidly their equilibrium mass fractions of 2.10^{-4}, 7.10^{-5}, 10^{-2} and 4.10^{-7} respectively. The NO cycle does not reach equilibrium: ^{16}O decreases steadily to 6.10^{-4} followed by ^{17}O whereas ^{12}C and ^{13}C follow the slight increase in the equilibrium abundance of ^{14}N which is due to the increase in temperature.

The major isotopes of the Ne-Na and Mg-Al cycles (^{20}Ne and ^{24}Mg) are not influenced by H burning. However, ^{21}Ne reaches its equilibrium abundance, which is very sensitive to temperature: at the end of H burning (central temperature $T_c \gtrsim 50.10^{6}$ °K) it goes from $1.5\ 10^{-5}$ down to 3.10^{-7}. There is also an increase in the abundance of ^{23}Na at the expense of ^{22}Ne, and of ^{26}Mg at the expense of ^{25}Mg. The abundance of ^{26}Al increases rapidely to $1.5\ 10^{-5}$ as a result of ^{25}Mg (p, γ), but after 10^6 years ^{26}Al starts β-decaying to ^{26}Mg and its abundance declines slowly to 5.10^{-6} at the end of core H burning. The composition of a large part of the mass of the star ($\sim 50\ M_\odot$ or 84%) is modified by the action of core H burning. A much smaller fraction of the stellar mass ($\sim 1\ M_\odot$) is processed through the subsequent phase of shell H burning, but its duration is too short ($\sim 0.1\ 10^5$ years) to bring any substantial alteration to the chemical composition of that region, despite the high temperatures prevailing there ($T \sim 4.10^7$ °K).

During the next phase of core He burning, ^{4}He is transformed first into ^{12}C and finally into ^{16}O. At the end of core He burning the stellar core ($M_{12} \sim 32\ M_\odot$) consists mainly of ^{16}O ($\sim 86\ \%$) and of a smaller amount of ^{12}C ($\sim 6\%$) and heavier isotopes. This high ^{16}O abundance is of course due to the use of the new $^{12}C\ (\alpha, \gamma)$ reaction rate (see also Prantzos and Arnould, 1983; Thielemann and Arnett, 1984).

At the beginning of He burning ^{14}N is very rapidly transformed into ^{18}O: for a very brief period (~ 1000 years) ^{18}O becomes the second major nuclide, its abundance rising spectacularly to 2%, but it is rapidly destroyed through $^{18}O\ (\alpha, \gamma)\ ^{22}Ne$. ^{13}C and ^{17}O left behind by the CNO tricycle also disappear at the beginning of He burning through (α, n) reactions. For a short period (~ 10 years) a "burst" of neutrons ($\sim 10^9$ neutrons/cm^3) is produced in the inner parts of the core, but their density goes down by a factor of 10^5 when ^{13}C, ^{17}O and ^{18}O disappear. A much more important neutron source is the $^{22}Ne\ (\alpha, n)$ reaction, operative at $T \sim 3.10^8$ K, producing neutron densities of 10^7 neutrons/cm^3 (see Prantzos et al., this volume, for the results of neutron-capture nucleosynthesis). The $^{16}O\ (\alpha, \gamma)\ ^{20}Ne$ and $^{20}Ne\ (\alpha, \gamma)\ ^{24}Mg$ reactions, also become operative at this point, leading to a substantial increasein the abundances of ^{20}Ne and ^{24}Mg up to 0.03 and 0.005 respectively). At the end of He burning ^{22}Ne is almost completely transformed through (α, n) and

(α,γ) reactions into ^{25}Mg and ^{26}Mg while the abundances of ^{21}Ne and ^{23}Na are also substantially enhanced by neutron captures on ^{20}Ne and ^{22}Ne respectively. ^{26}Al, which is not produced during He burning, is rapidly destroyed through (n,p) and (n,α) reactions, while the abundances of ^{27}Al and of the Si isotopes (^{29}Si and ^{30}Si) increase by a factor of 5.

Since the more energetically important reaction rates are known to within a few per cent, the lifetimes, structures and evolutionary paths of the stellar models are not influenced by the existing nuclear physics uncertainties, but the nucleosynthesis predictions might be considerably altered. In order to explore quantitatively the effects of these uncertainties, calculations have been performed with extreme values (0 and 1) of the uncertainty factor f and the results are compared (table 1) to the "standard" (f = 0.1) case. The resulting uncertainties in the abundances of some species are quite high, but, fortunately, none of them is very important in the context of this study (with the exception of ^{27}Al).

As already mentionned, the differences in the chemical evolution of the 4 studied stellar models (M_{ZAMS} = 50, 60, 80 and 100 $M_\odot$) are rather small. Fig.4 indeed shows that the concentrations of most of the species with X > 10^{-5} (^{16}O, ^{21}Ne, ^{23}Na, ^{25}Mg, ^{26}Mg, ^{27}Al, ^{28}Si, ^{29}Si, ^{30}Si) at the end of the He burning, are almost independent of the initial mass. This is not the case for ^{12}C and ^{22}Ne which are less abundant in stars with larger masses, whereas the opposite is true for ^{20}Ne and ^{24}Mg: they are 3-4 times more abundant in the 100 $M_\odot$ than in the 50 $M_\odot$ star. These results of course reflect the high sensitivity to temperature of the corresponding reaction rates. Of particular interest is the extremely high abundance of ^{16}O (86% in all our models) remaining at the end of He burning in these large stellar cores: in the case of the models with M_{ZAMS} = 100 $M_\odot$ and 80 $M_\odot$, the resulting, almost pure oxygen, cores with M > 45 $M_\odot$ are potential candidates for the formation of pair-creation supernovae (see Cahen et al., this volume).

4. THE EVOLUTION OF SURFACE ABUNDANCES AND THE COMPOSITION OF WR EJECTA

Mass losing stars offer a unique opportunity for a direct test of nucleosynthesis theories since they expose their core burning products to observation. Along with the H-R diagram, the surface composition of these stars provides a test for the models.

The evolution of the surface composition in our model stars is essentially the result of the interplay between 3 factors:
- the extent of the convective core;
- the mass loss rate, especially during the WR phase, as well as the moment of beginning of this phase;

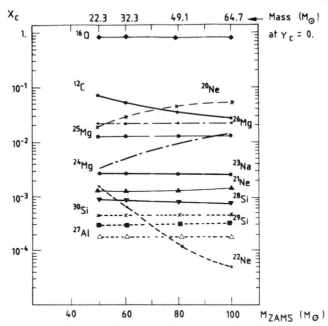

FIG. 4 : Core chemical composition at the end of He burning for our models with M_{ZAMS} = 50, 60, 80, and 100 $M_\odot$. The corresponding core masses are indicated at the top of the figure.

TABLE 1
Role of uncertainties in nuclear reaction rates
(tests for the M_{ZAMS} = 60 $M_\odot$ star)

Reaction	Rate ratios f = 0 : 0.1 : 1	Isotope affected	Mass fractions obtained at the end of the cycle		
			f=0	f=0.1	f=1
Hydrogen burning (T_6 = 50)					
$^{17}O(p,\alpha)$	0.05 : 1 : 10	^{17}O	3.7(-4)	3.5(-5)	3.5(-6)
$^{18}O(p,\alpha)$	10^{-2} : 1 : 10	^{18}O	1.1(-7)	1.5(-10)	1.8(-12)
		^{19}F	1.8(-8)	2.3(-11)	2.3(-13)
$^{21}Ne(p,\gamma)$	0.1 : 1 : 10	^{21}Ne	3.3(-7)	3. (-7)	2. (-7)
$^{23}Na(p,\alpha)$	0.1 : 1 : 10	^{22}Ne	5.5(-6)	3. (-7)	3. (-8)
$^{26}Mg(p,\gamma)$	10^{-3}: 1 : 10	^{26}Al	5.5(-6)	5.5(-6)	4.8(-6)
$^{27}Al(p,\gamma)$	10^{-4} : 1 : 10	^{27}Al	1.4(-4)	3.5(-5)	5.3(-6)
$^{27}Al(p,\alpha)$	10^{-5} : 1 : 10				
Helium burning (T_8 = 3)					
$^{24}Mg(\alpha,\gamma)$	$3\,10^{-3}$: 1 : 10	^{24}Mg	7. (-3)	4.9(-3)	7.7(-4)

The most spectacular result of the combination of these 3 factors is the discontinuities in the evolution of the surface abundances of most of the nuclides involved, an effect already noticed by Maeder (1983a). This is due to the fact that, when the convective core first reaches the surface, the abundance of each isotope is more or less different from its previous atmospheric one.

In the following we describe the principal features of the evolution of the surface abundances for our $M_{ZAMS} = 60\ M_\odot$ star. The evolution in the case of the models with $M_{ZAMS} = 50, 80, 100\ M_\odot$ is not very different.

When the H burning core first reaches the surface, the abundances of the isotopes that have been substantially modified at the very beginning of core H burning show a pronounced discontinuity. In particular, the abundance of ^{14}N jumps from 10^{-3} up to 10^{-2} and continues to rise slowly up to 2.10^{-2} for 10^6 years. After an initial fall down by a factor of ~ 50, the abundance of ^{12}C follows the slow rise of ^{14}N, being in its turn followed closely by ^{13}C. The He abundance increases slowly at the expense of H which goes down to 0.20 at the end of th Of phase. Thus, the H/He ratio shifts from its solar system value of ~ 3 down to 0.3. At the same time ^{16}O decreases slowly, and ^{17}O follows its decline, after a dramatic initial jump by 2 orders of magnitude. Thus the Of phase is characterized by almost constant ratios of C/N ($\sim 1.5\ 10^{-2}$; solar: ~ 3), $^{13}C/^{12}C$ (~ 0.3; solar: $1.5\ 10^{-2}$) and $^{17}O/^{16}O$ (~ 0.1; solar: 5.10^{-4}), while there is a variation in the ratios of O/N (from ~ 1 to 0.05, compared to the solar system value of ~ 9) and C/O (from 10^{-2} to 0.3, compared to the solar 0.4).

Very few of the heavier isotopes are affected during this phase, like ^{21}Ne (which increases by a factor of 2), ^{25}Mg and ^{26}Mg. However, since these are not the major isotopes of the corresponding elements, the elemental Ne and Mg compositions are not affected. In contrast, the abundance of the monoisotopic Na increases by a factor of 4 with respect to its solar system value. Most important is the appearance, during the Of phase of ^{26}Al the abundance of which increases slowly to 5.10^{-6}. This value is lower by a factor of ~ 3 than the maximum amount reached inside the core, due of course to the ^{26}Al decay in the stellar envelope for 2.10^6 years.

In the following, quite short, WN phase, the atmospheric He and N abundances are stabilized to ~ 0.90 and 2.10^{-2} respectively whereas H decreases steadily to 0.05. The particular shape of the ^{12}C and ^{13}C abundance curves at the end of the WN phase reflects the action of the, otherwise unimportant, shell H burning.

When the He burning products appear at the surface and the WC stage sets in, various species "disappear", like H, ^{13}C, ^{14}N and ^{17}O, the abundances of which drop by many orders of magnitude. The abundance of ^4He presents a small discontinuity (from 0.95 to 0.90) while the abundances of ^{12}C and ^{22}Ne rise more spectacularly, up to 5.10^{-3} and 10^{-2} respectively. In a very short time ($\sim 5\ 10^3$ years) they are stabilized to ~ 0.3 and 3.10^{-2} respectively, whereas ^{16}O continues to increase and becomes the dominant species $\sim 3.10^5$ years

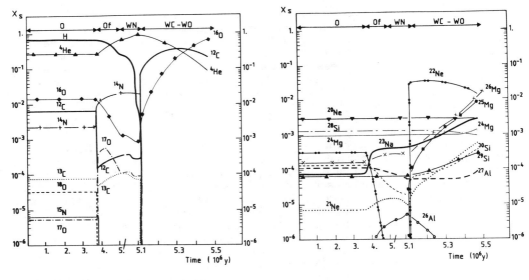

FIG. 5 : Surface composition of the 60 $M_\odot$ star (A < 20)   of the 60 $M_\odot$ star (A > 20)

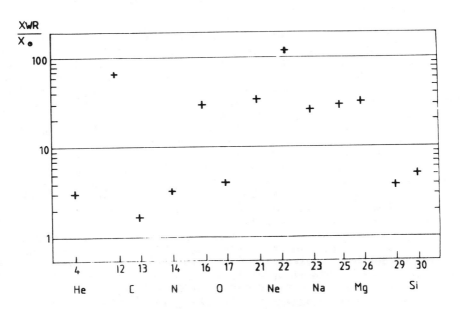

FIG. 7 : Overproduction factors of the principal nuclei (with respect to solar system) in the ejecta of WR stars.

after the beginning of the WR stage. This moment could be identified (rather arbitrarily) as the beginning of the WO phase. The abundances of ^{21}Ne, ^{23}Na, ^{25}Mg and ^{26}Mg also rise steadily in the WC phase, and at the end of He burning they are some 2 orders of magnitude larger than their respective solar system values. A more moderate increase (by a factor of 2-3) is observed for ^{29}Si and ^{30}Si. Thus, besides the C, O and Ne excess, the WC and especially WO stars should present an overabundance of Na and Mg.

The same qualitative features are present in all our models. Quantitatively, the discontinuities exhibited at the surface are slightly more pronounced in the stars with smaller masses: since their convective cores occupy a smaller fraction or the total stellar mass, and because of the smaller mass loss rate during the O and Of phases, the core burning products appear at the surface later (being thus more "processed") than in the heavier stars. The differences however do not exceed 20% between the 50 $M_\odot$ and the 100 $M_\odot$ star.

In spite of their short lifetimes, WR stars may have a significant contribution to the enrichment of the interstellar medium (ISM) in some heavy elements, because of their high mass loss rates. The mass input of WR winds to the ISM has been estimated to 5.10^{-5} $M_\odot$/year/kpc^2 (Abott, 1982).

In this section, the yield in the stellar winds of our WR models is presented. For each nuclide, the mass fraction in the WR wind is computed:

$$X_{i\ WR} = \int_{WR} X_{i\ S}\ \dot{m}_{WR}\ dt\ /\ M_{WR}$$

where: X_{iS} is the surface abundance of the ith nuclide, and $M_{WR} = \int_{WR} \dot{m}_{WR}\ dt$, is the total mass ejected during the WR phase.

The results are quite similar for the four stellar models, with differences of 20-30% obtained in some cases.

In order to estimate the global contributions of the ejecta of these stars ($50 < M_{ZAMS}/M_\odot < 100$), to the enrichment of the ISM, an initial mass function (IMF) in this mass range is used to calculate the average mass fraction:

$$X_i = \int_{50}^{100} X_{i\ WR}(M)\ \Phi(M)\ M_{WR}\ dM \bigg/ \int_{50}^{100} \Phi(M)\ M_{WR}\ dM$$

where $M = M_{ZAMS}$, and $\Phi(M) \propto M^{-2.5}$ is the IMF adopted from Humphreys and Mac Elroy (1984).

The resulting average enhancement factors (with respect to the solar system abundances: $f = X_i/X_{i\odot}$) are plotted in fig. 7 (Only nuclides with $f > 2$ are presented). Clearly ^{22}Ne is the most overproduced species (by a factor of $\sim$120) followed by ^{12}C ($f \sim 65$) and ^{16}O, ^{21}Ne, ^{23}Na, ^{25}Mg, ^{26}Mg.

These results confirm previous preliminary computations (Prantzos and Arnould, 1983), and they are not very different from the ones obtained by Maeder (1983b), except for a few cases (^{17}O, ^{21}Ne, ^{23}Na) which are more properly treated in our work or, for which revised nuclear data have been used (e.g. ^{12}C, ^{16}O).

5. CONCLUSION

We presented results of detailed nucleosynthesis computations for the core H and He burning phases of massive stars (50 M_{ZAMS}/M_O 100) with mass loss and overshooting. The main results may be summarized as following:

- The use of revised nuclear data (e.g. for the $^{12}C(\alpha,\gamma)^{16}O$ or the $^{22}Ne(n,\gamma)$ reaction rates) has a considerable influence on the core composition at the end of He burning, with important consequences for the later stages of the stellar evolution.

- Due to the combined effects of mass loss and overshooting, the core burning products appear at the stellar surfaces, which exhibit the characteristics of Of - WN - WC - WO stars.

- The enrichment of the ISM with the non-standard composition material of the WR winds may have interesting astrophysical implications for:
  - the early solar system isotopic composition (Arnould et al., this volume)
  - the gamma-ray line astronomy (Cassé et al., this volume)
  - the anomalous isotopic composition of galactic cosmic rays (Prantzos et al., this volume).

REFERENCES

- Abott, D.C. (1982): Astrophys.J., 263, 723
- Almeida, J., Kappeler, F. (1983): Astrophys.J., 265, 417
- Cameron, A.W.C. (1982): in "Essays in Nuclear Astrophysics", eds. C.Barnes, D.Clayton, D.Schramm, Cambridge Univ.Press
- Caughlan, G.R., Fowler, W.A., Harris, M.D., Zimerman, B.A. (1984): (Submitted to Atomic Data Nucl.Data Tables)
- Doom, C. (1982,a): Astron.Astrophys., 116, 303
- Doom, C. (1982,b): Astron.Astrophys., 116, 308
- Fowler, W., Caughlan, G., Zimerman, B.A. (1975): Ann.Rev.Astron. Astrophys., 13, 69
- Harris, M.J., Fowler, W.A., Caughlan, G.R., Zimerman, B.A. (1983): Ann.Rev.Astron.Astrophys., 21, 165
- Humphreys, R.M., Mc Elroy, D.B. (1984): Astrophys.J., 284, 565
- Maeder, A.(1983,a): Astron.Astrophys., 120, 113
- Maeder, A. (1983,b): Astron.Astrophys., 120, 130
- Prantzos, N., Arnould, M. (1983): in "Wolf-Rayet Stars: Progenitors of Supernovae?", eds.M.C.Lortet and A.Pitault, Meudon, France, p.II 33.
- Thielemann, K.F., Arnett, D.W. (1984): Preprint, Max Planck Institute
- Woosley, S.E., Fowler, W.A., Holmes, J.A., Zimerman, B.A. (1978): Atomic Data Nucl.Data Tables, 22, 371

NUCLEOSYNTHESIS IN MASSIVE STARS: WINDS FROM WR STARS AND ISOTOPIC ANOMALIES IN COSMIC RAYS

A. Maeder
Geneva Observatory
CH-1290 Sauverny
Switzerland

ABSTRACT: A short account of the main effects of mass loss on stellar nucleosynthesis is given.
  The problem of the isotopic anomalies of the galactic cosmic rays is considered as well. A new model is proposed. It is in good agreement with the observations and shows that the observed enhancement originates from the fact that cosmic rays are a probe of matter coming from inner galactic regions where metallicity is higher, and also from the fact that in these inner galactic regions the relative contribution of WC stars to the galactic enrichment is larger than in the solar neighbourhood.

1. MAIN EFFECTS OF MASS LOSS ON NUCLEOSYNTHESIS

Mass loss by stellar winds is of huge consequence concerning evolutionary tracks in the HR diagram and on the genetic relations between OB stars, Hubble-Sandage variables, blue and red supergiants and WR stars. It also influences strongly stellar nucleosynthesis (cf. Dearborn and Blake, 1979; Chiosi and Caimmi, 1979; Chiosi, 1979; Maeder, 1981b; Chiosi and Matteucci, 1982; Maeder, 1983ac, 1984a). Different kinds of effects are to be distinguished: the effects of mass loss on the overall stellar yields, on the status of supernovae precursors and also on the direct contribution of stellar winds to galactic enrichment. Let us summarize briefly the main results about these points.

1.1. Mass loss and the overall stellar yields

A common procedure, applied to estimate the effects of mass loss on nucleosynthesis, relies on the yields obtained by Arnett (1978) for constant mass evolution of bare helium cores of mass M_α. A relation $M(M_\alpha)$ between M_α and the initial stellar masses M is used to transfer the results of bare core calculations to those of standard stars. Mass loss reduces the size of convective cores and therefore the $M(M_\alpha)$ relation happens to be modified: in Arnett's calculations, a smaller mass M_α corresponds to a given initial mass M.

The procedure above has frequently been applied, but, as emphasized previously (Maeder, 1981b), it meets several difficulties. 1) The bare core calculations do not account for the new dredged-up elements in the external convective zone. 2) No account is given to the matter ejected into the winds. 3) M_α is considered as constant, whereas it does change considerably during the He burning-phase, particularly in the presence of mass loss. 4) Moreover, usually no difference is made between the newly synthetized and the original helium. Thus, the $M(M_\alpha)$ procedure is to be used with great caution.

A direct evaluation of all the various contributions of mass losing stars to the yields is preferable. This has been done to estimate the stellar yields in new helium and metals (Maeder, 1983c, cf. Fig. 1), and new models are now in progress in order to follow the evolution up to the pre-supernova and to obtain in case of mass loss the stellar

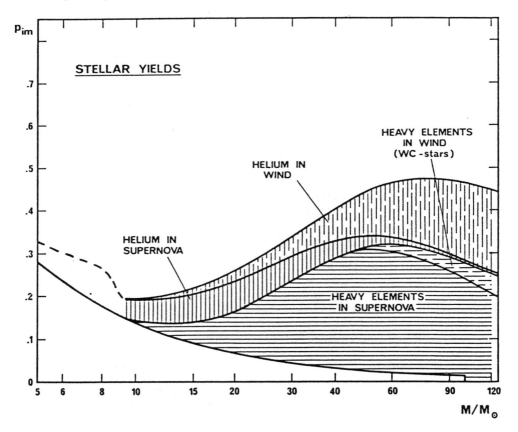

Fig. 1 *Representation of the stellar yields expressed in mass fractions. The helium and heavy elements ejected by the stellar wind are distinguished from those ejected in the supernova.*

yields for various interesting elements. We distinguish in Fig. 1 the yields due to the stellar winds and the yields induced from supernova explosions. For helium we note the increasing importance of the part ejected in the wind for large initial stellar masses, while the helium ejected by supernovae becomes for large masses quite negligible.

For heavy elements, Fig. 1 shows that the part ejected in stellar winds is in relation to the one in supernovae rather small, occuring only in WC stars from progenitors likely to be initially larger than about 50 $M_\odot$. The composition of these heavy elements, ejected in the winds, is highly peculiar and will be discussed below. The stellar yields in metals are reduced by heavy mass loss, an effect which is only significant for large initial masses (M $\geq$ 50 $M_\odot$). In this case, the fact that much helium is ejected and not transformed into metals evidently contributes to a large reduction in the metal yields. This is also the reason why the envelopes of the yield in heavy elements for large stellar masses are turned down in Fig. 1.

We see, from what preceeds, that the main change is due to mass loss and intervenes for large initial masses which are weakly weighted by the initial mass function (e.g. Miller and Scalo, 1979; Garmany et al. 1982). Therefore, when properly averaged by the initial mass function, the net yields are not very much modified by mass loss; between evolutions with constant mass and with mass loss (at the observed rates), there is only a decrease of about 10% for both the overall helium and the metal net yields, (Maeder, 1981). The following values were estimated (e.g. Maeder, 1983c): y_Y = 0.0189, y_Z = 0.0143.

We must emphasize that the above estimate also significantly depends on the limit M_{BH} for black hole formation and on the location of the mass cutoff between the ejected and collapsing parts. The above values were considered for M_{BH} = 150 $M_\odot$. However, a value of M_{BH} = 50 $M_\odot$ seems more reasonable on the basis of pulsars present in young associations (cf. Schild and Maeder, 1985). Supposing that stars with initial mases above 50 $M_\odot$ only contribute to galactic enrichment by their wind ejecta, we get values of y_Y = 0.0185 and y_Z = 0.0103. Cutting large masses reduces evidently much more the metal yield than the one of helium. The ratio y_Y/y_Z is 1.80 in this case; this corresponds in the framework of the closed galactic model with instantaneous recycling to a relative helium to metal enrichment $\delta Y/\delta Z$ = 1.35 (cf. Maeder, 1983c, 1984a).

1.2. Mass loss and the status of pre-supernovae

As a contrast to the very large effects of mass loss in the HR diagram, mass loss has in most cases only a small incidence on the course of central evolution, e.g. in the log Tc vs log ρc diagram (cf. Maeder and Lequeux, 1982). The physical reasons have been explained in the last reference. To give an example of the smallness of these effects, an initial 60 $M_\odot$ star containing only 25 $M_\odot$ at the end of the C-burning phase has exactly the same central conditions as its constant mass counterpart.

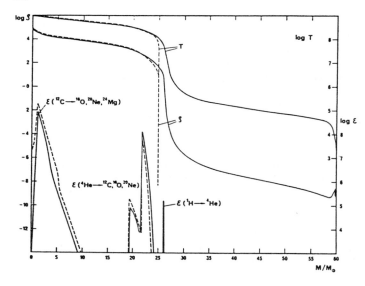

Fig. 2 *Example of the distribution of density, temperature T and rates of nuclear energy production ε in models from initial 60 $M_\odot$ stars at the end of the C-burning phase. The continuous lines refer to models without mass loss while the broken lines refer to the model with mass loss containing only 24.80 $M_\odot$ at the end of the C-burning phase.(cf. Maeder, 1981a).*

As shown in Fig. 2, the distribution of the internal temperature and density are very similar in common stellar parts and this is equally true for the distribution of the elements. Small differences appear only in extreme cases, namely when the central He + C/O core is eroded by mass loss. This occurs for WC stars, that is to say, for initial masses larger than 40 $M_\odot$ or 50 $M_\odot$ (cf. Schild and Maeder, 1984). And even in the case of WC stars with strong mass loss, the temperature deviations from the standard tracks in the log Tc vs log ρc diagram remains very limited (see values in Maeder and Lequeux, 1982).

Based on these facts, the last authors concluded that most mass losing stars undergo a phase of core collapse, and this in despite of the effect of mass loss by stellar winds. A major difference is however, that due to mass loss, stars with initial masses above 30 or 40 $M_\odot$ have hot blue stars as supernovae precursors. Supernovae from WR precursors are expected to exhibit characteristics analogous to those of Cas A (cf. Peimbert and van den Bergh, 1971; Lamb, 1978; Chevalier and Kirshner, 1978; Johnston and Yahil, 1984).

1.3. Direct contributions of stellar winds to galactic enrichment

Wolf-Rayet stars, which are currently identified with bare cores offer us most valuable nucleosynthetic tests, since products of CNO and He-burning reactions become visible at stellar surfaces. Detailed comparison between model predictions and observations have been performed for WR stars (cf. Maeder, 1983a). The general agreement which has been found strongly supports the advanced evolutionary stage of WR stars as leftover cores.

The contributions of the winds of WR stars to the enrichment of the interstellar medium have also been estimated (Abott, 1982; Maeder, 1981, 1983a). The yields due to stellar winds were estimated for the relevant elements. The result was that WC stars contribute to most of the galactic enrichment in ^{22}Ne (Maeder, 1983a). The contribution in ^{12}C is very significant as well, while the contribution in ^{16}O, ^{25}Mg and ^{26}Mg are very modest. Dearborn and Blake (1984) have also shown that WC stars may be an important source of ^{26}Al(see also M.Cassé in this conference).

II. ISOTOPIC ANOMALIES IN COSMIC RAYS

Several isotopic ratios in the galactic cosmic ray sources (GCRS) are enhanced with respect to the solar ratios (cf. Wiedenbeck, 1983), in particular ^{13}C/^{12}C, (^{18}O/^{16}O), ^{22}Ne/^{20}Ne, ^{25}Mg/^{24}Mg, ^{26}Mg/^{24}Mg, ^{29}Si/^{28}Si, ^{30}Si/^{28}Si. Various models have been proposed to explain the observed excesses (cf. Cassé, 1983).

Let us consider two of the presently most popular models. In the metal rich supernovae model (cf. Woosley, Weaver, 1983), GCRS are supposed to originate from inner galactic regions where the metallicity Z is higher than in the solar neighbourhood. The excess of isotopic ratio for secondary elements (2) with respect to primaries (1) scales like

$$E_{21}^{GCRS} \equiv \frac{\left(\frac{X2}{X1}\right)_{GCRS}}{\left(\frac{X2}{X1}\right)_{\odot}} \simeq \frac{Z(r_{int})}{Z(r_{\odot})} \qquad (1)$$

where r_{int} is the typical galactocentric distance from inner galactic regions, where cosmic rays originate; $r_{\odot}$ is the local galactocentric distance. This model suffers from the difficulty to predict correctly the ^{22}Ne/^{20}Ne excess which is the most significant one.

Another model is the WR model (cf. Cassé and Paul, 1982; Maeder, 1983b). In this model, the isotopic anomalies are supposed to be the result from injection by WC stars and the excesses are given by

$$E_{21}^{GCRS} = 1 + p \frac{X_2(WR)}{X_2(\odot)} \qquad (2)$$

where p is the ratio of the mass of matter injected by WR stars to that injected by supernovae. X_2(WR) and X_2 ($\odot$) are respectively the abundance of the considered secondary element in WR stars and in the Sun. The observed excesses of ^{12}C, ^{22}Ne/^{20}Ne, ^{25}Mg/^{24}Mg and ^{26}Mg/^{24}Mg could well be accounted for by an unique dilution factor p of three particles originating from WC stars among 100 in the GCRS. However, one difficulty is that the WC model cannot explain the excess of the ratios $^{29}Si/^{28}Si$ and $^{30}Si/^{28}Si$. Another difficulty is that it was (implicitly) assumed that WR stars, although they contribute to GCRS, make no significant contribution to the local galactic enrichment. Moreover, no account was given on the existing gradients in the galactic distribution of Wolf-Rayet stars and on the chemical gradients in the Galaxy.

A new model has been proposed (Maeder, 1984b). Hovever, this one is more general than the two previous ones, but contains them as limiting cases. It accounts for the fact that cosmic rays may propagate over a few kpc, so that there are at least two kinds of potential sources (supernovae and WR stars), and that both types of sources have a gradient in their galactic distribution and that the initial metallicity is locally different. The local ISM used as a basis of comparison is treated in the closed model with instantaneous recycling. We consider also that the fractions of the ejected nuclei of type (i), which, after being ejected are subsequently accelerated as cosmic rays, take a value λ_i^{SN} for the supernovae sources and λ_i^{WR} for the WR sources. The expression for the excess we get is,(cf. expr. 16, Maeder, 1984):

$$E_{21}^{GCRS} = \frac{Z(r_{int})}{Z(r_\odot)} \frac{\left(1+p \left(\frac{Z(r_{int})}{Z(r_\odot)}\right)^\alpha \frac{X_2(WR)}{X_2(SN)} \frac{\lambda_2^{WR}}{\lambda_2^{SN}}\right)}{\left(1+p \left(\frac{Z(r_{int})}{Z(r_\odot)}\right)^\alpha \frac{\lambda_1^{WR}}{\lambda_1^{SN}}\right)} \Bigg/ \left(\frac{1+p \frac{X_2(WR)}{X_2(SN)}}{1+p}\right) \quad (3)$$

In absence of any other indication, we can take $\lambda_i^{WR} = \lambda_i^{SN}$. From empirical relations between metallicity and the frequency of WR stars, α is estimated to be around 1.7, whereas p and Z have the same meaning as above.

We emphasize that this expression just reduces to expr. (1), if we ignore the contributions of WR stars, i.e. if we put p=0. It also reduces to expr. (2), if we ignore the contribution of WR stars to the local ISM, the metallicity gradient and the gradient in the distribution of WR stars. There are two unknown quantities in expr.(3), p and $Z(r_{int})/Z(r_\odot)$. The best agreement with observations (cf.Fig.3) is obtained for p=0.03 and $Z(r_{int})/Z(r_\odot)$=1.8. The value of p indicates that 3 particles over 100 in the local ISM originate from a WR star. This ratio is higher at r_{int}, i.e. $p(r_{int})$=0.08. We can estimate the value of $r_\odot-r_{int}$, i.e. of the distance from which the observed average GCRS originates. For a metallicity gradient of -.10 to -0.08 dex per kpc, the ratio $Z(r_{int})/Z(r_\odot)$=1.8 corresponds to a distance of 2.5 to 3 kpc towards the galactic interior.

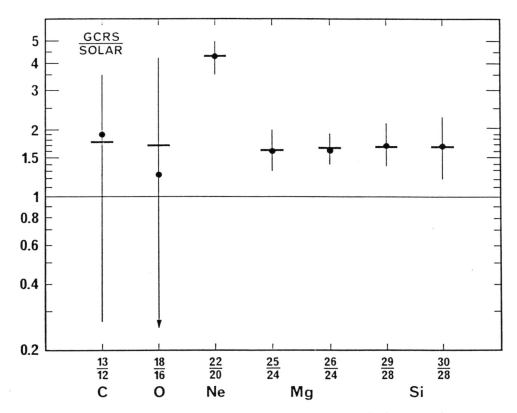

Fig. 3 *Comparisons of the observed excess of isotopic ratios represented by dots (cf. Wiedenbeck, 1983) and theoretical excesses represented by horizontal bars (cf. Maeder, 1984b).*

As a conclusion, we can say, that the basic idea of this model, which reproduces remarquably well the observations, is that the observed enhancements originate from the fact that cosmic rays are a probe of matter coming from inner galactic regions where metallicity is higher, and from the fact that in these inner galactic regions the relative contribution of WC stars to the galactic enrichment is larger than in the solar neighbourhood.

REFERENCES

Abbott D.:, *Astrophys. J.*, 263, 723
Arnett D.W.: 1978, *Astrophys. J.*, 219, 1008.
Cassé M.: 1983, in Composition and Origin of Cosmic Rays, Ed. M.M. Shapiro, *D. Reidel Publ. Co.*, p. 65.
Chevalier R.A., Kirshner R.P.: 1978, *Astrophys. J.*, 219, 931.

Chiosi C.: 1979, *Astron. & Astrophys.* 80, 252.
Chiosi C., Caimmi R.: 1979, *Astron. & Astrophys.* 80, 234.
Chiosi C, Mateucci F.: 1982, *Astron. & Astrophys.* 105, 140.
Dearborn D.S., Blake J.B.: 1979, *Astrophys. J.* 231, 193.
Dearborn D.S., Blake J.B.: 1984, *Astrophys. J.* 277, 783.
Garmany C.D., Conti P.S., Chiosi C.: 1982, *Astrophys. J.* 263, 777.
Johnston M.D., Yahil A.: 1984, *Astrophys. J.* 285, 587.
Lamb S.: 1978, *Astrophys. J.* 220, 186.
Maeder A.: 1981a, *Astron. & Astrophys.* 99, 97.
Maeder A.: 1981b, *Astron. & Astrophys.* 101, 385.
Maeder A., Lequeux J.: 1982, *Astron. & Astrophys.* 114, 409.
Maeder A.: 1983a, *Astron. & Astrophys.* 120, 113.
Maeder A.: 1983b, *Astron. & Astrophys.* 120, 120.
Maeder A.: 1983c, in Primordial Helium, ESO Workshop, Ed. P.A. Shaver, D. Kunth and K. Kjär, p. 89.
Maeder A.: 1984a, in Stellar Nucleosynthesis, 3rd Erice Workshop, Ed. C. Chiosi and A. Renzini, *Reidel Publ. Co.* p. 115.
Maeder A.: 1984b, *Advanced Space Res.* 4, 55.
Miller G.E., Scalo J.M.: 1979, *Astrophys. J. Suppl. Ser.* 41, 3.
Peimbert M, van den Bergh S.: 1971, *Astrophys. J.* 167, 233.
Schild H., Maeder A.: 1984, *Astron. & Astrophys.* 136, 237.
Schild H., Maeder A.: 1985, *Astron. & Astrophys.* 143, L7.
Widenbeck M.E.: 1983, in Composition and Origin of Cosmic Rays, Ed. M.M. Shapiro, *Reidel Publ. Co.* p.65.
Woosley S.E., Weaver T.A.: 1983, *Astrophys. J.* 234, 651.

EXPLOSIVE DISRUPTION OF STARS BY BIG BLACK HOLES

Jean-Pierre Luminet
Groupe d'Astrophysique Relativiste - CNRS
Observatoire de Paris, 92195 Meudon
France

ABSTRACT. The fate of a star deeply plunging in a strong external gravitational field may be described in terms of the "pancake" compression process during which explosive nucleosynthesis may be triggered in the core of the star. Consequences on the dynamics of stellar disruption, fuelling of active galactic nuclei and production of heavy elements are briefly examined.

As extensively discussed throughout this conference, it is widely recognized that heavy elements in the Universe are in most cases injected into the interstellar medium by supernovae explosions, which signal the catastrophic end of the thermonuclear burning stages in massive stars. In this lecture I would like to draw the attention to the fact that explosive nucleosynthesis may also be triggered in the core of quite peaceful "ordinary" stars (even little massive ones) when a strong external gravitational field acts as a detonator.

Such events requiring the interaction between stars and a supermassive compact object are likely to occur in galactic nuclei. The idea that at least active galactic nuclei are powered by accretion of gas onto a massive central body has now gained wide acceptance in the astrophysical community; current basic models involve generally a giant black hole surrounded by a dense cloud of stars, in which diffusion of orbits makes some stars to penetrate deeply in the region where black hole's gravity dominates. There, various violent processes are expected to occur so as to release huge amounts of energy and radiation : the Octopus Model (figure 1) summarizes in a funny way some of these processes, such as the ablation of stellar atmospheres by the external radiation field (associated for instance with a big accretion disk) and the more or less violent destruction of stars by tidal stresses or by high velocity interstellar collisions.

In this lecture I shall discuss the more extreme form of stellar disruption that can occur in the vicinity of a black hole, namely the squeezing process by which a star, instead of being continously decompressed and broken into pieces such as clouds, filaments or doughnuts, undergoes first a transitory short phase of strong compression whose effects on the dynamics of the stellar gas may be of primary importance.

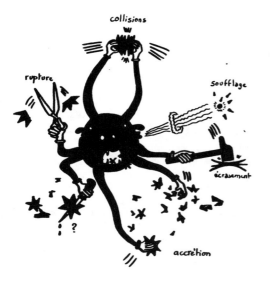

Figure 1 : The Octopus Model of Active Galactic Nucleus. Interactions between a giant black hole and individual stars such as tidal encounters, accelerated collisional disruptions, radiative ablation and squeezing are the major sources of gazeous accretion onto the black hole.

The "pancake" scenario of tidal disruption of a star by a large black hole was developed recently by Carter and Luminet (1982, 1983). Most of the previous studies of tidal disruption of stars had been based on the quite unrealistic incompressible model. Taking account of the high degree of compressibility of ordinary stars (for instance by assuming a polytropic equation of state), we have demonstrated that any star penetrating deeply inside the tidal radius of a large black hole was strongly compressed to a short-lived, flattened "pancake" configuration. This phenomenon is easily understood by a simple geometrical construction, as shown in fig.2. Any particle of the star enters in free-fall motion towards the black hole as soon as it penetrates inside the tidal radius, since the external tidal forces dominate rapidly internal forces such as pressure and selfgravitation. In the figure I have represented the (approximately parabolic) trajectory of a "North Pole" particle, lying in a plane passing through the black hole and intersecting the star's centre-of-mass trajectory slightly after the passage at periastron : at such a squeezing point the North Pole particle will have thus tendency to cross the orbital plane. Of course in the real situation the particles will not cross each other because when the volume of the star tends to zero the internal pressure will suddenly build-up and the star will bounce to a phase of expansion and ejection of its gas. Finally it is clear that the star passes through a fixed point at which it will look like a squeezed tube of toothpaste. This effect may also be viewed as a "rolling mill" effect or a pancake effect, in the sense that since the orbital velocity near the black hole (comparable with the velocity of light) is much greater than the internal sound speed, the squeezing can be considered as simultaneous all over the star. The approximate evolution of the shape of the star inside the tidal radius is shown in Figure 3. We recall that the instantaneous effect of a tidal field is extension along

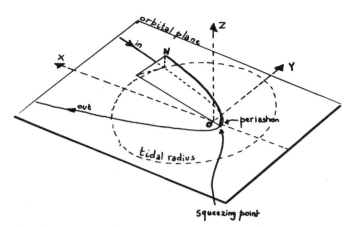

Figure 2 : Geometrical proof of the tidal squeezing.

the radial direction and compression along the two orthogonal directions ; now if we consider the effect of a *varying* tidal field such as occurs along an eccentric orbit, it is clear that the tidal field is always compressive along the direction orthogonal to the orbital plane (say the "vertical" direction), while along the two other principal directions (lying within the orbital plane) the compression and expansion tendencies will partially cancel out, due to the rotation of principal axes. This is the deep reason why the star will be strongly compressed *only* along the "vertical" direction.

One of the main results of a more detailed analysis of the pancake disruption of a star is that the amplitude of flattening at the instant of bounce is directly related to the maximum amplitude of the tidal field, or equivalently on the so-called penetration

Figure 3 : Shape of a star orbiting inside the tidal radius.
Left : projection into orbital plane.
Right : deformations along vertical direction.

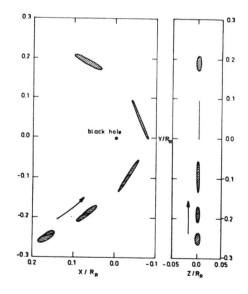

factor β, defined as the ratio between the tidal radius R_T and the periastron distance R_p. The most dynamical important quantity is the maximum compression velocity along the vertical direction, which is given merely by $v_{max} \simeq \beta v_*$, where v_* is the characteristic sound velocity in the star at spherical equilibrium (a few hundreds km/s for ordinary main-sequence stars). It follows that the maximum internal energy of the gas will be given by $U_{max} \simeq \beta^2 U_*$, where the value at equilibrium, U_*, is of order of 2 keV/nucleon in the core of main sequence stars. This traduces merely the fact that at the bounce the kinetic energy of vertical (free-fall) motion will be converted into heat. For non-degenerate equation of state, the maximum <u>pancake temperature</u> will thus be given by $T_{max} \simeq \beta^2 T_*$.

A more detailed specification of the equation of state, for instance non-relativistic polytropic, allow to fix the maximum density as well as the characteristic pancake duration (during which temperature remains maximum within a factor say two) : $\rho_{max} \simeq \beta^3 \rho_*$, $\Delta t_m \simeq \beta^{-4} \tau_*$ where τ_* is the internal timescale of the star at equilibrium (a few hundreds seconds), which from the definition of the tidal radius $R_T \simeq (M/\rho_*)^{1/3}$ turns out to be approximately the same as the orbital timescale for crossing the tidal sphere. Taking as an illustration thermodynamical conditions in a solar core, e.g. $T_* \simeq 10^7$ K, $\rho_* \simeq 100$ g/cc, $\tau_* \simeq 1000$ sec, we find for a penetration factor $\beta = 10$ the maximum pancake values $T_{max} \simeq 10^9$K, $\rho_{max} \simeq 10^6$g/cc, $\Delta t_m \simeq 0.1$ sec. Such conditions are obviously highly favourable for <u>explosive nucleosynthesis</u>. Nucleosynthesis in pancake stars will be discussed by B. Pichon (this conference). For my part, I will concentrate on some dynamical aspects and more general astrophysical implications of the pancake scenario.

To study in more detail the dynamics of violent disruption of a star, it is convenient to use as a first approximation the <u>affine star model</u> (Carter and Luminet, 1983, 1985 a,b), which allows compressibility, inhomogeneity, entropy generation by nuclear processes or viscosity and so on, but in which the layers of constant density are constrained to keep an ellipsoïdal shape. This model is likely to provide a good description of the behaviour of the main bulk of the star at least until the phase of bounce occurs (after what shock waves and significant non-linearities will be able to develop). The figure 4 shows the result of numerical integration of the affine equations of motion when the star's orbit is a parabola with penetration factor $\beta = 15$. The squeezing along the "vertical" direction is indeed very impressive. Still more impressive is the figure 5 which shows the result of integration of affine equations of motions in a <u>relativistic tidal field</u> (generated by a non-rotating black hole and described in the b ackground of the Schwarzschild curved space-time). The main difference with the previous newtonian treatment lies in the property that tight parabolic orbits in Schwarzschild space-time have a double point at finite distance. Geometrical arguments similar to those used above in euclidean space show that if the double point is located within the tidal radius, the particles of the star during the free-fall motion phase will have tendency to cross the orbital phane twice or more times, leading to the occurence of <u>several squeezing points and pancake flattenings</u> (Luminet and Marck, 1985). Furthermore we have shown that events involving two pancake compressions are fairly frequent (several % of all disruption

EXPLOSIVE DISRUPTION OF STARS BY BIG BLACK HOLES 219

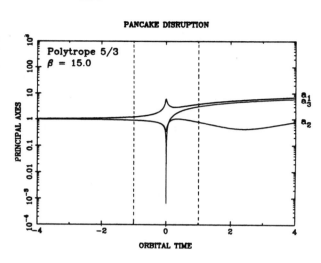

Figure 4 : Time variations of principal axes of the affine star model along parabolic orbit with penetration factor β = 15. Unit of time is the internal timescale $(G\rho_*)^{-1/2}$, origin is at periastron. Dashed vertical lines correspond to the passages of the star through tidal radius.

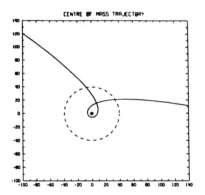
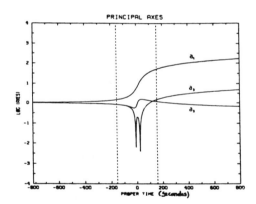

Figure 5 : The multi-pancake compression of a star in a Schwarzschild tidal field.

Left : trajectory of the centre of mass of the star with penetration factor β = 7 within the tidal radius (dashed circle).

Right : Time variation of principal axes. Unit of proper time is the second.

events). This has of course important consequences on nucleosynthesis, and since the multi-pancake compression by tidal field is a purely general relativistic effect, it might be used in the next future as a definite signature for the existence of large black holes in galactic nuclei.

Now it is time to say a word about the pancake effect in the context of <u>collisional</u> (instead of tidal) disruption of stars. In fact the relation $v_{max} \sim \beta v_*$ shows that the tidal encounter of a star with a black hole can be considered as a <u>collision of the star with itself</u> !. In the collision process, the ratio β between the relative collision velocity and the escape velocity v_* plays a role quite analogous to the penetration factor in the tidal encounter. In both cases, β appears as a <u>crushing factor</u> whose magnitude fixes the degree of maximum pancake compression. Since the pancake phenomenon associated with a tidal encounter requires high penetration inside the tidal radius without penetrating the black hole (inside which everything would be lost), it is expected to occur only in the vicinity of moderately massive black holes (say in the range $10^4 - 10^7$ M$_\odot$), i.e. in ordinary galactic nuclei, moderately active Seyfert nuclei, and perhaps some big globular clusters. On the contrary, the pancake phenomenon associated with interstellar collisions, which requires velocities widely in excess of ordinary stellar velocities, is expected to apply rather in giant elliptical and quasars which involve probably supermassive black holes (say $> 10^8$ M$_\odot$). Of course a crucial point is the frequency of pancake disruptions. It is easy to see that the mean number of stars suffering an event with crushing factor β divided by the mean number of disrupted stars (i.e. with $\beta > 1$) is of order β^{-1}. Thus, approximately 10% of all disrupted stars will likely undergo more or less violent pancake compression. Estimates for the Galactic Center give about 10^{-5} pancake events per year, while in quasars the rate might amount to 1 event per year. These rates may seem negligible compared with the rates of Supernovae, nevertheless I will conclude my talk by showing that the process of pancake disruption may have important astrophysical implications.

First of all, it is clear that a new scenario for active galactic nuclei arises. The previous scenario (see for instance Hills, 1975) was based on the fact that in the disruption of an incompressible body, the kinetic energy of the debris is of order of the initial binding energy of the star, therefore is negative. As a consequence the debris would remain bound to the black hole and would contribute entirely to feed the accretion process. In the present scenario, when compressibility and thermonuclear detonation are taken into account, the situation changes drastically since the (positive) nuclear energy injected may well exceed the (negative) binding gas energy ; in other words, a large fraction of ejected gas will be <u>unbound</u> to the black hole, and instead of accretion will take place an <u>outgoing wind</u> with velocities up to few thousands km/s.

An other obvious application, to which people at this meeting on nucleosynthesis will be presumably the most sensitive, is that the explosive disruption process provides <u>new sites for explosive nucleosynthesis and production of heavy elements.</u> For an individual event, the detailed thermonuclear processes depend sensitively on the crushing factor and on the initial chemical composition of the star (see Pichon, next talk), while global quantitative predictions will require averages over stellar statistics, mass spectrum of stars and black holes, differents types of chemical compositions and so on : this is clearly a hard, long-range work. I can however just sketch roughly two main trends. In population I stars, the dominant thermonuclear pancake processes are the accelerated proton captures on C, N, O ... seed elements (e.g. $(^{12}C(p,\gamma)^{13}N(e^+\nu)^{13}C$, etc...) which take place at about 10^8K, i.e. as soon as the crushing factor increases above 7. In small Population I & II stars, crushing factors greater than 15 may trigger the far more spectacular helium detonation in proton-rich medium. Production of γ-ray lines, neutrinos and proton-rich isotopes constitute of course the main astrophysical consequences of pancake nucleosynthesis.

All this suggests another, however more speculative, application of the pancake process in relation to the problem of primordial enrichment of population II stars. It is well known that the non-zero metallicity of the oldest population II halo stars observed until now raises a serious question about the origin of such metals. One is generally tempted to invoke a magic "population III" which does the job, for instance a first generation of massive stars which burnt and died quickly, leaving no trace except the right metals in the righ places. On the other hand, one would like to have also very numerous small primordial stars in order to relieve an other cosmological puzzle, namely the "missing mass" problem. The difficulty is that massive stars cannot be enough to solve the mass problem while the small stars cannot build heavy elements... unless they undergo accidentally the pancake process as described above ! Thus, the scenario of explosive disruption of a star in a strong external gravitational field may reconcile the idea of primordial small stars and primordial metal enrichment. But of course the price to pay is that you need a mixture of primordial massive black holes and small stars clustered around the holes. After all, this is no more speculative than the other ideas on the question, so I think that it deserves deeper investigation.

References

Carter B., Luminet J.P. (1982) Nature 296, 211
Carter B., Luminet J.P. (1983) Astron. Astrophys. 121, 97
Carter B., Luminet J.P. (1985a) Mon. Not. Roy. Astr. Soc. 212, 23
Carter B., Luminet J.P. (1985a) submitted to Astrophys. J.
Hills J.G. (1975) Nature 254, 295
Luminet J.P., Marck J.A. (1985) Mon. Not. Roy. Astr. Soc. 212, 56

NUCLEOSYNTHESIS IN PANCAKE STARS

Bernard PICHON
Groupe d'Astrophysique Relativiste
Département d'Astrophysique Fondamentale
Observatoire de Paris-Meudon
92195 Meudon Principal Cedex

ABSTRACT. When a star penetrates deeply into the tidal radius of a black hole, it undergoes a violent and large compression and heating which lead to examine the nuclear consequences, in particular the nucleosynthesis in stellar core. In this article, on the basis of the affine star model of Carter and Luminet, taking into account the variation of entropy, with the use of a new nucleosynthesis code including an optimized reaction network and a relativistic treatment of external gravitational field, we can confirm that the tidal disruption of stars provides really a new site of nucleosynthesis in galactic centers. This study has also pointed out the existence of a new nuclear flow called α- p as well as the possibility of tidal neutronization of a dwarf by a black hole.

By investigating the phenomenon of deep penetration of a star within the tidal radius of a black hole Carter and Luminet (see for example the contribution of J.P. Luminet in these procedings and references therein) have pointed out the possibility of a new site of nucleosynthesis. When the penetration factor β (defined as the ratio between the tidal radius and the periastron distance) is sufficently high (say β greater than 5.) the star undergoes compression to a short lived "pancake" configuration in which the density and temperature rise enough to burn some significant fraction of available nuclear material.

NUCLEOSYNTHESIS DURING PANCAKE PHASE.

Within the framework of the affine star model, the pancake phase can be caracterized by
- a high temperature $T_m \simeq \beta^2 T_*$ (eg $T9_m \geqslant 0.8$)

where $T9$ denotes the temperature in units of $10^9 K$, the subscript * refering to the initial central stellar value and the subscript m to the maximal central stellar value during the pancake phase.

- a large density $\rho_m \simeq \beta^3 \rho_*$ (eg $\rho_m \geqslant 10^5$ g/cm^3)

for a polytropic perfect gas of adiabatic index $\gamma = 5/3$

- a very short timescale $\Delta t_m \simeq \beta^{-4} \tau_*$, where τ_* is the internal timescale of the star given by $\tau_* = (G\rho_*)^{-\frac{1}{2}}$.

Such conditions are highly favourable for explosive nucleosynthesis, but such short timescales allow to neglect all weak decays during the pancake phase (but not later of course) and as a consequence there will be no possibility of combustion of hydrogen by pp chains or CNO cycles.

About the nucleosynthetic process, we have, in fact, a helium combustion in a proton rich medium with the following consequences :
- In addition to the triple alpha reaction, the principal reactions are (α,p) followed by (p,γ)
or (p,γ) followed by (α,p)
which are, in fact, equivalent to a (α,γ) reaction. Hence, only a very few hydrogen will be destroyed (mainly by $^{12}C(p,\gamma)$ $^{13}N(p,\gamma)$ ^{14}O or $^{25}Al(p,\gamma)$ ^{26}Si) during the pancake phase, thus it will still be available as fuel during post-pancake nucleosynthesis (if any).

- Thanks to this two-steps reaction, the rates of the equivalent $-(\alpha,\gamma)$ reactions are strongly enhanced, what is very important when the timescale of the nuclear process is driven by an external external dynamical effect.

- For the reasons above, such a process may be called the (explosive) α-p process.

For a temperature greater than $T_9 = 0.8$, the nuclear reaction network is shown in fig. 1.

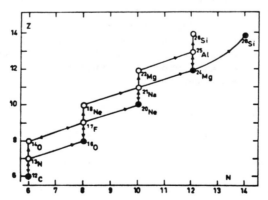

Figure 1 : the nuclear network used for the pancake nucleosynthesis calculation
white circles represent beta-unstable nuclei
black circles represent stable nuclei.

This network is "minimal" and correct to 2%, in the range $0.8 \leqslant T_9 \leqslant 3.5$ (that is relevant to our discussion), in the sense that we have systematically eliminated the reactions which contribute less than $0.02 \times (0.1)^2$ (the factor 0.1 warrants us for the possible inacurracy of the thermonuclear reaction rates) with respect of the other reactions for the creation or destruction of a given isotope. The thermonuclear reaction rates have been taken from numerous sources with priority to up-to-date rates. The two end-points of this network are ^{26}Si and ^{28}Si because beyond these isotopes, most of the required reaction rates are unknown, mainly because the major nuclear flow lies at the proximity of the line of proton instability of nuclei. In spite of this limitation, we have a general flow of nuclear species towards iron and/or nickel ; hence, we must consider, for example, the abundance of ^{26}Si as the sum of the abundance of itself and its different possible by-products.

For a while, let us fix our attention to the helium detonation (Pichon 1985a). We have plotted on fig. 2, for different values of initial central density, the fraction P of destroyed helium by the triple alpha reaction alone versus the penetration factor β (this leads, of course, to a lower bound for the nuclear energy release ε_N). The threshold, above which the dynamics of the disrupted star will be dominated by the helium detonation corresponds to the condition

$$\varepsilon_N + \varepsilon_g \geqslant \varepsilon_{3\alpha} + \varepsilon_g \geqslant 0$$

where $\varepsilon_g = -2$ keV per nucleon is the stellar self-binding energy and $\varepsilon_{3\alpha}$ is the energy released by the triple alpha reaction. Since we obtain $\varepsilon_{3\alpha} = 170$ P keV per nucleon, the energy condition is equivalent to $P \geqslant 0.0124$ as indicated by the dashed line of fig. 2.

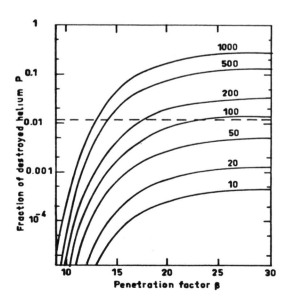

Figure 2 : fraction of destroyed helium by the triple alpha reaction alone versus the penetration factor. The curves are labelled with the initial central stellar density ρ_4 in g/cm^3. For the dashed line : see text.

This condition is fullfilled for a low mass star (central stellar density greater than 500 g/cm^3) with penetration factor greater than 13. As already been noticed by J.P. Luminet, when the condition $\varepsilon_N + \varepsilon_g$ positive is satisfied, ejected gas of the disrupted star will be unbound to the black hole and the standard scenario of accretion in AGN models will be modified.

As shown in figs 3 and 4, some isotopes processed in the pancake phase are unstable

e.g. ^{18}Ne with a period $T_{\frac{1}{2}} = 1.7$s)

^{22}Mg ($T_{\frac{1}{2}} = 4$s) or ^{26}Si ($T_{\frac{1}{2}} = 2.1$ s)

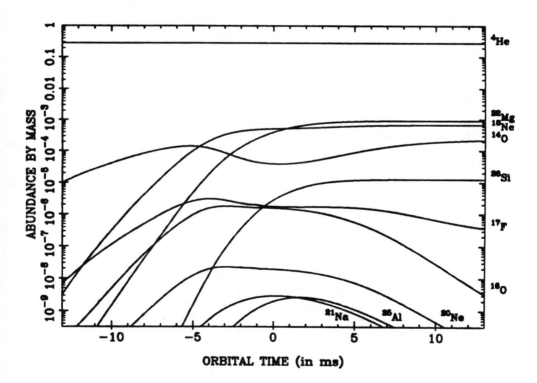

Figure 3 : time evolution of isotopic abundances during the pancake phase when the temperature exceeds $T_9 = 1$ for $\beta = 15$. The time axis has its origin at the instant of the maximum temperature.

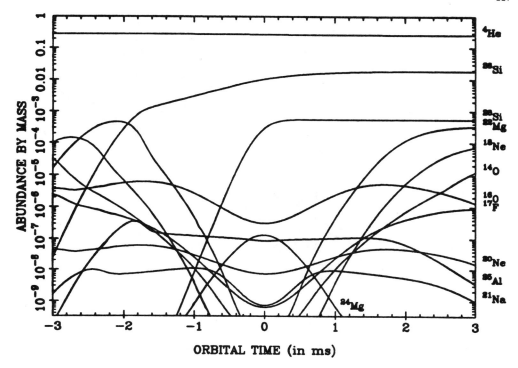

Figure 4 : the same as fig. 3 for β = 30.
We can notice that the results are, of course, very sensitive to the penetration factor.

By their decays, they provide a change in isotopic and chemical composition and an additional nuclear energy injected on longer time scale (on the order of τ_*). This leads to study the post-pancake nucleosynthesis.

POST-PANCAKE NUCLEOSYNTHESIS.

To do this, we need :
- An extended reaction network, including the α - p process, all weak decays of unstable isotopes (with the consideration of the emission of delayed particles for instance ^{20}Na($e^+\nu +\alpha$) ^{16}O, ^{21}Mg($e^+\nu +p$)^{20}Ne and ^{25}Si($e^+\nu +p$) ^{24}Mg) and, as a consequence, all subsequent reactions needed. Until A = 26, this new network contains 42 isotopes linked by only 44 thermonuclear reactions for a precision of 0.1 % (defined in the same way as previously) in the range $0.05 \leqslant T_9 \leqslant 3.5$ and since, beyond A = 26 or Z = 14 most of the required reaction rates are unknown, again, it is preferable in our opinion to stop there. The resulting network is shown in fig. 5.

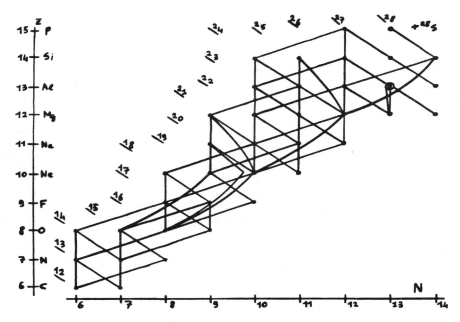

Figure 5 : the complete network used in the range $0.05 \leqslant T_9 \leqslant 3.5$.

Now we condider the variation of entropy, since the heat released by the exothermic reactions and decays is not negligible with respect to the internal energy of the star during the post-pancake phase. Although both contribute to the entropy generation, their respective parts are different. We can see on fig. 6 that

. for a little value of β (say $\beta = 10.$) there is only a little jump of entropy at the moment of the pancake. After, since in this case the part of beta-energy ε_β is greater or comparable to the nuclear energy ε_N, the entropy increases linearly and reaches its maximum after a few hundreds of seconds (when unstable isotopes have decayed). This saturation value is large because it is the beta-decay energy that contributes the more efficiently to the entropy generation.

. for a high value of β (say $\beta = 15.$ or $\beta = 30.$) there is a sudden and high jump of entropy that will drive formation of shock waves. But after, the entropy does not increase because the relative part of ε_β is well lesser than ε_N and it reaches quickly its saturation value.

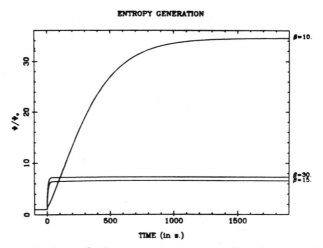

Figure 6 : time variation of the entropy generation factor Ψ/Ψ_* (where Ψ denotes the entropy by mass, the subscript $*$ refering to initial state value) for three penetration factors.

In all cases, the other principal effect of energy generation during post-pancake phase is to slow down the decrease of temperature in the cooling expansion phase as shown, for example, in fig. 7 for β = 15.

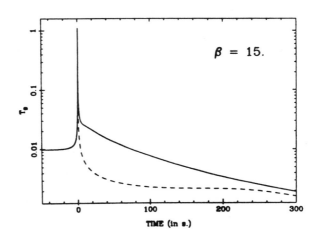

Figure 7 : time evolution of temperature (in units of 10^9 K) the dashed curve represents the adiabatic evolution of temperature whereas the solid curve represents the non adiabatic more realistic evolution of temperature.

This effect is, of course, very favourable for nucleosynthesis because the temperature remains greater than $T_9 = 0.02$ during 20 s instead of less than 1 s (in the case $\beta = 15.$).

When the temperature has decreased below $T_9 = 0.01$, after, few hundreds of seconds, the principal synthetised isotopes are shown in the table below :

β	$\epsilon/|\epsilon_g|$	^{24}Mg	^{25}Mg	^{27}Al	^{28}Si
15	5.5	0.011	0.0047	0.046	-
20	23.5	0.012	0.0038	0.052	-
30	40.	0.014	0.0025	0.050	0.008

The abundances are given by mass for three penetration factors (^{24}Mg and ^{27}Al may be overpruducted and hence observable). In addition in this table, the ratio between the total energy released ($\epsilon = \epsilon_N + \epsilon_\beta$) and the absolute value of self-binding energy of the star has also been indicated. One can remark that for sufficiently high β a large fraction of stellar debris will be blown away.

CONCLUSIONS AND WORK IN PROGRESS.

The results presented here show the existence of a new site of nucleosynthesis and hence might shed new lights on the isotopic enrichment of the interstellar medium near galactic centers. But quantitative results and predictions require averages over stellar statistics mass spectrum of stars (and black holes) and so on (see previous lecture by J.P. Luminet).

Besides this, we have already considered the inclusion of a relativistic treatment of the motion of the star around the black hole for which multi-pancake. compressions appear, separated by a few seconds. These successive compressions and consequently star reheatings (see fig. 8) have, of course, favourable effects on nucleosynthesis (Pichon and Luminet 1985).

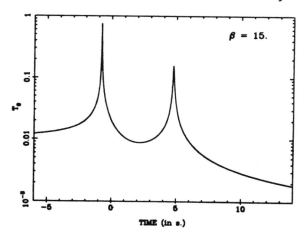

Figure 8 : time evolution of temperature in the case of stellar motion around Schwarzschild black hole. The origin of time scale is taken at the periastron.

Further work program contains the consideration of the effects of shock waves formation by a more performant hydrodynamical code as well as the possibility of theoritical calculation of unknown reaction rates to extend the present network.

In addition we have considered the influence of the use of an electronic semi-degenerate semi-relativistic equation of state to study the pancake effects on a white dwarf or a black dwarf. It appears that the pancake is now caracterized by

$T_m \simeq \beta^4 T_*$, $e_m \simeq \beta^6 e_*$ and $\eta_m \simeq \beta^{-2} \eta_*$ where η is the degeneracy parameter of the central core of the dwarf (the subscripts m and * refering still respectively to the extremum and initial values) (Pichon 1985b). We notice that electron degeneracy is strongly removed, leading with the aid of photodesintegration of stellar material at high temperature, to the neutronization of the pancake.

REFERENCES :

PICHON B. (1985a) Astron. Astrophys. 145 (1985) 387-390
PICHON B. (1985b) in preparation
PICHON B. and LUMINET J.P. (1985) Preprint, Observatoire de Meudon

OPTICAL SUPERNOVA REMNANTS

I.J. Danziger
European Southern Observatory
Karl-Schwarzschild-Str. 2
D-8046 Garching bei München, F.R.G.

ABSTRACT. A review of some unsolved problems in supernova remnant research is presented.

Before going directly to observational work which provides evidence for nucleosynthesis in SN events, it is perhaps worth noting some of the more general properties of supernova remnants and stressing where there are gaps in our understanding.

1. MORPHOLOGY OF REMNANTS

For those unfamiliar with the details of optical SNR, I want to begin by presenting my version of how one might classify optical SNR by their morphology. The following table summarizes this morphology with more detailed comments following.

Optical Morphology of SNR

1. Crab-like: filled centre, non-thermal optical continuum.

2. Filamentary
   a) Long thin filaments (Vela).
   b) Flocculi; geometry uncertain (quasi stationary flocculi in Cas A).
   c) High velocity blobs; roughly circular in shape very small (high velocity component of Cas A).
   d) Rings; usually of high velocity, not always a continuous entity (oxygen-rich filaments in N132D).

3. Compact stellar components.
   a) Pulsars (Crab).
   b) Binaries (W50 with SS433).
   c) Non-variable X-ray sources (RCW103).
   d) γ-ray burst sources (N49). This last association must at the moment be considered very uncertain.

1. The number of Crab-like remnants is increasing. N157B in the LMC may be another example. Apart from a non-thermal continuum these comparatively young objects contain elongated filamentary structures reminiscent of older SNR. In analysing the spectra of such objects for abundance effects one should be aware that both shock-heating and photo-ionization by a non-thermal source are possibly playing a role. It is particularly important to pursue studies of objects such as N157B as far as possible because it lies near the centre of the 30DorB HII region complex, which suggests it originates from a fairly massive star, a point of some debate for the Crab itself.

2.a) It seems increasingly likely that many of the long thin filaments seen in old SNR such as Vela and the Cygnus Loop are not rope-like structures, but are thin sheets seen edge-on. This can be tested by spectrophotometrically measuring the emissivity of these filaments. It may seem remarkable that filaments of this kind can maintain their identity and orientation to the line-of-sight over vast distances implied for example in the Cygnus Loop. Other more diffuse emissions seen in SNR have characteristic dimensions in the line-of-sight much less than those in the plane of the sky.

2.b) Nothing very quantitative is known about the flocculi seen in Cas A as the low velocity filaments, nor concerning those in Puppis A. Nevertheless the characteristic appearance is quite different from the elongated filaments discussed above. This very likely reflects a difference in the nature of the ISM into which the shock has penetrated - for example the presence or absence of higher density clouds embedded inside the lower density ISM.

2.c) The most striking characteristic of these blobs is that they are variable in brightness with time. Variability has always been observed in the sense of brightness incresing in Cas A and Kepler, although in the case of Kepler the filaments are apparently low-velocity nitrogen-enriched filaments near the limb of the SNR.

2.d) The only objects for which there is suggestive evidence of a ring structure from both a morphological and dynamical point of view, are the apparently young oxygen-line dominated remnants seen in our galaxy and in the Magellanic Clouds. One is also seen in the galaxy NGC 4449 but since it is not resolved it is impossible to say much about its structure.

3. Discussion of the origin and role of compact objects in SNR is beyond the scope of this talk. Although theoretical work may be on firmer ground there is little in the current observational work to show that compact objects reside preferentially in SNR of a given type and therefore originate, for example, in Type II rather than Type I SN. Although we know that the pulsar in the

Crab (and Vela) and SS433 in W50 are radiating prodigious amounts
of energy, it is unclear to what extent they affect what we see
as radiation from the filaments. Only in the case of Vela is it
clear that at the present epoch any effects must be small. The
nature of the non-variable X-ray sources is even less clear
because optical counterparts have not been identified.

The γ-ray burst source lying within the shell of the SNR N49 in
the LMC remains a mystery for the same reason. The association
with N49 is suggestive but it is worth remembering that current
estimates of its projected position place it ~5 parsecs from the
centre of the SNR, and therefore imply a very high proper motion
if it originated with the SN near the centre of the remnant.

2. HISTORICAL SN AND THEIR REMNANTS

If we consider the remnants originating from historical SN there
are in increasing order of youth only six. They are, together with the
associated remnant, as follows: SN 185 (RCW86), SN 1006 (P1459-41), SN
1054 (Crab), SN 1181 (3C58), SN 1572 (Tycho), SN 1604 (Kepler). Cas A
could also be included if it had been recorded by Flamsteed. In any
case the proper motions and the large radial velocities imply that it
resulted from a SN about 330 years ago.

All of these SNR, because of their linear size and known age
should have expansion velocities of several thousand km/sec or more.
The Crab and Cas A are the only historical remnants where radial
velocities seen in individual filaments are comparable to the implied
expansion velocity. In the case of Cas A we are probably seeing high
velocity filaments composed of material from the exploding star; with
the Crab the origin of the filaments is not so clear-cut, although
they contain at least an admixture of helium-enriched material.

A dynamical model of the Crab proposed by Clark et al. (1983)
from fairly complete velocity mapping, suggests that filaments tend to
lie on the inner or outer surface of an expanding thick shell (or it
could be 2 very thin shells). The velocity difference between the 2
surfaces is of the order of 270 km/sec. So far theoretical models have
not addressed these observations.

It has always been thought that Tycho, Kepler and SN 1006 all
originated from Type I SN. Yet the remnant of Kepler is completely
different from the other two. This may well be a result of different
environments surrounding the original exploding stars. Since the 3
remnants are situated reasonably far from the galactic plane, it has
not been obvious that this difference of the environment could be a
property of the unperturbed ISM but rather a difference in the history
of the mass loss process from the progenitor stars. But this reasoning
also leads to a conclusion that not all progenitors of Type I SN are
identical.

Although high velocity filaments are not a striking feature of young SNR in general there is a particular class of object, to which Cas A belongs, that is characterized by high velocities and abundance anomalies. These will be discussed in more detail later.

3. EXPANSION OF SUPERNOVA REMNANTS

In the past few years with the completion of surveys at X-ray, optical and radio wavelengths, particularly in the Magellanic Clouds, aimed at detecting SNR down to certain well defined flux limits, one has had the opportunity of investigating various statistical relationships. Here we do not describe in detail the well known 4 phases of SNR evolution proposed by Woltjer (1972). We want to consider only the first 2 phases, namely: 1. The free expansion phase, where material ejected from the star is moving outwards undecelerated by the interstellar medium. 2. The adiabatic or Sedov phase in which swept-up interstellar material is beginning to decelerate the outward expansion associated with material ejected from the star.

Under idealized conditions, assuming a unique density of the ISM and explosion energy, one would expect that for a population of SNR of different ages the following relationships would apply. If all SNR were in the free expansion phase the radius $R \propto t$ the age, and the accumulated number of remnants with radii $< R, N(<R) \propto R$. Whereas in the adiabatic phase

$$R \propto t^{2/5} \text{ and } N(<R) \propto R^{5/2}.$$

This latter relationship between N and R provides the opportunity to test these ideas if one has some sort of complete sample with accurately known radii. Such a sample exists in the Magellanic Clouds.

Long (1983) with X-ray data, Mills et al. (1983) with radio data and Mathewson et al. (1983) with optical data all found a relationship much less steep than the $R^{5/2}$ law, which led them to conclude that most of the remnants seemed to be in a dynamical state much closer to free expansion.

However, Fusco-Femiano and Preite-Martinez (1984) and Hughes et al. (1984) have shown that if one considers the more realistic possibility that there is a range of explosion energies and a range of interstellar densities into which SNR evolve, one has a range of populations of SNR each with a different cut-off in luminosity threshold for detectability. The combination of these different populations can then give rise to an N,R relationship close to the observed one. Hence the idea that most remnants are in the adiabatic phase is preserved. This does not necessarily apply to the very oldest or the very youngest SNR observed with this type of modelling.

Recently Danziger and Leibowitz (1985) have found that for a selection of SNR in the LMC the density Ne is not proportional to D^{-3}

(where D is the diameter), a relationship expected for SNR in the
adiabatic phase assuming a constant explosion energy. It remains to be
demonstrated that the observed relationship Ne $\propto$ D^{-1}, can be modelled
assuming the adiabatic phase and a realistic range of energy and
density. In principle, one could also derive a SN rate from the best
fit to the data, a number important among other reasons for the study
of chemical evolution of the LMC.

4. ABUNDANCE EFFECTS IN SNR

The following table provides a summary of information on SNR that
might identify where elements heavier than hydrogen could be produced.

SNR	Hist.	RV km/sec		Optical Abundance Peculiarities	X-Ray Abund.
CRAB	✓	1800-2000 Large PM		He variable × 5 N,O,Ni,S low-normal C normal (IR, UV)	-
Cas A HVF Q	(✓)	6000 Large PM	(ring)	O(50-70) S,Ar,Ca var. N × 10 He × 2-3	S,Ar,Ca 2-3×
LMC 132D G292.0+1.8 LMC 0540-69.3 SMC 0102-72.3 NGC 4449		2200 1600 1500 2000 3500	(ring) (ring)	O × 30 Ne O O O O	O/Fe×3
Puppis A LVF HVF LMC 0525-66.0 Kepler 3C 58	✓ ✓	300 300		N × 40 var. 1500 N,O,S N × 3-5 N	O,Ne O Si
RCW 86 RCW 102	✓	200 300		N × 3-4 N × 3-4	
Tycho SN 1006 4 LMC SNR	✓ ✓	200-1800 Large PM 200 Large PM (200)			Si

We have presented results obtained at both optical and X-ray wavelengths. It should be noted that in general, with one or two exceptions, the elements observable in the two wavelength ranges do not overlap very much and therefore reassuring consistency checks are lacking. It should also be noted that there is on-going discussion concerning the reliability of abundance determinations at X-ray wavelengths because of uncertainties concerning equilibrium versus non-equilibrium modelling. Non-equilibrium models in which the electron temperatures can differ from iron temperatures may cause huge differences in the derived abundances.

The table provides information on the observed radial velocities of filaments, and notes elements which seem to have a non-solar value. One general very significant observation can be made. Where oxygen overabundances exist, they are always associated with high velocity filaments, whereas nitrogen overabundances are confined to low velocity filaments. This in the past led to the idea that nitrogen enriched filaments were formed from circumstellar material ejected at comparatively low velocities prior to the SN explosion and subsequently shocked by the blast wave. The high velocity material is then the oxygen enriched material from the interior of an evolved star of mass 15-25 $M_\odot$.

In some cases such as Cas A both types of filaments exist (Chevalier and Kirshner, 1978). In other cases, for example N132D, oxygen-rich material exists without the nitrogen enrichment in low velocity filaments. Kepler on the other hand seems to be a case where nitrogen enrichment occurs but where high velocity filaments have not been detected.

The Crab. This object continues to provide puzzles. We have already noted the dynamical structure where there is evidence of a pattern of helium enrichment in which the enriched filaments tend to lie on the inner surface of a thick shell, while the more normal filaments lie on the outer surface. There have been claims in the past for underabundances of N, O, Ne, S but the most recent results of Péquignot (1983) who has used photoionized models, seem to suggest no significant under- or over-abundance of these elements. The UV spectra investigated by Davidson et al. (1982) indicate nothing very abnormal with the carbon abundance as does the IR spectrum of [CI] discussed by Dennefeld and Péquignot (1983). These latter authors have, together with Henry (1984), discussed the possibility that nickel is overabundant in the Crab. This possibility is especially interesting because of the possible production of iron through the nickel-cobalt-iron decay proposed to account for the energy release in Type I SN. At present, the reality of this anomaly in nickel must be considered uncertain both because of uncertain identification of lines and uncertain atomic collision strengths.

Cas A. This object can be considered together with other oxygen-rich remnants such as N132D in the LMC, SMC 0102-73.3, and the object in NGC 4449. There are hints of overabundances of other heavier elements that might be produced in regions of carbon and oxygen burning in a 20 $M_\odot$ star. The emission lines of most of these heavier elements appear in the spectra of much older SNR where one would not expect to see enrichment from the exploding star. It is therefore highly desirable to obtain independent evidence for the physical conditions in the regions where these lines appear. So far this has proved elusive.

A pattern that is emerging with these oxygen-rich SNR is the evidence that the filaments have a ring-like structure which is expanding at high velocity. This has been seen in the morphology and velocity pattern in N132D (Lasker, 1980), SMC 0102-72.3 (Tuohy and Dopita, 1983; Danziger, 1983) and may be apparent in Cas A from X-ray morphology. The case for G 292.0+1.8 is not nearly so convincing (Braun et al., 1983). These observations are sometimes claimed to be consistent with a model of a rotating massive star which has exploded as a SN (Bodenheimer and Woosley, 1983) and in which the hydrodynamical flow pattern produces oxygen rich material ejected and expanding outwards in the equatorial plane. At the moment the model calculations span a very small unit of time, and one needs to go to time scales long enough to see whether this configuration would be maintained over distances of several parsecs, which would correspond to what is actually observed.

Observationally, however, the matter is not yet completely without ambiguity. In some cases to interpret the data, one has to resort to the morphology of a highly warped ring. At a certain point it is not easy to discriminate observationally between a highly warped ring and an unfilled shell of which there are many examples.

Puppis A. This object is interesting because of the extremely high nitrogen abundances found in some filaments (40 × solar). In unpublished work Danziger and Dopita also found suggestions of overabundances of oxygen and helium. Recently Winkler and Kirshner (1985) have found oxygen enriched filaments with high velocities (~1800 km/sec). This reinforces an older idea that Puppis A is a more evolved version of Cas A, both of which derived from WN stars which have suffered mass loss before the SN explosion. Of all the SNR in the Magellanic Clouds the only one to show nitrogen overabundance effects is LMC 0525-66.0 (Danziger and Leibowitz, 1985). It seems to show other anomalies as well and may be a yet older counterpart of the Cas A phenomenon.

Kepler. This seems to be a genuine case where nitrogen enrichment has occurred without any manifestation of other overabundances. It remains to be seen whether this is consistent with expectations for a Type I SN. Most other apparent overabundance effects, particularly for nitrogen such as in RCW 86 and RCW 103, might just as easily be explained by abundance effects in the surrounding interstellar medium but varying from one to another because of the galactic abundance gradient.

Balmer Line Dominated SNR. At first these objects, such as Tycho, SN 1006 and 4 objects in the LMC (Tuohy et al., 1983) seem to have severe underabundance anomalies. However, their spectra have been successfully interpreted by Chevalier and Raymond (1978) to be the result of a fast collisionless shock propagating into partially neutral material where charge exchange is effective.

5. THE PROBLEM OF IRON IN SNR

The theoretical models for Type I SN seem to demand the production of considerable amounts of iron (0.4-0.8 $M_\odot$). There are pieces of evidence that iron should be present in large quantities. For example, blended emission lines of FeII and FeIII fit some features of the early spectra of Type I SN (Kirshner and Oke, 1975). The broad absorption features seen in spectra of Type I SN are explained by iron enriched material expanding at high velocities (Branch et al., 1985). There is also the requirement of the decay of Ni^{56} to Co^{56} and Fe^{56} proposed by Colgate and McKee (1969) and elaborated by Axelrod (1980) to produce the exponential light curve in Type I SN. Finally, Meikle et al. (preprint 1985) seem to have detected strong lines of [FeII] at 1.644 microns in the SN 1983n in M83, after it had become optically thin in the IR spectral region. Their modelling then requires ~0.3 $M_\odot$ of iron to be present. This last result is arguably the most convincing so far.

Nevertheless large overabundances of iron have never been apparent in real SNR, young or old. 0.5 $M_\odot$ of iron distributed through the volume of the Kepler SNR (if it originated from a genuine Type I SN) should be apparent as a huge overabundance. It is not visible at 10^6-10^7 °K in the X-ray region (nor with the FeXIV 5303 line); it is not visible at 10^5-10^6 °K region with optical lines of FeVII → FeX being accessible; and it is not visible at 10^4 °K in the visible filaments although there are many available [FeII], [FeIII] and [FeV] lines accessible to observation.

Is it possible that iron has been hidden in a temperature domain that is inaccessible at the present time? This would be a surprising coincidence if that temperature domain lay betwen 10^4 and 10^7 °K. Therefore one might expect it to be very hot or very cold. If it is very cold one might have hoped for a situation in a SNR where a reverse shock could have heated it. Until now the claim that FeII was seen in absorption with a large high velocity spread in the spectrum of a star behind the SN 1006 remnant has not so far been substantiated with better S/N UV spectra. The quandary remains.

REFERENCES

Axelrod, T.S.: 1980, Proceedings of the Texas Workshop on Type I SN, ed. J-C. Wheeler, Univ. of Texas.

Bodenheimer, P.B., Woosley, S.E.: 1983, Astrophys. J. __269__, 281.

Branch, D., Doggett, J.B., Nomoto, K., Thielemann, F. K.: 1985, Astrophys. J., in press.

Braun, R., Goss W.M., Danziger, I.J.: 1983, IAU Symposium 101, Supernova Remnants and Their X-Ray Emission, ed. I.J. Danziger, P. Gorenstein. Reidel, Dordrecht.

Chevalier, R.A., Kirshner, R.P.: 1978, Astrophys. J. 219, 931.

Chevalier, R.A., Raymond, J.C.: 1978, Astrophys. J. Lett. 225, L27.

Clark, D.H., Murdin, P.G., Wood, R., Gilmozzi, R., Danziger, I.J., Furr, A.W.: 1983, Mon. Not. Roy. astr. Soc. 204, 415.

Colgate, S.A., McKee, C.: 1969, Astrophys. J. 157, 623.

Danziger, I.J., Leibowitz, E.: 1985, Mon. Not. Roy. astr. Soc., in press.

Danziger, I.J.: 1983, IAU Symposium 101, Supernova Remnants and Their X-Ray Emission, ed. I.J. Danziger, P. Gorenstein. Reidel, Dordrecht.

Davidson, K., Gull, T.R., Maran, S.P., Stecker, J.P., Fesen, R.A., Parise, R.A., Hassel, C.A., Kafatos, M., Trimble, V.L.: 1982, Astrophys. J. 253, 696.

Dennefeld, M., Péquignot, D.: 1983, Astr. Astrophys. 127, 42.

Fusco-Femiano, R., Preite-Martinez, A.: 1984, Astrophys. J. 281, 593.

Henry, R.B.: 1984, Stellar Nucleosynthesis, eds. C. Chiosi and A. Renzini, Reidel, Dordrecht.

Hughes, J.P., Helfand, D.J., Kahn, S.M.: 1984, Astrophys. J. Lett. 281, L25.

Kirshner, R.P., Oke, J.B.: 1975, Astrophys. J. 200, 574.

Lasker, B.M.: 1980, Astrophys. J. 237, 765.

Long, K.S.: 1983, IAU Symposium 101, Supernova Remnants and Their X-Ray Emission, ed. I.J. Danziger, P. Gorenstein. Reidel, Dordrecht.

Mathewson, D.S., Ford, V.L., Dopita, M.A., Tuohy, I.R., Long, K.S., Helfand, D.J.: 1983, IAU Symposium 101, Supernova Remnants and Their X-Ray Emission, ed. I.J. Danziger, P. Gorenstein. Reidel, Dordrecht.

Meikle, W.P.S., Graham, J.R., Andrews, P.L.: 1985, preprint.

Mills, B.Y.: 1983, IAU Symposium 101, Supernova Remnants and Their X-Ray Emission, ed. I.J. Danziger, P. Gorenstein. Reidel, Dordrecht.

Péquignot, D.: 1983, IAU Symposium 101, Supernova Remnants and Their X-Ray Emission, ed. I.J. Danziger, P. Gorenstein. Reidel, Dordrecht.

Tuohy, I.R., Dopita, M.A., Mathewson, D.S., Long, K.S., Helfand, D.J.: 1983, IAU Symposium 101, Supernova Remnants and Their X-Ray Emission, ed. I.J. Danziger, P. Gorenstein. Reidel, Dordrecht.

Tuohy, I.R., Dopita, M.A.: 1983, IAU Symposium 101, Supernova Remnants and Their X-Ray Emission, ed. I.J. Danziger, P. Gorenstein. Reidel, Dordrecht.

Winkler, P.F., Kirshner, R.P.: 1985, Astrophys. J. Lett., in press.

Woltjer, L.: 1972, Ann. Rev. Astr. Astrophys. $\underline{10}$, 129.

LIGHT CURVES OF EXPLODING WOLF-RAYET STARS

S. Cahen[1], R. Schaeffer[2], and M. Cassé[1]
[1]Service d'Astrophysique, Centre d'Etudes Nucléaires de Saclay
France.
[2]Service de Physique Théorique, Orme des Merisiers, France.

ABSTRACT. The removal of the extended hydrogen envelope surrounding the dense core of preexplosive stars by stellar winds or mass transfer onto a companion star, obviously, has consequences on the optical manifestation of their final outbursts. The stripped models lead to faster and dimmer optical displays, with, of course, a lack of hydrogen lines. The effective temperature near maximum light is however predicted to be rather similar to that of classical Type II supernovae. Our calculations show that in principle it may be possible to distinguish Wolf-Rayet supernovae triggered by core collapse from those induced by pair production. Unfortunately, as shown by Maeder and Lequeux (1982), only 1 out 3 to 7 supernovae would be of this kind. These low statistics combined with the low luminosity and short duration of these events render their detection quite difficult with the present technology.

1. INTRODUCTION

Type II supernovae light curves owe their distinctive features - the presence of hydrogen lines and their long and bright plateau - to the fact that the dense core of the exploding star is buried in an extended envelope of the red supergiant type (for reviews, see Wheeler, 1981, and Trimble, 1982). A presupernova radius of the order of 1000 $R_\odot$ is required to explain these curves (Arnett 1971, Falk and Arnett 1973, Chevalier 1976a, Woosley and Weaver 1980). This huge envelope minimizes adiabatic losses during expansion, ensuring that internal energy is not totally converted into kinetic energy by expansion before it can be radiated (i.e. before the expanding medium becomes optically thin)(Colgate and White 1966, Chevalier 1976a, Woosley and Weaver 1982). Chevalier (1976b), and Woosley and Weaver (1982) already pointed out that removing the star's envelope would give dimmer light curves with shorter duration.
    It seems timely to elaborate on these exploratory calculations since consistent theories on stellar evolution in presence of mass loss and Roche lobe overflow in close binaries allow a fair estimate of the characteristics of the stripped object condamned to explode (e.g. Maeder 1984, de Loore et al. 1986, Prantzos et al. 1986, this conference).

2. MAIN EFFECTS OF MASS LOSS ON THE PRESUPERNOVA CONFIGURATION

Evolutionary models of mass losing stars yield contrasted results in comparison to conservative models.
i) Mass loss, specially during the Wolf-Rayet stage, leads to a strong reduction of the stellar radius. Instead of becoming red-supergiants ($R \simeq 10^{14}$ cm), stars with mass $\geqslant$ 45 $M_\odot$ become hot and compact objects ($R \simeq 10^{10}$ cm) (de Loore et al. 1986, Prantzos et al. 1986, this conference). Roche lobe overflow in massive close binaries (Doom et al., 1984) involving stars with $M_{ZAMS} \geqslant$ 15 $M_\odot$, and mass loss during the red-supergiant stage (Maeder 1982) affecting stars with 20 $M_\odot \lesssim M_{ZAMS} \lesssim$ 45 $M_\odot$ should also provide Wolf-Rayet stars.
ii) The stellar mass suffers continuous reduction. At the end of evolution the mass is only a fraction of the original mass. This fraction tion is unfortunately still uncertain. It depends critically for instance, on the moment at which the huge mass loss rate characterizing WR stars ($\dot{M} = 3.10^{-5}$ $M_\odot$ yr^{-1}) sets in (Cahen et al., in preparation).
iii) At helium exhaustion massive stars adopt a very simple configuration: a compact core in which oxygen predominates is capped with a thin shell of He, C, and O. After helium burning mass loss may become insignificant due to the accelerated evolution preceding explosion, for which two mechanisms are known.

For oxygen core masses above a critical mass M_{OC}, stars become unstable against electron-positron pair production (Fowler and Hoyle 1964). Recent estimates yield 48 $M_\odot$ for M_{OC} (Ober et al., 1983), corresponding to a ZAMS mass $M_i \simeq$ 90 $M_\odot$ (Prantzos et al., 1986, this conference). According to recent calculations the ^{56}Ni production seems to be negligible in this kind of events (Woosley and Weaver, 1982; Ober et al., 1983); and the energy released by the decay of this radionuclide should play a minor role in the late light curve, at variance with type I supernovae.

For $M_i \lesssim$ 90 $M_\odot$, massive stars undergo a core collapse following photodisintegration of iron and electron capture in the core. This collapse should be followed by a bounce triggering a shock wave which expells the mantle of the star. The energy in the shock is thought to be of the order of 10^{50} to 10^{51} ergs quite independently of the initial mass of the star.

3. TIME SCALES OF THE PROBLEM

We have extended Arnett's (1980, 1982) approach to take into account electronic recombination in the expanding plasma. To keep the model general we also include a variable contribution of radioactive ^{56}Ni. The detailed analytical model will be presented elsewhere (Schaeffer, Cassé and Cahen 1985, in preparation). In this short communication it is sufficient to indicate the relevant time scales entering the problem of light diffusion in an homologously expanding medium. These could be slightly modified in the final version of this work.

(1) The expansion time scale of the ejecta: τ_h
(2) The light (heat) diffusion time scale : τ_d
(3) Provided recombination can be neglected, the rate at which the heat content of the supernova is exhausted is given by $\tau_m = (2\tau_h\tau_d)^{1/2}$, which determines the width of the light curve.
(4) Recombination of free electrons on the last atomic orbits sets in at a time τ_ϵ, where the average temperature of the ejecta is comparable to the first ionization potential ϵ of the most abundant element, but is really achieved at a much later time of the order of $\tau_r \simeq 30\tau_\epsilon$. In the mean time recombination provides additional energy. At completion of recombination the opacity drops abruptly and the medium becomes transparent. Very rapidly, the energy content of the star is then radiated away. Practical expressions of the relevant time scales are:

$$\tau_h = 25 \left(\frac{E_o}{10^{51} \text{ergs}}\right)^{-1/2} \left(\frac{M_o}{10 M_\odot}\right)^{1/2} \left(\frac{R_o}{10^{10} \text{cm}}\right) \text{ sec} \quad (4)$$

$$\tau_d = 1.0 \; 10^{12} \left(\frac{M_o}{10 M_\odot}\right) \left(\frac{R_o}{10^{10} \text{cm}}\right)^{-1} \text{ sec} \quad (5)$$

$$\tau_m = 7 \; 10^6 \left(\frac{E_o}{10^{51} \text{ergs}}\right)^{-1/4} \left(\frac{M_o}{10 M_\odot}\right)^{3/4} \text{ sec} \quad (6)$$

$$\tau_r = 5 \; 10^6 \frac{Z^{1/3}}{A^{1/4}} \left(\frac{E_o}{10^{51} \text{ergs}}\right)^{-1/2} \left(\frac{M_o}{10 M_\odot}\right)^{3/4} \left(\frac{E_{Tot}}{10^{51} \text{ergs}}\right)^{1/16} \left(\frac{R_o}{10^{10} \text{cm}}\right)^{1/16} \text{ sec} \quad (7)$$

where E_o, M_o and R_o are the energy deposited by the explosion and the initial mass and radius, the total energy input is $E_{Tot} = E_{Th}(0) + E_o$, $E_{Th}(0) = \int_o^{R_o} aT^4 d^3r$ being the thermal energy stored in the star prior to explosion. Z and A are the mean atomic and mass numbers of the most abundant element in the expanding medium.

In this simplified scheme the fundamental parameters influencing the light curve of massive supernovae are:
i) the chemical composition of the bulk of the star through the energy released by recombination, which is much higher for heavy elements than for hydrogen,
ii) the initial radius of the exploding object (through all scales of the problem),
iii) the mass of the exploding star (through all time scales),
iv) the energy liberated by the explosion E_o (through τ_h, τ_m, and τ_r), and the thermal energy E_{Th} (through $E_{Tot} = E_{Th} + E_o$).

The main difference between ordinary type II supernova (dominated by hydrogen) and Wolf-Rayet supernova (dominated by oxygen) lies in fact in the total recombination energy (13.6 eV per proton against $\simeq$ 2 keV per O nucleus.

4. TYPICAL EXAMPLES

Table I shows the initial parameters of the two models computed based on presupernova structures obtained from evolutionary codes including mass-loss (Case A: 8 M_θ presupernova originating from a 40 M_θ star; Case B: 68 M_θ presupernova originating from a 100 M_θ star).

TABLE I

INITIAL PARAMETERS AND CHARACTERISTIC TIME SCALES OF THE EXPLODING MODELS

M_i/M_θ	Type[c] of explosion	M_0 (M_θ)	R_0 (cm)	E_0 (ergs)	E_{th} (ergs)	τ_h (s)	τ_r (s)	τ_m (s)	τ_d (s
40[a]	CC	8	$1.9\ 10^{10}$	10^{51}	$2.3\ 10^{51}$	41	$4.5\ 10^6$	$6.0\ 10^6$	$4.2\ 10^{11}$
100[b]	PP	68	$1.5\ 10^{10}$	10^{52}	$8\ 10^{52}$	30	$8.7\ 10^6$	$1.7\ 10^7$	$4.5\ 10^{12}$

(a) Model kindly supplied by André Maeder.
(b) Model obtained in the framework of the Brussels-Saclay collaboration (see Cassé 1984, Doom 1984, de Loore et al. 1986, Prantzos et al. 1986).
(c) CC : core collapse supernova - PP : pair production supernova.

Salient features of the computed light curves are the following (fig. 1 to 6).
(1) They are dominated by the recombination energy, since the other forms of energy (thermal and explosive) are strongly degraded by adiabatic losses in the early stage of expansion. ^{56}Ni plays a minor role unless the amount of ^{56}Ni gets close to 0.1 M_θ.
(2) The luminosity reaches respectively 3.10^{41} ergs s^{-1} and 10^{42}ergs^{-1} in cases A and B, against 5.10^{42} ergs s^{-1} in the canonical case (Arnett 1980). The absolute visual magnitudes at maximum are about -15 (case A) and -16 (case B).
(3) The width of the light curve is 4.10^6 s in case A and 8.10^6 s in case B, compared to 1.1 10^7 s in the canonical case.
We see that, apart from the possible existence of a spike in hard photons at early times (transient, see Arnett 1980), core collapse WR

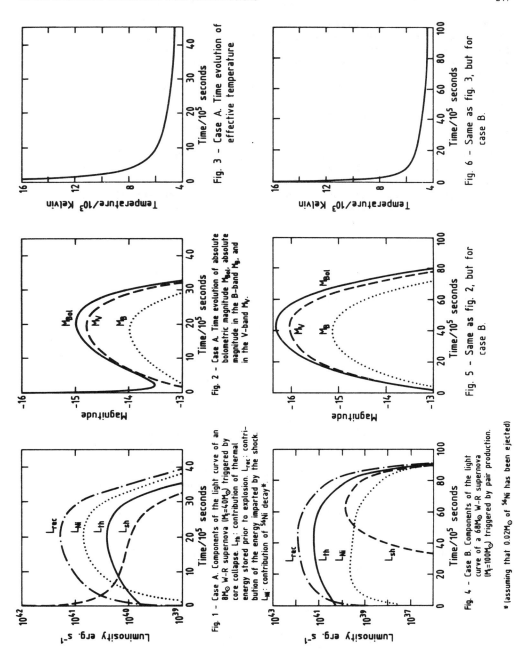

Fig. 1 - Case A. Components of the light curve of an 8M☉ W-R supernova (M_f=40M☉) triggered by core collapse. L_{th}: contribution of thermal energy stored prior to explosion. L_{rec}: contribution of the energy imparted by the shock. L_{Ni}: contribution of ^{56}Ni decay*

Fig. 2 - Case A. Time evolution of absolute bolometric magnitude M_{Bol}, absolute magnitude in the B-band M_B, and in the V-band M_V.

Fig. 3 - Case A. Time evolution of effective temperature

Fig. 4 - Case B. Components of the light curve of a 68M☉ W-R supernova (M_f=100M☉) triggered by pair production.

Fig. 5 - Same as fig. 2, but for case B.

Fig. 6 - Same as fig. 3, but for case B.

* (assuming that 0.02M☉ of ^{56}Ni has been ejected)

supernovae could be distinguished from pair production supernovae (if any) by a shorter duration and a lower luminosity. However the very low recurrence of WR outburst - 1 out of 3 to 7 supernovae, according to Maeder and Lequeux 1982 - render their detection problematic.
(4) The temperature levels off at about 4500 K, which corresponds to the temperature at which the last electron recombines in an oxygen dominated gas.

Finally, we would like to stress that our calculation applies strictly to WR stars isolated in the hot and dilute interstellar medium (runaway WR), since no circumstellar shell of the kind discussed by Falk and Arnett (1977) circumscribes the exploding objects. Work is in progress to investigate the case of WR exploding in stellar cavities (see e.g. Dorland, Montmerle and Doom, 1985).

More details will be given in a forthcoming article (Schaeffer, Cassé and Cahen, in preparation).

5. CONCLUSION

The removal of the hydrogen-rich envelope by mass loss and/or mass transfer onto a companion star has interesting consequences on the light curve generated by the final explosion.

(1) Obviously the spectra of supernovae descending from WR stars (i.e. with helium burning cores emerging at the stellar surface according to the picture generally admitted) should not exhibit hydrogen lines as normal type II supernovae. This kind of supernovae, however, must occur among extremely young objects at variance with the bulk of type I supernovae.

(2) Stripped models, as pointed out by Chevalier (1976b) and Woosley and Weaver (1982), give rise to dimmer and shorter optical events than clothed supernovae. The reduction in luminosity is however less dramatic than one may naïvely think, owing to the release of energy by recombination in the expanding plasma at a time where adiabatic losses are not any more devastating. The luminosity at maximum reaches 3.10^{41} ergs (for a 8 M_θ exploding WR star) and 10^{42} ergs (for a 6θ M_θ exploding WR star), against 5.10^{42} ergs for a typical type II supernova (Arnett 1982).

(3) The width of the light curve is slightly shorter than that of a typical type II supernova.

ACKNOWLEDGEMENTS

We are grateful to André Maeder for having generously offered his models and to C. Doom, C. de Loore, J.P. de Grève and M. Arnould and N. Prantzos for their constant help in stellar modelization. We are grateful to S.E. Woosley and J.C. Wheeler for useful discussions.

REFERENCES

Arnett, W.D., 1971, Ap. J., 163, 11.
Arnett, W.D., 1980, Ap. J., 237, 541.
Arnett, W.D., 1982, Ap. J., 253, 785.
Cahen, S., De Greve, J.P., and Doom, C., 1985, in preparation.
Cassé, M., 1984, in "Problems of Collapse and Numerical Relativity", eds. D. Bancel and M. Signore, NATO ASI Series, Reidel, Dordrecht, p.59.
Chevalier, R.A., 1976a, Ap. J., 207, 872.
Chevalier, R.A., 1976b, Ap. J., 208, 826.
Colgate, S A., and White, R.H., 1966, Ap. J., 143, 626.
de Loore, C., Prantzos, N., Doom, C., Arnould, M., 1986, this volume.
Doom, C., 1984, in "Problems of Collapse and Numerical Relativity", eds. D. Bancel and M. Signore, NATO ASI Series, Reidel, Dordrecht, p. 49.
Doom, C., De Greve, J.P., and de Loore, C., 1985, Ap. J., submitted.
Dorland, H., Montmerle, T.O., and Doom, C., 1985, Astron. Astrophys., submitted.
Falk, S.W., and Arnett, W.D., 1973, Ap. J. (Letters), 180, L65.
Falk, S.W., and Arnett, W.D., 1977, Ap. J. Suppl., 33, 515.
Fowler, W.A., and Hoyle, F., 1964, Ap. J. Suppl., 9, 201.
Maeder, A., 1982, Astron. Astrophys., 105, 149.
Maeder, A., 1984, in "Observational Tests of the Stellar Evolution Theory", IAU Symposium n°105, eds. A. Maeder and A. Renzini, Reidel, Dordrecht, p. 299.
Maeder, A., and Lequeux, J., 1982, Astron. Astrophys., 114, 409.
Ober, W.W., El Eid, M.F., and Fricke, K.J., 1983, Astron. Astrophys., 119, 61.
Prantzos, W., Doom, C., Arnould, M., and de Loore, C., 1986, this volume.
Trimble, V., 1982, Rev. Mod. Phys., 54, 1183.
Wheeler, J.C., 1981, Rep. Prog. Phys., 44, 85.
Woosley, S.E., and Weaver, T.A., 1980, Ann. N.Y. Acad. Sci., 336, 335.
Woosley, S.A., and Weaver, T.A., 1982, in "Supernovae: A Survey of Current Research", eds. M.J. Rees and R.J. Stoneham, Reidel, Dordrecht, p. 79.

III – S PROCESS

s-PROCESS NUCLEOSYNTHESIS -
STELLAR ASPECTS AND THE CLASSICAL MODEL

F. Käppeler
Kernforschungszentrum Karlsruhe GmbH
Institut für Kernphysik
Postfach 3640
D-7500 Karlsruhe
Federal Republic of Germany

ABSTRACT. In this contribution the sensitivity and the limits of s-process analyses with the classical model are discussed with respect to the interpretation of abundance patterns in the atmospheres of red giant stars. In particular, the effects of neutron capture cross sections, mean neutron exposure, and neutron density on the resulting s-process overabundances are considered.

1. INTRODUCTION

The classical s-process model has been shown to be an attractive tool for phenomenological studies, complementing stellar model calculations: It is easy to handle, contains only a few parameters (seed abundances, mean neutron exposure τ_0), and yields very accurate results (σN-curve, s-process abundances, neutron density, etc.). For a detailed discussion of this model see, for example, Käppeler et al. (1982) or Mathews and Ward (1985).
   The corresponding analyses were based on a fit of the model parameters to the s-only isotopes along the s-process synthesis path. If the classical model is to be applied to abundance patterns observed in stars, there is the problem that only element abundances can be derived from stellar spectra (except for very cool stars, where isotope assignments can be made via molecular lines). This feature complicates the interpretation of stellar abundances, as the clear s-process signature on the isotopic pattern is washed out by the averaging conversion into element abundances. This is illustrated in Fig. 1 which shows the s-process contributions to solar matter (solid line). The outstanding abundance peaks at the neutron magic isotopes are completely diluted except for barium which is one of the few elements - together with Sr, Y, Zr, and Ce - with practically negligible r-process contributions. These elements are therefore known to be good s-process indicators.
   Despite the difficulties of observation and analysis, abundances from red giants are a fascinating alternative to solar matter because of

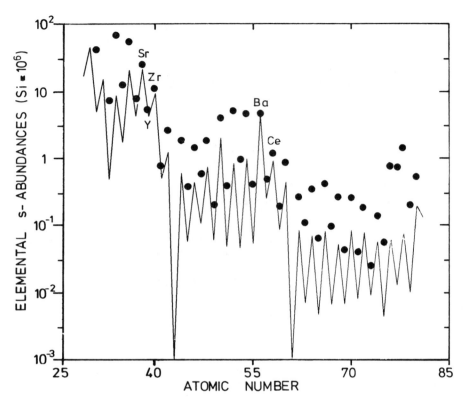

Figure 1. s-process contributions (solid line) to solar abundances (dots).

the strong overabundances of s-processed material, which result most likely from nucleosynthesis in just these stars. While the origin of the solar s-process component is rather unclear, the direct observation of s-enhancements in particular stars provides an immediate link to the s-process site and hence facilitates the comparison to stellar models.

Recent improvements in observational techniques (low noise, high resolution spectra obtained with multi-diode arrays) allowed for the determination of remarkably complete sets of stellar abundances (Tomkin and Lambert 1983; Holweger and Kovács 1984). Based on these results one might expect sufficiently detailed information for the future that reliable s-process analyses can be made for particular stars.

The classical s-process model was systematically applied for the analysis of stellar spectra by Cowley and Downs (1980). At that time not only the observations were limited but also the model itself was hampered by the lack of accurate input data. Meanwhile, many new neutron capture cross sections became available and the model was also shown to account quantitatively for the various branchings, in particular for the Kr-85 branching which is most important with respect to stellar abundances (Walter 1984, Walter et al. 1985). Therefore, it seems worth while to discuss the sensitivity and the limits of the classical s-process model in deriving information from stellar abundance patterns.

2. SOLAR ABUNDANCES AND STELLAR ATMOSPHERES

The most accurate information on solar abundances is obtained from analyses of primitive meteorites, the carbonaceous chondrites, which are believed to represent the original composition of the solar nebula. These abundances show typical uncertainties of 5 to 10% (Anders and Ebihara 1982), and hence are about a factor 2 more accurate than the results derived from the spectrum of the sun (Holweger 1979, Ross and Aller 1976). Another advantage of meteorite data is that they represent a complete isotope pattern. Only the abundances of the noble gases have to be interpolated from neighboring elements.

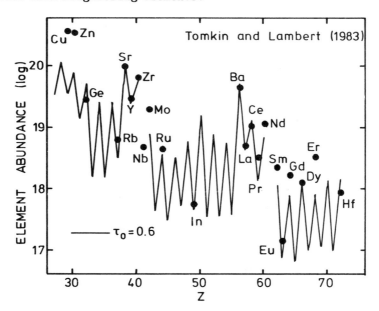

Figure 2. Observed s-process enhancements in HR774 (Tomkin and Lambert 1983) fitted with the classical model (solid line).

Analyses of the solar spectrum, on the other hand, yield element abundances only. In addition, a number of elements can not be observed in the spectrum of the sun (As, Se, Kr, Te, Xe, I, and Ta), either because they are rare or because their lines are masked by more abundant elements. Among these, Se and Kr would be important for s-process analyses (section 3.4).

Compared to the sun, the spectra of stars are more difficult to analyze although low noise spectra can be recorded with the improved techniques. Problems are due to uncertain parameters characterizing the stellar atmosphere. Consequently, stellar abundances are reported with uncertainties of typically 2 to 4 tenth of a decade. A nice example for a relatively complete set of abundances in the red giant HR774 is given in Fig.2 (Tomkin and Lambert 1983). Inclusion of Rb in these data is especially important for the s-process as we shall see in section 3.3 .

In the following section we discuss whether meaningful information on the s-process can be deduced from the still relatively uncertain stellar element abundances.

3. ELEMENTAL s-ABUNDANCES

3.1 Influence of Cross Sections

Since the work of Cowley and Downs (1980) a number of important cross sections were significantly improved. Because the product of cross section and abundance, σN, is the characteristic quantity of the s-process, which is known to be a smooth function of mass number, any cross section change propagates directly to the resulting isotopic abundance. Therefore, it was interesting to compare the present results with the

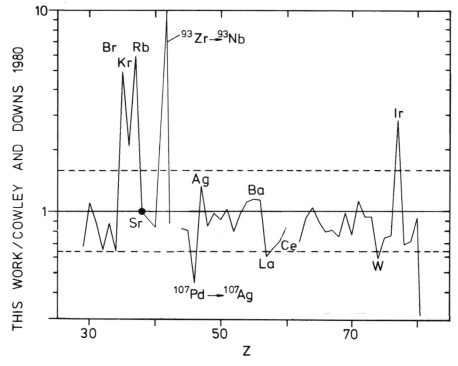

Figure 3. The influence of cross sections and branchings on the s-process abundances.

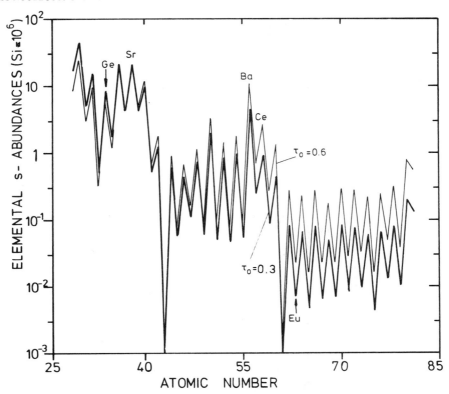

Figure 4. s-process abundances derived for different mean neutron exposures. The distributions are normalized at Sr.

abundances of Cowley and Downs. Fig. 3 shows the ratio of the two abundance sets, normalized at Sr. The dashed lines indicate the minimal uncertainties which could at best be expected for present observations.

One finds a number of significant differences in the two calculations although the pronounced isotopic discrepancies are washed out in the summation to element abundances. Nevertheless, with only a few exceptions these are smaller than the uncertainties in the observations. In all cases where the dashed limit is exceeded, the difference is primarily caused by details in the s-process path which were neglected in the work of Cowley and Downs. This holds especially for the s-process branchings at Se-79 and Kr-85 which determine the Br, Kr, Rb abundances. There, the neutron capture branch runs through isotopes with significantly smaller cross sections compared to those in the beta decay branch, thus building up higher abundances (see Tomkin and Lambert 1983 and also section 3.3).

The differences in the case of Pd/Ag and Ir/Pt are mainly due to the fact that Cowley and Downs did not account for the decay of the unstable isotopes Pd-107 and Pt-193 which are almost 'stable' on the time scale of the s-process but decay lateron. The only important difference which is

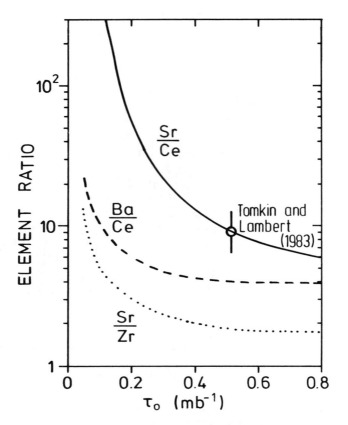

Figure 5. Determination of the mean neutron exposure τo from the most sensitive element ratios.

exclusively due to a cross section effect is the change in the Ba/La ratio.

3.2 Mean Neutron Exposure

In the classical model an exponential distribution of neutron exposures

$$\rho(\tau) \propto \exp(-\tau/\tau o)$$

is assumed, where the mean neutron exposure τo is a model parameter which is fitted to reproduce the observed abundances. It determines the number of neutrons captured per seed and hence is important for the neutron balance and the associated question for the neutron source in the s-process.

Fig. 4 illustrates the effect of the mean neutron exposure on the element abundances. The value of 0.3 is that obtained from solar abundances whereas 0.6 corresponds to the abundances in HR774 (Tomkin and

Lambert 1983). One finds that the larger neutron exposure leads to a flatter abundance distribution. The differences between the two curves are rapidly growing at magic neutron numbers, where the small capture cross sections act as a bottle neck for the s-process flow. This means that abundance ratios of elements situated below and above magic neutron numbers are most sensitive to the mean neutron exposure. The best cases in this respect are plotted in Fig. 5. The Sr/Zr and Ba/Ce ratios represent the crossing of magic neutron numbers 50 and 82, both exhibiting a very similar dependence on τ_0. At higher values of τ_0 the combined effect of both examples in form of the Sr/Ce ratio yields better sensitivity. Comparison with the observed ratio of Tomkin and Lambert (1983) shows that the slope of the curve is steep enough to allow for meanigful results up to $\tau_0 \sim 0.6$, but that more accurate observations would be needed to define larger exposure parameters. Consequently, it is important to include the crucial elements Sr, Zr, Ba, and Ce in the analysis of stellar spectra.

It should be noted here, that the usual method of determining s-process enhancements from stellar abundances might lead to avoidable systematic uncertainties. In order to account for the composition of the stellar envelope before it was enriched with s-processed material, one normally subtracts a 'solar system component' from the observed abundances. This component is normalized via those elements which have negligible s-process contributions, e.g. Ge and Eu (marked by arrows in Fig. 4). However, if the initial envelope abundance was not solar but characterized by $\tau_0=0.6$ as in HR774, then Fig. 4 shows that elements below Sr would be overcorrected while the corrections are too small at higher Z. If one goes back to Fig. 2 one finds that Ge and Eu follow such a trend with respect to the calculated curve. This point could be checked more consistently by inclusion of other elements with negligible s-process contribution such as Cs, Sb, Tb, Ho or Tm.

3.3 Neutron Density

It was shown in section 3.1 that the branchings of the s-process path at Se-79 and Kr-85 have a strong effect on the element abundances of Br, Kr, and Rb. As Br and Kr are not accessible to observations in stars, the Kr-85 branching manifests itself only through the Rb abundance. Tomkin and Lambert (1983) have shown how this can be used for estimating the neutron density during the s-process.

The s-process flow through the mass region 75<A<90 is sketched in Fig.6. At both branching points, Se-79 and Kr-85, there is almost equal probability for neutron capture and beta decay. The numbers associated to the isotopes in Fig. 6 are the respective neutron capture cross sections in (mb). One clearly finds that the isotopes on the neutron capture branch have much smaller cross sections compared to those on the beta decay branch. The most extreme case are the two Rb isotopes. As their abundances are almost proportional to the inverse of their cross sections, the Rb abundance is expected to change by a factor 30 between the extremes of very low and very high neutron density. This change is reduced to a factor of ~10 if the population of the short lived isomer in Kr-85 is taken into account which decays promptly to Rb-85.

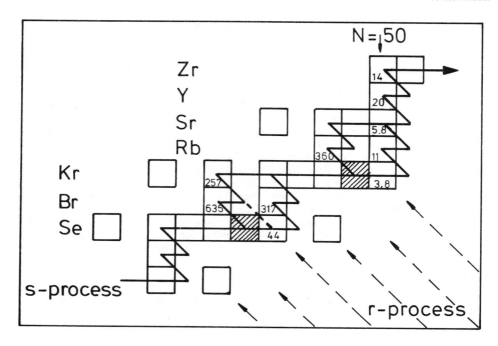

Figure 6. Significant branchings in the s-process path which allow for estimates of the neutron density during the s-process.

In Fig. 7 the Sr/Rb and Y/Rb ratios are plotted versus the neutron density. Obviously, neutron capture starts to dominate for neutron densities above $2 \cdot 10^8$ cm^{-3}, so that reasonable estimates can only be obtained below this limit. The arrows in Fig. 7 indicate the ratios observed by Tomkin and Lambert (1983) leading to a neutron density of about $2 \cdot 10^7$ cm^{-3}. All other branchings do not significantly influence the element distribution for neutron densities typical for solar s-process material. But for neutron densities $> 10^9$ cm^{-3} Mathews (1986) has found that Ce-141 becomes an important branching point which then affects the Ce/Nd ratio.

3.4 Temperature and Time Scales

The branching at Se-79 is characterized by the strong temperature dependence of the Se-79 half life. For a recent application of this s-process thermometer to solar matter see Walter (1984). With respect to stellar element abundances inspection of Fig. 6 shows that the Br abundance could be changed by this branching at most by a factor two. Indeed one finds that the Kr/Br ratio

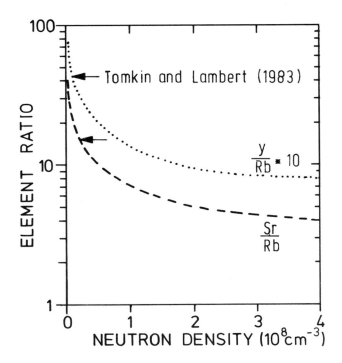

Figure 7. The s-process neutron density versus the abundance ratios resulting from the Kr-85 branching.

changes from 7 to 16 in the range of thermal energies 10<kT<50 keV which is characteristic for the s-process. However, this is of no use until the elements Br and Kr can actually be observed.

Another important aspect of the s-process concerns the time scales of the synthesis itself and for the transport of s-processed material to the surface of the star. Information on these features can be obtained via the unstable isotopes Zr-93 and Tc-99. A discussion of this problem is presented in another contribution to this conference (Mathews 1986).

4. SUMMARY

(i) The analysis of s-process enhancements in the atmospheres of red giants is of key importance because it provides direct access to the s-process site.

(ii) The present status of input data for the classical s-process model appears to be sufficient with respect to the analysis of stellar abundance patterns.

(iii) The quality of present observations allows for meaningful estimates of mean neutron exposures and neutron densites, (at least in certain ranges of these parameters) provided that the abundance pattern includes the important key elements Rb, Sr, Y, Zr, Ba, Ce.

(iv) For improvements in this field more accurate and comprehensive observations are called for. Element patterns need to be as complete as possible including not only a few important species. The more elements are determined the better the analysis can be made, including cross checks for consistency of abundance ratios and corrections for the initial envelope composition. For reliable s-process analyses a few complete abundance patterns are certainly of more help than a survey of many stars which includes only a few elements.

REFERENCES

Anders, E., Ebihara, M. 1982, Geochim. Cosmochim. Acta, **46**, 2363
Cowley, C.R., Downs, P.L. 1980, Ap. J., **236**, 648
Holweger, H. 1979, in Les Élements et leurs Isotopes dans l'Univers
    (Liège: University of Liège, Institute of Astrophysics) p.117
Holweger, H., Kovács, N. 1984, Astr. Ap., **132**, L5
Käppeler, F., Beer, H., Wisshak, K., Clayton, D.D., Macklin, R.L.,
    Ward, R.A. 1982, Ap. J., **257**, 821
Mathews, G.J. 1986, contrib. to this conference
Mathews, G.J., Ward, R.A. 1985, Reports on Progress in Physics, in print
Ross, J.E., Aller, L.H. 1976, Science, **191**, 1223
Tomkin, J., Lambert, D.L. 1983, Ap. J., **273**, 722
Walter, G. 1984, Report KfK-3706, Kernforschungszentrum Karlsruhe
Walter, G., Beer, H., Käppeler, F., Penzhorn, R.-D. 1985, Astr. Ap.,
    in print

s-PROCESS NUCLEOSYNTHESIS BELOW A=90

Hermann Beer
Kernforschungszentrum Karlsruhe, IK III
P.O.B. 3640
D-7500 Karlsruhe 1
Federal Republic of Germany

ABSTRACT. A complete s-process analysis of heavy elements is carried out in the frame of the classical model with three exponential neutron exposure distributions. Special emphasis is placed on the mass region below A=90 where the main and weak s-process component are effective. The derived astrophysical parameters of this model are able to reproduce heavy s-process element abundances correctly and consistently with the other processes of nucleosynthesis.

1. INTRODUCTION

In the mass range A=90-200 s-process nucleosynthesis is well described in the frame of an s-process model with one exponential distribution of neutron exposures [1,2]. However, at the beginning and at the termination of the s-process path problems exist. Significant s-process abundances are severely underproduced. In order to meet these difficulties a superposition of exponential distributions of neutron exposures was suggested by Ward et al. [2], where the additional fluence distributions, one for the isotopes of lead and bismuth and one for the nuclei below A=90, can be adjusted so that they contribute to the abundances significantly only in the required mass regions. This was demonstrated for the s-process termination already recently [3]. The subject of the present paper is a thorough study of the s-process nuclei including also s-process production of rare isotopes in the sulfur and calcium region. Previous studies [4,5,6,7,8] were constrained to partial solutions and/or disposed of less accurate capture cross sections.

2. THEORY

Mathematically the s-process can be formulated as a system of linear differential equations where the iron peak nuclei from explosive oxygen and silicon burning (e-process) [9] act as seed material for the heavy elements beyond. The abundance change $dN(A)/dt$ of an isotope A on the synthesis path is given by the equation:

$$dN(A)/dt = \lambda(A-1)N(A-1) - \lambda(A)N(A) \qquad (1)$$

Modifications of this equation occur at a radioactive nucleus A' where the neutron capture rate $\lambda(A')$ is in competition with the β-decay rate $\lambda_-(A')$.

$$dN(A')/dt = \lambda(A'-1)N(A'-1) - [\lambda_-(A') + \lambda(A')]N(A') \qquad (2)$$

and the stable isobar A" of A' obtained by β-decay from A' is given by

$$dN(A")/dt = \lambda_-(A')N(A') - \lambda(A")N(A") \qquad (3)$$

The occurring neutron capture rate, λ, is defined in the usual way as the product of neutron density, n, times the capture cross section σ averaged over a Maxwell-Boltzmann distribution of velocities v:

$$\lambda = n<\sigma v> = nv_T <\sigma v>/v_T \qquad (4)$$

$<\sigma v>/v_T$: Maxwellian averaged capture cross section

This Maxwellian averaged capture cross section, in the following simply designated as σ, was defined by introducing the thermal velocity v_T.

In general, the solutions of the equations of type (1), (2) and (3) are functions of the time dependent neutron density and temperature of the s-process environment. In the s-process model with an exponential distribution of neutron exposures [2] neutron density and temperature of the s-process are assumed constant and a superposition of solutions from eqs (1),(2),(3) with an exponential exposure distribution

$$\rho(\tau) = Go \exp(-\tau/\tau o) = [N(56)/\tau o] \exp(-\tau/\tau o) \qquad (5)$$

is chosen, where N(56) is the required iron seed abundance, τ the time integrated neutron density and τo the average of this quantity. Under these conditions the system of differential equations is solved recursively. For eqs (1),(2) and (3) we obtain:

$$\sigma N(A) = \{1 + 1/[\sigma(A)\tau o]\}^{-1} \sigma N(A-1) \qquad (6)$$

$$\sigma N(A') = \{1/[1-f] + 1/[\sigma(A')\tau o]\}^{-1} \sigma N(A'-1) \qquad (7)$$

$$\sigma N(A") = [f/(1-f)] \sigma N(A') \qquad (8)$$

These formulae demonstrate that the characteristic quantity of the s-process for an isotope A is the Maxwellian averaged capture cross section σ times the s-process abundance N. The factor f has to be introduced to specify a branching:

$$f = \lambda_-(T)/[\lambda_-(T) + \lambda(n)] \qquad (9)$$

f is always a function of the neutron density n via λ and can be additionally dependent on temperature, T, (and electron density) if the

half-life of the radioisotope is sensitive on temperature (and electron density).
The extension of this one component model to a three component model to account for s-process nucleosynthesis below A=90 and at Pb and Bi is straightforward using the following exposure distribution function:

$$\rho(\tau) = G_0 \exp(-\tau/\tau_0) + G_1 \exp(-\tau/\tau_1) + G_2 \exp(-\tau/\tau_2) \quad (10)$$

with $G_2 \ll G_0 \ll G_1$ and $\tau_2 \gg \tau_0 \gg \tau_1$

Due to the different strength of τ_0, τ_1, and τ_2 the corresponding components are termed as main, weak and strong.
Finally, the total $\sigma N(A)$ value of a nucleus A is simply given by the sum of the individual components:

$$\sigma N(A) = \sigma N^0(A) + \sigma N^1(A) + \sigma N^2(A) \quad (11)$$

3. EMPIRICAL $\sigma N(A)$-VALUES

s-process calculations are greatly simplified through the existence of s-only isotopes which are shielded from r-process contributions by stable neutron richer isobars. The empirical distribution of $\sigma N(A)$ values for these nuclei can be specified as the solar abundance is identical with the s-process abundance. This distribution is supplemented by s-process dominant isotopes where an r-process correction can be carried out sufficiently accurate. Minor additional corrections of the s-only nuclei are due to:
--A p-process contribution, estimated from neighboring p-only isotopes.
--Neutron capture on thermally populated excited states leading to an effective capture cross section different from ground state capture.
The empirical $\sigma N(A)$ data points are plotted in Fig.1(above). They form a smooth over large parts rather flat distribution with steps at the magic neutron numbers 50, 82, and 126. Other irregularities can be interpreted as s-only or preferentially s-process isotopes located in a branching of the synthesis path. Therefore, they are partially bypassed in the nucleosynthesis and lie distinctly below the distribution of the other data points. This occurs for Kr80, Kr86, Rb87, Gd152, Er164, Yb170, Hf176, Lu176, Os186, and Pt192. Not shown is Ta180 which also might be partially an s-process isotope [10,11,12].

4. s-PROCESS CALCULATION

The fit of empirical $\sigma N(A)$ values below A=90 is a two step calculation. The s-process abundances consist of a mixture of two components, the main component and the weak component. The main component is fitted in the mass region A=90-200 and then the computer calculation is extended below A=90 using the derived astrophysical parameters (Table I [13]). Neutron density and temperature has been adjusted via the s-only isotopes in the branchings at Sm151, Dy163, Tm170, W185, and Ir192. This main component fit forms the basis for an adjustment of the weak component below A=90. The parameters of the weak component (Table II) account for the rapid increase of the σN(A) curve required to meet the empirical σN(A) values below A=90. The crucial mass region for the analysis is the range between Se and Sr. At lower mass number overlap with the e-process nucleosynthesis already occurs. After the calculation of the contribution of the main component to the Se79 and Kr85 branchings they are used to determine neutron density and temperature of the weak component.

The Kr85 branching is only sensitive to the neutron density as the Kr85 half-life (10.7yr) is practically independent from temperature effects in the range of interest. An additional complication is created by an isomeric state at 305 keV excitation energy which is populated directly in the capture process [(52±5)%] [14] and decays chiefly [78.8%] to Rb85. Therefore, the s-process flow bypasses the Kr85 ground state to 41%. This effect was taken into account.

The temperature information of the weak component is derived from the Se79 branching. The terrestrial Se79 half-life is drastically changed under s-process conditions by allowed beta decay from thermally populated excited states, especially the 1/2(-) level at 96 keV.

Fig.1(above) shows the final result of our computer calculation. At the radionuclei with branchings the synthesis path becomes two or sometimes even threefold. If no information about the branching can be deduced because of the lack of an s-only nucleus on one of the branches, the sum of the individual σN-values for each mass number is plotted. In cases where an s-only isotope A contains a radiogenic s-process abundance from an unstable isobar A' the sum

$$\sigma N(A) + [\sigma(A)/\sigma(A')] \sigma N(A') \tag{12}$$

is calculated for reasens of comparison with the respective empirical value of A. This kind of calculation is for instance necessary for Sr86 and Yb170 which contain contributions from Rb86 and Tm170, respectively.

For the presentation of the sum of weak and main s-process nucleosynthesis in one σN(A) plot the σN(A)-values of the weak component have been normalized to the temperature of the main component. This

---

Fig.1 →
(above). The product of capture cross section σ and s-process abundance Ns as a function of mass number. The symbols correspond to empirical data for s-only isotopes or to s-process dominant isotopes near magic neutron shells. Symbols in full black were used to adjust the main component. Significant branchings were identified due to the low empirical σN values Kr80, Kr86, Rb87, Gd152, Er164, Yb170, Os186, and Pt192. The branching at Lu176/Hf176 was not treated here [13]. Below A=90 the influence of the weak component becomes apparent. The upper curve is the sum of main, weak and strong component. The strong component yields significant contributions only for Pb206,207,208, and Bi209.
(below). r-process residuals calculated as the difference between solar (ref.[15]) and s-process abundances. Below A=70 the steep rise indicates the onset of e-process. Data points with an arrow symbolize upper limits only. Full black symbols are the pure r-process isotopes. At Pb206,207,208 and Bi209 the analysis includes also a correction for the r-process of short-lived transbismuth progenitors and radiogenic Th232 and U235,238.

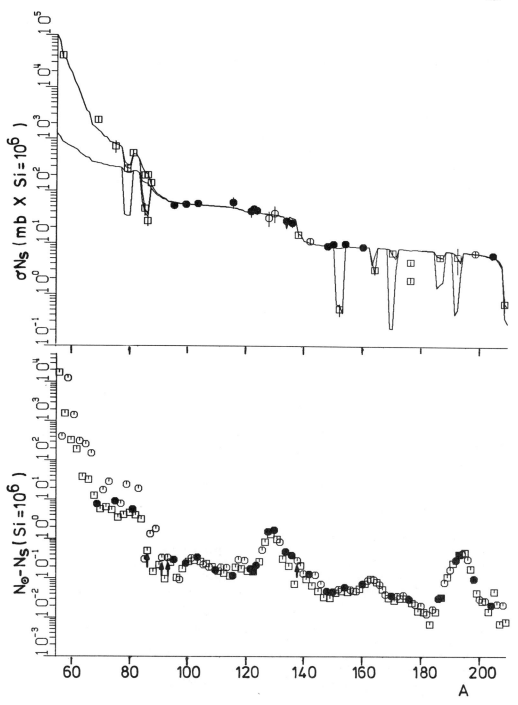

leads to no significant distortions of the curve as the temperature dependence of the capture cross sections is frequently close to $\sigma \sim 1/\sqrt{kT}$.

The parameters of the various components are summarized in Table I. The much smaller value of $\tau 1$ compared to $\tau 0$ yields the desired effect of a rapid increase torwards iron.
The allowed ranges of temperature and neutron density for the weak component are plotted in Fig.2 together with the respective result for the main component [8]. There exists no common range of values. This is a strong indication that weak and main component are two independent s-process nucleosyntheses.

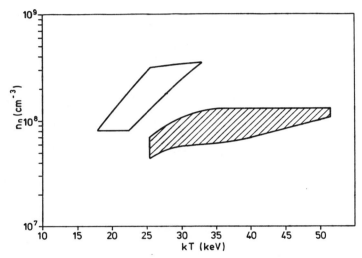

Fig.2 s-process neutron density as a function of s-process temperature. Consistent values for the main and weak component lie within the indicated areas [8]. The area of the weak component is hatched. The fact that there is no common range of values for the two components is interpreted as a strong evidence for their independence.

5. r-PROCESS RESIDUALS

The complete description of the s-process from Fe56 to Bi209 (an analysis of Pb,Bi in terms of the strong component was given in ref.[3]) allows finally the calculation of the r-process residuals as the difference of solar abundance minus total s-process abundance. Fig.1(below) shows the distribution curve of these nuclides. The black points are the pure r-process nuclei. The other points are the result of the s-process subtraction. The good agreement between the distribution of black points and the calculated residuals can be taken as a confirmation that our concept to treat heavy elements is correct. The accuracy of this determination of r-process residuals strongly depends on the accuracy and size of the s-process abundance. This is the reason why in four cases (Rb87, Zr92,94, Ba138) only an upper limit of the r-process abundance could be deduced. Below A=70 a steep rise of the abundances indicates the onset of e-process nucleosynthesis. Another hint for the influence of the e-process still at mass number 70 is supplied by Ge70, an r-process shielded isotope the abundance of which is only to 60% accounted for by the s-process. The rest is ascribed to the e-process.

Table I. Parameters for the flux distributions of the main, weak, and strong components

s-process component	Fraction of solar iron seed abundance	Average time integrated neutron flux at kT=30keV	Average number of neutrons per iron seed
Main	$(0.048\pm0.003)\%$	(0.30 ± 0.01)/mb	11.2 ± 0.7
Weak	$(1.6\pm0.2)\%$	(0.06 ± 0.01)/mb	1.4 ± 0.4
Strong	$(1.2\pm0.7)10^{-4}\%$	≥ 6/mb	≥ 150

6. s-PROCESS OF LIGHT ISOTOPES

An important criterion for the consistency of our analysis is that no overproductions of stable isotopes are generated. Except for pure s-process isotopes other species should always have an s-process abundance at least not larger than the solar abundance to allow for possible contributions of other nucleosyntheses (r-process, e-process...). This requirement is critical if the s-process is the dominant contribution of the solar abundance. In the domain of the main s-process component crucial isotopes are Rb87, Kr86, Sr88, Ba138 and Ce140.
For the weak s-process certain light isotopes represent a crucial check. They can be easily produced by neutron capture on highly abundant progenitor isotopes which act as seed material. Table II summarizes the studied isotopes. As the comparison with solar abundances shows there is no case where we have found an overproduction. The most critical case appears to be Fe58 where about 84% of the solar abundance has to be ascribed to the s-process.

7. CONCLUSIONS

We have found that the s-process abundances of isotopes below A=90 can be consistently described by a two component s-process model with exponential neutron exposures. The two components seem to be two different s-process sytheses because s-process neutron density and temperature have no common range of values. No inconsistencies were found by checking the s-process production of certain neutron rich light isotopes. The e-process nucleosynthesis appears to be effective up A=70.
Besides the assumption of a constant neutron density and temperature during the synthesis which is certainly an oversimplification, we have not treated the influence of a pulsed s-process. Influences of the pulse conditions are expected at the branchings. The analysis shows that the effects are negligible if the pulse duration $\Delta t \gg t_n$, the neutron capture time. This condition is, for instance, not satisfied for Kr85

where one estimates $t_n=13.5$ yr for a neutron density of $10^8/cm^3$. This capture time is not anymore small compared to the pulse duration of current stellar models.

Table II. Rare neutron rich light isotopes produced preferentially by the weak s-process via abundant progenitors as seed nuclei

Seed Nuclei	Nucleus	Abundances [Si=10^6]		$N/N_\odot$ [%]
		s-process N	solar $N_\odot$	
S32	S33	113.9	3860	3
"	S36	46.2	88	52
Ar36,38	Cl37	396.7	1270	31
" "	K41	21.5	253.7	9
Ca40	Ca42	11.6	395	3
"	Ca43	3.0	82.5	4
Ca40,44	Ca46	0.04	2.17	1.6
" "	Sc45	1.8	33.8	5
Ti48,49	Ti50	13.2	125	10.6
Cr50,52,53	Cr54	52.2	316	16.5
Fe56,57	Fe58	2203	2610	84
Fe56,57,Ni58,60	Ni64	136.1	449	30
" " " "	Cu63	155.2	356	43.6
" " " "	Cu65	120.5	158	76.3
" " " "	Zn67	19.2	51.7	37.1
" " " "	Ga69	10	22.7	44
" " " "	Ge70	13.6	21.7	62.2
" " " "	Ga71	9.3	15.1	61.6

REFERENCES

[1] P.A.Seeger, W.A.Fowler, D.D.Clayton, Ap.J.Suppl. 97(1965)121
[2] R.A.Ward, M.J. Newman, D.D.Clayton, Ap.J.Suppl. 31(1976)33
[3] H.Beer, R.L.Macklin, Phys.Rev. C (in press)
[4] R.A.Ward, M.J.Newman, Ap.J. 219(1978)195
[5] L.D.Hong, H.Beer, F.Käppeler, Proc.Int.Astrophys. Colloquium of Liege(1978)p.79
[6] F.Käppeler, H.Beer, K.Wisshak, D.D.Clayton, R.L.Macklin, R.A.Ward, Ap.J. 257(1982)821
[7] G.Walter, H.Beer, F.Käppeler, R.-D.Penzhorn, Astron. Astrophys. (in press)
[8] G.Walter, Report KfK 3706(1984), Kernforschungszentrum Karlsruhe
[9] S.E.Woosley, W.D.Arnett, D.D.Clayton, Ap.J.Suppl. 26(1973)231
[10] H.Beer, R.A.Ward, Nature 291(1981)308
[11] H.Beer, R.L.Macklin, Phys.Rev. C26(1982)1404
[12] K.Yokoi, K.Takahashi, Nature 305(1983)198
[13] H.Beer, G.Walter, R.L.Macklin, P.J.Patchett, Phys.Rev. C30(1984) 464
[14] R.-D.Penzhorn, G.Walter, H.Beer, Z.Naturforschung 38a(1983)712
[15] E.Anders, M.Ebihara, Geochim. et Cosmochim. Acta 46(1982)2363

PULSED-NEUTRON-SOURCE MODELS FOR THE ASTROPHYSICAL S-PROCESS

W. M. HOWARD, G. J. MATHEWS, K. TAKAHASHI, and R. A. WARD
University of California
Lawrence Livermore National Laboratory
Livermore, CA 94550

ABSTRACT. The astrophysical s-process is a sequence of neutron-capture and beta-decay reactions on a slow time scale compared to beta-decay lifetimes near the line of stability. We systematically study this detailed sequence of neutron capture, continuum and bound-state beta decay, positron decay, and electron-capture reactions that comprise the s-process for a broad range of astrophysical environments. Our results are then compared with the solar-system abundances of heavy elements to determine the range of physical conditions responsible for their nucleosynthesis.

1. INTRODUCTION

It has been clear for some time[1] that an exponential distribution of neutron exposures is required to fit the solar-system s-process σN curve (neutron capture cross section times abundance) as a function of atomic mass. Ulrich[2,3] showed that an exponential distribution of exposures could be achieved in a single star which subjected initial seed material to periodic neutron exposures followed by dredge up of some fraction of the irradiated material to the stellar surface. This s-process scenario has been explored in a series of papers[4-7] based on a $^{22}Ne(\alpha,n)^{25}Mg$ neutron source for the s-process during thermal pulses of asymptotic giant branch stars. There is sufficient uncertainty in the stellar models, however, that a different approach is warranted, i.e. to utilize the observed solar-system s-process σN curve to define the constraints on any stellar model for the s-process. This is the subject of the present work. This study reveals that the s-process is best fit with conditions similar to those expected for relatively low-mass red-giant stars near the end of their lifetime. Deviations of the fit from the observed solar-system values highlights the need for improved nuclear data.

2. S-PROCESS CALCULATION

In the classical s-process (without beta-decay branching) the abundance of an isotope is given simply by the solution to the set of coupled differential equations,

$$dN_A/d\tau = \sigma(A-1)N_{A-1} - \sigma(A)N_A \, , \quad (1)$$

where τ is the time integrated neutron flux (i.e. neutron exposure) and σ the Maxwellian averaged neutron capture cross section.

At equilibrium, a single exposure would lead to a constant σN value for all isotopes in the s-process. The solar-system σN curve (see Fig. 1), however, requires an exponential distribution of neutron exposures, i.e. the probability, $\rho(\tau)$ for a given exposure is taken to be,

$$\rho(\tau) \propto \exp(-\tau/\tau_o) \, . \quad (2)$$

In a periodic s-process operating in a single star[2,3], the mean exposure, τ_o, is simply related to the average exposure per pulse, $\Delta\tau$, and the fraction of material, r, which remains after each dredge up, i.e. $\tau_o = -\Delta\tau/\ln(r)$. Note, that the mixing fraction and exposure per pulse are not independent parameters.

In dynamic stellar environments, such as thermally-pulsing red giants, the simplicity of the classical s-process (Eq. 1) is lost due to a break down of the assumption that neutron captures are slow compared to beta decay. We therefore, compute the nucleosynthesis in a network:

$$\frac{dN(Z,A)}{dt} = N(Z,A-1)n_n\sigma_{n,\gamma}(Z,A-1) + N(Z-1,A)\lambda_\beta(Z-1,A)$$
$$- N(Z,A)[n_n\sigma_{n,\gamma}(Z,A) + \lambda_\beta(Z,A)] \, . \quad (3)$$

The time-dependence of the flux and temperature dependence of the cross sections and beta rates are included. In a few cases, positron, electron-capture, or alpha decay must also be added to Eq. (3). We utilize experimental neutron-capture cross sections when available[8-16]. Cross sections for unstable or unmeasured nuclei were taken from Hauser-Feshbach estimates[17] along with the temperature dependence of all cross sections. Decay rates from thermally populated excited states were calculated assuming thermal equilibrium and with appropriately choosen ft values[18]. The initial seed abundances were taken from solar-system values[19].

3. RESULTS

We have adjusted the parameters to minimize χ^2 for the fit to 23 s-only nuclei with $Z \geq 40$. Lighter nuclei were omitted because of possible contribution from other sources[9]. The most sensitive parameters in the fit are τ_o, and n_n, although the temperature, pulse shape, and interpulse period also enter.

Figure 1 is an example of a good fit ($\nu^2 \sim 2.3$) for a mean exposure of $\tau_o = 0.277$ mb^{-1}, a constant density ($n_n = 1.0 \times 10^8$ cm^{-3}) and temperature (T = 0.348×10^9K), with a long

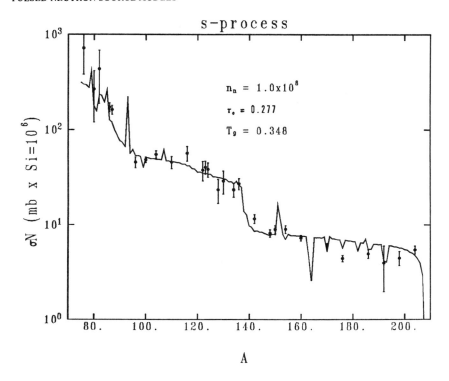

Figure 1 Fit to the observed σN curve.

interpulse interval ($\sim$460 y). The χ^2 quickly increases for higher fluxes, but increases more slowly as the flux is decreased. From this study we conclude that the best-fit parameters are $n_n = 1.0(^{+0.6}_{-0.5}) \times 10^8$ cm^{-3}, $T = 0.30(\pm 0.04) \times 10^9$K), and $\tau_0 = 0.28 \pm 0.01$ (mb^{-1}), corresponding to a pulse duration of $500(^{+300}_{-100})$ yrs. This is consistent with previous estimates[3,14] of the neutron density based on one or two branches alone. For higher densities than 10^8 cm^{-3} many branches (all above A = 140) are bypassed, while for low densities the σN curve approximates the nominal s-process. What is most interesting is that these stellar parameters, (derived solely from input nuclear data and observed abundances) are very similar to the conditions thought to exist[4] in relatively low-mass (M $\lesssim$ 3 M_O) red-giant stars as they ascend the asymptotic giant branch.

The best fit for the solar-system σN curve exhibits pronounced deviations from the smooth monotonically decreasing classical s-process curve. The reason for this is that many stable nuclei are actually produced as beta-unstable progenitors due to neutron captures on beta-unstable nuclei. If the progenitor has a larger (or smaller) cross section there will be a dip (or peak) in the σN curve when σN for the stable daughter is calculated.

This is the reason for the pronounced peaks on Fig. 1 at ^{93}Nb (partially produced as ^{93}Zr) and ^{151}Eu (partially produced as ^{151}Sm).

For a broad range of average neutron densities the temperature minimum occurs for T_9 = 0.30 ($\pm$ 0.04) (kT = 26$\pm$4 keV). This is due both to the change in cross sections and the fact that certain beta decays are too slow at low temperature to provide adequate competition with neutron capture at certain key branch points (e.g. ^{79}Se, ^{134}Cs, and ^{154}Gd). At low temperatures the fact that ^{80}Kr, ^{134}Ba, and ^{154}Gd are bypassed is particularly evident. The fact that ^{154}Gd is bypassed is at first surprizing since the precursor is ^{154}Eu with a terrestrial half life of only 8 yrs. The capture cross section for ^{154}Eu is extremely large (8 barns at kT = 10 keV), however, so that capture occurs at these neutron densities before beta decay. The poorer fit at high temperature is due primarily to changes in the neutron capture cross sections; also the increased beta-rates at high temperatures tends to washout the branches at ^{170}Yb and ^{186}Os. Thus, we find that the best solution is for kT = 26 keV, n_n = 10^8cm^{-3}, and τ_o = 0.28 mb^{-1}. The rather narrow range of allowed temperatures for a broad range of densities is significant because these temperatures are characteristic of the ^{22}Ne(α,n)^{25}Mg source reaction. Temperatures in this range are just what is required to give the correct neutron exposure from this reaction. Most other possible s-process sources such as ^{13}C(α,n)^{16}O and carbon burning would normally be expected to occur at temperatures considerably lower or higher than the ranges which give the best fit to the σN curve. It is also encouraging that this completely independent means of estimating the temperature (which makes no assumption about the stellar site other than that it experiences sequential neutron irradiations) comes so close to the traditionally assumed value for kT = 30 keV based on the assumption that this process should occur at typical helium-burning temperatures.

ACKNOWLEDGEMENT

Work performed under the auspices of the U.S. Department of Energy by the Lawrence Livermore National Laboratory under contract number W-7405-ENG-48.

REFERENCES

1. P. A. Seeger, W. A. Fowler, D. D. Clayton, Astrophys. J. Suppl., 97, 121 (1965).
2. R. K. Ulrich, in "Explosive Nucleosynthesis", eds. D. N. Schramm and W. D. Arnett (Univ. Texas Press, Austin, 1973) p.139.
3. R. K. Ulrich, in "Essays in Nuclear Astrophysics", eds. C. A. Barnes, D. D. Clayton, and D. N. Schramm (Cambridge Univ. Press, N. Y., 1982) p. 301.
4. I. Iben, Jr., Ap. J., 217, 788 (1977).
5. J. W. Truran and I. Iben, Jr., Ap. J., 216, 197 (1977).
6. I. Iben, Jr., and J. W. Truran, Ap. J., 220, 980 (1978).

7. K. R. Cosner, I. Iben, and J. R. Truran, Ap. J. Lett., 238, L91 (1980).
8. B. J. Allen, J. H. Gibbons, and R. L. Macklin, Adv. Nucl. Phy., 4, 205 (1971).
9. F. Käppeler, H. Beer, K. Wisshak, D. D. Clayton, R. L., Macklin, and R. A. Ward, Ap. J., 257, 821 (1982).
10. M. J. Newman, Ap. J., 219, 676 (1978).
11. H. Beer and R. L. Macklin, Phys. Rev., C26, 1404 (1982).
12. H. Beer, F. Käppeler, G. Reffo, and G. Ventorini, Ap. Space Sci., 97, 95 (1983).
13. R. R. Winters, F. Käppeler, K. Wisshak, A. Mengoni, and G. Reffo, (submitted to Ap. J., 1984).
14. H. Beer, F. Käppeler, K. Yokoi, and K. Takahashi Ap. J., 278, 388 (1984).
15. H. Beer and G. Walter, Astron. Ap. (1984 in press).
16. G. J. Mathews and F. Käppeler, Ap. J., 286, 810 (1984).
17. J. A. Holmes, S. E. Woosley, W. A., Fowler, and B. A. Zimmerman, Atom. Nucl. Data Tables, 18, 306 (1976).
18. K. Takahashi and K. Yokoi (to be published).
19. E. Anders and M. Ebihara, Geochim. Cosmochim. Acta, 46, 2263 (1982).

STELLAR S-PROCESS DIAGNOSTICS

G. J. Mathews, R. A. Ward, K. Takahashi, and W. M. Howard
University of California
Lawrence Livermore National Laboratory
Livermore, CA 94550

ABSTRACT. We argue that the solar-system σN curve can be best understood if the s-process is largely produced by stars in the mass range of $M \sim 2-4~M_\odot$. Several observations are then studied as indicators of the validity of this hypothesis. We find that isotopic Zr abundances and elemental (Ba/Sr) vs. (Ba/Nd) ratios are consistent with these conditions. The elemental (Tc/Nb) ratio is found not to be an indicator of temperature as has been suggested but rather a measure of the age of an AGB star in the third dredge-up phase.

1. INTRODUCTION

As we have already heard at this conference, the σN curve for solar-system material seems to imply s-process conditions of $n_n \sim 1.0(^{+0.2}_{-0.6}) \times 10^8 \mathrm{cm}^{-3}$, $T = 0.30 \pm 0.04 \times 10^9$ K and $\tau_0 = 0.28 \pm .01~\mathrm{mb}^{-1}$ corresponding to a thermal pulse duration of $\sim 500^{+300}_{-100}$ yrs (Mathews, et al. 1984ab; Howard, et al. 1985).
    These parameters seem to be telling us something about the nature of the astrophysical site for the s-process. In particular, the empirically derived temperature is remarkably close (Almeida and Käppeler 1983) to the optimum temperature for a $^{22}\mathrm{Ne}(\alpha,n)^{25}\mathrm{Mg}$ neutron source, which operates in the thermally pulsing phase for AGB stars (Iben 1977) with core masses > 1.0 $M_\odot$. On the other hand, the pulse duration, is more characteristic of AGB stars with lower mass cores. In this paper we make the point that this apparent contradiction can be resolved on the basis of more recent calculations (Becker 1985) which show that the temperatures for the full amplitude pulses for low-mass stars are sufficiently high ($T_9 \sim 0.3$) that the $^{22}\mathrm{Ne}(\alpha,n)^{25}\mathrm{Mg}$ source can be significant. Thus, a consistent argument can be constructed that the most likely site for the s-process is in the thermally-pulsing phase of relatively low-mass (M $\sim 2-4~M_\odot$) AGB stars.
    Since stars in this mass range also correspond to most observed S- and Ba stars, we next investigate the consistency of this interpretation with observed s-process abundances in these stars. In particular, we find that a measure of the neutron exposure can be

obtained from the Sr/Ba elemental ratio, and an upper limit to the neutron density can be inferred from the Ba/Nd ratio. Isotopic abundances from ZrO lines (Zook 1978) can also be used to determine ranges of allowed values for n_n and τ_o. Both of these constraints are consistent with relatively low-mass AGB stars as the site for the s-process.

Finally, we briefly consider quantitative observations (Smith and Wallerstein 1983) of the Tc abundance on S-stars which have been suggested as a probe of the s-process temperature. We show that the Tc abundance is largely independent of temperature due to the fact that the neutron production is even more temperature sensitive than the ^{99}Tc beta half life. We then compute the expected Tc/Nb and Tc/Mo ratios and show that these quantities can be used as a measure of the lifetime of a star in the thermally pulsing third dredge-up phase.

2.   IN SEARCH OF THE SITE FOR s-PROCESS NUCLEOSYNTHESIS

In a number of papers (Iben 1977; Truran and Iben 1977; Iben and Truran 1978; Cosner, Iben and Truran 1980) it has been argued that the s-process occurs in thermally pulsing AGB stars with relatively massive ($M_c \sim 1-1.4 \, M_\odot$) electron-degenerate carbon-oxygen cores. This conclusion was reached on the basis of the fact that the ^{22}Ne(α,n)^{25}Mg reaction seems to give the neutron exposure required to fit the solar-system σN curve.

The neutron densities corresponding to these core masses, however, are too high to give a satisfactory fit to the solar-system s-only nuclei. We have investigated two possible remedies to this dilemma. One is to postulate (Clayton 1984) a weak interpulse exposure ($\Delta\tau \sim .01 - 0.10 \, mb^{-1}$) which is insignificant compared with the total exposure but which can heal the dips in the σN curve caused by the high neutron density. The problem we have found with this approach is that, by the time that enough exposure is introduced to heal the σN curve, branching at the low-temperature exposure leads to a poor fit to the σN curve.

The scenario which we prefer is based on the calculations of Becker (1981; 1985). These calculations show that, for the low-mass cores, the thermal pulses continue to heat up for about the first 20 pulses to a maximum temperature which is very close to the empirically derived temperature of $T_9 \sim 0.30$. For these temperature, the pulses endure long enough to provide sufficient neutron exposure without an excessively high neutron density. In calculating the exposure we have used the analytic relations of Iben and Truran (1978) and a constant peak temperature of $T_9 = 0.31$ for $M_c < 0.96$ $M_\odot$. This is a reasonable approximation to the actual numerical output from the stellar models.

Figure 1 is an example of the reduced χ_r^2 for the fit as a function of core mass for these models with increased temperature. The fit is quite good for $M_c \sim 0.65 \, M_\odot$, which is shown in Fig. 3. This initial core mass corresponds to roughly a 3 $M_\odot$ star (Iben

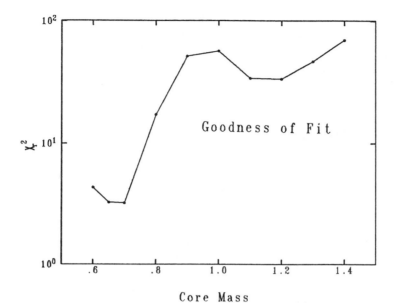

Fig. 1 Reduced χ_r^2 for fit to solar-system σN curve as a function of core mass (units of $M_\odot$).

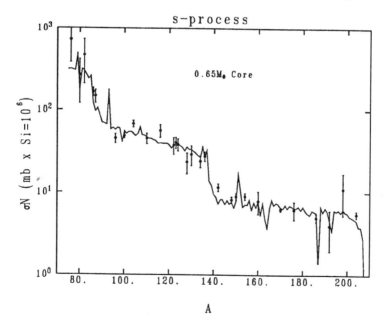

Fig. 2 Fit to solar-system σN curve for a star with a 0.65 $M_\odot$ carbon-oxygen core.

and Truran 1978). The implication is that the solar system s-process material has been largely produced by relatively low-mass AGB stars since the fit is significantly worse if an average over the entire mass range is used.

This is consistent with the suggestion (Scalo and Miller 1981) that the third dredge-up may not occur in the higher-mass stars and the lack of observation of Tc lines in AGB stars with $M \geq 3\ M_\odot$.

3. STELLAR s-PROCESS DIAGNOSTICS

This leads us to suggest that it would be useful to consider observations of s-process abundances in low-mass AGB stars to see if these data are consistent with the implications from the solar system ON curve. It would, of course, be most useful to have good isotopic abundance measurements. Unfortunately these are not available for most elements. However, Zook (1978) has obtained isotopic Zr abundances from ZrO lines in three S-stars. These data are shown in Fig. 3 along with calculated Zr isotopic abundances as a function of core mass. As can be seen, the best fit is for the low core masses.

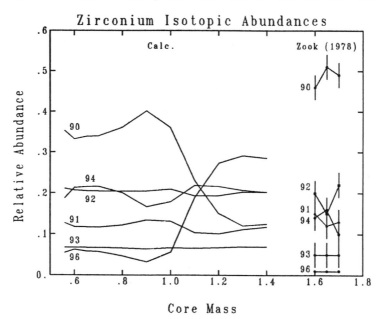

Fig. 3 Relative Zr isotopic abundances calculated for stellar models. The data are from the observations of Zook (1978).

What about elemental abundances? From variation of neutron exposure at a constant neutron density, we find that there is a dramatic variation of a stepwise behavior of the elemental abundances. From this study it appears that the Sr/Ba ratio is a good indicator of neutron exposure. Variation of the neutron-density over 11 orders of magnitude for a fixed exposure produces little change in the relative abundances. In particular, the Ba/Sr is invariant to a good approximation. This elemental ratio is, therefore, a good measure of the neutron exposure.

There is, however, one significant change. The Nd abundance decreases by about a factor of two. The reason for this is that the Nd abundance is normally dominated by the s-only isotope, ^{142}Nd. For neutron densities $n_n \geq 10^9$ cm^{-3}, however, ^{142}Nd is bypassed due to neutron captures on the unstable isotope, ^{141}Ce. Thus, the Ba/Nd ratio is a measure of the neutron density when the $n_n > 10^9$ cm^{-3}.

Figure 4 shows predicted behaviors for the Sr/Ba and Ba/Nd ratios as a function of neutron density and neutron exposure. The observations (Cowley and Downs 1980) for several of Ba stars are also indicated. Although the uncertainties are large, it is clear that, at least for the stars indicated, the densities and exposures tend to be consistent with the low neutron densities associated with lower core masses.

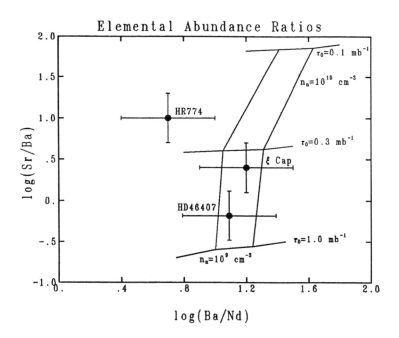

Fig. 4   Correlation of (Sr/Ba) and (Ba/Nd) elemental ratios. Lines correspond to constant neutron density and exposure. The data are from Cowley and Downs (1980).

4. TECHNETIUM ABUNDANCES

Another possible probe of the stellar environment is suggested by the quantitative observations of unstable Tc (probably ^{99}Tc) on stellar surfaces (Smith and Wallerstein 1983). Because the beta-decay halflife for ^{99}Tc is expected (Cosner, Despain and Truran 1984) to be drastically diminished at stellar temperatures ($\tau_{1/2} \sim 1$ yr. rather than $\tau_{1/2} \sim 2.1 \times 0^5$ yr. terrestrially), it has been suggested that Tc abundances on stellar surfaces may indicate low-temperatures for nucleosynthesis. We find (Mathews et al. 1985), however, on the basis of our detailed network calculations, that a significant fraction of the Tc abundance (70-90%) survives to the end of the convective shell, almost independent of core mass in these models. The reason is simply that the abundance of ^{99}Tc is determined by both the beta-decay rate and its neutron capture rate. Because the capture cross section for ^{99}Tc is so large (854 mb (Macklin 1984)), and the ^{22}Ne$(\alpha,n)^{25}$Mg, neutron source increases so rapidly with temperature, the σN value for ^{99}Tc is very close to the value it would have if there were no beta decay.

Since the ^{99}Tc abundance is not particularly temperature sensitive, it is useful to see what the observations (Smith and Wallerstein 1983) are telling us about the stars producing ^{99}Tc. Figure 5, (from Mathews et al. (1985)) shows the predicted (Tc/Nb) and (Tc/Mo) ratios as a function of the lifetime of a star in the thermally pulsing third dredge-up phase. This figure is based on a simple analytical two-reservoir model (Anders 1958; Peterson and Wrubel 1966). The points are two S-stars measured by Smith and Wallerstein

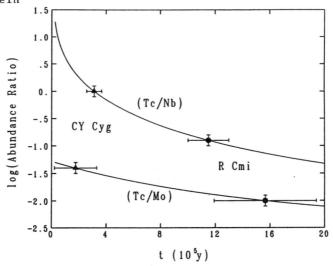

Fig. 5   Predicted behavior of the surface (Tc/Nb) and (Tc/Mo) ratios as a function of the age of an AGB star in the thermally-pulsing third dredge-up phase.

(1983). From this figure it can be seen that this approach can be used to give a good indication of the age of a star in this phase of evolution.

5. CONCLUSION

The basic conclusion of this work is that, from a number of different ways of looking at the problem, the thermally-pulsing phase of AGB stars in the mass range of 2-4 $M_\odot$ appears to the most promising site for the s-process. We suggest that it will be useful for further study to compile Zr isotopic abundances as well as (Sr/Ba), (Ba/Nd), and (Tc/Nb) elemental ratios for a number of stars as a best means to test this hypothesis.

6. ACKNOWLEDGEMENT

Work performed under the auspices of the U.S. Department of Energy by the Lawrence Livermore National Laboratory under contract number W-7405-ENG-48.

7. REFERENCES

Almeida, J. and Käppeler, F. 1983, Ap. J., 265, 417.
Anders, E. 1958, Ap. J., 127, 355.
Becker, S. A. 1981, in "Physical Processes in Red Giants", I. Iben Jr., and A. Renzini (eds.), (Reidel, Dordrecht) pp. 141-146.
_____. 1985 (Priv. Comm.).
Clayton, D. D. 1984 (Priv. Comm.)
Cosner, K. R., Iben, I., and Truran, J. R. 1980, Ap. J. Lett., 238, L91.
Cowley, C. R. and Downs, P. L. 1980, Ap. J., 236, 648.
Howard, W. M., Mathews, G. J., Takahashi, K., and Ward, R. A. 1985 (Submitted to Ap. J.).
_____. 1977, Ap. J., 217, 788.
Iben, I. Jr., and Truran, J. W. 1978, Ap. J., 220, 980.
Macklin. R. L. 1984, Nucl. Sci. Eng., 81, 520.
Mathews, G. J., Howard, W. M., Takahashi, K., and Ward, R. A. 1984a, in "Neutron-Nucleus Collisions as a Probe of Nuclear Structure", J. Rapaport, R. W. Finlay, S. M. Grimes, and F. S. Dietrich (eds.), (Burr Oak, Ohio 1984) (American Institute of Physics, New York), p. 511.
_____. 1984b, in "Capture Gamma-Ray Spectroscopy and Related Topics-1984", S. Raman (ed.), (Knoxville, Tenn.) (American Institute of Physics, New York), p. 766.
Mathews, G. J., Takahashi, K., Ward, R. A., and Howard, W. M. 1985, (submitted to Ap. J. Lett.).
Mathews, G. J. and Ward, R. A. 1985, Rep. Prog. Phys. (in press).

Petersen, V. L. and Wrubel, M. H. 1966, in "Stellar Evolution", R. F. Stein and A. G. W. Cameron, eds., (Plenum Press; New York) p. 419.
Scalo, J. M. and Miller, G. E. 1981, Ap. J., 246, 251.
Smith, V. V. and Wallerstein, G. 1983, Ap. J., 273, 742.
Takahashi, K., Yokoi, K. 1984, to be published in At. Nucl. Data Tables.
Truran, J. W. and Iben, I. Jr. 1977, Ap. J., 216, 197.
Zook, A. C. 1978, Ap. J., 221, L113.

PRODUCTION AND SURVIVAL OF ^{99}Tc IN He-SHELL RECURRENT THERMAL PULSES

K.Takahashi, G.J.Mathews and R.A.Ward
University of California
Lawrence Livermore National Laboratory
Livermore, CA 94550
USA

and

S.A.Becker
University of California
Los Alamos National Laboratory
Los Alamos, NM 87545
USA

ABSTRACT. After a brief introduction to the present state of art of nuclear beta-decay studies in astrophysics, we report our recent work on the long-standing ^{99}Tc problem. Having combined a detailed study of the recurrent He-shell thermal-pulse, third dredge-up episodes in a 2.25 M$_\odot$ star and an s-process network calculation, we show that a substantial amount of ^{99}Tc can be produced by the s-process and can survive to be dredged up to the stellar surface. We stress that the factual observation of ^{99}Tc at the surface of certain stars does not necessarily preclude the ^{22}Ne(α,n)^{25}Mg reaction from remaining as the neutron source for the s-process. The calculated surface abundances of ^{99}Tc and elements with neighboring atomic numbers are compared with observations.

1. INTRODUCTION

1.1. Beta-Decays in Astrophysics

The main difficulty in most studies of nuclear beta-decays of astrophysical interest is to reliably evaluate beta-decay nuclear matrix elements for terrestrially unknown transitions, either individually or collectively as a strength function.

Our knowledge of beta-strength functions has in recent years increased drastically (at least for allowed β^- transitions), mainly because of a series of (p,n) experiments elucidating the long-suspected Gamow-Teller giant resonance (e.g. Takahashi 1983 for an introductory chronicle). This is further reinforced as large-basis shell-model calculations have become feasible. Indeed, such a calculation has been carried out by Fuller and Bloom (1985) to evaluate the electron capture rates for Fe-group nuclei which are extremely important in supernova problems (e.g. Fuller, Fowler and Newman 1980, 1982; Woosley 1986). It still remains to be seen if the method can be successfully applied to study beta-decay properties of unknown nuclei such as those involved in the r-process, and if the stability of its predictions with respect to different choices of the nuclear Hamiltonian and model space is high enough, so that one can, without hesitation, discard the values from simpler models currently on the market (Takahashi, Yamada and Kondoh 1973; Klapdor, Metzinger and Oda 1984). In addition, how to cope with deformed nuclei might remain as a crucial question (Krumlinde and Møller 1984). Now that the search for the astrophysical site(s) for the r-process is in chaos (e.g. Schramm 1983; Mathews and Ward 1985), it is of extreme importance to challenge the refinements of nuclear input data including beta-strength functions (e.g. Mathews 1983)

The difficulty increases when individual beta-transitions come into the game as in s-process branchings which often constrain certain astrophysical conditions (e.g. temperature, neutron density) for the s-process (Cameron 1959; Ward, Newman and Clayton 1976; Käppeler et al. 1982; Ulrich 1983; Howard et al. 1985). Again, the possibility of large-basis shell-model calculations, as recently performed in conjunction with the solar neutrino detector (Mathews et al. 1985) and with the ^{99}Tc problem (Takahashi, Mathews and Bloom 1985), leaves good hopes that for at least some selected spherical nuclei we will have predictions in the near future that are better than those available now (Cosner and Truran 1981; Yokoi and Takahashi 1985).

The story does not end here since in some cases a detailed study of atomic aspects stemming from ionization is needed (Takahashi and Yokoi 1983). In particular, the roles played by bound-state β^--decays have appeared to be very important in the ^{187}Re-^{187}Os chronometry (Clayton 1969; Perrone 1971; Yokoi, Takahashi and Arnould 1983) and in the possible ^{205}Pb-^{205}Tl chronometry (Yokoi, Takahashi and Arnould 1985; cf. Blake, Lee and Schramm 1973). It should be noted here that the question whether the bound-state β^--decay contribution on neutral ^{187}Re atoms shows up as a difference between its decay rates measured in meteorites and in the laboratory (electron measurements) has been discussed (Dyson 1972; Williams, Fowler and Koonin 1984).

1.2 ^{99}Tc "Problem"

The existence of Tc (most probably ^{99}Tc) at the surface of certain (e.g. type S) stars (Merrill 1952; Iben and Renzini 1983 and references therein; Smith and Wallerstein 1983) is one of the strongest supports for the idea of nucleosynthesis of heavy elements via slow neutron capture (e.g. Mathews and Ward 1985 and references therein). On the other hand, various studies of the solar abundance curve for s-process nuclides (e.g. Beer 1986; Käppeler 1986; Howard et al. 1985) as well as the existing scenarios for the astrophysical site for the s-process (e.g. the He-shell recurrent thermal-pulses in asymptotic-giant-branch intermediate-mass stars with the ^{22}Ne$(\alpha,n)^{25}$Mg neutron source: Iben 1975, 1977; Iben and Truran 1978; Cosner, Iben and Truran 1980) suggest a typical temperature of 3×10^8 K at the site.

A crucial question first raised by Cameron (1959) is whether ^{99}Tc can survive such a hot environment until it is dredged up and observed at the stellar surface. Indeed, the ^{99}Tc beta-decay half-life at such temperatures will most certainly be much much shorter than its terrestrial value of 2.1×10^5 yr. This is expected because the thermal excitation of the $7/2^+$ (141 keV) and $5/2^+$ (181 keV) states opens the channel for the Gamow-Teller allowed transitions (Fig.1): the effective half-life at a temperature of 3×10^8 K might be as short as the order of years (Cameron 1959; Cosner and Truran 1981; Schatz 1983; Yokoi and Takahashi 1985). If most ^{99}Tc consequently decays into ^{99}Ru, then it would contradict the factual observation of substantial amounts of ^{99}Tc at certain stars. This dilemma has since long remained open as the "^{99}Tc problem". In particular, this apparent conflict has tempted some authors (e.g. Smith and Wallerstein 1983) to argue that the ^{22}Ne(α,n)^{25}Mg reaction is not a favorable neutron source for the s-process as it requires temperatures of $\sim 3 \times 10^8$ K to be effective. [An alternative choice for the neutron source is ^{13}C(α,n)^{16}O reaction, which operates at lower temperatures of $\sim 10^8$ K. The corresponding astrophysical scenario has, however, its inherent difficulty: Ulrich 1983.]

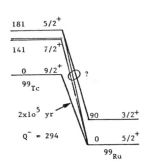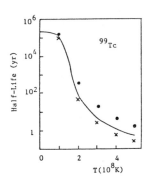

Figure 1. Beta-decay of ^{99}Tc in stellar interiors. Energies are in keV. The Gamow-Teller transitions of astrophysical interest are indicated with a question mark. The beta decay of the $1/2^-$ isomeric state at 143 keV is known to be slow (Lederer and Shirley 1978). Predicted effective beta-decay half-lives are plotted on the right-hand-side against temperature T : Cosner and Truran 1981 (crosses), and Yokoi and Takahashi 1985 (curve) have used Gamow-Teller matrix elements for the unknown transitions obtained from empirical systematics, while Takahashi, Mathews and Bloom 1985 (dots) have performed a shell-model calculation. We will adopt the second calculation as 'conservative' estimates for the half-life. [Note that the longer the half-life the more favorable for the ^{99}Tc survival.]

Recently, Mathews et al. (1985) have performed a detailed s-process network within a framework of schematized versions of the thermal pulse model (Iben 1977; Iben and Truran 1978; Cosner, Iben and Truran 1980) to show that the ^{99}Tc depletion due to the enhanced beta-decay rates is largely compensated by its production due to even more enhanced ^{22}Ne $(\alpha,n)^{25}$Mg rates such that a high temperature does not necessarily mean a low abundance of ^{99}Tc or vice versa. Based on a simple two-reservoir model (Anders 1958; Peterson and Wrubel 1966), in addition, they have resurrected the classical idea (Anders 1958; Cameron 1957) that the observed abundances of ^{99}Tc and elements with neighboring atomic numbers can be a good indicator of recurrent stellar mixing timescales in opposed to the idea of a stellar temperature (e.g. Smith and Wallerstein 1983).

2. A THERMAL PULSE MODEL FOR A 2.25 $M_\odot$ STAR

Our aim is to examine the compatibility of thermal-pulse s-process models with the observed surface abundances of Tc and elements with neighboring atomic numbers. In the present work, we rely on the detailed characteristics of the 18th pulse in a 2.25 $M_\odot$ star (Becker 1981), and simulate the other pulses with the aid of the simple analytic expressions given by Iben and Truran (1978). It should be noted here, however, that the maximum temperatures at the base of the convective shell, $T_{base,max}$ for low C-O core mass M_c ($\leq 0.96\ M_\odot$) turned out to be considerably higher (Becker 1981) than previously thought (Iben and Truran 1978) and almost independent of M_c there. We have therefore adopted $T_{base,max}=2.8\times10^8$ K, the value for the 18th pulse, and kept it constant for all the pulses ($M_c \leq 0.8\ M_\odot$ for our 2.25 $M_\odot$ star). The temperature profile during the 18th pulse can be well approximated to be linear as a function of the mass coordinate as well as in time (until near the end of the lifetime of the convective shell). The density profile is also taken from the 18th pulse results which show $\rho \propto T^n$ with $n=2$. We have assumed that these dependences hold for every pulse. For simplicity, furthermore, we have approximated the shape of the convective shell as triangular after its maximum amplitude (Fig.2).

As for the other physical quantities, we have utilized the simple analytic approximation formulas (Iben and Truran 1978) as functions of M_c in order to follow the recurrent pulses, but with slight readjustments of the parameter values so as to be consistent with the results for the 18th pulse. For simplicity, we have also assumed that the third dredge-up operates from the first pulse and takes place at a time of ~160 yr (the value for the 18th pulse) after each convective shell dies.

3. s-PROCESS AND INPUT DATA

We have performed an s-process network for nuclei heavier than ^{22}Ne for short time intervals during each pulse. Since the timescale of the convection evaluated from the mixing length theory is as short as 5 days for the 18th pulse, we assume that matter is throughly mixed during the time interval, and calculated the relevant nuclear reaction rates by averaging over the above-mentioned temperature and density profiles for

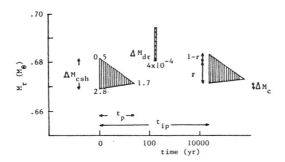

Figure 2. Schematic model adopted to describe the He-shell recurrent thermal pulses in a 2.25 $M_\odot$ star. The left triangle represents the convective shell characteristics at the 18th pulse after its maximum amplitude is reached. The attached numbers are the temperatures in 10^8 K The other physical quantities are ΔM_{csh}: the maximum mass of the convective shell, r: the overlap of the consecutive pulses, ΔM_c: the net increase of the core mass, ΔM_{dr}: the dredged-up mass, t_p: the pulse duration, and t_{ip}: the interpulse period.

each time step. This procedure, especially the averaging the neutron density, may not be strictly valid (Arnould 1985) if one considers a much shorter timescale: A then tiny portion of matter near the top of the convective shell feels only low temperatures and thus low neutron densities, and will be left above the shell before the mixing. To follow all this requires a detailed, time-consuming multi-mass-zone calculation with a fine time step. Fortunately, however, this may not cause serious errors with respect to the abundances in the dredged-up material for the following reason. Since the heavy elements to be dredged up experience the pulse for only a short period (see Fig.2), their composition will be essentially the same as that just before the pulse, namely a mixture of surface composition and the one resulted from the previous pulse (and interpulse) in which most matter had enough time to be mixed by the convection.

We have assumed that the radioactive elements left above the convective shell decay at respective temperature, and if dredged up they undergo free decays with the terrestrial half-lives. Judging from the composition at the 18th pulse, we take the ^{22}Ne abundance to be 1% throughout each pulse. The initial surface composition of the heavier isotopes for the first pulse is assumed to be solar (Anders and Ebihara 1982). The ^{22}Ne(α,n) ^{25}Mg rates are taken from Fowler, Caughlan and Zimmerman (1975), while the adopted neutron capture cross sections and beta-decay rates are those summarized in Howard et al. (1985) and Yokoi and Takahashi (1985), respectively.

4. RESULTS AND DISCUSSION

The resultant <u>surface</u> abundances after some 70 pulses (chosen simply because M_c reaches an estimated WD remnant mass) are displayed in Fig 3 for the mass $70 \leq A \leq 100$ region in comparison with the solar abundances of a few pure s-process nuclides, which show systematic overabundances by a factor up to ~5. Figure 4 shows the evolution of Zr, Nb, Mo and Tc abundances at the surface during the third dredge-up, along with the growth of the core mass. It can be seen that a substantial amount of ^{99}Tc could indeed survive the s-process environment and be dredged up to the surface. The gradual decrease of the Tc abundance is due to the fact that the ^{99}Tc produced by preceding pulses have decayed at the surface.

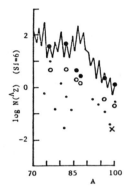

Figure 3. Calculated surface abundances for the mass number $70 \leq A \leq 100$ region in $Si=10^6$ units. The points for pure s-process nuclides are highlighted, the solar values for which are displayed by open circles. The cross corresponds to ^{99}Tc.

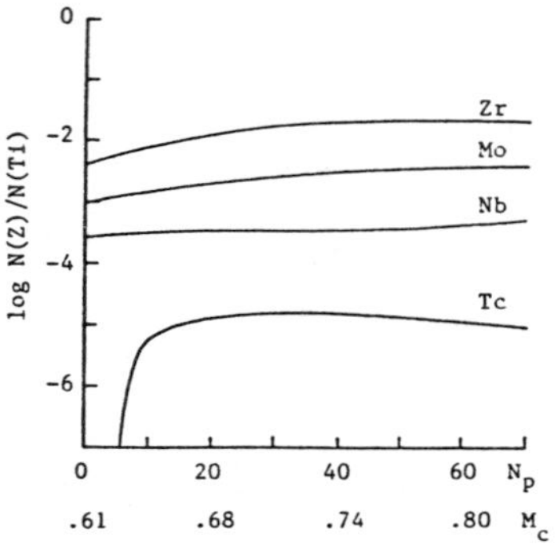

Figure 4. Calculated elemental Zr, Nb, Mo and Tc abundances at the surface relative to the Ti abundance as a function of the number of pulses N_p and the corresponding core mass M_c (in $M_\odot$). The initial values are solar.

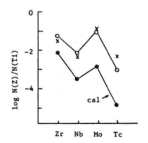

Figure 5. Calculated surface abundances after 20 pulses in comparison with the values observed in R Cmi (open circles) and in CY Cyg (crosses) [Smith and Wallerstein 1983]. The values are relative to the Ti abundance.

Figure 5 compares the calculated abundances (after 20 pulses) and the values observed in two stars by Smith and Wallerstein (1983). It shows that the model cannot produce enough of these elements, although a fair agreement is achieved with respect to the relative abundances. Of course, there is hardly a good reason to expect that this specific model will explain the observation in those specific stars, and we consider that the qualitative agreement of this sort gives us some hopes for further challenges for the eventual understanding of the observations in terms of the s-processing during recurrent thermal pulses in intermediate-mass AGB stars. In particular, the present results of systematically low abundances can be understood as follows. The temperature at the convective shell is relatively low in such low mass stars as we have considered, and the consequent relatively low neutron density is not enough to transform light seed nuclei to heavier nuclei and could easily result in the systematic underproduction seen in Fig.5.

With this background in mind, we are planning a more stringent calculation for a 3 $M_\odot$ star. Nevertheless, it may not be unfair to conclude that significant amounts of ^{99}Tc can be produced via the s-process triggered by the ^{22}Ne$(\alpha,n)^{25}$Mg reaction at the He-shell thermal pulse phase in intermediate mass stars. The ^{99}Tc "problem" may not be much of a problem in the classical sense, but much remains to be worked out along with further observations such as performed by Smith and Wallerstein (1983) before we eventually decipher what Nature is telling us about her principles on ^{99}Tc.

ACKNOWLEDGMENTS This work has benefited from an unpublished note by D.D.Clayton on the thermal pulse model (1984). We also thank M.Arnould, S.D.Bloom, W.M.Howard and K.Yokoi for various collaborations in the background of this report. Work performed under the auspices of the U.S. Department of Energy by the Lawrence Livermore National Laboratory under contract number W-7405-ENG-48. Work supported in part by the Lawrence Livermore National Laboratory Institute for Geophysics and Planetary Physics.

REFERENCES

Anders, E., 1958, Astrophys.J. **127**, 355
Anders, E. and Ebihara, M., 1982, Geochim.Cosmochim.Acta **46**, 2363
Arnould, M., 1985, private communication
Becker, S.A., 1981, in Physical processes in red giants, eds.
    I.Iben, Jr. and A.Renzini (Reidel), p.141
Beer, H., 1986, this meeting
Blake, J.B., Lee, T. and Schramm, D.N., 1973, Nature **242**, 98
Cameron, A.G.W., 1957, Chalk River report CRL-41
_____, 1959, Astrophys.J. **130**, 452
Clayton, D.D., 1969, Nature **224**, 56
Cosner, K. and Truran, J.W., 1981, Astrophys.Space Sci. **78**, 85
Cosner, K., Iben, I.Jr. and Truran, J.W., 1980, Astrophys.J. **238**, L91
Dyson, F.J., 1972, in Aspects of quantum theory (Cambridge Univ.), p.213
Fowler, W.A., Caughlan, G.R. and Zimmerman, B.A., 1975, Ann.Rev.
    Astron.Astrophys. **13**, 69

Fuller, G.M. and Bloom, S.D., 1985, submitted to Nucl.Phys.A
Fuller, G.M., Fowler, W.A. and Newman, M.J., 1980, Astrophys.J.Suppl. 42, 447
_____, 1982, Astrophys.J. **252**, 715
Howard, W.M., Mathews, G.J., Takahashi, K. and Ward, R.A., 1985, submitted to Astrophys.J.
Iben, I.Jr., 1975, Astrophys.J. **196**, 525
_____, 1977, Astrophys.J. **217**, 788
Iben, I.Jr. and Truran, J.W., 1978, Astrophys.J. **220**, 980
Iben, I.Jr. and Renzini, A. 1983, Ann.Rev.Astron.Astrophys. 21, 271
Käppeler, F., 1986, this meeting
Käppeler, F., Beer, H., Wisshak, K., Clayton, D.D., Macklin, R.L. and Ward, R.A., 1982, Astrophys.J. **257**, 821
Klapdor, H.V., Metzinger, J. and Oda, T., 1984, Atom.Nucl.Data Tables 31, 81
Krumlinde, J. and Møller, P., 1984, Nucl.Phys. **A417**, 419
Lederer, C.M. and Shirley, V.S. (eds.), 1978, Table of isotopes, 7th ed. (Wiley)
Mathews, G.J., 1983, Proc.NEANDC specialists meeting on yields and decay data of fission product nuclides, Brookhaven: BNL 51778, p.485
Mathews, G.J. and Ward, R.A., 1985, Rep.Prog.Phys., in the press
Mathews, G.J., Takahashi, K., Howard, W.M. and Ward, R.A. 1985, submitted to Astrophys.J.
Mathews, G.J., Bloom, S.D., Fuller, G.M. and Bahcall, J.N., 1985, submitted to Phys.Rev.C
Merrill, P.W., 1952, Science **115**, 484
Perrone, F., 1971, Ph.D. thesis, Rice University, unpublished
Peterson, V.L. and Wrubel, M.H., 1966, in Stellar evolution, eds. R.F.Stein and A.G.W.Cameron (Plenum), p.419
Schatz, G., 1983, Astron.Astrophys. **122**, 327
Schramm, D.N., 1983, in Essays in nuclear astrophysics eds. C.A.Barnes, D.D.Clayton, D.N.Schramm (Cambridge Univ.), p.325
Smith, V.V. and Wallerstein, G., 1983, Astrophys.J. **273**, 742
Takahashi, K. 1983, Proc.NEANDC specialists meeting on yields and decay data of fission product nuclides, Brookhaven: BNL 51778, p.157
Takahashi, K. and Yokoi, K., 1983, Nucl.Phys. **A404**, 578
Takahashi, K., Mathews, G.J. and Bloom, S.D., 1985, to be published
Takahashi, K., Yamada, M. and Kondoh, T., 1973, Atom.Nucl.Data Tables 12, 101
Ulrich, R.K., 1983, in Essays in nuclear astrophysics, eds. C.A.Barnes, D.D.Clayton, D.N.Schramm (Cambridge Univ.), p.301
Ward, R.A., Newman, M.J. and Clayton, D.D., 1976, Astrophys.J.Suppl. 31, 33
Williams, R.D., Fowler, W.A. and Koonin, S.E., 1984, Astrophys.J. **281**, 363
Woosley, S.E., 1986, this meeting
Yokoi, K. and Takahashi, K., 1985, KfK report 3849 Kernforschungszentrum Karlsruhe, and to be published
Yokoi, K., Takahashi, K. and Arnould, M., 1983, Astron.Astrophys. **117**, 6
_____, 1985, Astron.Astrophys., in the press

NEUTRON CAPTURE NUCLEOSYNTHESIS IN MASSIVE STARS

N. Prantzos[1], J.-P. Arcoragi[2,3], and M. Arnould[3]
[1] Service d'Astrophysique, CEN Saclay, France
[2] Département de Physique, Université de Montréal, Canada
[3] Institut d'Astronomie et d'Astrophysique
Université Libre de Bruxelles, B-1050 Bruxelles, Belgium

ABSTRACT. Neutron-capture nucleosynthesis during core He burning in massive ($50 \leq M_{ZAMS}/M_\odot \leq 100$) mass losing stars is studied in the framework of new stellar models based on the latest available nuclear data. The resulting stellar surface abundances as well as the composition of the stellar winds corresponding to the WC phase of Wolf-Rayet stars are presented. Some implications for the isotopic composition of galactic cosmic rays are also discussed.

1. INTRODUCTION

The $^{13}C(\alpha,n)^{16}O$ and $^{22}Ne(\alpha,n)^{25}Mg$ reactions have long been considered as the most efficient neutron sources for s-process nucleosynthesis /1,2/. Core He burning in massive stars has been identified as a promising site for neutron production by the latter reaction /3/, ^{22}Ne being produced abundantly through $^{14}N(\alpha,\gamma)^{18}F(e^+\nu)^{18}O(\alpha,\gamma)^{22}Ne$, while $^{22}Ne(\alpha,n)^{25}Mg$ becomes operative at temperatures $T \gtrsim 2.5 \; 10^8$ K. Subsequent studies /4,5/ have shown that only s-nuclides with mass number up to A $\approx$ 90 could be substantially overproduced with respect to their solar system abundances in the cores of $M \gtrsim 10 M_\odot$ stars. These limitations arise because neutrons are mainly captured by $^{25,26}Mg$ and other nuclides lighter than ^{56}Fe.

In the past few years, new models for the evolution of massive stars have become available, including important new ingredients, like high mass loss rates and overshooting (e.g. /6,7/). If at least one of those effects is large enough, the He burning products can even appear at the stellar surface during a phase which is identified as the WC-WO stage of Wolf-Rayet (WR) stars. This offers an opportunity for a direct test of nucleosynthetic theories, and may also have other important astrophysical implications /8,9/.

On the other hand, new nuclear data concerning He burning and neutron capture reactions have become available recently. As a few important examples, we could mention the $^{12}C(\alpha,\gamma)^{16}O$ reaction rate which is $\approx$3-5 times higher than previously thought /10-12/, or the $^{22}Ne(n,\gamma)^{23}Ne$ cross-section, which is $\approx$20 times larger than the one used in previous computations /13/. In such conditions, ^{22}Ne could be an

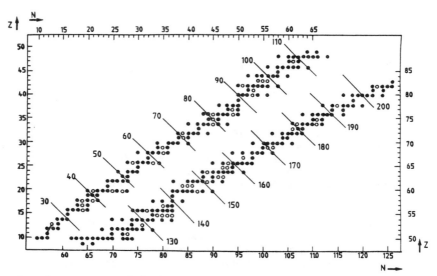

Figure 1. Isotopes used in our s-process network. Full and open circles represent stable and unstable nuclei, respectively

important neutron poison, thus reducing considerably the build-up of heavy nuclides.

In the light of these recent nuclear data and stellar models, a reexamination of the whole s-process in massive stars appears to be called for. Sect. 2 describes the nuclear reaction network, the adopted nuclear data and some computational details. In Sect. 3, we present the evolution of neutron densities, the final abundances, and a comparison with previous computations. Sect. 4 deals with the composition of the stellar ejecta. Some implications for the isotopic composition of the galactic cosmic rays (GCR) are also discussed.

2. NETWORK AND NUCLEAR DATA

The nuclear reaction network used in our computations includes some 340 nuclei (fig. 1) linked by about 450 reactions, comprising α-induced reactions (from 3α up to $^{26}Mg(\alpha,\gamma)^{30}Si$), (n,γ) reactions (from ^{20}Ne up to ^{210}Po), as well as (n,α) or (n,p) reactions, e^--captures, and β-decays.

Charged particle reaction rates are taken from /14-16/. Rates for (n,γ) reactions are adopted from recent compilations or measurements (e.g. /13,17/). In absence of experimental data, the Hauser-Feshbach predictions of /18/ for 30 keV (n,γ) cross sections are selected. The (n,α) or (n,p) rates are derived from /19/. The complete set of our neutron induced reaction rates is given in /20/. Finally, the temperature and density dependent β-decay rates of /21/ are adopted for nuclei beyond the iron peak, while laboratory values are used for lighter nuclei.

Preliminary computations /22/ have shown that neutron concentrations

$n_n < 10^8$ n/cm^3 are probably characteristic of core He burning in $M_{ZAMS} \gtrsim 60$ M⊙ stars. In order to cope with such typical conditions, all unstable nuclei with β-decay half-lives in excess of about 1 year are included in our network. Shorter-lived species are allowed to decay instantaneously.

Our s-process calculations are performed for physical conditions (initial chemical composition, temperature and density as a function of mass coordinate and time) derived from detailed evolutionary sequences for stars in the mass range $50 \leq M_{ZAMS}/M\odot \leq 100$ /6,7,23/. During core He burning, those stars are predicted to develop huge convective cores (≈95% of the total stellar mass) with masses in the range 20-60 M⊙ at the end of He burning. Since convection homogeneizes the chemical composition on time scales ($\tau_c \approx 10^6$ sec) much smaller than a typical integration time step ($\Delta t \approx 10^{10}$ sec), an average reaction rate $\langle \sigma v \rangle_{av} = \int_{M_c} \langle \sigma v \rangle_m \, dM/M_c$ must be used over the whole convective region of mass M_c, $\langle \sigma v \rangle_m$ being the reaction rate at the shell with mass coordinate m. Species with nuclear lifetimes $\tau_N > \tau_c$ are considered to be homogeneously distributed over the whole convective region. In contrast, those for which $\tau_N < \tau_c$ have to be considered in local equilibrium at each shell. This is always the case for neutrons, for which $\tau_N \approx 10^{-4}$ sec under He burning conditions.

3. RESULTS

As stated above, computations are performed for stars with M_{ZAMS} = 50, 60, 80, and 100 M⊙. Because of mass losses during the O and Of phases, the corresponding masses at the beginning of He burning are reduced to M = 38, 45, 60.5, and 76 M⊙, respectively. During the He burning phase, a constant WR mass loss rate of 3 10^{-5} M⊙/yr is applied, resulting in masses M = 22, 31.5, 48, and 64.5 M⊙, respectively, at the end of the He burning phase. Figs. 1-3 of /7/ describe the temporal variations of the stellar and convective core masses of a M_{ZAMS} = 60 M⊙ star, the correspondind evolution of the central temperature and density, as well as the convective core abundances of the light (A < 30) nuclides. The evolution characteristics of the other models are qualitatively similar.

Of great importance for our study is the evolution of the neutron density, presented in fig. 2 for a few mass shells of the 60 M⊙ star. At the very beginning of He burning, a "burst" of neutrons (characterized by a maximum value of the order of 3 10^8 n/cm^3 in the central regions of the considered star) is produced as a result of the $^{13}C(\alpha,n)^{16}O$ reaction, ^{13}C being one of the ashes of the CNO burning. Those burst conditions can be handled quite satisfactorily by the adopted s-process network, as it includes nuclei with β-decay half-lives in excess of about 1 year. However, ^{13}C, as well as ^{17}O and ^{18}O, are very rapidly consumed by α-captures, so that the neutron density goes down to about 10 n/cm^3 after ≈ 1000 years.

With time, ^{22}Ne builds up to a mass fraction of the order of 3 10^{-2} through the $^{14}N(\alpha,\gamma)^{18}F(\beta^+)^{18}O(\alpha,\gamma)^{22}Ne$ reaction chain. However, ^{22}Ne becomes an efficient neutron source for T $\gtrsim$ 2.5 10^8 K only, that is close to the end of He burning. At that stage, peak neutron densities of ≈10^7 n/cm^3 are reached at the stellar center. Since the $^{22}Ne(\alpha,n)^{25}Mg$ reaction

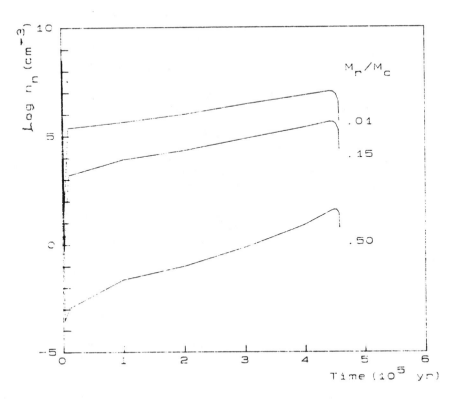

Figure 2. Evolution of the neutron density in different mass shells of the core of a M_{ZAMS} = 60 M☉ star. The mass shells are identified by M_r/M_c (mass inside a sphere of radius r normalized to convective core mass)

rapidly outwards (fig. 2). Consequently, the neutron-capture nucleosynthesis is efficient only in a very small fraction of the total stellar mass contained in the innermost regions. The subsequent dilution of that processed material through convective mixing in the whole stellar core reduces considerably the efficiency of the s-process nucleosynthesis.

The higher the mass of the star, the greater is the neutron production efficiency of the $^{22}Ne(\alpha,n)^{25}Mg$ reaction, since higher central temperatures are attained. For example, the peak central neutron density (excluding the early neutron burst) in the M_{ZAMS} = 100 M☉ star is ≈2 10^7 n/cm³, that is ≈2.2 times higher than the one obtained in the M_{ZAMS} = 50 M☉ star. However, the larger the mass of the convective core, the greater is the quantity of matter where essentially no neutrons are present, and into which the centrally processed material is diluted. It turns out that these two effects cancel, so that the final abundance patterns are very similar for all the studied models. In fact, most of the abundances differ by less than 50% betweeen the M_{ZAMS} = 50 and 100 M☉ stars.

Fig. 3 displays the enhancement factors $X_{i,FINAL}/X_{i,☉}$ obtained for

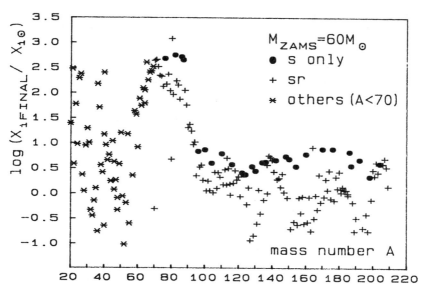

Figure 3. Overabundances in the convective core at the end of core He burning in a $M_{ZAMS} = 60$ M⊙ star

the M_{ZAMS} = 60 M⊙ star, $X_{i\,FINAL}$ and $X_{i\,\odot}$ being the mass fraction of nucleus i in the convective core at the end of central He burning (or, equivalently, the WR phase) and in the solar system, respectively. For 30 ≲ A ≲ 50 nuclei, substantial enhancements factors ($X_{i\,FINAL}/X_{i\,\odot}$ ≳ 10) are obtained for ^{36}S, ^{37}Cl, ^{40}Ar, ^{40}K, ^{50}Ti, and ^{54}Cr. (For A < 30, see /7/). Some A ≲ 80 nuclei are also significantly overproduced, especially ^{65}Cu, ^{67}Ga, ^{71}Ga, ^{70}Ge, ^{72}Ge, ^{76}Se, ^{80}Kr, ^{82}Kr, and ^{87}Sr. In contrast, for A ≳ 90, even "pure" s-nuclei are not substantially overproduced, even if they are in general more enhanced than the other neighboring nuclei.

A direct comparison of these results with the ones presented in /5/ is not easy, since different (constant) stellar masses are considered there. In addition, a number of reaction rates (including ^{12}C(α,γ)^{16}O and ^{22}Ne(α,n)^{25}Mg) are different from the ones used in our computations. Globally, the two studies predict substantial enhancements for some nuclei up to A ≈ 90, and an overproduction by a factor ≲ 10 for heavier s-only nuclei. However, it is important to note that our results for M_{ZAMS} = 60 M⊙ correspond more to the 15 M⊙ model of /5/, which has a helium core of only M_α ≈ 4 M⊙. This means that the "efficiency" of our central He burning s-process is substantially reduced with respect to previous expectations.

In principle, ^{12}C(α,γ)^{16}O competes with ^{22}Ne(α,n)^{25}Mg for the depletion of α-particles, and our use of a higher α-capture rate by ^{12}C should reduce the neutron production efficiency. However, neutrons are liberated only at the end of He burning, when very few ^{12}C is left in our models. In such conditions, the increase of the rate has only a very small net effect on the neutron production efficiency. In contrast, the increase by a factor of ≈ 20 of the $^{22}\text{Ne}(n,\gamma)^{23}\text{Ne}$ cross section is largely responsible for the reduction of the neutron density. A more detailed analysis of these results is given elsewhere /20/.

4. THE COMPOSITION OF THE WC EJECTA AND GCR ISOTOPIC ANOMALIES

According to our models, the high WR mass loss rate sooner or later strips the star of its He-rich envelope, and the core He-burning products appear at the surface (WC phase). They are subsequently ejected by the strong stellar wind into the interstellar medium (ISM), which gets enriched with material of non-standard composition.

For each of our models, the composition of the WC ejecta is computed from $X_{i,WC}^{(W)} = \int_{WC} X_i^{(S)}(t)dt/\tau_{WC}$, where τ_{WC} is the duration of the WC phase, and $X_i^{(S)}(t)$ is the WC surface mass fraction of nucleus i at time t. The ejected mass of nuclide i is thus given by $M_{i,WC}^{(W)} = X_{i,WC}^{(W)} \Delta M_{WC}$, ΔM_{WC} being the total mass lost during the WC phase (ΔM_{WC} = 12.8, 12.6, 11.6, 10.8 M☉ for the stars with M_{ZAMS} = 50, 60, 80, 100 M☉, respectively).

As in the case of the final stellar abundances X_{FINAL}, there is an overall similarity (within a factor of ≈ 3) between the $M_{i,WC}^{(W)}$ values for the different models. There is also a qualitative similarity in the enhancement patterns between the final core and stellar wind compositions. In particular, $30 \lesssim A \lesssim 90$ nuclei are substantially enhanced ($A \lesssim 30$ nuclei are subject to α-induced reactions, and show different patterns in the two cases). Quantitatively, however, the stellar wind abundance enhancements are smaller than those of the core because of the delay between the nucleosynthesis in the core and the appearance of the products at the surface. Indeed, as stated previously, neutron capture nucleosynthesis essentially takes place close to the end of He burning, and the products have no time to appear at the surface by the end of He burning (which corresponds to the endpoint of our calculations). Thus, the enhancement factors in the WC ejecta are typically ≈ 20 times lower than the corresponding ones in the final composition of the core. Of course, the s-process products made near the end of the He burning could be ejected at later times in the interstellar medium, either through stellar winds or through more violent mass ejection (which could be due in particular to a supernova explosion of the pair-creation type).

In order to have a global estimate of the composition of WC ejecta for stars in the $50 \lesssim M_{ZAMS}/M_☉ \lesssim 100$ range, we make an average over the initial mass function ($\phi(M) \alpha M^{-2.5}$, according to /24/). The resulting averaged enhancements factors $f_i = \langle X_{i,WC}^{(W)}/X_☉\rangle$ are given in Table I for nuclei with $f_i > 5$. As already mentioned, these results may have important astrophysical implications concerning

(i) the isotopic composition of the early solar system /8/;

(ii) γ-ray line astronomy, since many radioactive species are produced in the interior of WR stars (for the question of the ^{26}Al 1.8 MeV line, see /9/). However, only species with lifetimes greater than $\approx 10^4$ years have any chance to reach the surface before substantial decay in the ≈ 1 M☉ stellar envelope. Most of the surviving species decay directly to their ground state. Among the few exceptions, the most interesting one seems to be ^{60}Fe (half-life $\approx 3 \cdot 10^5$ yr), which decays via an excited state of

Table I. 12 ≤A≤ 90 nuclei for which the average enhancement factor $f_i > 5$

Nucleus	f	Nucleus	f	Nucleus	f
^{12}C	55.4	^{57}Fe	6.1	^{69}Ga	13.3
^{16}O	31.2	^{58}Fe	40.2	^{71}Ga	15.2
^{21}Ne	50.3	^{59}Co	20.3	^{70}Ge	15.7
^{22}Ne	94.3	^{61}Ni	23.7	^{72}Ge	9.5
^{23}Na	24.0	^{62}Ni	11.6	^{73}Ge	6.2
^{25}Mg	38.5	^{64}Ni	10.1	^{74}Ge	5.3
^{26}Mg	39.5	^{63}Cu	20.9	^{76}Se	11.3
^{30}Si	52.7	^{65}Cu	36.8	^{79}Se	5.2
^{36}S	31.0	^{64}Zn	8.4	^{80}Kr	25.5
^{37}Cl	20.7	^{66}Zn	12.0	^{82}Kr	12.2
^{40}K	88.5	^{67}Zn	15.6	^{86}Sr	12.5
^{54}Cr	8.3	^{68}Zn	11.7	^{87}Sr	11.5

^{60}Co emitting 59 keV photons. Unfortunately, the quantities of ^{60}Fe obtained in our computations (≈1000 times lower than the corresponding ^{26}Al amount) are far too low for the resulting γ-ray lines to be detectable by present day instruments;
(iii) the isotopic composition of the Galactic Cosmic Rays (GCR). According to /25/, the strong stellar winds of WC stars might be at the origin of the anomalies observed in the otherwise standard GCR isotopic composition (e.g. /26/).

Indeed, the observed GCR anomalies in ^{22}Ne/^{20}Ne, ^{25}Mg/^{24}Mg and ^{26}Mg/^{24}Mg could readily be explained if only one out of ≈35 GCR particles comes from the winds of our WR stellar models (the remaining 34 may have

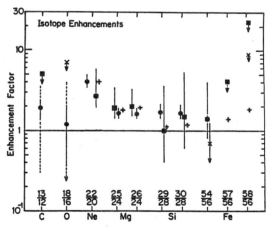

Figure 4. Observed isotopic anomalies in GCR (data taken from /27/) compared with the theoretical predictions from our WR model (+)

an essentially solar-system like composition), as can be seen in fig. 4. On the other hand, our computations show that the WC winds are not able to explain the $^{29}Si/^{28}Si$ and $^{30}Si/^{28}Si$ ratios, but the only relevant observation has not been confirmed. Clearly, more measurements are needed in this case. Some other interesting predictions are made by this WR scenario concerning the $^{36}S/^{34}S$ ratio (predicted to be enhanced in GCR by a factor $f \approx 1.8$), $^{40}K/^{39}K$ ($f \approx 3.5$), or $^{61}Ni/^{60}Ni$ ($f \approx 1.6$).

5. CONCLUSION

The main results of this work can be summarized as follows:
(i) at the end of core He burning in massive stars ($M_{ZAMS} \gtrsim 50 M\odot$), ≈ 1-$2\ 10^7$ neutrons/cm^3 are released through the $^{22}Ne(\alpha,n)^{25}Mg$ reaction;
(ii) the capture of these neutrons on seed nuclei produces some light ($A \lesssim 56$) nuclei and a "weak" s-process (nuclei with $A \lesssim 100$);
(iii) the ejection of these isotopes into the ISM by the WR winds may have interesting implications for γ-ray line astronomy, early solar system isotopic composition, or GCR anomalies.

The work presented in this paper has been performed within the Collaboration on the evolution of massive stars involving the Institut d'Astronomie et d'Astrophysique of the Université Libre de Bruxelles, CEN Saclay, and the Astrophysisch Instituut of the Vrije Universiteit Brussel. The Collaboration is supported by the Belgian Fund of Joint Fundamental Research (FRFC) under Contract Nr. 2.9002.82. M. Arnould is Chercheur Qualifié F.N.R.S. (Belgium)

REFERENCES

1. Cameron, A.G.W.: 1955, *Ap.J.* **121**, 144
2. Cameron, A.G.W.: 1960, *Astron.J.* **65**, 485
3. Peters, J.G.: 1968, *Ap.J.* **154**, 225
4. Couch, R.G., Schmiedecamp, A.B., Arnett, W.D.: 1974, *Ap.J.* **190**, 95
5. Lamb, S.A., Howard, W.M., Truran, J.W., Iben, I.Jr.: 1977, *Ap.J.* **217**, 213
6. de Loore, C., Prantzos, N., Arnould, M., Doom, C.: this volume
7. Prantzos, N., de Loore, C., Doom, C., Arnould, M.: this volume
8. Arnould, M., Prantzos, N.: this volume
9. Cassé, M., Prantzos, N.: this volume
10. Kettner, K.U., Becker, H.W., Buchmann, L., Görres, J., Krawinkel, H., Rolfs, C., Schmalbrock, P., Trautvetter, H.P., Vlieks, A.: 1982, *Z. Phys.* **A308**, 73
11. Descouvemont, P., Baye, D., Heenen, P.-H.: 1984, *Nucl. Phys.* **A430**, 426
12. Trautvetter, H.P.: this volume
13. Almeida, J., Käppeler, F.: 1983, *Ap.J.* **265**, 417
14. Fowler, W.A., Caughlan, G.R., Zimmerman, B.A.: 1975, *Ann. Rev. Astron. Astrophys.* **13**, 69

15. Harris, M.J., Fowler, W.A., Caughlan, G.R., Zimmerman, B.A.: 1983, *Ann. Rev. Astron. Astrophys.* **21**, 165
16. Caughlan, G.R., Fowler, W.A., Harris, M.J., Zimmerman, B.A.: 1985, *Atomic Data Nucl. Data Tables* **32**, 197
17. Käppeler, F., Beer, H., Wisshak, K., Clayton, D.D., Macklin, R.L., Ward, R.A.: 1982, *Ap.J.* **257**, 821
18. Harris, M.J.: 1981, *Astrophys. Space Sci.* **77**, 357
19. Woosley, S.E., Fowler, W.A., Holmes, J.A., Zimmerman, B.A.: 1978, *Atomic Data Nucl. Data Tables* **22**, 371
20. Prantzos, N., Arcoragi, J.-P., Arnould, M.: in preparation
21. Yokoi, K., Takahashi, K.:1985, Report KfK-3849, Kernforschungszentrum Karlsruhe
22. Prantzos, N., Arnould, M.: 1983, in *Wolf-Rayet Stars: Progenitors of Supernovae?*, eds. M.C. Lortet, A. Pitault, Observatoire de Paris-Meudon, p.II.33
23. Prantzos, N., Doom, C., Arnould, M., de Loore, C.: 1985, in preparation
24. Humphreys, R.M., McElroy, D.B.: 1984, *Ap.J.* **284**, 565
25. Cassé, M., Paul, J.: 1982, *Ap.J.* **258**, 860
26. Meyer, J.-P.: 1985, *Ap.J. Suppl.*, **57**, 173
27. Simpson, J.A.: 1983, *Ann.. Rev. Nucl. Part. Sci.* **33**, 323

A PARAMETRIZED STUDY OF THE $^{13}C(\alpha,n)^{16}O$ NEUTRON SOURCE

A. Jorissen, M. Arnould
Institut d'Astronomie et d'Astrophysique
Université Libre de Bruxelles, C.P. 165
Av. F.D. Roosevelt 50
B-1050 Bruxelles, Belgium

ABSTRACT. After briefly reviewing the astrophysical sites where protons could be mixed into He- and C-rich layers, we present the main features of the neutron processing resulting from that mixing. Those are used as a guideline for an analytical investigation of the conditions required for $^{13}C(\alpha,n)^{16}O$ to be an efficient neutron source when $T \lesssim 1.5\ 10^8$ K. We analyze the sensitivity of the results to some reaction rate uncertainties

1. INTRODUCTION

The relative importance of the $^{13}C(\alpha,n)^{16}O$ and $^{22}Ne(\alpha,n)^{25}Mg$ neutron sources for the synthesis of the heavy elements in specific stars or at a galactic level has been largely debated, and still remains more or less controversial in many instances.

The $^{13}C(\alpha,n)^{16}O$ reaction has raised much interest recently following various studies pointing out the possibility for protons to be mixed into He- and C-rich zones at temperatures $T \gtrsim 10^8$ K ($T_8 \gtrsim 1$). Under such conditions, the $^{12}C(p,\gamma)^{13}N(e^+\nu)^{13}C(\alpha,n)^{16}O$ chain can indeed develop, producing a substantial amount of energy along with neutrons. A preliminary study of this scenario can be found in /1/.

The interest for the ^{13}C α-captures has been amplified further because the $^{22}Ne(\alpha,n)^{25}Mg$ neutron source might have difficulties in explaining various observations, namely the s-process enrichment in certain classes of relatively low-mass stars (e.g./2/), or even the solar system s-process content (e.g./3/).

After a brief review of the various stellar sites where protons are expected to be mixed into He- and C-rich layers (Sect.2), we reanalyze the ability of the $^{13}C(\alpha,n)^{16}O$ reaction to produce neutrons. This work is motivated by the fact that the studies already devoted to that problem
(i) rely too heavily on predictions of specific stellar models, many features of which are still very uncertain. We adopt instead a parametrized approach which is hoped to encompass a large variety of physically plausible situations, and is therefore more illuminating than specific studies;
(ii) are essentially concerned with the production of heavy elements, and

at most with a (too) limited number of light nuclei. Our study deals instead with all the nuclei between ^{12}C and ^{28}Si. We examine if the observational determination of the abundances of certain of those species could unravel specific signatures of the considered nucleosynthesis process;
(iii) make use of nuclear reaction rates for which new data or evaluations are available. On the other hand, some reactions which might be important in certain relevant conditions have been neglected in previous studies. Finally, and in contrast to the existing work in the field, we analyze the sensitivity of our predictions to changes in certain key reaction rates which are still very uncertain.

2. A BRIEF REVIEW OF THE SITES IN WHICH ^{13}C$(\alpha,n)^{16}$O COULD OPERATE

As mentioned in Sect.1, we limit ourselves to sites in which the ^{13}C$(\alpha,n)^{16}$O reactions result from a proton mixing into He-rich layers. We do not consider here the occurence of that reaction at the beginning of He burning, where the available ^{13}C comes from the CNO cycle (e.g./4/), or during C burning in certain massive stars (e.g./5/).

2.1. Central He flash in M $\lesssim$ 2 M⊙ stars

The possibility has been raised of a proton mixing at that stage with a concomitant neutron production (which some authors even consider to lead to a r-type neutron processing (e.g./6/)). However, the modeling of the central He flash encounters many difficulties /7/, and the occurence of the required mixing is still very uncertain.

2.2. Asymptotic giant branch (AGB) evolution of single low and intermediate mass stars

The question of mixing in such circumstances has been studied by several authors (e.g./8-10/). They demonstrate that the entropy barrier between the H-burning shell and the recurrently "thermally unstable" He-burning layer characterizing the considered stars is large enough to prevent the mixing of protons into the He-burning zone. It is also demonstrated that the mixing obtained in /11/ is just an artefact of the neglect of radiation pressure.
However, it has been proposed that the mixing of interest might take place as a result of semiconvection in rather low mass AGB stars (e.g./2/ for a review). Unfortunately, such a possibility cannot be firmly established in view of the difficulties of modeling semiconvection.

2.3. Nuclei of planetary nebulae on their way to white dwarfs

The possibility has been raised by several authors (e.g./12,13/) of the occurence of a "last" thermal pulse in such stars, leading possibly to quite spectacular effects on the evolutionary tracks in the H-R diagram, as well as on the surface composition, resulting in particular from the mixing of protons into the nuclearly active He zone. The occurence of such a mixing in the considered stars, which have a low mass envelope and a strong nuclear shell source, agrees with the mixing criteria established in /10/.

However, many problems remain in the structural and nucleosynthesis modeling of the last pulse. For example, if the proton injection rate in the He-burning layer is as high as 2.10^{45} protons s^{-1} /12/, the stability of that zone against convection could be substantially altered /14/. On the other hand, no detailed treatment of the element production during that pulse has ever been attempted.

2.4. Massive star evolution

Calculations indicate that some H-rich material could mix into He layers at the onset of shell C-burning in a 9 Me star /15/. This might be characteristic of stars in the approximate $8 \lesssim M/M_\odot \lesssim 10$ range. However, no calculation of the ensuing nucleosynthesis has been performed.

On the other hand, some proton mixing into the He burning core is found at the beginning of the WN phase in certain of the massive mass losing stars modeled in /16/. The exact circumstances under which such a phenomenon could occur and its eventual consequences will be discussed elsewhere.

2.5. Accretion of H-rich material on a white dwarf

As indicated by many studies (e.g. /17-19/), this scenario might lead, under certain circumstances, to a double shell structure similar to that of the AGB stars, and to some H-He mixing if the theoretical criteria mentioned in /10/ are fulfilled. However, no detailed study of such a possibility has been conducted to-date in this context. Also note that this accretion scenario (and the possible concomitant nucleosynthesis) might apply to certain symbiotic stars /20/.

3. PRELIMINARY ANALYSIS OF LIGHT NUCLIDE ABUNDANCES

3.1 Network Calculations

Our network involves all the stable nuclei in the $12 \leq A \leq 28$ mass range, as well as the unstable nuclei ^{13}N, ^{14}C, ^{22}Na, and ^{26}Al. Those species are linked by proton-, neutron-, α-captures and β-decays which can be of importance in the temperature range $1 \leq T_9 \leq 3$. The nuclear reaction rates are taken from /21-28/. We also include a neutron sink nucleus. Its mass fraction is kept constant at the value $X_{sink} = \Sigma_i {}^{56}_{29} X_i e$, where $X_i e$ is the solar system mass fraction of species i /29/, while its neutron capture cross section is taken equal to $\sigma_{sink} = \Sigma_i {}^{56}_{29} \sigma_i X_i e$, where σ_i is the 30 keV Maxwellian-averaged radiative neutron capture cross section of nucleus i, taken from /30/. The initial mass fractions are representative of a mid-He burning composition ($X(^4He)=.49$, $X(^{12}C)=.47$, $X(^{13}C)=3.5\ 10^{-4}$, $X(^{14}N)=1.0\ 10^{-8}$; low ^{15}N, ^{17}O content; high ^{16}O, ^{18}O, ^{20}Ne, ^{22}Ne content). One proton per 10 ^{12}C nuclei is injected in this material ($X(^1H)=3.9\ 10^{-3}$).

The time evolution of several abundances when $T_9=1$ and $T_9=3$ ($\rho=500$

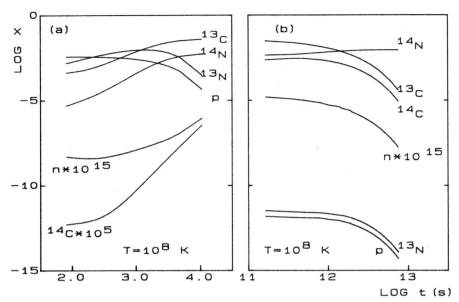

Figure 1. Time variation of the mass fractions of various nuclei, resulting from the injection at t=0 of one proton per 10 ^{12}C nuclei in a material with mid-He burning composition (see text). Temperature and density are assumed to remain constant at $T=10^8$ K and $\rho=500$ gcm^{-3}. Parts (a) and (b) correspond to phases characterized by ^{13}C production and destruction, respectively. They are separated by a ^{13}C stationary stage

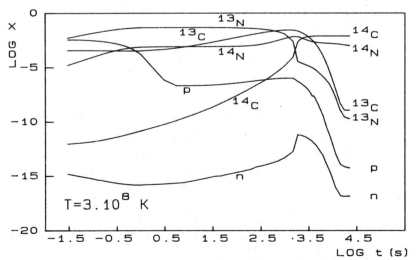

Figure 2. Same as Fig. 1, but for $T=3\ 10^8$ K. In this case, the ^{13}C production and destruction phases [parts (a) and (b) of Fig. 1] are not separated by a ^{13}C stationary stage

Table I : Schematic presentation of the main features of the nuclear processing resulting from the mixing of protons in He- and C-rich matter (based on caculations for $\rho=500$ gcm^{-3}, $p(o)/^{12}C(o)=.1$)

$T_8 =$	1.0	1.5	2.0	2.5	3.0
I	\multicolumn{4}{l}{^{13}C synthesis phase separated from neutron liberation phase by ^{13}C stationary stage}	no ^{13}C stationary stage
II	^{13}N at quasi-equilibrium with ^{12}C through $^{12}C(p,\gamma)^{13}N(e^+\nu)$ $^{13}N(p,\gamma)^{14}O$		^{13}N governed by $^{13}N(\alpha,p)^{16}O$ and $^{13}N(n,p)^{13}C$ and if $\rho X(^1H)$ sufficently high		
III	protons described by $p=p(0)\exp(-t/\tau_p)$		proton, neutron, ^{13}N, ^{13}C abundances strongly coupled		
	\multicolumn{5}{l}{ε_{peak} (erg g^{-1} s^{-1})=}
	8.2 10^{12}	8.5 10^{13}	1.0 10^{15}		2.5 10^{16}
	\multicolumn{5}{l}{E_{tot} (erg g^{-1})=}
	2.2 10^{16}	2.4 10^{16}	2.6 10^{16}		5.8 10^{16}
	\multicolumn{5}{l}{$\Delta t(s)=$}
IV	1.0 10^4	2.2 10^3	1.6 10^3		1.0 10^3
V	\multicolumn{5}{l}{n- and p- captures coupled through main neutron poison(s): $^{14}N(n,p)^{14}C$ \hspace{2em} $^{13}N(n,p)^{13}C$ and $^{14}N(n,p)^{14}C$}
	\multicolumn{5}{l}{N_n^{max} (cm^{-3})=}
	6.0 10^8	8.0 10^9	7.0 10^{12}		2.0 10^{15}
	\multicolumn{5}{l}{Total integrated neutron exposure (mb^{-1})=}
VI	.59	.87	1.04		.62
	\multicolumn{5}{l}{Main ashes of the process:}
VII	s-process		intermediate between s- and r-process		
VIII	$^{14}N(\alpha,\gamma)^{18}F$ inefficient		^{14}N if matter removed from He-burning zone before $^{14}N(\alpha,\gamma)^{18}F$ takes place		
IX	$^{14}C(\alpha,\gamma)^{18}O$ inefficient		^{14}C if matter removed from He-burning zone before $^{14}C(\alpha,\gamma)^{18}O$ proceeds		

gcm^{-3} in both cases) are given in Figs. 1 and 2, while the main nuclear features (independent of any stellar model input) are outlined in Table I. Features I to IV and V to IX refer to the ^{13}C synthesis and neutron liberation phases, respectively. Those features are used as a guideline for the construction of approximate analytical descriptions of the nuclear processing that provide a useful way to analyze the results, and in particular the conditions required for the ^{13}C$(\alpha,n)^{16}$O reaction to be an efficient neutron source.

Since the situation around $T_8=1$ is quite different from that around $T_8=3$, two different analytical approaches are needed. Sect. 3.2 outlines the main features of the formalism which is well suited for $T_8=1$. The analytical treatment of the $T_8=3$ case will be presented elsewhere.

3.2 Analytical approximation for $T_8 \lesssim 1.5$

On the basis of items I to III of Table I, and assuming that the ^{12}C abundance remains constant in time (valid if $p(o)/^{12}C(o) \lesssim 1$), we find that the number ratio of ^{13}C at the end of the proton capture phase to the initial ^{12}C is given by Eq.(1):

$$\frac{^{13}C(\infty)}{^{12}C(o)} = f_{13N}(T,\rho X_p(o)) \frac{\langle\sigma v\rangle_{12C,p\gamma}}{\langle\sigma v\rangle_{13C,p\gamma}} [1-\exp(-\tau_p/\tau_{13C})] + \exp(-\tau_p/\tau_{13C})\frac{^{13}C(o)}{^{12}C(o)}$$

where

$$f_{13N}(T,\rho X_p(o)) = [\tau_+(^{13}N) \rho X_p(o) N_{Av}\langle\sigma v\rangle_{13N,p\gamma} + 1]^{-1}$$

is the branching ratio between $^{13}N(e^+\nu)^{13}C$ and $^{13}N(p,\gamma)^{14}O$, $\tau_+(^{13}N)$ being the ^{13}N β-decay lifetime, $\langle\sigma v\rangle_{12C,p\gamma}/\langle\sigma v\rangle_{13C,p\gamma}$ is just the $^{12}C/^{13}C$ equilibrium value (depending on T only), and where the factor in square brackets is related to the proton abundance decrease during the ^{13}C synthesis (this factor is absent in classical CNO studies, where p>>CNO). In our analysis, a central role is played by the ratio of the proton lifetime against capture to the ^{13}C lifetime against proton capture

$$\frac{\tau_p}{\tau_{13C}} = \frac{\langle\sigma v\rangle_{13C,p\gamma}}{\langle\sigma v\rangle_{12C,p\gamma}} \frac{p(o)}{^{12}C(o)} [1 + \Sigma_J \frac{\langle\sigma v\rangle_{J,p}}{\langle\sigma v\rangle_{12C,p\gamma}} \frac{N_J(o)}{^{12}C(o)}]^{-1},$$

where j denotes all p-poisons other than ^{12}C and ^{13}C (essentially $^{15}N(p,\alpha)^{12}C$ and $^{18}O(p,\alpha)^{15}N$).

Eq.(1) allows us to express the conditions required for $^{13}C(\alpha,n)^{16}O$ to be efficient at $T_8 \lesssim 1.5$:
i) $^{13}N(e^+\nu)^{13}C$ must be more rapid than $^{13}N(p,\gamma)^{14}O$. This sets a limitation on $\rho X_p(o)$. Of course, this constraint depends upon the $^{13}N(p,\gamma)^{14}O$ rate, which is still somewhat uncertain, and has been revised recently /31/;
ii) $p(o)/^{13}C(o) > 1$, for the ^{13}C synthesis to be useful;
iii) low $^{18}O/^{12}C$ during He-burning. High values of this ratio may reduce significantly the ^{13}C production efficiency;
iv) $p(o)/^{12}C(o) < 1$, in order for protons to make ^{13}C rather than ^{14}N.

From Eq.(1), we may calculate the efficiency $^{13}C(\infty)/^{12}C(o)$ of the ^{13}C synthesis as a function of $p(o)/^{12}C(o)$. The results for $T_8=1$ are

Table II. ^{13}C synthesis efficiency at $T_8=1$ as a function of $p(o)/^{12}C(o)$

$p(o)/^{12}C(o)$	1	.1	.01	.001
$^{13}C(\infty)/p(o)$	.28	.84	.98	.99
$[^{14}N(\infty)-^{14}N(o)]/^{12}C(o)$	.36	$8.\ 10^{-3}$	$1.\ 10^{-4}$	$5.\ 10^{-6}$

presented in Table II, which shows that most of the protons injected initially are used to synthesize ^{13}C only when $p(o)/^{12}C(o) < 0.01$. In the other cases, those protons serve to build up a large amount of ^{14}N, which can be calculated to a good approximation from the requirement of baryon number conservation. This leads to

$$^{14}N(\infty) = {}^{14}N(o) + \tfrac{1}{2}\, p(o)\, (1 - {}^{13}C(\infty)/p(o)) \ . \qquad (2)$$

Thus, ^{14}N acts as a strong neutron poison during the neutron liberation phase through $^{14}N(n,p)^{14}C$ (due to that proton production, a correct analysis of this phase must involve proton capture reactions; see the discussion about $^{17}O/^{18}O$ in Sect. 3.3). Eqs. (1) and (2) allow also to estimate the maximum neutron density which can be achieved (Table I):

$$N_n = \frac{\langle \sigma v \rangle_{^{13}C,\alpha n}\ ^4He\ ^{13}C(\infty)}{\langle \sigma \rangle_{^{12}C,n} v_T\ ^{12}C(o) + \langle \sigma \rangle_{^{13}C,n} v_T\ ^{13}C(\infty) + \langle \sigma v \rangle_{^{14}N,np}\ ^{14}N(\infty) + \langle \sigma \rangle_{sink} v_T N_{sink}} ,$$

where v_T is the most probable relative neutron velocity at 30 keV, while the other symbols have their usual meaning.

Even more important is the fact that the presence of ^{14}N (its destruction by α-captures is not efficient on the relevant time scales at $T_8=1$) and of ^{14}C (if $t < \tau_-({}^{14}C)$; Fig.1) together with heavy elements could be a signature of the $^{13}C(\alpha,n)^{16}O$ neutron source.

3.3. Sensitivity of the results to reaction rates

Up to now, we have looked for possible alterations of the previous results due to cross section uncertainties only at $T_8=1$, and by varying the $^{12}C(n,\gamma)^{13}C$, $^{14}N(n,\gamma)^{15}N$, and $^{14}N(n,p)^{14}C$ rates.

An increase by a factor of 10 of the ^{12}C radiative neutron capture cross section value given in /23/ leads to a time delay in the neutron irradiation rise. This is due to the fact that more neutrons are trapped in the $^{12}C(n,\gamma)^{13}C(\alpha,n)^{16}O$ chain. However, the neutron production process as a whole is slightly more efficient (total integrated neutron exposure of 0.64 mb^{-1}, instead of 0.59 mb^{-1} in the standard case; see Table I).

More dramatic consequences result from changes of the $^{14}N(n,p)^{14}C$ and $^{14}N(n,\gamma)^{15}N$ rates by factors 0.5 and 10, respectively. The most evident implication is the production of a considerable ^{15}N amount. At the end of the process at $T_8 \approx 1$, a ratio $(^{14}N+^{14}C)/^{15}N = 10$ (i.e. $X_{15} = 10^{-3}$) is obtained, while this quantity reaches a value of ≈ 1100 in the standard case!

Another direct consequence is that protons are now underabundant by more than a factor of 10 relative to the standard case. This means that the neutron source efficiency is slightly lowered, because the ^{13}C regeneration that occurs even during the neutron liberation phase through $^{12}C(p,\gamma)^{13}N(\beta^+)^{13}C$ is now reduced.

A further interesting change, which is quite unexpected at first glance, concerns $^{17}O/^{18}O$. This ratio reaches an equilibrium value of 230, compared with 1200 in the standard case. This can be understood in terms of an equality between the $^{14}C(\alpha,\gamma)^{18}O$ and $^{18}O(p,\alpha)^{15}N$ flows and of the equilibrium between ^{17}O and ^{18}O through $^{16}O(n,\gamma)^{17}O(n,\alpha)^{14}C$ and $^{16}O(p,\gamma)^{17}F(\beta^+)^{17}O(n,\alpha)^{14}C$ (those two chains are roughly equally probable). Thus, the decrease of $\sigma_{n,p}(^{14}N)$ by a factor of 2 reduces the number of available protons, and consequently the ^{17}O production. At the same time, the ^{18}O production is increased through $^{14}C(\alpha,\gamma)^{18}O$, while its destruction by $^{16}O(p,\gamma)^{19}F$ is reduced.

4. CONCLUSIONS

The results reported in this paper represent the first step towards a more thorough reinvestigation of the $^{13}C(\alpha,n)^{16}O$ neutron source, and of the ensuing heavy element synthesis. Such a nuclear processing could be activated in various astrophysical sites, especially as a result of the mixing of protons into He- and C-rich layers at rather high temperatures.

We stress the interest of a parametrized approach to the question, as well as of the development of analytical approximations. Those greatly help understanding the main features of the rather complex nuclear pattern possibly characterizing the H-He mixing considered here, and can in particular provide an "educated appraisal" of the consequences of certain reaction rate uncertainties.

Our study will be extended in various directions. In particular, more systematic numerical investigations will be conducted, involving the light ($A \leq 28$) nuclei already considered here, as well as the heavier ($28 < A \leq 210$) species, the abundances of which could be affected by the more or less strong neutron irradiation resulting from the operation of $^{13}C(\alpha,n)^{16}O$. Such calculations will be performed with the aid of an extension of the already large network used in /4/, and will always put special emphasis on the role of key nuclear reaction rate uncertainties.

As a necessary complement to that numerical work, the development of analytical approximations will be pursued, particularly for $T \gtrsim 1.5 \, 10^8$ K.

REFERENCES

1. Sanders, R.H.: 1967, *Ap.J.* **150**, 971
2. Iben, I.Jr., Renzini, A.: 1983, *Ann. Rev. Astron. Astrophys.* **21**, 271
3. Howard, W.M.: this volume
4. Prantzos, N., Arcoragi, J.-P., Arnould, M.: this volume
5. Arnett, W.D., Thielemann, F.-K.: 1984, preprint
6. Cowan, J.J., Cameron, A.G.W., Truran, J.W.: 1984, in *Stellar Nucleosynthesis*, eds. C. Chiosi, A. Renzini, Reidel, p. 151
7. Deupree, R.G.: 1984, *Ap.J.* **282**, 274; and *Ap.J.* **287**, 268
8. Despain, K.H., Scalo, J.M.: 1976, *Ap.J.* **208**, 789

9. Iben, I.Jr.: 1976, *Ap.J.* **208**, 165
10. Fujimoto, M.: 1977, *Pub. Astron. Soc. Japan* **29**, 331
11. Schwarzschild, M., Härm, R.: 1967, *Ap.J.* **150**, 961
12. Schönberner, D.: 1979, *Astron. Astrophys.* **79**, 108
13. Iben, I.Jr., Kaler, J.B., Truran, J.W., Renzini, A.: 1983, *Ap.J.* **264**, 605
14. Sweigart, A.V.: 1974, *Ap.J.* **189**, 289
15. Weaver, T.A., Axelrod, T.S., Woosley, S.E.: in *Proceedings of the Texas Workshop on Type I Supernovae*, ed. J.C. Wheeler, University of Texas and McDonald Observatory, p. 113
16. de Loore, C., Prantzos, N., Arnould, M., Doom, C.: this volume
17. Sienkiewicz, R.: 1975, *Astron. Astrophys.* **45**, 411
18. Nomoto, K., Sugimoto, D.: 1977, *Pub. Astron. Soc. Japan* **29**, 765
19. Iben, I.Jr.: 1981, *Ap.J.* **243**, 987
20. Paczynski, B., Rudak, B.: 1980, *Astron. Astrophys.* **82**, 349
21. Wagoner, R.V.: 1969, *Ap.J. Suppl.* **18**, 247
22. Macklin, R.L.: 1971, *Phys. Rev.* **C3**, 1735
23. Fowler, W.A., Caughlan, G.R., Zimmerman, B.A.: 1975, *Ann. Rev. Astron. Astrophys.* **5**, 525
24. Fowler, W.A., Caughlan, G.R., Zimmerman, B.A.: 1975, *Ann. Rev. Astron. Astrophys.* **13**, 69
25. Woosley, S.E., Fowler, W.A., Holmes, J.A., Zimmerman, B.A.: 1978, *Atomic Data Nucl. Data Tables* **22**, 371
26. Görres, J., Becker, H.W., Buchmann, L., Rolfs, C., Schmalbrock, P., Trautvetter, H.P., Vlieks, A.: 1983, *Nucl. Phys.* **A408**, 372
27. Harris, M.J., Fowler, W.A., Caughlan, G.R., Zimmerman, B.A.: 1983, *Ann. Rev. Astron. Astrophys.* **21**, 165
28. Mathews, G.J., Dietrich, F.S.: 1984, preprint UCRL-90349
29. Anders, E., Ebihara, M.: 1982, *Geochim. Cosmochim. Acta* **46**, 2363
30. Almeida, J., Käppeler, F.: 1983, *Ap.J.* **246**, 52
31. Mathews, G.J., Dietrich, F.S.: 1984, *Ap.J.* **287**, 969

IV – CHEMICAL EVOLUTION AND CHRONOMETERS

TYPE I SNe FROM BINARY SYSTEMS:
CONSEQUENCES ON GALACTIC CHEMICAL EVOLUTION

F. Matteucci
European Southern Observatory
Karl-Schwarzschild-Str. 2
D-8046 Garching bei München, F.R.G.

L. Greggio
Dipartimento di Astronomia
C.P. 596
I-40-100 Bologna, Italy.

ABSTRACT. The binary model for type I SN precursors is assumed, and the abundances relative to the Sun of several chemical elements (^{12}C, ^{16}O, ^{24}Mg, ^{28}Si, ^{56}Fe) are computed by means of a chemical evolution model for the solar vicinity. The contribution to the galactic enrichment by type II SNe is also taken into account. We conclude that, the binary model for type I SNe together with the available nucleosynthesis computations and an initial mass function constant in space and time can account for the observational behaviour of most of the studied elements.

1. INTRODUCTION

Both observations and theory suggest that type I SNe are likely to be important contributors to the iron and, to some extent, to the elements from ^{12}C to ^{28}Si in galaxies. In fact, SNI light curves have been successfully explained in terms of $^{56}Ni \rightarrow {}^{56}Co \rightarrow {}^{56}Fe$ radioactive decay, requiring an amount of ^{56}Ni in the range 0.25-0.7 $M_\odot$ (Arnett, 1979, 1982; Colgate et al., 1980; Weaver et al., 1980). In addition, the presence of iron and elements from ^{12}C to ^{28}Si has been found in the early and late time SNI spectra (Branch et al., 1983).

Progenitors of type I SNe can, in principle, be envisaged both in binary systems and in single stars. White dwarfs in binary systems which accrete matter from a companion can ignite carbon under degenerate conditions, when their mass reaches the Chandrasekhar limit, and explode without leaving any remnant. Single stars with initial masses in the range 4-8 $M_\odot$ could also explode by C-deflagration, when their core reaches the Chandrasekhar limit, but this depends crucially on the rate of mass loss that they have experienced before. In fact, if the mass loss process reduces the mass of the star below the Chandrasekhar limit, carbon will never be ignited in the core and the star will end its life as a white dwarf. It is worth noting that in both cases, i.e. white dwarf in binary system and single C-deflagrating star, the explosive nucleosynthesis products are the same, since the mechanism of the explosion is C-deflagration occurring in a core of ~1.4 $M_\odot$. Nucleosynthesis computations predict, in this case, that a substantial amount of iron, as well as elements from C to Si, are produced (Nomoto et al., 1984).

The effects of C-deflagration in single stars (type I1/2 SNe) on galactic chemical evolution have already been explored by Matteucci and Tornambè (1985). Their conclusion was that the main source of iron in the Galaxy must be envisaged in stars with intermediate masses, no matter what is the assumed model for type I SN precursors, which instead influences the temporal behaviour of the type I SN rate. However, since the binary model for type I SN precursors seems to be the most likely in order to explain light curves and spectra of type I SNe (see Matteucci and Greggio, 1985, and references therein), it seems worthwhile to take into account this model in the computation of the type I SN rate.

For the binary model for type I SN precursors we have some possibilities: i) a He or C-O white dwarf, accreting material rich in H and He from a companion filling its Roche lobe (Whelan and Iben, 1973) or ii) a C-O white dwarf accreting matter from another C-O white dwarf (Iben and Tutukov, 1984). Although model ii) is the most likely explanation for type I SNe (see Iben and Tutukov, 1984), we have assumed here model i), for the sake of simplicity. The main difference between model i) and model ii) consists in the clock for the explosion. In the former, the clock for the explosion is directly given by the evolutionary lifetime of the secondary component, which is defined as the primordially less massive star in the system. In the latter, a further delay is introduced by the time taken by the gravitational wave radiation to cause the final merge of the system (from 10^5 to 10^{10} years). In computing the chemical evolution model we followed the formulation for the type I SN rate given by Greggio and Renzini (1983) in the frame of model i).

2. THE CHEMICAL EVOLUTION MODEL

The main features of the chemical evolution model are:
1) one-zone, with the instantaneous mixing of gas,
2) no instantaneous recycling approximation (i.e. stellar lifetimes are taken into account),
3) the evolution of several chemical elements (^{12}C, ^{16}O, ^{24}Mg, ^{28}Si, ^{56}Fe) due to stellar nucleosynthesis, stellar mass ejection and infall of primordial gas, is followed in detail.

If G_i is the fractional mass of an element i with abundance by mass X_i, for each element we can write the following equation:

$$\dot{G}_i(t) = -X_i(t)\psi(t) + \int_{M_L}^{M_{Bm}} \psi(t-\tau_M) Q_{Mi}(t-\tau_M)\phi(M)dM +$$

$$+ A \int_{M_{Bm}}^{M_{BM}} \phi(M_B) \left\{ \int_{\mu_{inf}}^{0.5} d\mu\, f(\mu)\psi(t-\tau_{M_2}) Q_{M_{1i}}(t-\tau_{M_2}) \right\} dM_B \quad (1)$$

$$+ (1-A) \int_{M_{Bm}}^{M_{BM}} \phi(M_B)\psi(t-\tau_{M_B}) Q_{M_{Bi}}(t-\tau_{M_B}) dM_B$$

$$+ \int_{M_{BM}}^{M_U} \psi(t-\tau_M) Q_{Mi}(t-\tau_M)\phi(M)dM + [\dot{G}_i(t)]_{infall}$$

The first five terms on the right side of eq. (1) represent, respectively, the rate at which the gas is astrated by star formation, and the rate at which the matter is restored to the interstellar medium (ISM) by: (i) single stars with masses between M_L, which is the lowest mass contributing to the galactic enrichment ($\sim 0.8\ M_\odot$), and M_{Bm}, which indicates the minimum mass for a binary system which is able to produce a type I SN; (ii) binary systems producing type I SNe, the mass limits of which, M_{Bm} and M_{BM}, are respectively 3 and 16 $M_\odot$ as indicated in Greggio and Renzini's (1983) model for a C-O white dwarf. $f(\mu)$ is the distribution function of the mass fraction of the secondary, $\mu=M_2/M_B$, and is taken from Greggio and Renzini (1983), their case $\gamma=2$. The parameter A indicates the proportion of binary systems in the mass range $M_{Bm}-M_{BM}$ able to produce type I SNe, and is obtained by imposing that the type I SN rate be equal to the type II SN rate at the present time; (iii) single stars in the mass range $M_{Bm}-M_{BM}$; (iv) single stars in the mass range $M_{BM}-M_U$, where M_U is the maximum mass contributing to the galactic enrichment (100 $M_\odot$). Finally, the last term represents the rate of infall of primordial gas and is expressed in the form of an exponential law (see Matteucci and Greggio 1985 for details). The quantities $\phi(M)$ and $\psi(t)$ represent the initial mass function and the star formation rate, respectively. For the initial mass function we assumed the Salpeter (1955) slope (x=1.35) for a mass range of 0.1-100 $M_\odot$. For the rate of star formation we have used a law based on the formulation of Talbot and Arnett (1975) based, in turn, on the Schmidt (1958) law of star formation (e.g. proportional to the second power of the gas density), the description of which can be found in Matteucci and Greggio (1985). The term $Q_{Mi}(t-\tau_M)$, the so-called "production matrix" (Talbot and Arnett, 1973), represents the fraction of mass ejected by a star of mass M in the form of an element i. The subscripts 1 and 2 for the quantity M indicate the mass of the primary and secondary, respectively. The term $Q_{M_{1i}}(t-\tau_{M_2})$ indicates that, in the case of the binary systems producing type I SNe, the primary mass is responsible for the chemical enrichment whereas the secondary is responsible for the clock of the explosion, given by its lifetime τ_{M_2}. The stellar lifetimes used here are the same as described in Matteucci and Greggio (1985).

3. NUCLEOSYNTHESIS PRESCRIPTIONS

3.1. Single Stars

a) For low and intermediate mass stars ($0.8 < M/M_\odot < 8$) we used Renzini and Voli's (1981) nucleosynthesis prescriptions (Model with $\eta=0.33$ and $\alpha=1.5$).

b) For massive stars ($M > 8\ M_\odot$) we adopted Arnett's (1978) nucleosynthesis computations but we used a relationship between the initial mass M and the mass of the He-core, M_α, taken from Maeder (1981, 1983), his case B with mass loss. It is worth noting that the $M(M_\alpha)$ relationship of Maeder does not substantially differ from the original $M(M_\alpha)$ relationship given by Arnett (1978). For the iron and silicon production, we assume that they represent 64% and 36%, respectively, of the yield of silicon plus iron, as given by Arnett (1978). This prescription seems to be in agreement with detailed nucleosynthesis computed for a 25 $M_\odot$ by Woosley and Weaver (1982).

3.2. Binary Stars

We adopted the Renzini and Voli (1981) nucleosynthesis prescriptions for the evolution of the primary until the WD phase. The explosive nucleosynthesis products are taken from the Nomoto et al. (1984) computations for the case of an original C-O white dwarf which accretes matter at a rate of $4 \cdot 10^{-8}\ M_\odot\ yr^{-1}$ (their case W7). They predict that a supernova produces 0.6 $M_\odot$ of ^{56}Fe, 0.023 $M_\odot$ of ^{24}Mg, 0.16 $M_\odot$ of ^{28}Si, 0.14 $M_\odot$ of ^{16}O and 0.032 $M_\odot$ of ^{12}C.

4. MODEL RESULTS

The theoretical type I and II SN rates as functions of time are shown in Fig. 1. The two rates reach almost the same value at the present time, as required by the observations ($0.01-0.03\ pc^{-2}\ yr^{-1}$; Tammann, 1982), and this is obtained with the particular choice of the parameter A=0.09 in eq. (1). This value depends on the assumed initial mass function and star formation rate. The temporal behaviour of the two rates is quite different, since that of type II SNe reaches a maximum early on and declines afterwards, whereas that of type I SNe starts with a certain delay and reaches a maximum much later than that of type II. This is due to the different progenitors of the two types of supernovae and accounts for the observational behaviour of several chemical elements. For example, in Fig. 3 is shown the [O/Fe] ratio as a function of [Fe/H] (the symbol [ ] indicates the abundances relative to the Sun), which agrees quite well with the observations (Clegg et al., 1981). The overabundance of oxygen at low [Fe/H], and the subsequent decline of the [O/Fe] toward the solar value with increasing [Fe/H], is a direct consequence of the two SN rates. Therefore, the detailed computation of the type I SN rate, by taking into account the binary nature of type I SN progenitors, accounts quite naturally for

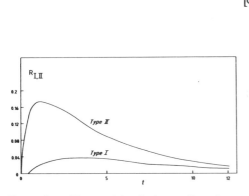

Fig. 1. Theoretical type I and II SN rates, in units of pc^{-2} Gyr^{-1}, as functions of time, in units of Gyr. The present time value for both the rates is in the range 0.01-0.03 pc^{-2} Gyr^{-1} (Tammann, 1982).

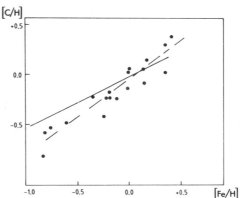

Fig. 2. Theoretical relative abundance of carbon as a function of the theoretical relative abundance of iron. The dashed line represents the best fit of the data as given by Clegg et al. (1981).

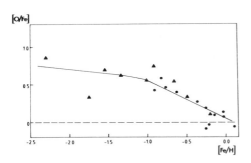

Fig. 3. Theoretical [O/Fe] as a function of [Fe/H]. The data points are from Clegg et al. (1981) (circles), and from Sneden et al. (1979) (triangles). The dashed line represents the solar ratio, [O/Fe]=0.

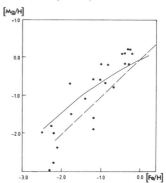

Fig. 4. Theoretical relative abundance of magnesium as a function of the relative abundance of iron. The data points are from Gratton and Ortolani (1984) and the dashed line represents the solar ratio [Mg/Fe]=0.

the temporal behaviour of oxygen and iron in the solar neighbourhood. Fig. 6 shows the [Fe/H] as a function of time compared with the data of Twarog (1980) and the agreement is excellent. Since the behaviour of iron in time is mostly due to the combined effect of the SN rates and infall rate, this means again that the predicted type I SN rate is quite realistic.

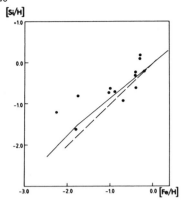
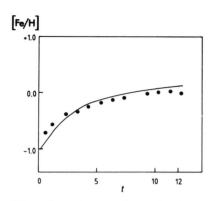

Fig. 5. The same as Fig. 4 for silicon. The data points are from Gratton and Ortolani (1984) and the dashed line represents the solar ratio [Si/Fe]=0.

Fig. 6. Temporal evolution of the [Fe/H] ratio. The time is in units of Gyr. The data points are from Twarog (1980).

For the relative abundance of carbon as a function of the relative abundance of iron, the agreement with observations is not so good, as shown in Fig. 2. We predict, in fact, an overabundance of carbon with respect to iron going towards low [Fe/H] values, whereas the observational data of Clegg et al. (1981) show a solar ratio for C/Fe among all disk stars. In addition, a recent study by Laird (1985) has shown that this trend is valid also for very metal poor stars. The disagreement between theory and observations is due to the fact that the ratio between the predicted amounts of carbon produced by type II SNe and by single intermediate mass stars is greater than the ratio between the amounts of iron produced by type II and type I SNe. In this case, we could suggest that either the yield of carbon from massive stars of Arnett (1978) is too high, and this could be the case (see Woosley, this conference), or the yield of iron from type I SNe, as given by Nomoto et al. (1984), is too high. In the latter case, better agreement would also be reached between the predicted and observed solar iron abundance.

In fact, from table I, where are reported the predicted and observed solar ratios for the studied elements, it turns out that the model prediction for iron is higher by a factor of two with respect to the observed value. On the other hand, part of iron, ejected by type I SNe, could escape from the Galaxy (see Wheeler, this conference) or condense into grains. However, we want to stress the point that we do not find any dramatic iron problem, in spite of the fact that we allow also type II SNe to be iron contributors.

In any case, an amount of iron of ~ 0.3 $M_{\odot}$ per SN of type I could reconcile the predicted and the observed solar iron abundance.

Finally, in Figs. 4 and 5 we show the predictions for magnesium and silicon compared with the data of Gratton and Ortolani (1984). Magnesium and silicon are found to be overabundant with respect to iron in metal poor stars by ~0.5 and 0.3 dex, respectively. This seems to be qualitatively in agreement with the observations. In fact, even if the data of Figs. 4 and 5 do not show any noticeable overabundance (the observational uncertainty is 0.2-0.3 dex), a number of other studies (Luck and Bond, 1981, 1985; Edvardsson et al., 1984) seem to indicate average overabundances of the order of 0.4-0.5 dex in metal poor stars for both Mg and Si. The difference between the predicted overabundance of magnesium and that of silicon is due to the different proportions in which these elements are produced by type I and II SNe, and only extremely accurate data could confirm or exclude this prediction.

Table I. Predicted and observed solar ratios.

	C/O	Mg/O	Si/O	Fe/O
Cameron (1982)	0.42	0.08	0.095	0.17
Present work	0.32	0.071	0.105	0.28
Allen (1973)	0.38	0.06	0.089	0.214

5. CONCLUSIONS

We have shown that the type I SN rate computed in the frame of the binary model for type I SN precursors (Whelan and Iben, 1973), together with the nucleosynthesis prescriptions of the C-deflagration model and a constant IMF, can account for the behaviour of several chemical elements in the solar neighbourhood.

In particular: i) We have found excellent agreement between the predicted and the observed [O/Fe] ratio both in halo and disk stars. The agreement requires that a substantial fraction of iron comes from type I SNe but that, in the meantime, a non-negligible contribution from type II SNe be present (~30%). ii) Magnesium and silicon are predicted to behave similarly to the oxygen with respect to iron, but their overabundances with respect to iron in metal poor stars are lower than the one of oxygen (~0.6-0.7 dex) and of the order of 0.5 and 0.3 dex, respectively. iii) Carbon is predicted to be overabundant with respect to iron in metal poor stars, at variance with the observations. Less carbon from massive stars or less iron from type I SNe could improve the agreement. However we cannot exclude the possibility that part of the iron could escape from the Galaxy or condense into grains. In addition, a smaller amount of iron restored to the ISM from type I SNe, would improve the agreement between observed and predicted solar iron abundance.

Finally, we want to remind that the double degenerate model for type I SN precursors seems to be a more likely explanation for type I

SNe (Iben and Tutukov, 1984). In this case the type I SN rate computed here would become the birthrate of the systems which will eventually produce type I SN events.

In fact, the time of explosion of the systems will be further delayed by a time interval of the order of the timescale for shortening the separation of the two white dwarfs, as a result of gravitational wave radiation. This time delay will depend mainly on the initial separation between the two white dwarfs. Our prediction is that, if the distribution function of the separations is a smooth function, the net result of the additional temporal delay will consist in an enhancement of the SNI rate at late epochs and, correspondingly, a reduction of its maximum reached at early epochs. However, only a detailed model, like the one presented here, could predict the effects of this SNI rate on galactic chemical evolution.

REFERENCES

Allen, C.W.: 1973, Astrophysical Quantities (3d ed; London Athlone Press).
Arnett, D.W.: 1978, Ap.J. 219, 1008.
Arnett, D.W.: 1979, Ap.J. Lett. 230, L37.
Arnett, D.W.: 1982, Ap.J. 253, 785.
Branch, D., Lacy, C.M., McCall, M.L., Sutherland, P.G., Nomoto, A., Wheeler, J.C., Wills, B.J.: 1983, Ap.J. 270, 123.
Cameron, A.G.W.: 1982, in "Essays in Nuclear Astrophysics", eds. C.A. Barnes, D.D. Clayton, and N.D. Schramm (Cambridge Univ. Press), p. 377.
Clegg, R.E.S., Lambert, D.L., Tomkin, J.: 1981, Ap.J. 250, 262.
Colgate, S.A., Petscheck, A.G., Kriese, J.T.: 1980, Ap.J. Lett. 237, L81.
Edvardsson, B., Gustafsson, B., Nissen, P.E.: 1984, ESO Messenger No. 38.
Gratton, R., Ortolani, S.: 1984, Astron. Astrophys. 137, 6.
Greggio, L., Renzini, A.: 1983a, Astron. Astrophys. 118, 217.
Iben, I., Tutukov, A.: 1984, Ap.J. Suppl. Ser. 54, 335.
Laird, J.B.: 1985, Ap.J. 289, 556.
Luck, R.E., Bond, A.E.: 1981, Ap.J. 244, 919.
Luck, R.E., Bond, A.E.: 1985, preprint.
Maeder, A.: 1981, Astron. Astrophys. 101, 385.
Maeder, A.: 1983, in "Primordial Helium", ESO Workshop, eds. P.A. Shaver, D. Kunth, K. Kjär, p. 89.
Matteucci, F., Tornambè, A.: 1985, Astron. Astrophys. 142, 13.
Matteucci, F., Greggio, L.: 1985, preprint.
Nomoto, K., Thielemann, F.K., Yokoi, K.: 1984, Ap.J. 286, 644.
Renzini, A., Voli, M.: 1981, Astron. Astrophys. 94, 175.
Salpeter, E.E.: 1955, Ap.J. 121, 161.
Talbot, R.J., Arnett, D.W.: 1973, Ap.J. 186, 51.
Talbot, R.J., Arnett, D.W.: 1975, Ap.J. 197, 551.
Tammann, G.A.: 1982, in "Supernovae: a Survey of Current Research", eds. M. Rees and R. Stoneham, p. 371.

Twarog, B.A.: 1980, Ap.J. 242, 242.
Weaver, T.A., Axelrod, T.S., Woosley, S.E.: 1980, in "Type I Supernovae", ed. J.C. Wheeler, p. 113.
Wheeler, J.C.: 1986, this conference.
Whelan, J.C., Iben, I. Jr.: 1973, Ap.J. 186, 1007.
Woosley, S.E.: 1986, this conference.
Woosley, S.E., Weaver, T.A.: 1982, in "Supernovae: a Survey of Current Research", eds. M. Rees and R. Stoneham, p. 79.

MAGNESIUM ISOTOPES AND GALACTIC EVOLUTION

B. Barbuy
Universidade de Sao Paulo, Dept Astronomia
C.P. 30627, Sao Paulo 01051, Brazil
and
UA 173, Université de Paris VII, Observatoire
de Paris-Meudon, 92195 Meudon Cedex Principale,
France

ABSTRACT: The carbon-burning products ^{24,25,26}Mg in halo stars and super-metal-rich stars are studied. The isotopic composition as a function of metallicity is derived and a comparison of results to nucleosynthesis theories is done.

I. Introduction

The species ^{23}Na, ^{24,25,26}Mg, ^{27}Al, ^{29,30}Si are products of carbon burning. The C-burning can be hydrostatic or explosive, and which phase is dominant in the nucleosynthesis is still cause of discussion. The neutron-rich species ^{23}Na, ^{25,26}Mg and ^{27}Al can be studied for a nucleosynthesis test.

As a matter of fact, if these isotopes are mainly produced by an explosive C-burning, they would be deficient in metal-poor stars and overabundant in super-metal-rich stars, relative to ^{24}Mg, the relevant even-Z species, due to the so-called odd-even effect. Starting from ^{12}C, ^{16}O and ^{22}Ne (or ^{25}Mg depending on stellar mass), (α,γ) reactions will produce ^{20}Ne, ^{24}Mg and (α,n), (n,γ) reactions will produce the ^{23}Na, ^{25,26}Mg, ^{27}Al, ^{29,30}Si isotopes, during C-burning (Arnett, 1971; Truran and Arnett, 1971). These latter isotopes would be thus produced in abundances depending on the neutron excess η(Arnett, 1971) and on the metallicity [M/H] and their underabundances should scale with the even-Z elements abundances.

On the other hand, the production of these elements during hydrostatic phases may be important (Arnett and Wefel, 1978; Arnett and Thielemann, 1983; Weaver and Woosley, 1980). It seems that the outburst would heat only 10% of the shell to T = 2 10(9) and the C-burning would proceed mostly during hydrostatic activity at T = 1.0 - 1.3 10(9).

The time scale of hydrostatic burning permits the excess of neutrons over protons to raise and the neutron-rich species would be produced in proportions less correlated to [M/H] (whereas about half of the neutron-rich Mg isotopes survive from core helium burning).

As regards in particular the ^{25,26}Mg isotopes, the intermediate mass stars also play a role. Their production is discussed here in the context of heavy s-elements production. Truran and Iben (1977) present a mechanism of production of s-elements in thermally pulsing stars ($1.5 < M/M_o < 8$) where ^{22}Ne(α,n)^{25}Mg is the neutron source. The production of ^{25,26}Mg comes here as a consequence of the production of neutrons. Relative to massive stars, the contribution of intermediate mass stars to the enrichment of these isotopes in the interstellar medium should not be negligible, possibly accounting for 50% of these isotopes abundance (Truran, 1985).

II. Observational results

Observational results are available in the literature mostly for ^{23}Na and ^{27}Al. The results from different authors appear to be controversial regarding [^{23}Na,^{27}Al/Fe] ratios but agreement is found for [^{23}Na,^{27}Al/Mg] ratios to be below solar values in the halo; the degree of deficiency and the law of variation with deficiency remain controversial (Spite and Spite, 1980; Peterson, 1981; Tomkin et al., 1985; Edvardsson et al., 1984; François, 1984, 1985).

The present observations concern the isotopic ratios ^{25,26}Mg/^{24}Mg. MgH lines should be convenient for a test from the fact that, in the determination of ^{25}MgH, ^{26}MgH/^{24}MgH ratios, possible errors coming from model atmospheres, non-LTE effects, uncertainties in oscillator strengths, are cancelled out.

The MgH lines studied are at $\lambda 5135$ Å, belonging to the $A^2\Delta - X^2\Pi$ green system. The isotopic separations are $\Delta\lambda = 0.06$ to 0.1 Å for these lines. The same lines have been observed by Tomkin and Lambert (1980) in the halo dwarf Gmb 1830, and by a few other authors in the disk stars.

In the present observations, halo stars (Barbuy, 1985a) and super-metal-rich (SMR) stars (Barbuy, 1985b) are considered. All observations were obtained at La Silla (ESO) at the 1.4 m CAT telescope using the Coudé Echelle Spectrometer and a Reticon as detector. A wavelength resolution $\Delta\lambda = 0.03$ Å (for the brighter stars) to 0.06 Å and a dispersion of 2 Å/mm are achieved. The limiting magnitude is $V \simeq 7$ mag.

The calculations on spectrum synthesis are done in LTE, the input data required are a model atmosphere, microturbulent velocities and elemental abundances. Model atmospheres are obtained by interpolation in the grids of metal-poor stars (Gustafsson et al., 1975, 1981), an super-metal-rich stars (Gustafsson, 1984).

We noticed the following problems: (1) the presence of C_2 lines disturbs the determinations in some cases, (2) the microturbulent velocity is generally not known and is determined from a fit to atomic lines; uncertainties may affect the determinations in the few more evolved stars and (3) the instrumental convolution step is obtained from a fit to a narrow FeI line; it is found to be equal to 0.03 to 0.06 Å (FWHM). The precise value in each case is important, specially in super-metal-rich, C_2 rich stars, where one has to rely on the $\lambda 5136.014$ Å line ($\Delta\lambda_{isotopic}$ = 0.06 Å) width.

Calculations were carried out for each star with the admixtures $^{24}Mg : ^{25}Mg : ^{26}Mg$ equal to 79:10:11 (terrestrial-meteoritic proportions), 100:0:0 (no $^{25,26}Mg$ isotopes) and 60:20:20 (maximum of $^{25,26}Mg$ yet detected).

III. Objects observed

Halo stars from the catalogue by Carney (1980) and super-metal-rich stars from the catalogue by Cayrel de Strobel et al. (1983, 1985) were selected. We consider here a simplistic way of looking to galactic evolution by studying the evolution of chemical species with metallicity. In that sense the super-metal-rich stars are extreme, although appearing to be old in fact: from their kinematics, typical of old disk, ages of 8-10 billion years are expected, whereas from fits to isochrones the SMR stars show ages between 2 to 15 billion years, indicating that they formed continuedly since the formation of the galactic disk (Cayrel de Strobel, 1985).

IV. Results and discussion

A plot of $^{25}Mg/Mg$ or $^{26}Mg/Mg$ (where $Mg \equiv ^{24}Mg + ^{25}Mg + ^{26}Mg$), given in percentages, vs. [Fe/H] is shown in figure 1.

From figure 1 the following conclusions are drawn:

(1) A discontinuity at [Fe/H] = -0.5 seems to be present. This might be symptomatic of relatively low-mass long-lived stars giving a delayed enrichment of certain elements.

(2) $[^{25,26}Mg/^{24}Mg]$ is somewhat deficient in the more metal-poor stars but the deficiency seems to be constant with decreasing [Fe/H]. This behaviour does not follow a purely explosive nucleosynthesis prediction. More points for more extreme stars would be necessary in order to ascertain whether hydrostatic phases dominate or not.

(3) $[^{25,26}Mg/^{24}Mg] = 0$ in the super-metal-rich stars studied here. The odd-even effect is thus not present here; however, the overabundance in metals is maybe not high enough ([Fe/H] < +0.5) so as to detect the effect.

(4) $[^{25,26}Mg/^{24}Mg] > 0$ is found for some giants. This might reflect a mixing with He-burning material at this stage of evolution.

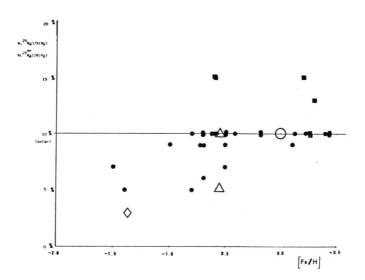

Figure 1 - Ratios $[^{25}Mg/^{24}Mg] \equiv [^{26}Mg/^{24}Mg]$, given in percentages, vs. [Fe/H]. For simplicity the 11% of ^{26}Mg in solar-type mixtures are plotted as 10% similarly to ^{25}Mg. Symbols: o: sun, ●: dwarfs and subgiants, ■: giants, Δ : Arcturus (where ^{25}Mg and ^{26}Mg are different (Tomkin and Lambert, 1976), ◊ : Gmb 1830 (Tomkin and Lambert, 1980).

References

Arnett, W. D.: 1971, Ap. J. **166**, 153
Arnett, W. D., Wefel, J. P.: 1978, Ap. J. **224**, L139
Arnett, W. D., Thielemann, F. K.: 1983, in *Stellar Nucleosynthesis*, eds C. Chiosi, A. Renzini, Reidel: Dordrecht, Holland
Barbuy, B.: 1985a, Astron. Astrophys., in press
Barbuy, B.: 1985b, in preparation
Bell, R. A., Branch, D.: 1970, Ap. Letters **5**, 203
Carney, B. W.: 1980, *A Catalogue of field population II stars*, unpublished

Cayrel de Strobel, G.: 1985, in *La composition chimique des étoiles
    dans le voisinage solaire*, Strasbourg
Cayrel de Strobel, G., Bentolila, C.: 1983, Astron. Astrophys. 119, 1
Cayrel de Strobel, G., Bentolila, C., Hauck, B., Duquennoy, A.: 1985,
    Astron. Astrophys. Suppl. 59, 145
Edvardsson, B., Gustafsson, B., Nissen, P. E.: 1984, The Messenger 37,33
François, P.: 1984, C. R. Acad. Sci. Paris 299, 195
François, P.: 1985, in preparation
Gustafsson, B., Bell, R. A., Eriksson, K., Nordlund, Å.: 1975, Astron.
    Astrophys. 42, 407
Gustafsson, B.: 1981, unpublished
Gustafsson, B.: 1984, unpublished
Peterson, R.: 1981, Ap. J. 244, 989
Spite, M., Spite, F.: 1980, Astron. Astrophys. 89, 118
Tomkin, J., Lambert, D. L.: 1976, Ap. J. 208, 436
Tomkin, J., Lambert, D.L.: 1980, Ap. J. 235, 925
Tomkin, J., Lambert, D. L., Balachandran, S.: 1985, Ap. J. 290
Truran, J. W., Arnett, W. D.: 1971, Astrophys. Spa. Sci. 11, 430
Truran, J. W., Iben, I.: 1977, Ap. J. 216, 797
Truran, J.: 1985, private communication
Weaver, T. A., Woosley, S. E.: 1980, Ann. N. Y. Acad. Sci. 336, 335

^{26}Al EXPERIMENTAL RESULTS

Philippe Durouchoux

Service d'Astrophysique, Centre d'Etudes Nucléaires de Saclay, France

ABSTRACT

Recent experimental results have, for the first time, demonstrated the existence of extra-solar nuclear gamma ray lines.

These lines, due to the desexcitation of ^{24}Mg* in SS 433 and the beta decay reaction of ^{26}Al in the interstellar medium provide useful information on nuclear reactions which might take place in a peculiar astrophysical site and nucleosynthesis of supernova and stellar winds of O and WR stars.

We will summarize and discuss these experimental results and present a new generation of gamma ray line experiment with a high energy resolution and good spatial resolution.

INTRODUCTION

Observations of spectral lines in the high energy range, 10 keV to 10 MeV, are directly related to the understanding of many classes of objects. Neutron stars, black holes, supernova remnants, the interstellar medium, the galactic nucleus and galactic active nuclei are known or predicted to be sources of spectral lines which can be studied with high energy spectroscopy.

The specific information is contained in the intensity of the line, its variability, the centroid energy with respect to a rest frame, and the width and profile of line as well as line-to-line and line-to-continuum ratios. This information relates directly the theory of the line-forming processes, and when coupled to source models, provides critical tests of models which describe the macroscopic nature of the source. A possibility of detection of a narrow line at the energy 1809 keV was suggered by Ramaty and Lingenfelter (1977). This line results from the beta decay of ^{26}Al to the first excited state of ^{26}Mg. Because of its long lifetime ($\tau_{\frac{1}{2}}$ = 7.4 x 10^5 yr), ^{26}Al decays and produces gamma rays after it has been mixed with the interstellar medium; so the line was expected to be spatially diffuse and narrow. The predicted intensity was 5.10^{-5} photons/cm^2.s.rad and width $\leqslant 3$ keV.

EXPERIMENTAL RESULTS

In 1979 NASA launched the satellite HEAO 3. It carried a gamma ray spectrometer, consisting of four germanium detectors in a cesium iodide shield. The instrument, designed by the Jet Propulsion Laboratory to operate for a year, scanned most of the sky to look for gamma ray lines in the range 50 keV-10 MeV.

A scan of the galactic plane took place two times: during fall 1979 and spring 1980.

Some difficulties raised during the data reduction, principally the variation of the background along the orbit, the background lines and the variation versus time of the energy resolution. For example, the ^{26}Al decay line at 1809 keV, was close to two background lines (1764 keV ^{238}U and 1779 keV ^{28}Si desexcitation).

An other problem is the great penetration power of 1.8 MeV gamma rays in the 6cm thick shield which becomes transparent at these energies : in other words, the response of the instrument pointing in the direction of a diffuse source is not only a flux coming through the 42° F.W.H.M. field of view but also includes 20% of the gamma rays which penetrate the sides of the shield and make an interaction in the central detectors.

So it is needed to assume a <u>source distribution</u> in order to determine the net flux emitted by a diffuse source. The distribution shown in fig 1 (which is the COS B 70 MeV- 6 GeV distribution) was taken in the data reduction and the intensity of the 1.809 MeV line was found to be =(4.8 $\pm$ 1.0) 10^{-4} ph/cm^2.s.rad.

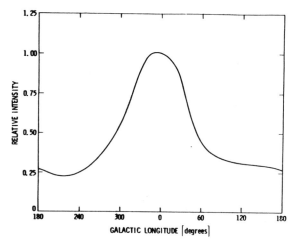

Fig. 1: COS B (70 MeV-5 GeV) gamma ray distribution used for deconvolution of 1809 MeV line with HEAO 3 (from Mahoney et al 1984).

A part of the spectrum of the galactic plane is shown on fig 2

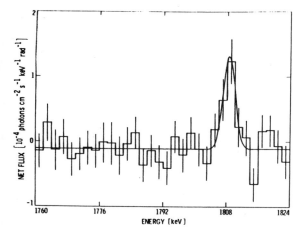

Fig. 2: Total net diffuse galactic plane emission near 1809 keV. (from Mahoney et al 1984).

A confirmation of this results came from the Solar Maximum Mission Satellite experiment. A gamma ray spectrometer on board, pointed in 1980-1981 and 1982 toward the galactic center region and detected the 1.8 MeV line. The measured flux is $(4.0 \pm 0.4)\ 10^{-4}$ ph/cm^2.s.rad. These results are summarized on table 1.

TABLE 1

SPACECRAFT	OBSERVING PERIOD	ENERGY (KeV)	INTRINSIC WIDTH FWHM(keV)	APERTURE FWHM	INTENSITY PH/cm^{2}s.rad.	REFERENCES
HEAO C	1979 Sep26-Oct.8 1980 March4-Apr.4	1808.49±0.41	3.0 (1σ)	42°	$(4.8±1.0)10^{-4}$	Mahoney et al (1984)
SMM	1980-1981-1982	1804 ± 4	38 $^{21}_{38}$	130°	$(4.0±0.4)10^{-4}$	Share et al (1985)

OTHER LINES

Other lines than ^{26}Al decay were observed during the past decade. They are summarized in table 2

TABLE 2 Observed Gamma-Ray Lines

Object or Phenomenon	Process	Energy keV	Maximum Flux ph/cm^2-sec	Maximum Detection significance of sigma*	References
Galactic Center Region	(e^-,e^+) 2γ	510.90±.25	1.8×10^{-3}	9	1,2
	$^{26}Al \rightarrow ^{26}Mg$	1809	4.8×10^{-4}	4.8	3,21
SS 433	^{24}Mg	1500	1.5×10^{-3}	6	22
X-ray Pulsator		1200	1.1×10^{-3}	5	22
Her X-1	Cyclotron	44-58	3×10^{-3}	5	4,5,6
4U 0115+63	Cyclotron	20	2×10^{-3}	6.2	7
GX 1+4	Cyclotron	42	3×10^{-3}	4.8	8
Crab Pulsar	Cyclotron	73-77	4×10^{-3}	3.4	9,10,11
10 June 1974 Transient	(e^-,e^+) 2γ $^1H(n,\gamma)^2H$ $^1H(n,\gamma)^2H$ $^{56}Fe(n,\gamma)^{57}Fe$	413±2** 1790±6** 2219±6 5947±4**	7×10^{-3} 3×10^{-2} 1.5×10^{-2} 1.5×10^{-2}	3.4 4.4 3.1 3.2	12,13
Gamma-Ray Burst	Cyclotron (e^-,e^+) 1γ $^{56}Fe(x,x'y)^{56}Fe$	20-100 400-460** 738±10**	~ 3 73 ~ 4	18 9.4 3.5	14 14,15,16 15,16
Solar Flare	(e^-,e^+) 2γ $^1H(n,\gamma)^2H$ $^{12}C(x,x'\gamma)^{12}C$ $^{16}O(x,x'\gamma)^{16}O$	511 2225±1 4439 6129	9×10^{-2} 1.0 ~ 1 ~ 1	29 16 ~ 40 ~ 40	17,18 17,18,19,20 17,18,19 17,18

*From Matteson et al (1982)
** Gravitationally redshifted

One should notice that a few only are nuclear gamma ray lines, except in solar flares and gamma ray bursts.

With more sensitive spectrometers, the other lines most likely to be detected are 0.847 and 1.240 MeV from the decay of ^{56}Co ; such observations would give information on galactic and extragalactic supernovae younger than 10 years old.

From galactic supernovae younger than 300 years old, 1.156, 0.078 and 0.068 MeV lines from ^{44}Ti decay are also expected.

Finally the detection of a line at 58.6 keV from ^{60}Fe $\longrightarrow$ ^{60}Co decay, and the measurement of its distribution could be correlated to galactic supernovae events over last 10^6 years. The predicted fluxes are shown in table 3.

TABLE 3

Predicted Fluxes*

GAMMA-RAYS PREDICTED FROM EXPLOSIVE NUCLEOSYNTHESIS

DECAY (Parent Daughter)	γ-RAY ENERGY (keV) BRANCHING (%)	HALFLIFE	YIELD/SN EVENT (nuclei) a) b)		FLUX/SN EVENT PEAK ($cm^{-2}-s$)$^{-1}$ c)	AVERAGE GALACTIC ($cm^{-2}-s$)$^{-1}$ d)	GAMMA-RAYS INDICATE:
^{57}Ni → ^{57}Co	127(14),1370(86)	36h	2(53)	5(52)	–		Galactic and extragalactic supernovae 10 years old, evolution of SN shell, measurement of nucleosynthesis yield
^{57}Co → ^{57}Fe	14.4(89),122(89),136(11)	270d	2(53)	5(52	.5 to 2 × 10^{-1}		
^{56}Ni → ^{56}Co	163(85),276(34),427(34), 748(51),812(85)	6.1d	4(54)	2(54)	–		
^{56}Co → ^{56}Fe	847(100),1030(16),1240(67) 1760(14),2600(17),e$^+$(20)	77d	4(54)	2(54)	.5 to 1 × 10^0		
^{22}Na → ^{22}Ne	1275(100),e$^+$(90)	2.6y	2(51)	1(52)	2 to 8 × 10^{-3}		
^{44}Ti → ^{44}Sc	68(100),78(100)	48y	3(51)	4(51)	2 × 10^{-4}	5 × 10^{-4}	Galactic SN 300 years old, nucleosynthesis yield
^{44}Sc → ^{44}Ca	1156(100),e$^+$(94)	3.9y	3(51)		2 × 10^{-4}	5 × 10^{-4}	
^{60}Fe → ^{60}Co	58.6(100)	3(5)y	4(50)	4(50-51)	.2 to 2x10^{-8}	5x10^{-4}	Distribution of galactic SN events over last 10^5-10^6 years, product of nucleosynthesis yield and supernovae frequency
^{60}Co → ^{60}Ni	1170(100),1330(100)	5.3y	4(50)	4(50-51)	.2 to 2x10^{-8}	5x10^{-4}	
^{26}Al → ^{26}Mg	1809(100),1130(4),e$^+$(85)	7.4(5)y	5(49)	5(50)	.1 to 1x10^{-3}	7x10^{-5}	

a) Calculated from Woosley and Axelrod (1980). Assumes a 25 Mo progenitor except for ^{60}Fe which is based upon 15 Mo.
b) Based on production of a galactic mass of the elements at the known abundances (Clayton, 1980).
c) Calculated at 300 days subsequent to explosion with no scattering. Assumed distance is 10kpc.
d) Assumes a disk distribution of SN events, scale height 200pc equivalent to n average source distance of 7kpc.

*From Matteson et al (1982).

FUTURE

An instrument with a sufficient sensitivity, energy and angular resolution, would provide information on gamma rays predicted from explosive nucleosynthesis, and accompanying annihilation line. Moreover, the measurement of the distribution of ^{26}Al in the galactic plane and relation with type II supernovae and WR distribution would provide critical tests of models of stellar evolution (Deaborn et al 1985, Cassé et al 1986). So, it was decided (UCSD/UCB/CESR/CEN-Saclay) to built a balloon borne experiment having high energy resolution capability and good spatial resolution (HIREX) fig 3.

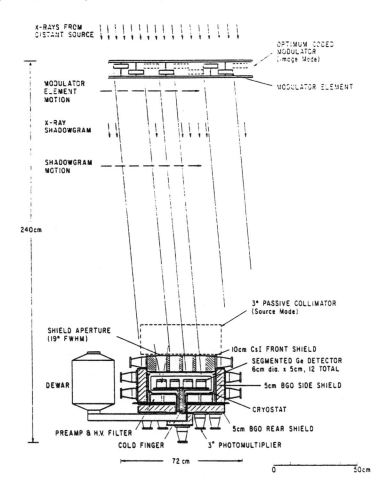

Fig.3 High Resolution Spectrometer (HIREX)

The high energy resolution is given by an array of 12 hyperpur germanium detectors, operating at the liquid nitrogen temperature. On the other hand, the good angular resolution is achieved with an optimum coded modulator, moving 2 meters above the detector plane. The narrow line sensitivity of the instrument (few times 10^{-5} photons/cm^2.s. at 1 MeV) is shown on fig 4, for different operating modes.

A comparison with HEAO 3 experiment (30 day scan) is shown on fig 5.

The engeneering model of HIREX is scheduled to fly in fall 1985 and the scientific payload in 1986.

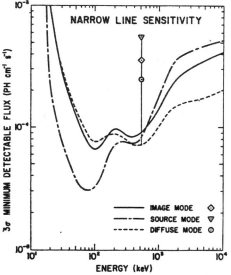

Fig.4: The 3 sigma narrow line sensitivity calculated for 6 hours observation.

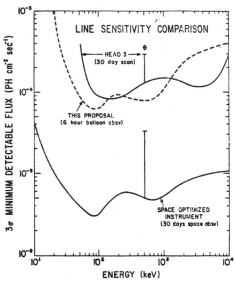

Fig.5: Image mode line sensitivity compared with HEAO3 and predicted space instrument.

CONCLUSION

The HIREX experiment was principally developed to demonstrate the instrument capabilities and the new techniques used in the gamma ray astronomy. A program of balloon flights, starting in 1986 might provide important scientific results. But the instrument also has features required for future space missions. A space optimized version, on board the spacelab or the space station, with a sensitivity of the order of few times 10^{-6} photons/cm^2.s. would be the most exciting prospect for gamma ray line astronomy.

REFERENCES

Cassé, M. and Prantzos, N., 1986 this volume.
Chupp, E.L., 1982, in Gamma Ray Transients and Related Astrophysical Phenomena, Lingenfelter, R.E. , Hudson, H.S., and Worrall, D.M., eds., AIP, New York,p.363.
Chupp, E.L., 1982, Presented at Cospar XXII, Ottawa, May 1982
Dearborn, D.S.P., and Blake, J.B. 1985, Ap.J.(Letters), 288,L21.
Gruber, D.E., Matteson, J.L., Nolan, P.L., Knight, F.K., Baity, W.A., Rothschild, R.E., Peterson, L.E., Hoffman, J.A., Scheepmaker, A., Wheaton, W.A., Primini, F.A., Levine, A.M., and Lewin, W.H.G. 1980, Ap.J. (Letters) 240, L127.
Hudson, H.S., Bai T., Gruber, D.E., Matteson, J.L., Nolan, P.L., and Peterson, L.E. 1980, Ap.J. (Letters) 236, L91.
Jacobson, A.S., Ling, J.C., Mahoney, W.A., and Willett, J.B. 1979, in Gamma Ray Spectroscopy in Astrophysics, T.L. Cline and R.Ramaty, eds., NASA T.M. 79619, 228.
Lamb, R.C., Ling, J.C., Mahoney, W.A., Riegler, G.R., Wheaton, W.A., and Jacobson, A.S., 1983, Nature 305, 37.
Leventhal, M. , McCallum, C.J., and Watts, A.C. 1977b, Ap.J. 216, 491.
Leventhal, M., McCallum, C.J., and Stang, P.D. 1978, Ap.J. (Letters), 223, L11.
Ling, J.C., Mahoney, W.A., Willett, J.B., and Jacobson, A.S. 1979, Ap.J. 231, 896.
Ling, J.C., Mahoney, W.A., Willett, J.B., and Jacobson, A.S. 1972, in Gamma Ray Transients and Related Astrophysical Phenomena, R.E.Lingenfelter, H.S. Hudson and D.M. Worrall, eds. AIP, New York, 143.
Mahoney, W.A., Ling, J.C., Wheaton, WM.A., and Jacobson, A.S. 1984, Ap.J. 286, 576.
Manchanda, R.K., Bazzomo, A., La Padula, C.D., Polcaro, V.F., and Ubertini,P., 1982, Ap.J. 252, 172.
Matteson, J.L., Pelling, M.R., Lin, R.P., Hurley, K.C., 1982, Proposal submitted to National Aeronautics & Space Administration Washington, D.C. 20546.
Mauer, G.S., Johnson, W.N., Kurfess, J.D., and Strickman, M.S. 1982, Ap.J. 254, 271.
Mazts, E.P., Golenetskii, S.V., Aptekar', R.L., Gur'yan, Yu.A., and Il'inskii,V.N., 1981, Nature 290, 378.
Pelling, M.R., Matteson, J.L., Worrall, D.M., Lin, R., Pehl, R., Hurley, K., Durouchoux, Ph., 163rd AAS Meeting Las Vegas, 1984.
Prince, T.A., Ling, J.C., Mahoney, W.A., Riegler, G.R., and Jacobson,A.S., 1982, Ap.J. 259, 392.
Ramaty, R., and Lingenfelter, R.E. 1977, Ap.J.(Letters) 213, L5.
Riegler, G.R., Ling, J.C., Mahoney, W.A., Wheaton, W.A., Willett, J.B., and Jacobson, A.S. 1981, Ap.J. (Letters) 248, L13.
Share, G.H. 1985, private communication.

^{26}Al PRODUCED BY WOLF-RAYET STARS AND THE 1.8 MeV LINE EMISSION OF THE GALAXY

M. Cassé and N. Prantzos
Service d'Astrophysique, Institut de Recherche Fondamentale,
C E N Saclay

ABSTRACT. Massive stars (50 to 100 M_θ initially), both synthesize and eject significant quantities of ^{26}Al in their Of and WR stages. We estimate the collective contribution of these extreme population I objects to the 1.8 MeV line emission of the galactic plane.

1. INTRODUCTION

The recent satellite observation of the 1.8 MeV line from the decay of ^{26}Al (HEAO 3: Mahoney et al., 1984, SMM: Share et al., 1984, private communication), has given a new impetus to the study of the nucleosynthesis of ^{26}Al (e.g. Clayton 1984 and Fowler 1984 for a review). In this communication we discuss the production and ejection of ^{26}Al by massive mass-losing stars (Of and WR stars), in the light of recent stellar models (see also Dearborn and Blake 1984, 1985). We also derive the longitude distribution of the ^{26}Al γ-ray line produced by the galactic collection of WR stars, based on various estimates of their radial distribution. This longitude profile provides i) a specific signature of massive stars on the background of other potential ^{26}Al sources, as novae (see e.g. Clayton 1984), supernovae (Truran and Cameron 1978, Arnett and Wefel 1978, Woosley and Weaver 1980, Arnould et al., 1980), certain red-giants (Norgaard 1980) and possibly AGB stars (Cameron 1984) and ii) a possible tool to improve the data analysis of the HEAO 3 and SMM experiments.

2. THE PRODUCTION AND EJECTION OF ^{26}Al BY Of and WR STARS

An evolutionary model of massive stars (initial mass from 50 to 100 M_θ), including mass loss and extended internal mixing, has been recently developed, aimed at following O stars through their subsequent evolution into the Of, WN, WC and WO stages (de Loore et al., 1986, Prantzos et al., 1986). This mass range seems to correspond to most of the WR progenitors (Humphreys et al., 1985). The nucleosynthesis of all species up to ^{30}Si is closely followed thanks to a detailed nuclear network supplied with updated nuclear data relevant to the H and He-burning phases (Fowler et al., 1975, Ward and Fowler 1980, Harris et al., 1983, Caughlan et al., 1984, Almeida and Kappeler, 1983). For detail of the network interesting specifically ^{26}Al, see Cassé and Prantzos 1985 and Prantzos et al., 1986. ^{26}Al is produced and homogenized in the stellar convective core during H-burning, through the reaction ^{25}Mg(p,γ), and is destroyed at the onset of He-burning through (n,α) and (n,p) reactions. This nuclide is also $β^+$ unstable with a mean lifetime $τ_{26} \cong 1$ million years. It appears at the stellar surface when the hydrogen-rich envelope is dispersed by the intense stellar wind, i.e. during the Of and WN phases, and disappears at the beginning of the WC phase, when it is the turn of He-burning products to emerge at the surface (fig. 1). The ^{26}Al dispersed by the wind in the interstellar medium still decays long after the final explosion of the Wolf-Rayet star. Fig. 2 displays the history of ^{26}Al (in mass fraction) in the core and at the surface of a 100 $M_\odot$ star (initially), together with the resulting γ ray-line luminosity, $L_γ$ (in photons sec^{-1}) and its time averaged values over the Of and WR phases.

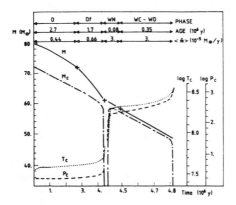

Figure 1. Structural evolution of a massive star
The evolutionary model includes mass loss and overshooting from the Schwarzschild convective core (Prantzos et al. 1986). The dash-dotted line delineates the boundary of the convective region. When the H- and He-burning products appear successively at the surface, the star enters first the Of-WN and then the WC-WO phase.
This example concerns a 80 $M_\odot$ (initial mass) star.

The quantity of ^{26}Al ejected is found to increase with <u>mass</u> and should depend linearly on <u>metallicity</u> of the stellar progenitor (see also Dearborn and Blake 1984, 1985). The ^{26}Al yield and the corresponding γ-line luminosity, averaged over i) the initial mass function of Humphreys and McElroy 1984 (assumed to be uniform throughout the galactic disk, see however Boissé et al., 1983) and ii) the radial metallicity gradient derived by Shaver et al., 1983 (extrapolated up to $\cong 4$ kpc from the galactic center) amount to $Y_{26} = 1.1 \ 10^{-4} \ M_\odot$ and $L_γ = 1.3 \ 10^{38}$ photons sec^{-1} respectively. These values should be characteristic of an average galactic WR star.

Assuming a steady state abundance (e.g. Clayton 1984), the total mass of live ^{26}Al scattered in the whole galaxy at present time is $M_{26} = N_T \cdot Y_{26}$, where N_T is the total number of WR stars that have contributed to the galactic ^{26}Al production in the last lifetime ($\tau_{26} \cong 10^6$ years). N_T, in turn, is proportional to the WR birthrate, $B_{WR} = n_{WR}/\tau_{WR}$, n_{WR} being the present number of WR stars in the galaxy and τ_{WR} their average lifetime. Current models (e.g. Maeder and Lequeux 1982, Prantzos et al., 1986) predict that $\tau_{WR} \cong 3$ to $5 \; 10^5$ years, at least for solar metallicity models. We assume provisionally that this number applies to the whole galaxy as well. The error introduced by this simplification is expected to be small compared to the uncertainty on n_{WR}, which is, as we shall see, considerable.

3. THE NUMBER OF WR STARS IN THE GALAXY

A reasonable estimate of the total number of WR stars present in the galaxy is difficult, but it must, in our opinion, include one of the two following factors or both.

a) The increase of the Star Formation Rate (SFR) with Decreasing Galacto-Centric Distance (R):

Since the WR catalogs are complete only up to 2.5 kpc from the sun (e.g. Hidayat et al., 1982, Conti et al., 1983), we must rely on qualitative tracers of star formation to derive the radial distribution of young and massive stars inward, including the very central region.

b) The increase of the Ratio WR to O Stars (N_{WR}/N_O) with Metallicity Z:

From counts of WR and O stars in the Magellanic clouds and in different regions of the Milky way, Maeder (1984) derived the relation $N_{WR}/N_O \alpha Z^{1.7}$. We assume that this relation still holds for $Z > 0.03$ (i.e. in the inner galaxy and in the very central region where $Z \cong 0.09$, Güsten and Ungerechts, 1985). Indeed an increase of N_{WR}/N_O with Z, presumably due to an increase of the mass loss rate of O stars with Z, is not unexpected (Maeder 1982) and can be understood, at last qualitatively, in the framework of radiation driven wind models of O stars (e.g. Abbot 1982).

Both effects tend to increase significantly the γ-line luminosity of the inner galaxy. Three different cases have been considered to illustrate their relative importance (fig. 3).

A. Following Maeder and Lequeux (1982), we assume that WR stars follow the distribution of giant HII regions, as given by Guibert et al., (1978). In this case $n_{WR} \cong 1000$ and $M_{26} \cong 0.4 \; M_\odot$ (leaving aside the very central region of the galaxy), much less than the mass required to sustain the 1.8 MeV line at the observed intensity ($\cong 3 \; M_\odot$, Mahoney et al., 1984).

B. The WR surface density $\sigma(WR)$ has been scaled to that of molecular hydrogen $\sigma(H_2)$, using the radial distribution of Sanders et al. 1985, recalibrated to the local $\sigma(H_2)$ of Dame and Thaddeus 1985, which seems more appropriate for nearby regions (Lebrun, private communication). This is equivalent to assume that the formation rate of the WR progenitors is proportional to the gas density, at large scale, and that the N_{WR}/N_O ratio is uniform across the galactic disk. n_{WR} is then 3000 (2000 in the disk, 1000 in the center) and $M_{26} \cong 1.3\ M_\odot$.

C. Applying to distribution B the metallicity correction discussed above, we get distribution C, n_{WR} is now $\cong$ 8000 (6000+2000) and $M_{26} \cong 3.2\ M_\odot$. This last case is encouraging, but remember that it rests on a a rather speculative basis. Dedicated models of metal-rich WR stars are needed to substantiate these ideas.

4. LONGITUDE DISTRIBUTION OF THE WR γ-LINE EMISSION

Knowing the typical luminosity L_γ of individual sources and their galactic distribution, it is a matter of numerical integration to calculate the longitude distribution of the arriving γ-ray line flux (e.g. Harding 1981). The fluxes resulting from radial profiles A, B and C are shown in fig. 4 a,b.,(a) normalized to the peak value, to emphasize the center/anticenter contrast, b) in absolute flux). Only in case C, as expected, the flux from the galactic center direction is comparable to the one derived from the HEAO 3 data ($4.8 \pm 1 \times 10^{-4}$ photon cm^{-2} sec^{-1} rad^{-1}, between b = $\pm$ 30°, Mahoney et al., 1984).

Note that the three proposed profiles are sharper than the COSB one, which served as a reference in the HEAO 3 and SMM data treatment. For consistency, it would be desirable to reiterate the data analysis on the basis of theoretical profiles A, B and C.

5. CONCLUSION

We have tentatively estimated the contribution of WR stars to the 1.8 MeV line emission of the galactic plane on the basis of recent stellar evolution models. These seem to be interesting candidates, but because of i) large uncertainties in their galactic distribution, and ii) the lack of dedicated metal-rich WR models, it would be premature to conclude that they are the unique source of ^{26}Al in the galaxy. Future experiments with improved spatial resolution ($\lesssim 5°$) will help to identify the most generous ^{26}Al sources, galaxy-wise. A present, it would be desirable to refine the data analysis of the HEAO 3 and SMM satellites on the basis of theoretically derived longitude distribution, such as distributions A, B and C for instance.

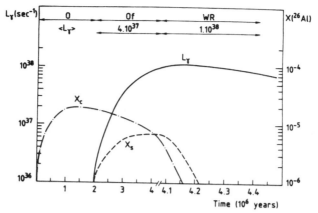

Figure 2. *Abundance (by mass) of ^{26}Al in the stellar core, at the stellar surface (versus time) and the corresponding luminosity in the 1.8 MeV γ-ray line*
^{26}Al is produced and homogenized in the stellar core at the beginning of H-burning and appears at the surface later on, during the so-called Of and WN phases, when the overlying layers are dispersed by the intense stellar wind. ^{26}Al decays in the ejected material, giving rise to a time-dependent γ-ray line luminosity, L_γ.
This example concerns a 100 $M_\odot$ (initial mass) star.

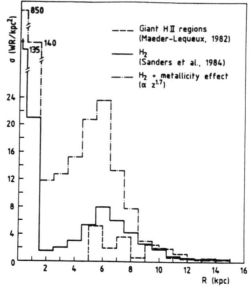

Figure 3. *Three possible galactocentric distribution of WR stars*
Their radial distribution is assumed to follow that of the extreme population I tracers as i) the giant HII regions (Guibert et al., 1978, Maeder and Lequeux, 1982) (Case A) or ii) the molecular gas (Sanders et al. 1985, renormalized to the local surface density given by Dame and Thaddeus, 1985) (Case B). In addition it is assumed that the number of WR stars relative to O stars varies radially with $Z(R)^{1.7}$ (Maeder, 1984) (Case C).

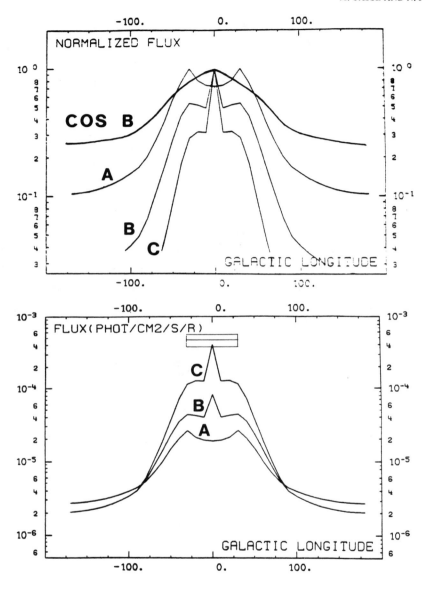

Figures 4 a,b. Longitude profiles derived from the three possible WR distributions
a) Normalized at peak value to show the center/anticenter contrast, of interest for the observers. The COS B profile is shown for comparison.
b) Expected flux (in photons cm^{-2} s^{-1} rad^{-1}) versus longitude in the A, B and C cases. The shaded area indicates the measured value ($4.8 \pm 1 \cdot 10^{-4}$ photons cm^{-2} s^{-1} rad^{-1}, between $\pm$ 30° in longitude, Mahoney et al., 1984).

ACKNOWLEDGEMENTS

We thank our colleagues C. Doom, M. Arnould, C. de Loore and J.P. de Grève for constant help and support. We are grateful to A. Maeder for very useful discussion on the number and distribution of Wolf-Rayet stars.

REFERENCES

Abbot, D.C., (1982), Ap. J., 259, 282.
Almeida, J. and Kappeler, F., (1983), Ap. J. 265, 417.
Arnett, D.W. and Wefel, J.P., (1978), Ap. J. (Letters), 224, L139.
Arnould, M., Norgaard, H., Thielemann, F.-K., and Hillebrandt, W., (1980), Ap. J. 237, 931.
Boissé, P., Gispert, R., Coron, N., Wijnbergen, J.J., Serra, G., Ryter, C. and Puget, J.P., (1981), Astron. Astrophys., 94, 265.
Cameron, A.G.W., (1984), Icarus, 60, 416.
Cassé, M. and Prantzos, N., (1985), in preparation.
Caughlan, G.R., Fowler, W.A., Harris, J. and Zimmermann, B.A., (1984), preprint.
Clayton, D.D. (1984), Ap. J., 280, 144.
Conti, P.S., Garmany, C.D., de Loore, C. and Vanbeveren, D., (1983), Ap. J. 274, 302.
Dame, T.M. and Thaddeus, P., (1985), preprint.
Dearborn, D.S.P. and Blake, J.B., (1984), Ap. J. 277, 783.
Dearborn, D.S.P. and Blake, J.B., (1985), Ap. J. (Letters), 288, L21.
de Loore, C., Prantzos, N., Doom, C. and Arnould, M., (1986), Moriond Symposium "Nucleosynthesis and its applications on nuclear and particule physics", to be published.
Fowler, W.A. (1984), Rev. Mod. Phys., 56, 149.
Fowler, W.A., Caughlan, G.R. and Zimmerman, B.A., (1975), Ann. Rev. Astron. Astrophys., 13, 113.
Guibert, J. Lequeux, J. and Viallefond, F., (1978), Astron. Astrophys., 68, 1.
Güsten, R. and Ungerechts, H., (1985), Astron. Astrophys., 145, 241.
Harding, A.K., (1981), Ap. J., 247, 639.
Harris, M.J., Fowler, W.A., Caughlan, G.R. and Zimmerman, B.A., (1983), Ann. Rev. Astron. Astrophys., 21, 198.
Hidayat, B., Supelli, K. and van der Hucht, A.K., (1982), IAU Symp. 99, eds. C. de Loore and A. Willis, Reidel, Dordrecht, p. 27.
Humphreys, R.A., Nichols, M. and Massey, P., (1985), Astron. J., 90, 101.
Humphreys, R.A. and McElroy, D.B., (1984), Ap. J., 284, 565.
Maeder, A., (1984), Adv. Space Res., 4, 55.
Maeder, A. and Lequeux, J., (1982), Astron. Astrophys., 114, 409.
Mahoney, W.A., Ling, J.C., Wheaton, W.A. and Jacobson, A.S., (1984), Ap. J. 286, 278.
Norgaard, H., (1980), Ap. J., 236, 895.
Prantzos, N., Doom, C., Arnould, M. and de Loore, C., (1986), Moriond Symposium "Nucleosynthesis and its applications on nuclear and particle physics", to be published.

Shaver, P.A., McGee, R.X., Newton, L.M., Danks, A.C. and Pottasch, S.R., (1983), M.N.R.A.S., 204, 53.
Sanders, D.B., Solomon, P.M. and Scoville, N.Z., (1984), Ap. J., 276, 182.
Truran, J.W. and Cameron, A.G.W., (1983), Ap. J., 219, 226.
Ward, R.A. and Fowler, W.A., (1980), Ap. J. 238, 266.
Woosley, S.E. and Weaver, T.A., (1980), Ap. J., 238, 1017.

^{26}Al-destruction in neutron rich environments

H.P. Trautvetter
Inst. f. Kernphysik der Universität Münster,
West-Germany

I. Introduction

Indications for live ^{26}Al in the ISM were given in recent years for two reasons:
i) The isotopic anomaly of ^{26}Mg which was found to be correlated with ^{27}Al in inclusions of the Allende-meteorit and was therefore identified to originate from ^{26}Al which decayed in situ to ^{26}Mg [Wa 82]. Typical ^{26}Al/^{27}Al-ratios were found to be of the order of 5×10^{-5} but could also reach high values up to 10^{-3} [Wa 82].
ii) From γ-ray astronomy (HEA03-satellite) one observes the $E_\gamma = 1.809$ MeV-line [Ma 82] from the decay of ^{26}Al with the half live of 7.2×10^5 y. From its intensity Clayton estimates [Cl 84] that about 4.2 $M_\odot$ of ^{26}Al are distributed in the ISM.

The question what kind of astrophysical scenario produces this amount of ^{26}Al is of great interest and was raised by several authors ot these proceedings [Wi 85, Wo 85, Du 85, Ca 85, Ar 85, Pr 85]. Clearly, any model-calculation which tries to give quantitave results must take into account proper reaction rates for the ^{26}Al production as well as for its destruction. This contribution is concerned with ^{26}Al-destruction by neutrons as they are produced e.g. in explosive nucleosynthesis.

II. Experiment

Inspection of fig. 1 shows that neutron-capture by ^{26}Al leads to high energies in ^{27}Al of ~ 13 MeV where the level density in ^{27}Al should be high such that a statistical approach for obtaining the reaction rate seems to be justified. However, the spin of the ground state in ^{26}Al is $J^\pi = 5^+$ and hence with low neutron energies (s-wave) only high spin states in ^{27}Al can be populated. The density of

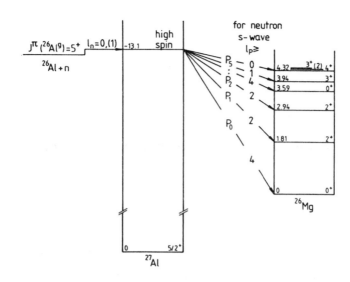

Fig. 1: Relevant parts of the level diagram involving the ^{26}Al(n,p)^{26}Mg reaction. s-wave neutron capture dominates for low energies hence high spin compound states in ^{27}Al will be populated from which the minimal values for the orbital angular momentum l_p of the reaction protons are deduced.

these high spin states is much lower so that individual structure for the ^{26}Al(n,p)^{26}Mg reaction could be expected. Indeed, this has been seen by [Sk 83] who studied the reverse reaction ^{26}Mg(p,n)^{26}Al. However, from this work, applying the method of detailed balance, only the rate for the p_0-group of the reaction ^{26}Al(n,p_0)^{26}Mg could be determined. This p_0-transition is, however, greatly inhibited by the high centrifugal barrier involved since the final state in ^{26}Mg has a $J^\pi = 0^+$ assignment. It is therefore expected, that the p_1-transition to the first excited state ($J^\pi = 2^+$) in ^{26}Mg should be much stronger. It's intensity could only be determined by a direct measurement.

The production of ^{26}Al-targets has been described in detail by [Bu 84]. Two targets have been used with $(1.3\pm0.1) \times 10^{15}$ atoms/cm^2 called MIII and $(2.3\pm0.2) \times 10^{15}$ atoms/cm^2 called MIV. The experimental arrangements are shown in fig. 2. In fig. 2a and b the neutrons were produced by the ^{7}Li(p,n)^{7}Be and the T(p,n)^{3}He reaction, which gave neutron energy distributions having nearly Maxwell-Boltzmann shape [Be 80] with kT=31 and 71 keV, and a nearly homogeneous distribution with E_n = 310 ± 40 keV. In fig. 2c the set up at the ILL-reactor at Grenoble is shown (for full experimental details see [Tr 84,85]).

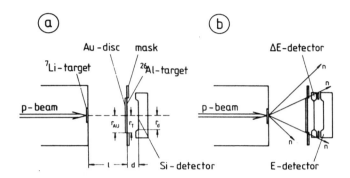

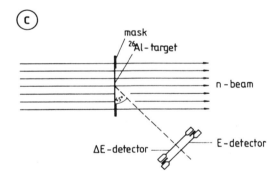

Fig. 2: Three typical experimental set ups
a) using a single particle detector and
b) a ΔE-E telescope where the neutrons are produced by nuclear reactions.
c) Set up of the ΔE-E telescope at the ILL-reactor in Grenoble.

Using first the simple set up of fig. 2a by placing a 100 μm thick Si-SB-detector in front of the ^{26}Al-target we received a spectrum shown in fig. 3. This detector was too thin to stop and hence observe the p_0-group, but by using a thick enough detector for the p_0-group the neutron induced background increased enormously so that no detection of any particle group was possible. The identification of the peak in fig. 3 with the p_1-group was based on i) its energy and ii) its intensity correlation by changing the ^{26}Al-targets with different ^{26}Al content (M III and M VI).

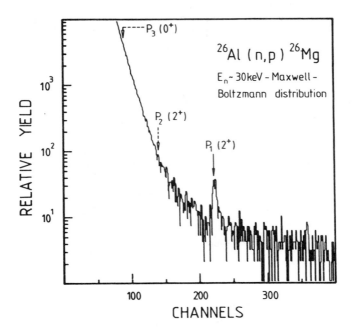

Fig. 3: Typical spectrum for the set up of fig. 2a using a 100 μm-thick Si-detector (for details see text).

Improvement of the signal to noise ratio could be obtained by choosing a stack of thin detectors as well as coincidence technique, which can also be used for particle identification if the energy loss of a particle going through the ΔE-detector is recorded against the total energy $\Delta E + E$, where the particle is stopped in the E-detector. Such a detection system was checked out at the ILL-reactor in Grenoble (fig. 2c), where it was possible to set up the detectors outside the neutron beam (this was not possible for the set ups of fig. 2a and b because of the method for neutron production. The exposure of the particle telescope to the neutrons, which was unavoidable, will destroy the expensive detectors in relatively short time periods).

Examples of 3D-spectra obtained at Grenoble are shown in fig. 4, where the presentation was rotated by 180° as compared to standard presentations in order to get a better survey. The p_1- and p_0-group could be clearly identified

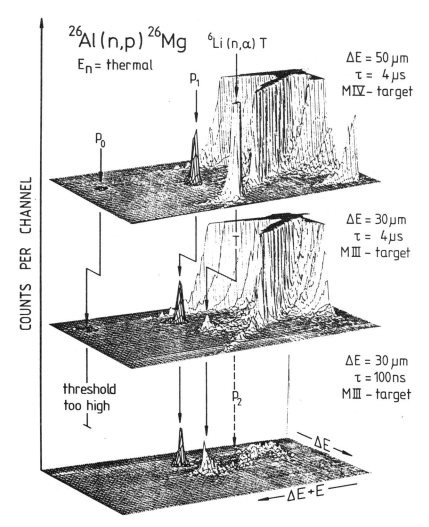

Fig. 4: Typical 3D-spectra as obtained at the ILL-reactor in Grenoble. The spectra are rotated by 180° compared to standard representation in order to gain a better survey (for details see text).

according to their expected coordinates. By reducing the thickness of the ΔE-detector (middle spectrum) the particle groups move accordingly to lower ΔE-values. At the lower spectrum a tighter coincidence requirement (Δt ~ 100 ns) between the ΔE- and the E-signal served as a gate and "cleaned up" the spectrum to allow extraction of an upper limit for the p_2-group. Similar spectra were then recorded for higher neutron energies and are shown and discussed in [Tr 84,85]. From the peak intensities M.B. averaged cross sections [Be 80] were deduced which were subsequently converted into reaction rates.

III. Discussion

In fig. 5 our results are shown as experimental points in a graph where $N_A <\sigma v>$ [cm^3 $mole^{-1}$ sec^{-1}] is given versus the temperature T_9. The data points for the p_0-group can be compared with the curve obtained by [Sk 83] from the reverse reaction $^{26}Mg(p,n_0)^{26}Al$ and are in good agreement with this curve. Our data points for the p_1-group lie about an order of magnitude higher then those for the p_0-group and the sum of both is shown as open circles.

In previous stellar model calculations the curve of [Wo 78] was adopted which is the result of Hauser-Feshbach calculations.

In the new compilation of [CFHZ 84] this curve was lowered by a factor 3.3 on the basis of level density arguments concerning ^{27}Al (fig. 5). Since our measurements were not sensitive to the higher proton-groups and only an upper limit for the p_2-transition of $\leq 10\%$ compared with the p_1-transition could be deduced at thermal neutron energies, our data must be considered as a lower bound. Judging from the results at thermal neutron energies, and implying that the p_2-transition is always lower than 10%, we would recommend a reduction in the $^{26}Al^g(n,p)^{26}Mg$ reaction rate of a factor 4 as compared to [Wo 78].

We would like to stress one additional point. If the statistical model for the $^{26}Al(n,p)^{26}Mg$ reaction would be fully applicable, then the cross section should follow the 1/v-law over a wide range of energies and the resulting reaction rate would be a constant versus temperature. This is obviously not the case when examining fig. 5. It would be therefore dangerous to measure the cross section just at thermal neutron energies and then extrapolate to stellar energies using the 1/v-law. It should also be pointed out for the same reasons that for $4.6 \times 10^{-7} \leq T_9 \leq 0.2$ the reaction rate is experimentally uncertain. Future experiments with targets of greater ^{26}Al-density will hopefully clarify all remaining open questions.

Woosley [Wo 85] has used in his recent nucleosynthesis network of a 25 $M_\odot$ star going through SN-explosion the new

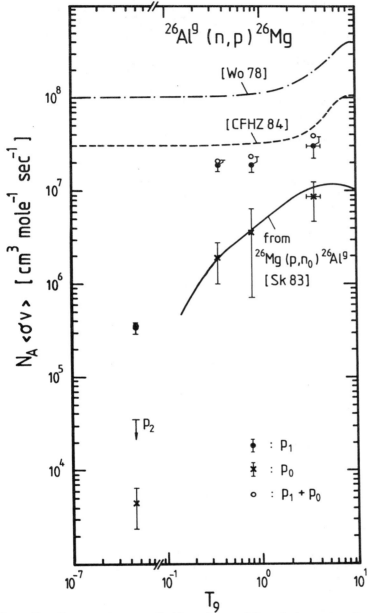

Fig. 5: Comparison of the directly determined reaction rate for $^{26}Al(n,p)^{26}Mg$ with the rate for the p_0-group as obtained by [Sk 83] from the inverse reaction $^{26}Mg(p,n)^{26}Al$ and with Hauser-Feshbach calculations [Wo 78], CFHZ 84] (for details see text).

[CFHZ 84] rate for ^{26}Al(n,p)^{26}Mg and can now explain about 10% of the ^{26}Al-content in the ISM. It would be interesting to see the results for an even lower ^{26}Al(n,p)^{26}Mg-rate.

Wiescher et al. [Wi 85] has shown that in new network calculations for novae the ^{26}Al output can drop by a factor of 4 to 100 depending on the nova model. These results are in conflict with the explanation [Cl 84] that all the ^{26}Al in the ISM comes from novae.

Cassé [Ca 85] argues that a significant contribution of ^{26}Al comes from W.R.-stars.

What ever the final answer to the question of the origin of ^{26}Al in the ISM will be, our impression is, that basic nuclear physics input data to the various stellar model calculations have to be further improved before conclusive statements can be made.

References

[Ar 85] M. Arnould: proceedings Moriond (1985)
[Be 80] H. Beer and F. Käppeler: Phys.Rev. C21 (1980) 534
[Bu 84] L. Buchmann, H. Baumeister and C. Rolfs:
    Nucl. Instr. Meth. B4 (1984) 132
[Ca 85] M. Cassé: proceedings Moriond (1985)
[CFHZ 84] G.R. Caughlan, W.A. Fowler, M.J. Harris and
    B. Zimmermann: Atomic Data Nucl. Data
    Tables   (194)
[Cl 84] D.D. Clayton: Ap. J. 280 (1984) 144
[Du 85] P. Durouchoux: proceedings Moriond (1985)
[Ma 82] W.A. Mahoney, J.C. Ling, A.S. Jacobson and R.E.
    Lingenfelter: Ap. J. 262 (1982) 742
[Pr 85] N. Prantzos: proceedings Moriond (1985)
[Sk 83] R.T. Skelton, R.W. Kavanagh and D.G. Sargood:
    Ap. J. 271 (1983) 404
[Tr 84,85] H.P. Trautvetter: Habilitationsschrift Univ.
    Münster (1984) and to be published
[Wa 82] G.J. Wasserburg and D.A. Papanastassiou: Essays in
    Nuclear Astrophysics, ed. C.A. Barnes, D.D. Clayton
    and D.N. Schramm (Cambridge Press), (1982) 77
[Wi 85] M. Wiescher et al.: proceedings Moriond (1985)
[Wo 78] S.W. Woosley, W.A. Fowler, J.A. Holmes and B.A.
    Zimmerman: Atomic Data and Nucl. Data Tables
    22 (1978) 371
[Wo 85] S.E. Woosley: proceedings Moriond (1985)

Certainties and Uncertainties in Long-Lived Chronometers

Bradley S. Meyer and David N. Schramm
Department of Astronomy and Astrophysics
University of Chicago
Chicago, Ill. 60637

Abstract. Evidence is provided for the confirmation of the long-held suspicion that ^{232}Th/^{238}U is not a very long-lived chronometric pair. It is argued, however, that it nevertheless provides a firm lower limit to the age of the Galaxy. It is also shown that ^{187}Re/^{187}Os is a very long-lived chronometric pair; thus, the Re/Os pair gives an upper limit to T_{Gal}, the Galaxy's age. From current data, our limits give a model independent range for T_{Gal} of 6.4 Gyr $\lesssim T_{Gal} \lesssim$ 22.6 Gyr with a best fit model of $T_{Gal} \approx$ 15 Gyr. Any conclusions about needing a cosmological constant to fit age constraints are thus very model dependent. A discussion of the current uncertainties in the input parameters is also presented.

1. Introduction

By virtue of their long lifetimes, the chronometric pairs ^{232}Th/^{238}U and ^{187}Re/^{187}Os are useful tools for the investigation of the age of the Galaxy. Schramm and Wasserburg (1970) have even shown that if the chronometers are very long-lived, they give constraints on the Galaxy's age independent of the model of Galactic evolution. The promise of long-lived nucleocosmochronology is tarnished, however, by the uncertainties that plague the Th/U and Re/Os pairs. Furthermore, the Th/U pair may not even be very long-lived since the ^{238}U lifetime may be less than the duration of nucleosynthesis prior to solar system formation. In this paper, we investigate whether we can put firm constraints on the Galaxy's age despite these uncertainties.

2. Basic Equations

The general equation governing the abundance N_i of a nuclear species with decay rate λ_i is (Schramm and Wasserburg 1970)

$$\frac{dN_i(r,t)}{dt} = -\lambda_i N_i(r,t) + B(r,N_i,t) \tag{1}$$

where $B(r,N_i,t)$ is a generalized production function. (It also allows for destruction by means other than radioactive decay.) Equation (1) can be linearized by suitable choice of galactic evolution model (Tinsley 1975, Hainebach and Schramm 1977) to give

$$\frac{dN_i}{dt} = -\lambda_i N_i - \omega N_i + P_i \psi(t) \qquad (2)$$

where ω is the rate of movement of material out of the gas for reasons other than radioactive decay, for example, by trapping in white dwarfs, neutron stars, or black holes or by enriched galactic infall, P_i is the number of nuclei i produced per unit mass going into stars, and $\psi(t)$ is the time rate of mass going into stars. In general, we expect ω to be a function of time.

We solve equation (2) by integrating over a time T and then allowing for free decay over an interval Δ. This corresponds to a model of solar system formation in which the solar system nebula receives no new nucleochronometer material after T. The interval Δ corresponds to the time required for formation of solid bodies in the solar system where decay products are retained. The solution of (1) under this scenario is given by Schramm and Wasserburg (1970) and Symbalisty and Schramm (1981) as

$$N_i(T+\Delta) = P_i T <\psi> \exp(-\lambda_i \Delta)\exp[-\nu(T)]\exp[-\lambda_i(T-t_\nu)][1+\delta_i] \qquad (3)$$

where

$$\nu(t) \equiv \int_0^t \omega(\tau)d\tau, \qquad (4)$$

$$<\psi> \equiv \frac{1}{T}\int_0^T \psi(t)e^{\nu(t)}dt, \qquad (5)$$

$$t_\nu \equiv \frac{1}{<\psi>T}\int_0^T t\psi(t)e^{\nu(t)}dt, \qquad (6)$$

$$\delta_i \equiv \sum_{n=2}^{\infty} \frac{\lambda_i^n}{n!}\mu_n, \qquad (7)$$

and

$$\mu_n \equiv \frac{1}{<\psi>T}\int_0^T (t-t_\nu)^n \psi(t)e^{\nu(t)}dt. \qquad (8)$$

If we then take the ratio of two nuclide abundances i and j and expand δ_i and δ_j, we obtain the result

$$T-t_\nu = \Delta_{ij}^{\max} - \Delta + \frac{(\lambda_i+\lambda_j)}{2}\mu_2 + \frac{(\lambda_i^2+\lambda_j^2+\lambda_i\lambda_j)}{6}\mu_3 + \dots, \qquad (9)$$

where

$$\Delta_{ij}^{\max} = \frac{\ln[(P_i/P_j)/(N_i(T+\Delta)/N_j(T+\Delta)]}{\lambda_i-\lambda_j}. \qquad (10)$$

We note in equation (9) that, since μ_n is proportional to T^n, if $\lambda_i T$ and $\lambda_j T$ are $\ll 1$, we can make the approximation

$$T - t_\nu = \Delta_{ij}^{\max} - \Delta. \tag{11}$$

This result is extremely nice for two reasons. First, it is independent of detailed models of galactic evolution; that is, it requires no knowledge of ω or $\psi(t)$. Second, it has an easy interpretation. $\Delta_{ij}^{\max}$ in equation (10) is simply the time required for decay of two nuclei i and j, produced in a single event in the abundance ratio P_i/P_j, to reach an abundance ratio at $T+\Delta$ of N_i/N_j. This makes sense since the longest-lived chronometer pairs see nucleosynthesis as a single event. Two chronometric pairs that may satisfy this long-lived criterion are ^{232}Th/^{238}U and ^{187}Re/^{187}Os. In section 5 we will see that equation (11) is satisfactory for ^{187}Re/^{187}Os but probably not for ^{232}Th/^{238}U.

Knowledge of $T - t_\nu$ allows us to put limits on the age of the Galaxy. Figure 1 shows two simple, extreme models of Galaxy evolution. From these models we can conclude that

$$(T-t_\nu)+\Delta+t_{ss} \lesssim T_{\text{Gal}} \lesssim 2(T-t_\nu)+\Delta+t_{ss}, \tag{12}$$

where $t_{ss} = 4.6$ Gyr is the age of the solid bodies in the solar system. The longest-lived chronometers satisfy equation (11); hence, for them, equation (12) becomes

$$\Delta_{ij}^{\max}+t_{ss} \lesssim T_{\text{Gal}} \lesssim 2(\Delta_{ij}^{\max}-\Delta)+\Delta+t_{ss} \tag{13}$$

For the longest-lived chronometers, then, it is simply a matter of determining (P_i/P_j) and $(N_i(T+\Delta)/N_j(T+\Delta))$ to get limits on T_{Gal}. N_i/N_j is derived from meteoritic, terrestrial, and lunar studies. P_i/P_j comes from nuclear calculations and is discussed next.

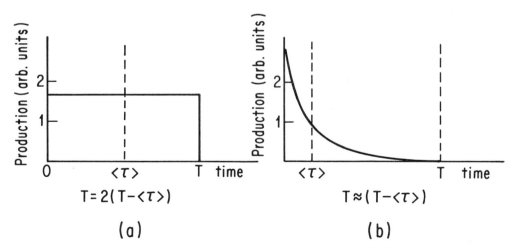

Figure 1. Two simple, extreme models of Galaxy evolution. (a) is the steady synthesis model.

3. Production Ratios

The basic method of calculating P_i/P_j is not difficult. If we follow the prescription by Seeger and Schramm (1970), we first assume equal abundances per isobar since there is generally no shell structure near typical progenitors of our chronometers. Next we correct for decay back effects forcing material out of the path of decay to the nucleus in question. Finally, we simply add up progenitors. As an example, consider the calculation of P_{232}/P_{238}. ^{232}Th has six alpha decaying β-stable progenitors: (232), (236), (240), (244), (248), and (252). The progenitors of ^{238}U are (238), (242), (246), and 1/10 of (250). The other 9/10 of (250) fissions. The production ratio, then, is simply $6/3.1 \approx 1.9$.

There are two caveats that must be mentioned in the calculation of P_{232}/P_{238}. The first is that Wene (1975) and Thielemann et al. (1983) have pointed out that β-delayed fission may be enhanced for some of the higher A progenitors of ^{232}Th. This would tend to lower P_{232}/P_{238}, thereby increasing $\Delta^{max}_{232,238}$. The number Thielemann et al. (1983) give for P_{232}/P_{238} is 1.4, so the effect can be quite large.

The second caveat is that the assumption of equal abundances per isobar may not be valid. There may be a bump in the abundance vs. A curve for the actinides. The effect of this bump would tend to counter the effect of the delayed fission because the bump leads to a lower relative abundance at higher A so that we knock out fewer progenitors of thorium by delayed fission. Since the value 1.9 was calculated for no delayed fission, we will regard it as an upper limit on P_{232}/P_{238}. It will thus provide us with a lower limit on $\Delta^{max}_{232,238}$, which in turn will give a lower limit on T_{Gal} via equation (13) since the μ_2 term in equation (9) will tend to increase $T - t_\nu$. Furthermore, since the bump counters the delayed fission, we might regard 1.4 as a lower limit for P_{232}/P_{238}. This would provide an upper limit on $\Delta^{max}_{232,238}$.

4. Re/Os

The other long-lived pair is ^{187}Re/^{187}Os ($\tau_{1/2}$ for ^{187}Re $\approx 43 \times 10^9$ yr, ^{187}Os is stable), first proposed for study by Clayton (1964). Because ^{187}Os is stable and has no direct contribution from the r-process since it is shielded from β-decay from below by ^{187}Re, the formulae required to derive $\Delta^{max}_{187,187}$ are different from those of other chronometers. The present ^{187}Os abundance results from an s-process part, $(^{187}\text{Os})_s$, and from a cosmoradiogenic part $(^{187}\text{Os})_c$, resulting from β-decay from ^{187}Re. The number $R_{187,187}$ is then given by (Schramm 1974)

$$R_{187,187} = 1 + \frac{(^{187}\text{Os})_c}{^{187}\text{Re}}, \quad (14)$$

where

$$\Delta^{max}_{187,187} = \frac{\ln R_{187,187}}{\lambda_{187}}. \quad (15)$$

We get the cosmoradiogenic abundance of ^{187}Os from ^{186}Os since it is not shielded and, consequently, not changing after nucleosynthesis:

$$\frac{(^{187}\text{Os})_c}{^{186}\text{Os}} = \frac{^{187}\text{Os}}{^{186}\text{Os}} - \frac{(^{187}\text{Os})_s}{^{186}\text{Os}} \quad (16)$$

or

$$\frac{(^{187}\text{Os})_c}{^{186}\text{Os}} = \frac{^{187}\text{Os}}{^{186}\text{Os}} - \left(\frac{\sigma_{186}}{\sigma_{187}}\right)_{\text{lab}} \times f, \qquad (17)$$

where the σ's are neutron capture cross sections and f is a factor that converts lab cross sections appropriate for conditions inside stars, that is, at kinetic energies of approximately 30 keV. Equation (17) is derived from the so-called local approximation for the s-process:

$$(^{187}\text{Os})_s(\sigma_{187})_{30\text{keV}} \approx (^{186}\text{Os})_s(\sigma_{186})_{30\text{keV}}. \qquad (18)$$

There are a number of uncertainties in the Re/Os pair. The first is that the half-life of ^{187}Re β-decay inside stars may be much different from the half-life in the lab. Two effects conspire to make this so. The first is that, although in the lab ^{187}Re β-decay goes to the continuum 99% of the time (Williams et al. 1984), in stars outer electron orbits are vacant so that β-decay can easily go to bound states (Arnould et al. 1984). This increases the rate of decay. The second effect is that more low lying nuclear excited states are populated at higher temperatures than in the lab. When one sums up the decay rates (weighted by occupation numbers) from the different excited states, one finds a significant increase in λ_{187} with increasing temperature (Cosner and Truran 1981). Clearly, then, the half-life of ^{187}Re decreases dramatically with temperature; thus, we must find an effective λ_{187} to use with equation (15) to get $\Delta^{\max}_{187,187}$. We expect this effective λ_{187} to be larger than the lab λ_{187}, consequently, we expect it to decrease $\Delta^{\max}_{187,187}$.

A second uncertainty in the Re/Os pair is that the local approximation (equation (18)) may not be valid in the region of interest for this chronometer. Arnould et al. (1984) have raised such doubts. They claim that since some of the s-process ^{186}Re may neutron capture before β-decaying, we must question the use of equation (17) in solving for $(^{187}\text{Os})_c$. Truran (1985), however, has calculated $\Delta^{\max}_{187,187}$ using various values of $\frac{(^{187}\text{Os})_s}{^{186}\text{Os}}$ and has found it to be quite insensitive to the assumed value of $\frac{(^{187}\text{Os})_s}{^{186}\text{Os}}$. Furthermore, it is unimportant in equation (14) whether the abundance of ^{187}Re is r only or r+s. From these facts, we conclude that, even though there may be branching at ^{186}Re, its effect on $\Delta^{\max}_{187,187}$ is negligible.

The final uncertainty in the Re/Os pair is in the factor f. It has been claimed (Arnould et al. 1984) that a low-lying nuclear excited state in ^{186}Os may lead to a large uncertainty in f. Recent experimental work by Winters (1984), however, yields a small range for f: $0.80 \leq f \leq 0.83$. We thus assume that the uncertainty in $\Delta^{\max}_{187,187}$ due to f is negligible, at least compared to the uncertainty due to the effect of astration on the β-decay of the ^{187}Re.

Our presumption is thus that the uncertainty in the Re/Os pair due to the effect of astration on ^{187}Re β-decay is the dominant of the three uncertainties and has the largest effect. To calculate the magnitude of this effect, we must know the length of time ^{187}Re spends at various temperatures inside stars, a quantity which is clearly model dependent. We conclude from the existence of deuterium, which is easily destroyed inside stars, that much material never goes into stars. Furthermore, since the abundance by mass of metals is only $Z \approx 0.02$, we conclude that not a lot of stellar processing has occurred. From these two observations, we might infer that the time ^{187}Re spends inside stars may be relatively small so that our basic conclusions may require little modification. On the other hand, Yokoi et al. (1983) find in their detailed Galactic evolution models an effective ^{187}Re half-life of $\tau \approx 35 \times 10^9$ yr. This translates into an approximately 20% decrease in $\Delta^{\max}_{187,187}$ from the $\Delta^{\max}_{187,187}$ calculated from the lab half-life. This 20% decrease is significant. The effective ^{187}Re half-life question obviously needs more work. What we can say at present, however, is that since the dominant uncertainty in the Re/Os pair is the ^{187}Re β-decay rate, and since this uncertainty always increases the decay rate, $2(\Delta^{\max}_{187,187} - \Delta)$ (see equation (13)) calculated from the lab λ_{187} should represent a good upper limit to T. This is an important point and deserves attention.

5. Results

We now present results for $\Delta^{\max}$ for Th/U and Re/Os. From the current range of values for P_{232}/P_{238} (1.4 to 1.9) and for values of N_{232}/N_{238} derived from consistency with lead isotopes (2.3 to 2.7, Symbalisty and Schramm, 1981), with best values of $P_{232}/P_{238} \approx 1.6$ and $N_{232}/N_{238} = 2.5$, we get $\Delta^{\max}_{232,238} = 4.2^{+2.5}_{-2.4} \times 10^9$ yr. From this we conclude $T_{\text{Gal}} \gtrsim 6.4 \times 10^9$ yr.

From Winters' (1984) values for f (0.8 to 0.83) and $\dfrac{\sigma_{186}}{\sigma_{187}}$ (0.478 ± 0.022) and from Luck et al.'s (1980) values for $(\dfrac{^{187}\text{Os}}{^{186}\text{Os}})_{T+\Delta}$ (0.805 ± 0.011) and $(\dfrac{^{187}\text{Re}}{^{187}\text{Os}})_{T+\Delta}$ (3.20 (± 10%)), we get $\Delta^{\max}_{187,187} = 7.5^{+1.5}_{-1.2} \times 10^9$ yr. This $\Delta^{\max}_{187,187}$ has been calculated with $\tau_{1/2}$ for ^{187}Re $= 43 \times 10^9$ yr; therefore, $2(\Delta^{\max} - \Delta) + \Delta + t_{ss}$ represents an upper limit to T_{Gal}. This limit is 22.6 Gyr.

We now investigate the effect of higher moment terms on T. For this analysis we consider only a steady synthesis model, that is, one for which $\psi e^{\nu} = $ constant. To zeroth order in μ, equation (11) applies and $T = 2(\Delta^{\max} - \Delta)$. Such a model is supported by the work of Reeves and John (1976) and Hainebach and Schramm (1977). Table 1 shows values of T calculated to order μ_2, the resulting values of T_{Gal}, and the percent change δ in T, where

$$\delta(\%) = \frac{T - 2(\Delta^{\max} - \Delta)}{2(\Delta^{\max} - \Delta)} \times 100\% \qquad (19)$$

for various values of $\Delta^{\max} - \Delta \approx \Delta^{\max}$ (for long-lived chronometers).

As is clear from the table, corrections to T for Th/U can be large (>10%). Our best estimate for T_{Gal} from Th/U is pushed up from 13.0 Gyr to 14.8 Gyr. Corrections due to μ_3 are zero (μ_3 is an odd integral about t_{ν} for steady synthesis) and the correction due to μ_2^2 and μ_4 amounts to only 0.85% for $\Delta^{\max}_{232,238} = 4.2 \times 10^9$ yr. For Re/Os, on the other hand, the corrections $\delta(\%)$ are small. In the steady synthesis model, then, Re/Os is a long-lived chronometer and 22.6 Gyr does indeed represent a good upper limit to T_{Gal}.

Pair	$\Delta^{max}-\Delta$ (Gyr)	T (Gyr)	T_{Gal} (Gyr)	$\delta(\%)$
Th/U	1.8	3.9	8.5	8.3
	4.2	10.2	14.8	21.4
	6.2	17.8	22.4	43.5
Re/U	6.3	12.8	17.4	1.6
	7.5	15.3	19.9	2.0
	9.0	18.5	23.1	2.8

Table 1. This table shows T and T_{Gal} calculated from Th/U and Re/Os for a steady synthesis model including order μ_2 in equation (9) for the indicated values of $\Delta^{max}-\Delta \approx \Delta^{max}$. $\delta(\%)$ (see equation (19)) represents the percent increase of the calculated T over the very long-lived expression for T, namely $2(\Delta^{max}-\Delta)$.

6. Conclusions

Although the study of the Galactic age through the use of long-lived chronometers is at present fraught with uncertainties, there are two important conclusions we can nevertheless draw before these uncertainties are resolved. First, we can get a firm, model-independent lower limit to the age of the Galaxy from the Th/U pair. In particular, $T_{Gal} \gtrsim \Delta^{max}_{232,238} + t_{ss}$. Our other conclusion is that the dominant uncertainty in the Re/Os chronometric pair is in the effective half-life of ^{187}Re. As we have argued, however, this uncertainty always decreases $\Delta^{max}_{187,187}$; hence, we expect $\Delta^{max}_{187,187}$ calculated from the lab λ_{187} to provide us with an upper limit to T_{Gal}: $T_{Gal} \lesssim 2(\Delta^{max}_{187,187} - \Delta) + \Delta + t_{ss}$.

Acknowledgements

We would like to thank the hospitality of those who made the Fifth Moriond Astrophysics Meeting an enjoyable and fruitful experience. We would also like to thank Jim Truran for helpful discussions and for his results on the s-process branching at ^{186}Re. This work was supported in part by NSF grant AST 8313128. One of us (B.S.M.) would like to acknowledge the support of a National Science Foundation Graduate Fellowship.

References

Arnould, M., Takahashi, K., and Yokoi, K. 1984, *Astr.Ap.*, **137**, 51.
Clayton, D. D. 1964, *Ap.J.*, **139**, 637.
Cosner, K., and Truran, J.W. 1981, *Astrophys. Space Sci.*, **78**, 85.
Hainebach, K., and Schramm, D.N. 1977, *Ap.J.*, **212**, 347.
Luck, J.M., Birck, J.L., and Allegre, C.J. 1980, *Nature*, **283**, 256.
Reeves, H., and Johns, O. 1976, *Ap.J.*, **206**, 958.
Schramm, D.N. 1974, *Ann. Rev. Astr. Ap.*, **12**, 303.
Schramm, D.N., and Wasserburg, G.J. 1970, *Ap.J.*, **162**, 57.

Seeger, P.A., and Schramm, D.N. 1970, *Ap.J.*, **160**, L157.
Symbalisty, E.M.D., and Schramm, D.N. 1981, *Rep.Prog.Phys.*, **44**, 293.
Thielemann, F.-K., Metzinger, J., and Klapdor, H.V. 1983, *Astr.Ap.*, **123**, 162.
Tinsley, B.M. 1975, *Ap.J.*, **198**, 145.
Truran, J.W. 1985, *private communication*.
Wene, C.O. 1975, *Astr.Ap.*, **44**, 233.
Williams, R.D., Fowler, W.A., and Koonin, S.E. 1984, *Ap.J.*, **281**, 363.
Winters, R.R. 1984, in AIP **124**, *Neutron-Nucleus Collisions: A Probe of Nuclear Structure*, eds. J. Rapaport, R.W. Finlay, S.M. Grimes, and F.S. Dietrich, pp. 495-497.
Yokoi, K., Takahashi, K., and Arnould, M. 1983, *Astr.Ap.*, **117**, 65.

MORE ABOUT NUCLEOCOSMOCHRONOLOGY: THE RELIABILITY OF THE LONG-LIVED
CHRONOMETERS, AND THE PRODUCTION OF EXTINCT RADIOACTIVITIES

M. Arnould
Institut d'Astronomie et d'Astrophysique
Université Libre de Bruxelles, Belgium
N. Prantzos
Service d'Astrophysique, CEN Saclay, France

ABSTRACT. After reassessing the opinion that the ^{232}Th-^{238}U, ^{235}U-^{238}U and ^{187}Re-^{187}Os pairs do not appear to be very reliable galactic age indicators, we examine a model for the origin of extinct radionuclides calling for massive mass losing Of and WR stars. We show that these stars may be responsible for the inferred presence of ^{26}Al, ^{107}Pd and possibly also other short-lived species in certain meteoritic condensates.

1. INTRODUCTION

In nucleocosmochronological studies, use is made most classically of the r-process radionuclides ^{129}I, ^{232}Th, ^{235,238}U, and ^{244}Pu. This set is sometimes complemented with the ^{187}Re-^{187}Os pair, which is very often regarded as one of the best candidates for providing an estimate of the "age" of the r-process nuclei, and thus a lower limit to the age of the Galaxy (e.g./1/).
   The development of other chronometers has recently been the subject of many experimental and theoretical studies. This excitement is largely related to the discovery of meteoritic isotopic anomalies which are attributed to the in-situ decay of now extinct radionuclides /2-4/. In this field, most of the recent experimental data concern ^{26}Al and ^{107}Pd. In general, they are interpreted as strong evidence for the presence of those short-lived nuclei in the early solar system. In such views, those radionuclides may provide key informations about the time span between the last events that were able to modify the composition of the solar nebula and the formation of solar system solid bodies.
   This work deals with two quite different questions:
(1) in Sect. 2, we want to reassess briefly some quite pessimistic views expressed in great detail elsewhere /5,6/ concerning the reliability of the predicted age of the Galaxy based on the use of the long-lived chronometric pairs ^{232}Th-^{238}U, ^{235}U-^{238}U, and ^{187}Re-^{187}Os. This is at variance with the optimism expressed by some authors (e.g. /7,8/), which, however, appears to have subsided very recently /9/;
(2) in Sect. 3, we propose a new model for the origin of some extinct radionuclides. More specifically, we show that Of and Wolf-Rayet stars

could have contaminated the solar nebula with ^{26}Al, and also with other short-lived radionuclides.

The main conclusions of this work are summarized in Sect. 4.

2. CAN 232Th-238U, 235U-238U, AND 187Re-187Os REALLY PROVIDE RELIABLE ESTIMATES FOR THE AGE OF THE GALAXY ?

This question has already been debated in great detail in /5,6/. Here we just summarize some of the main points stressed in those studies:
A) The cosmochronological estimates are performed in the framework of a model for the chemical evolution of the solar neighborhood which satisfies various observational constraints, and in particular the age-metallicity relation and metallicity distribution /10-12/;
B) Physically acceptable solutions to our chemical evolution model are found for galactic ages in the approximate range $11 \lesssim T_G \lesssim 15$ Gyr (this result is obtained without any reference to nucleocosmochronometers);
C) In that model, the chronometric virtues of the ^{232}Th-^{238}U or ^{235}U-^{238}U pairs are not clearly evident, especially once the many astrophysical, meteoritic or nuclear physics uncertainties are taken into account. Those uncertainties concern in particular the r-process production ratios, the evaluation of which suffers from the lack of precise identification of the r-process site, as well as from the poor reliability of the predicted properties of very neutron-rich nuclei by the available nuclear models. In addition, some abundance ratios at the time T_{SOLID} of solidification in the solar system are still not known precisely enough.

Figs. 6a and b of /4/ are quite suggestive of the difficulty of estimating T_G reliably with the aid of the above mentioned pairs¹. This situation can be explained by the loss of information about the early galactic epoch, resulting from the expected weak time dependence of the stellar birth rate. Our conclusions contradict quite common views in the field (e.g. /7,8/), based on the use of a simplified (exponential-type) model for the chemical evolution of the Galaxy which, in particular, does not take into account the observational constraints mentioned in A);
D) The chronological virtues sometimes attributed to the ^{187}Re-^{187}Os pair are not evident yet. The difficulties encountered in the use of that clock concern the possible enhancement of the ^{187}Re and ^{187}Os β-transmutation rates in stars /5,15/, as well as the exact $N_s(^{187}Os)/N_s(^{186}Os)$ s-process production ratio. This latter question is discussed in /6/, where it is shown that some of its aspects have been largely overlooked.

The considerations summarized above lead us to the conclusion that the ^{232}Th-^{238}U, ^{235}U-^{238}U, and ^{187}Re-^{187}Os pairs may not be as faithful galactic clocks as commonly imagined. One has at least to be aware of the many uncertainties involved in their use!

---

¹Fig. 6b of /4/ displays a value of 2.5 ± 0.2 (taken from /1/) for the ^{232}Th/^{238}U abundance ratio at T_{SOLID}. From /3/, we derive 2.3 ± 0.2, while values from 1.7 to 3.0 are obtained from /14/. The adoption of this latter range would of course reinforce our views concerning the rather poor reliability of the ^{232}Th-^{238}U pair

3. THE PRODUCTION OF 26Al AND OTHER SHORT-LIVED RADIONUCLIDES BY MASSIVE MASS-LOSING STARS DURING THE Of AND WOLF-RAYET STAGES

3.1. Generalities

There is now strong experimental evidence for the signature of the in-situ decay of ^{26}Al and ^{107}Pd in meteorites (e.g. /4, 16/). Such observations are generally interpreted in terms of the injection of live ^{26}Al and ^{107}Pd (possibly along with other short-lived species) in the solar nebula, followed by their trapping into condensing solids. In such views, important informations can be gained on the time Δ^* elapsed between the last astrophysical event(s) able to affect the composition of the solar nebula and the solidification of some of its material.

The very nature of such (an) event(s) is still debated. In particular, a single nearby supernova that exploded within a few million years of solar system formation has been envisioned /17/. Another appealing model calls for the birth of the solar system in an OB association, where the contamination could be due to stellar wind mass losses of the constitutive massive stars and/or their explosion as supernovae /18,19/. Low mass stars ($M \approx 1 M\odot$) in their asymptotic giant branch (AGB) evolutionary stage have also been proposed as a promising source of the short-lived radioactivities in the early solar system /20/.

In contrast, some models do not recognize any chronometric virtue to the short-lived radionuclides. In particular, Clayton suggests that their decay did not take place in the solar system itself, but instead in alien presolar grains incorporated into meteorites without much alteration (e.g. /2,3,21/). Of course, such a model does not require (a) nucleosynthetic event(s) just preceding the solar system formation. On the other hand, a *local* production in the solar system by an extremely active young Sun has sometimes been advocated.

In this work, we propose that massive mass losing stars during their Of and Wolf-Rayet (WR) stages are potentially appealing candidates for the production of ^{26}Al, ^{107}Pd, and other short-lived radionuclides, as well as for the contamination of the solar nebula with such species (either live in gas and/or grains, or extinct in grains). This contamination would be especially efficient if the solar system has been born in an OB association /18,19/, and still more so if certain at least of the envisioned massive stars would experience a supernova explosion.

We base our study on the massive mass losing star models described in /22,23/. The nucleosynthesis during the O or Of phases (associated to central H burning) and WR stage (corresponding to central He burning) is followed with the aid of the detailed networks described in /23,24/

3.2. ^{26}Al yields from Of and WR stars

The ^{26}Al mass (expressed in $M\odot$) present in the interstellar medium (ISM) as a result of the wind ejection by our star models between the zero-age main sequence (ZAMS; t=0) and time t is defined by

$$M_{26}^{(W,O)}(t) = \int_0^t \dot{m}(t')X_{26}^{(S)}(t')\exp(-t'/\tau_{26})dt', \quad (1)$$

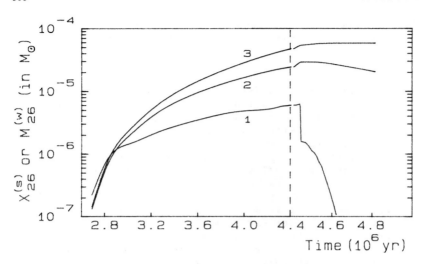

Figure 1. Values of the ^{26}Al mass fraction at the stellar surface $X_{26}^{(s)}$ (curve 1), and of the ^{26}Al masses $M_{26}^{(w,D)}$ (curve 2) or $M_{26}^{(w,ND)}$ (curve 3) in the wind of the M_{ZAMS} = 80 M⊙ model during the Of and WR stages, which correspond to $t \lesssim 4.4\ 10^6$ yr and $t \gtrsim 4.4\ 10^6$ yr, respectively (note the change of scale in abscissa)

if the ^{26}Al β-decay in the ISM is taken into account. In this expression, $\dot{m}$ is the mass loss rate in M⊙/yr, τ_{26} is the ^{26}Al lifetime, while $X_{26}^{(s)}(t)$ is the stellar surface ^{26}Al mass fraction at time t. A similar quantity, $M_{26}^{(w,ND)}$, can also be defined, which neglects the ISM ^{26}Al β-decay (i.e. $\tau \longrightarrow \infty$ in Eq. (1); of course, the ^{26}Al decay is properly taken into account in the stellar core and envelope).

Fig. 1 displays $X_{26}^{(s)}$, $M_{26}^{(w,D)}$ and $M_{26}^{(w,ND)}$ for our M_{ZAMS} = 80 M⊙ star model. It is seen that the ^{26}Al mass increases rapidly during the Of stage, as a result of the ^{26}Al production through the MgAl cycle. In contrast, it increases much more slowly at the beginning of the WN phase. Later on, $M_{26}^{(w,ND)}$ reaches a constant value, while $M_{26}^{(w,D)}$ levels off. $X_{26}^{(s)}$ indeed decreases very rapidly after the start of He burning, as a result of the ^{26}Al destruction by ^{26}Al(n,p)^{26}Mg and ^{26}Al(n,α)^{23}Na.

From Fig. 1, it can be concluded that the star under consideration can be an important contributor to the ISM ^{26}Al. Similar calculations have been performed for M_{ZAMS} = 50, 60, and 100 M⊙, and indicate that $M_{26}^{(w)}$ increases slightly with M_{ZAMS}. In fact, $M_{26}^{(w,D)}$ reaches maximum values (in M⊙) ranging from $1.5\ 10^{-5}$ for M_{ZAMS} = 50 M⊙ to $5.4\ 10^{-5}$ for M_{ZAMS} = 100 M⊙. For those same stellar masses, $M_{26}^{(w,ND)}$ has "asymptotic" values of $2.3\ 10^{-5}$ and $1.2\ 10^{-4}$ M⊙, respectively.

These results may have interesting consequences for the interpretation of the observed ISM ^{26}Al γ-ray line /25/, as well as of certain Mg isotopic anomalies found in some chondrites. This latter problem is examined below.

3.3. The ^{26}Al-related Mg isotopic anomalies in meteorites, and the Of evolutionary phase

If attributed to the in-situ decay of ^{26}Al, the ^{26}Mg excess (^{26}Mg*) found in certain Ca-Al-rich inclusions of some chondrites requires an abundance ratio (^{26}Al/^{27}Al)$_o$ ≈ 5 10^{-5}, where the subscript o refers to the time of solidification T_{SOLID} in the solar system. However, there appears to be a large spread around that "canonical" value. In fact, values in the range $0 \lesssim (^{26}$Al/^{27}Al)$_o \lesssim 10^{-3}$ have been reported (e.g. /4/).

Up to now, no undisputable correlation has been found between ^{26}Mg* and other isotopic anomalies. In particular, the associated Ca isotopic composition and ^{25}Mg/^{24}Mg ratio are found to be essentially solar. This puts constraints on the ^{26}Al production models, which call for red giants /20,26/, novae /27-29/, or supernovae /30-32/. All those models encounter various astrophysical or nuclear physics difficulties (e.g. /26,27,33/).

Even if it is not free from uncertainties, the ^{26}Al source model proposed in this work (just calling for the MgAl cycle during the main-sequence phase!) appears to be relatively more secure than models relying on complicated evolutionary phases (AGB or explosions) and/or still rather uncertain nuclear physics. It remains to be seen if our model does not predict a complement of other elements which would have a sufficiently non-solar isotopic composition to be detected as anomalies in the ^{26}Mg*-bearing inclusions. This important question will be discussed in detail elsewhere /34/. Here, let us simply summarize two of the main points relevant to the Of phase: (i) for $(^{26}$Al/^{27}Al)$_o$ = 5 10^{-5}, it is found that $(^{25}$Mg/^{24}Mg)$_o$ departs from the solar value by less than about 1 per mil for high enough M_{ZAMS} ($\gtrsim$ 80 M⊙) and short enough times Δ^* ($\lesssim$ 5 10^5 yr), and (ii) the possibly accompanying Si, Ca or Ba are essentially solar, in agreement with the observations. Of course, these results are obtained by applying the same "dilution" factor to all the nuclear species, fractionation effects being ignored.

From this brief summary, it can be concluded that ^{26}Mg* observed in certain inclusions might be the result of a ^{26}Al ejection into the ISM during the Of phase. In addition, reasonable conditions can be found for which the possibly accompanying Mg, Si, Ca or heavier elements have essentially a solar isotopic composition.

3.4. Contribution of the WR stage to the yields of ^{26}Al and other very short-lived radionuclides

During the WR (WN + WC/WO) stage, the neutrons produced in the He-burning core /24/ are responsible for the ^{26}Al destruction (see Fig. 1), as well as for a "mini" s-process. Here, we want to concentrate on the role of that processing for the synthesis of various short-lived radionuclides, and for the related ISM enrichment.

Following a method similar to the one used in Sects. 3.2 and 3.3 for the evaluation of $(^{26}$Al/^{27}Al)$_o$, the abundance ratio $(R/S)_o$ of a radionuclide R to a stable nucleus S corresponding to the matter ejected during the whole WR phase can be evaluated straightforwardly for given M_{ZAMS} and Δ^* values (Δ^* being interpreted here as the time span between the end of the WR phase and T_{SOLID}). Note that the decay of R between its ejection

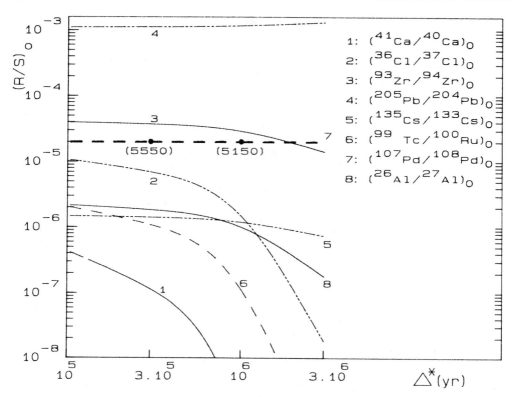

Figure 2. Abundance ratios $(R/S)_0$ at the time of solidification T_{SOLID} for various values of Δ^*, R and S representing a radionuclide and a stable nucleus, respectively. The curves are calculated from the total masses of R ($M_R^{(W,ND)}$) and S ejected into the ISM during the whole WR stage of a M_{ZAMS} = 80 M$\odot$ star. The data are normalized to $(^{107}Pd/^{108}Pd)_0$ = 2 10^{-5}. The corresponding normalization factors, indicated on the Pd curve for two Δ^* values, are assumed to be the same for the elements which exhibit in chondrites a refractory (solid lines), siderophile (dashed lines; Tc is put into this category), or volatile behavior (dot-dashed lines)

and the end of the WR phase will be neglected in the following. This approximation is quite acceptable for the s-process radionuclides, in view of the fact that they are ejected rather close to the end of the WR phase /24/, but is somewhat less satisfactory for ^{26}Al, a fraction of which is ejected in the early stages of the WR phase.

The above described procedure is used to construct Fig. 2, which displays several $(R/S)_0$ ratios calculated for $10^5 \leq \Delta^* \leq 3 \; 10^6$ yr from the composition of the material ejected during the WN + WC/WO phase of our M_{ZAMS} = 80 M$\odot$ model. All the results are normalized to $(^{107}Pd/^{108}Pd)_0$ = 2 10^{-5}, which is a "canonical" value derived from the analysis of various iron meteorites (e.g. /16/) (there is of course some spread around that value). That observed ratio can be easily accounted for by our model, as indicated by the high dilution factors which are marked on the $(^{107}Pd/^{108}Pd)_o$ curve, and which have to be applied in order to normalize our theoretical predictions to the above mentioned value of 2 10^{-5}.

Using these same dilution factors in order to construct the other $(R/S)_o$ curves of Fig. 2 (i.e. neglecting fractionation effects between the refractory, siderophile and volatile elements), it appears that various short-lived nuclei of potential interest can be produced in significant amounts. In particular, let us emphasize that

(1) in addition to the Of phase, the WR stage also contributes to the ISM ^{26}Al, even if the corresponding $(^{26}Al/^{27}Al)_o$ ratio is lower than its canonical value, at least for the adopted normalization. In this respect, it has to be recalled that the Al data of Fig. 2 do not include any contribution from the Of phase, and that a large spread is observed in the $(^{26}Al/^{27}Al)_o$ ratio. In addition, it has to be stressed that no correlation has been discovered to-date between $^{26}Mg^*$ and the ^{107}Ag excess resulting from the ^{107}Pd decay;

(2) for low enough Δ^*, the $^{36}Cl/^{37}Cl$ and $^{41}Ca/^{40}Ca$ ratios can exceed the *very uncertain* limits $(^{36}Cl/^{37}Cl)_o \lesssim 10^{-8}$ /35/ and $(^{41}Ca/^{40}Ca)_o \lesssim 10^{-8}$ /36,37/ put by preliminary analyses of experimental data. It has also to be stressed that a ^{41}K excess $(^{41}K^*)$ due to the ^{41}Ca decay has been very tentatively identified in an Allende $^{26}Mg^*$-bearing inclusion /37/. The experimental confirmation of the existence of a correlation between ^{26}Al and ^{41}Ca in certain meteoritic material would be of interest for constraining Δ^* and models for the origin of the short-lived radionuclides;

(3) $(^{99}Tc/^{100}Ru)_o$ ratios in excess of about 10^{-7} are obtained for $\Delta^* \lesssim 10^6$ yr. A possible excess of ^{99}Ru ($^{99}Ru^*$) resulting from the in-situ decay of ^{99}Tc has been searched for /38/ in six Allende and Leoville samples (including LEO-1, which might contain some $^{41}K^*$ /39/). No clear evidence for such an excess has been obtained to-date. A further search for $^{99}Ru^*$, and the confrontation of its (either positive, or negative) results with the theoretical expectations, as those reported in this work, would be very interesting;

(4) $(^{205}Pb/^{204}Pb)_o$ ratios as high as about 10^{-3} are predicted, and exceed the experimental upper limit of 9 10^{-5} /40/. The particular interest of the ^{205}Pb-^{205}Tl pair has been discussed in detail elsewhere /41/. Let us just recall that it might very usefully complement other extinct radionuclide data, as ^{205}Pb might be a pure s-process nucleus. In addition, it is shown in /41/ that other studies (e.g. /42/) have underestimated the s-process $^{205}Pb/^{204}Pb$ production ratio, in particular as a result of a drastic underestimate of the ^{205}Pb effective lifetime. This relates to the neglect of the ^{205}Tl *bound-state β-decay*, which can effectively hinder the ^{205}Pb destruction in a quite large variety of astrophysical environments (see also /15/). This work still reinforces the view expressed in /41/ that the ^{205}Pb-^{205}Tl pair is not necessarily a farfetched s-process chronometer, and still gives more credit to a *plead for a renewed search for extinct ^{205}Pb in meteorites*;

(5) other short-lived nuclei are produced in more or less large amounts by the WR stars, like ^{93}Zr or ^{135}Cs. However, the decay of the first one to ^{93}Nb cannot lead to observable isotopic anomalies in view of the mono-

isotopic nature of Nb. On the other hand, let us just remark that ^{135}Ba and ^{137}Ba excesses and a ^{135}Ba deficiency are reported for the FUN inclusions EK-1-4-1 and C-1, respectively /43/, the latter observation being sometimes interpreted as a *deficit* of ^{135}Cs decay. We will not attempt to relate those ^{135}Ba anomalies to our WR model, which is anyway unable to account for a ^{137}Ba excess. Let us also note that the small amounts of ^{60}Fe emerging from the M_{ZAMS} = 80 M⊙ star ((^{60}Fe/^{58}Fe)$_o$ $\lesssim 10^{-8}$) are probably insufficient to have some interest for γ-ray line astronomy /24/, or for the generation of ^{60}Ni anomalies in meteorites (see also /44/).

The results obtained for the other calculated M_{ZAMS} models are not drastically different from those displayed in Fig. 2. In fact, the most mass-dependent results are found for (^{26}Al/^{27}Al)$_o$ and (^{135}Cs/^{133}Cs)$_o$, which increases and decreases with M_{ZAMS}, respectively, and for (^{60}Fe/^{58}Fe)$_o$, which is quite variable with mass (it can reach values close to 10^{-7} for $M_{ZAMS} \approx 60$ M⊙).

4. CONCLUSION

This work first reassesses the opinion (presented in detail in /5,6/) that the ^{232}Th-^{238}U, ^{235}U-^{238}U and ^{187}Re-^{187}Os pairs do not appear to be very reliable clocks for the age of the r-process nuclei. At least, they seem unable at the present state of the art to further constrain estimates of the age of the Galaxy derived from other means. These conclusions are based on a model for the chemical evolution of the solar neighborhood which satisfies various observational constraints, and on the consideration of astrophysical, meteoritic and nuclear physics uncertainties;

We then examine a model for the origin of some extinct radionuclides centered on the evolution of massive mass losing stars. In this field, our main results may be briefly summarized as follows:
(1) Of and WR stars can eject substantial amounts of short-lived radionuclides into the ISM. More specifically, rather large quantities of ^{26}Al can emerge from the Of stage, while the WR phase may be responsible for an ISM enrichment of ^{26}Al, ^{107}Pd, ^{36}Cl, ^{41}Ca, ^{205}Pb, as well as some other short-lived species;
(2) those massive stars could thus be responsible for the inferred presence of ^{26}Al and ^{107}Pd (possibly along with other short-lived nuclei) in certain meteoritic material, especially if the solar system has been born in an OB association. The alien component containing those radionuclides might have been injected into the solar system either in the form of gas, or/and grains (which are known to form around WR stars).

However, the considered stars are unable to produce other short-lived radionuclides of interest, like ^{129}I or ^{146}Sm. In this respect, it has to be noted that a ^{129}I production by the last event(s) able to modify the solar system composition does not seem to be strictly required (e.g. /4/). On the other hand, it has to be emphasized that this work deals only with the central H and He burning evolutionary phases. The further evolution of the considered stars, which could possibly end (at least for some of them) with a supernova explosion, might bring its share of isotopic anomalies and of short-lived radionuclides. We hope to be able to provide information on this important question in a near future.

The work presented in this paper has been performed within the Collaboration on the evolution of massive stars involving the Institut d'Astronomie et d'Astrophysique of the Université Libre de Bruxelles, CEN Saclay, and the Astrophysisch Instituut of the Vrije Universiteit Brussel. The Collaboration is supported by the Belgian Fund of Joint Fundamental Research (FRFC) under Contract Nr. 2.9002.82. M. Arnould is Chercheur Qualifié F.N.R.S. (Belgium)

REFERENCES

1. Symbalisty, E.M.D., Schramm, D.N.: 1981, *Rep. Prog. Phys.* **44**, 293
2. Clayton, D.D.: 1979, *Space Sci. Rev.* **24**, 147
3. Begemann, F.: 1980, *Rep. Prog. Phys.* **43**, 1309
4. Wasserburg, G.J., Papanastassiou, D.A.: 1982, in *Essays in Nuclear Astrophysics*, eds. C.A. Barnes, D.D. Clayton, D.N. Schramm, Cambridge University Press, p. 77
5. Yokoi, K., Takahashi, K., Arnould, M.: 1983, *Astron. Astrophys.* **117**, 65
6. Arnould, M., Takahashi, K., Yokoi, K.: 1984, *Astron. Astrophys.* **137**, 51
7. Thielemann, F.-K., Metzinger, J., Klapdor, H.V.: 1983, *Astron. Astrophys.* **123**, 162
8. Thielemann, F.-K.: 1984, in *Stellar Nucleosynthesis*, eds. C. Chiosi, A. Renzini, Reidel, p. 389
9. Thielemann, F.-K.: this volume
10. Twarog, B.A.: 1980, *Astrophys. J.* **242**, 242
11. Pagel, B.E.J., Patchett, B.E.: 1975, *M.N.R.A.S.* **172**, 13
12. Pagel, B.E.J.: 1976, in *The Galaxy and the Local Group*, RGO Bull. No. 182, eds. R.J. Dickens, J.E. Perry, Herstmonceux, p. 65
13. Anders, E., Ebihara, M.: 1982, *Geochim. Cosmochim. Acta* **46**, 2363
14. Meyer, J.-P.: 1979, in *Les Eléments et leurs Isotopes dans l'Univers*, eds. A. Boury, N. Grevesse, L. Remy-Battiau, Univ. de Liège, p. 153
15. Takahashi, K.: this volume
16. Chen, J.H., Wasserburg, G.J.: 1983, *Geochim. Cosmochim. Acta* **47**, 1725
17. Cameron, A.G.W., Truran, J.W.: 1977, *Icarus* **30**, 447
18. Reeves, H.: 1978, in *Protostars and Planets*, ed. T. Gehrels, Univ. of Arizona Press, Tucson, p. 399
19. Olive, K.A., Schramm, D.N.: 1982, *Ap.J.* **257**, 276
20. Cameron, A.G.W.: 1984, *Icarus* **60**, 416
21. Clayton, D.D.: 1983, *Ap.J.* **268**, 381
22. de Loore, C., Prantzos, N., Arnould, M., Doom, C.: this volume
23. Prantzos, N., de Loore, C., Doom, C., Arnould, M.: this volume
24. Prantzos, N., Arcoragi, J.-P., Arnould, M.: this volume
25. Cassé, M., Prantzos, N.: this volume
26. Nørgaard, H.: 1980, *Ap.J.* **236**, 895
27. Arnould, M., Nørgaard, H., Thielemann, F.-K., Hillebrandt, W.: 1980, *Ap.J.* **237**, 931
28. Hillebrandt, W., Thielemann, F.-K.: 1982, *Ap.J.* **255**, 617
29. Clayton, D.D.: 1984, *Ap.J.* **280**, 144
30. Truran, J.W., Cameron, A.G.W.: 1978, *Ap.J.* **219**, 226
31. Arnett, W.D., Wefel, J.P.: 1978, *Ap.J.* **224**, L139
32. Woosley, S.E., Weaver, T.A.: 1980, *Ap.J.* **238**, 1017

33. Wiescher, M.: this volume
34. Arnould, M., Prantzos, N.: in preparation
35. Jordan, J., Pernicka, E.: 1981, *Meteoritics* **16**, 332
36. Begemann, F.: 1985 (private communication)
37. Huneke, J.C., Armstrong, J.T., Wasserburg, G.J.: 1981, *Lunar Planet. Sci. Conf. XII*, Abstracts, p. 482
38. Poths, H., Schmitt-Strecker, S., Begemann, F.: 1983, *46th Meeting Meteoritical Soc. (Mainz)*, Abstracts, p. 162
39. Stegmann, W., Specht, S.: 1983, *46th Meeting Meteoritical Soc. (Mainz)*, Abstracts, p. 185
40. Huey, J.M., Kohman, T.P.: 1972, *Earth Planet. Sci. Letters* **16**, 401
41. Yokoi, K., Takahashi, K., Arnould, M: 1985, *Astron. Astrophys.* **145**, 339
42. Blake, J.B., Schramm, D.N.: 1975, *Ap.J.* **197**, 615
43. McCullogh, M.T., Wasserburg, G.J.: 1978, *Ap.J.* **220**, L15
44. Kohman, T.P., Robison, M.S.: 1980, *Lunar Planet. Sci. Conf. XI*, Abstracts, p. 564

CHRONOMETER STUDIES WITH INITIAL GALACTIC ENRICHMENT

F.-K. Thielemann* and J.W. Truran
Department of Astronomy, University of Illinois,
1011 W. Springfield Avenue, Urbana, IL 61801, USA

ABSTRACT. The long-lived actinide chronometers ^{232}Th/^{238}U and ^{235}U/^{238}U have been used to determine the galactic age. Emphasis is put on constraints for the nucleosynthesis production function from galactic evolution models and astronomical observation. The short-lived pairs ^{129}I/^{127}I and ^{244}Pu/^{238}U were not included, because of large uncertainties in the meteoritic ratios of the first and the r-process production ratios of the latter. The results are strongly dependent on the amount of initial enrichment and vary between 12.6 and 24.6 billion years. The inclusion of the ^{187}Re/^{187}Os pair within the same framework shows the existence of consistent solutions, but the intrinsic uncertainties in the Re/Os pair cannot lead to a more precise age determination. Thus, at present, nucleocosmochronology cannot provide smaller uncertainties than the analysis of globular cluster ages or determinations of the Hubble constant.

1. INTRODUCTION

Various attempts have been made in the past to determine the age of the Galaxy from considerations of nucleocosmochronology, with particular emphasis on long-lived nuclei like ^{187}Re ($t_{1/2}$=4.5×10^{10}y), ^{238}U (4.46×10^{9}y), ^{232}Th (1.405×10^{9}y), and ^{235}U (7.03×10^{8}y). Nuclei with shorter half-lives like ^{244}Pu (8.26×10^{7}y) and ^{129}I (1.57×10^{7}y) have been used to obtain information about last nucleosynthesis events before the formation of the solar system. All of the above mentioned nuclei are products of r(rapid neutron capture)-process nucleosynthesis.

These studies were performed either with an exponential model, which assumes that the time dependence of the nucleosynthesis production

---

*On leave from Max-Planck-Institut für Physik und Astrophysik, Garching b. München, FRG

follows an exponential behavior (Fowler, Hoyle 1969; Truran and Cameron 1971; Fowler 1972, 1978), or with a so-called model-independent approach (Schramm, Wasserburg 1970; for a more recent review see Symbalisty and Schramm 1981). The latter treatment determines only the mean ages of the elements, but the interpretation in terms of a total age of the Galaxy is again model dependent (Tinsley 1977, 1980; Meyer 1986).

This means that, besides the necessity of knowing both production ratios of various isotopes in nucleosynthetic processes and their abundance ratios at the formation of the solar system (meteoritic values), the time dependence of galactic nucleosynthesis also has to be known. This led Tinsley (1980) among others to the conclusion that there is a large uncertainty in any age estimate of the Galaxy, derived from nucleochronology. An attempt to perform a consistent galactic evolution and chronology study has recently been undertaken by Yokoi, Takahashi, Arnould(1980; see also Truran and Cameron 1971). Uncertainties associated with the $^{187}Os/^{187}Re$ chronometric pair unfortunately did not allow them to reach precise conclusions (Arnould, Takahashi, Yokoi 1984).

We therefore continue to utilize the method of a model ansatz which allows us to determine the free parameters in the model. This has been performed again with the use of an exponential model, but with the consistent inclusion of the $^{187}Os/^{187}Re$ pair together with the other r-process chronometers. Such an attempt had not previously been undertaken. In addition, we consider the possibility of an initial enrichment of metals, which is incorporated in many galactic evolution models to account for pre-disk populations of stars (Pop II and Pop III). It is of special interest to determine whether such an initial enrichment can reduce the relatively large ages obtained with the standard exponential model, when the reduced production ratios for $^{232}Th/^{238}U$ and $^{235}U/^{238}U$ were applied, which resulted from the inclusion of β-delayed fission and neutron emission in the prediction of r-process production ratios (Thielemann, Metzinger, Klapdor 1983; Thielemann 1984). Those results ($t_G = (17.6 \mp 4) \times 10^9 y$) differed significantly from earlier determinations (Fowler 1978, $t_G = (10.9 \mp 2) \times 10^9 y$).

Before presenting the current model and the results obtained, we want to discuss the relationship between the paramters of such a model ansatz and observational quantities, like the star formation rate $\Psi(t)$, its time dependence, the rate of infall, and the rate of injection of new nucleosynthesis products into the interstellar gas. Those observational quantities should also enable us to put constraints on the results of cosmochronology studies.

2. THE EXPONENTIAL MODEL AND OBSERVATIONAL CONSTRAINTS

When applying a galactic evolution model with instantaneous recycling, the abundance of nucleus A in the interstellar gas is governed by the differential equation (Tinsley 1980 or also Thielemann, Metzinger, Klapdor 1983b)

(1) $\dot{N}_A = -\omega N_A + P_A \Psi$ $(-\lambda_A N_A$ if A is radioactive)

with P_A describing the production of nucleus A due to stellar evolution and ejection into the interstellar medium and Ψ denoting the star formation rate (mass of interstellar gas being transformed into stars per unit time). ω is given by

(2) $\omega = -\dot{m}_g/m_g + f/m_g (1-Z_f/Z)$

and for the case of metal-free infall ($Z_f = 0$), ω describes the rate with which gas is turned into stars. Under the assumptions ω = const and $\Psi(t) = \Psi_0 \exp(-\mu t)$ the solution is

(3) $N_A(t) = P_A \Psi_0 \exp(-\omega t)/(\lambda_R - \lambda_A) \left(\exp(-\lambda_A t) - \exp(-\lambda_R t) \right)$

with $\lambda_R = \mu - \omega$. This leads to

(4) $N_A(t)/N_B(t) = P_A/P_B \, f(\lambda_R, \lambda_A, \lambda_B, t)$

which is only dependent on λ_R or $\exp(-\lambda_R t)$. The latter function has the same time dependence as $\Psi(t)\exp(\omega t)$, under the mentioned assumptions. Therefore it is called the "effective nucleosynthesis rate" and cosmochronological studies describe its time dependence rather than the time dependence of the nucleosynthesis production $P_A \Psi(t)$. For a more general derivation see Tinsley (1980).

One of the major constraints on simplified chronometer studies is the need to find solutions which give a time dependence of the effective nucleosynthesis rate (ENR) which is in accordance with limits derived from galactic evolution models and astronomical observations. This means that $\Psi(t)$ and ω have to be known. ω is related to the more commonly used gas consumption time scale τ_g by $\omega = 1/\tau_g$. Tinsley (1977) gives limits to the above mentioned quantities

$$0.125 \leq \Psi(present)/\Psi_{av} \leq 1 \text{ and } 0.6 \leq \tau_g/10^9 y \leq 10$$

which are updated by an extensive study of Miller and Scalo (1979) to

$$0.18 \leq \Psi(present)/\Psi_{av} \leq 2.5 \text{ and } 1 \leq \tau_g/10^9 y \leq 3.$$

The latter limits are not independent; for a decreasing $\Psi(t)$, $\tau_g = 3 \times 10^9$ y holds true (see tables 5 and 10 in Miller and Scalo 1979). Under the following assumptions we can draw conclusions for the time dependence of the ENR

$$\begin{align}(5) \quad \text{ENR}(t_G) &= \Psi(t_G)\exp(t_G/\tau_g)\\ &= (\Psi(t_G)/\Psi_{av})(\Psi_{av}/\Psi_o)\exp(t_G/\tau_g)\\ &= (\Psi(t_G)/\Psi_{av})^2 \exp(t_G/\tau_g)\, \text{ENR}(0).\end{align}$$

If we use the extreme limits of 0.18 and 3×10^9y for the ratio of present to average star formation rate and the gas consumption time scale, respectively, and a minimum age of the Galaxy of $t_G = 10^{10}$y, we find that $\text{ENR}(t_G)/\text{ENR}(0) = 0.91$. Such a value indicates an almost constant ENR. Miller and Scalo (1979) argue that they give extreme limits for the behavior of the star formation rate, which should be rather reduced by a factor of 2. Taking 0.36 instead of 0.18 for the lower limit for $\Psi(t_G)/\Psi_{av}$ results in $\text{ENR}(t_G)/\text{ENR}(0) = 3.63$; applying a galactic age of 15 rather than 10 billion years yields 19.23.

As discussed before, the ENR can also be expressed in terms of $\text{ENR}(t) = \Psi_o \exp(-\lambda_R t)$. The three cases 0.91, 3.63, and 19.23 translate into values for λ_R of 9.6×10^{-12} y^{-1} ($\cong 0$), -1.3×10^{-10}, and 1.97×10^{-10}. Thus, vanishing or negative values are allowed while positive values are excluded. This means we have a constant ENR which was the conclusion drawn from a variety of realistic galactic evolution models by Hainebach and Schramm (1977), or even a rising ENR.

3. GENERALIZATION OF THE EXPONENTIAL MODEL WITH INITIAL ENRICHMENT

The original form of the exponential model of galactic nucleosynthesis (Fowler 1972), contained four parameters: Δ (duration of galactic nucleosynthesis); λ_R (coefficient governing the time dependence of the effective nucleosynthesis rate $\Psi(t)\exp(t/\tau_g)$, with Ψ being the star formation rate and τ_g the gas consumption time scale); S_Δ (contribution of a final spike of nucleosynthesis, i.e. the last passage through a spiral arm); and δ (free decay period before the formation of meteorites in the solar system). Solving the chronometer equations with such a model ansatz for the most recent chronometric production ratios and meteoritic abundance ratios resulted in relatively large galactic ages (Thielemann, Metzinger, Klapdor 1983ab; Thielemann 1984).

Many galactic evolution models also include an initial enrichment of metals (Truran and Cameron 1971; see also Tinsley 1980 and references therein) caused by pre-disk populations (Pop II and Pop III?). We might note that an initial spike in the present context is equivalent to r-process production on a halo collapse time-scale ($\sim 10^8$-10^9y) which is consistent with the fact that metal-poor stars tend to show r-process abundances (Truran 1980; Sneden and Pilachowski 1985). The question arises, as to whether the inclusion of an initial enrichment can alter these results (Fowler and Meisl 1985). Therefore, we generalize the exponential model to a form which includes an initial spike and we add a chronometric equation for the ^{187}Os/^{187}Re pair to be solved consistently with the other equations. For these assumptions, Eq.(5a) in Thielemann et al. (1983b) changes to

(6) $N_A(\Delta+\delta) = \Psi_0 \exp(-\omega\Delta) P_A f(\Delta,\lambda_R,S_0,S_\Delta,\delta;\lambda_A)$ with
$f = \exp(-\lambda_A\delta) \left((1-S_0-S_\Delta)/(\lambda_R-\lambda_A) \right.$
$(\exp(-\lambda_A\Delta)-\exp(-\lambda_R\Delta))$
$+ S_0/\lambda_R (1-\exp(-\lambda_R\Delta)) \exp(-\lambda_A\Delta)$
$\left. + S_\Delta/\lambda_R (1-\exp(-\lambda_R\Delta)) \right)$.

with S_0 and S_Δ indicating an initial and a final nucleosynthesis spike, respectively, and $\lambda_A = 1/\tau_A = \ln2/t_{1/2}$ denoting the decay rate of radioactive nucleus A. For two nuclei which decay by means of entirely independent decay chains, the chronometric equation results

(7) $\dfrac{N_A(\Delta+\delta)}{N_B(\Delta+\delta)} = \dfrac{P_A}{P_B} \dfrac{f(\Delta,\lambda_R,S_0,S_\Delta,\delta;\lambda_A)}{f(\tilde{},\lambda_R,S_0,S_\Delta,\delta;\lambda_B)}$

which relates the (r-process) production ratios P_A/P_B and the meteoritic ratios $N_A(\Delta+\delta)/N_B(\Delta+\delta)$.

When a nucleus D (daughter) is a non-radioactive decay product of a radioactive nucleus P (parent) which was produced in the r-process, and it is also not directly produced in the r-process, then the differential equation Eq.(3) in Thielemann et al. (1983b) changes to

(8) $\dot{N}_D = -\omega N_D + \lambda_P N_P$

with the production term $\lambda_P N_P$ (λ_P being the decay rate of nucleus P). Then Eq.(4a) in Thielemann et al (1983b) changes to

(9) $N_D(t) = \exp(-\omega t) \int_0^t \lambda_P N_P(t') \exp(\omega t') \, dt'$ $t \leq \Delta$

and $N_D(\Delta+\delta) = N_D(\Delta) + N_P(\Delta)(1-\exp(-\lambda_P\delta))$. The result of some algebra is

(10) $N_D(\Delta+\delta) = \Psi_0 \exp(-\omega\Delta) P_P g(\Delta,\lambda_R,S_0,S_\Delta,\delta; \lambda_P,\lambda_P^*)$

with

(11) $g = (1-S_0-S_\Delta)/(\lambda_R-\lambda_P^*) \left(1-\exp(-\lambda_P^*\Delta-\lambda_P\delta) \right.$
$+ \exp(-\lambda_R\Delta-\lambda_P\delta)-\exp(-\lambda_R\Delta) -\lambda_P^*/\lambda_R(1-\exp(-\lambda_R\Delta)) \left. \right)$
$+ S_0/\lambda_R(1-\exp(-\lambda_R\Delta))(1-\exp(-\lambda_P^*\Delta -\lambda_P\delta))$
$+ S_\Delta/\lambda_R(1-\exp(-\lambda_R\Delta))(1-\exp(-\lambda_P\delta))$.

Here we have included the possibility that the parent nucleus has a different effective decay rate λ_P^* in stellar environments (see Yokoi, Takahashi, Arnould 1983 for the case of ^{187}Re) during the galactic evolution period $0 \leq t \leq \Delta$, than in the laboratory (λ_P). The latter is only applied in the free decay period δ. The chronometric equation then reads

$$(12) \quad \frac{N_D(\Delta+\delta)}{N_P(\Delta+\delta)} = \frac{g(\Delta,\lambda_R,S_o,S,\delta;\lambda_P,\lambda_P^*)}{f(\Delta,\lambda_R,S_o,S,\delta;\lambda_P,\lambda_P^*)} .$$

$f(..;\lambda_P,\lambda_P^*)$ changes compared to $f(..;\lambda)$ in Eq. (6) by replacing λ_A everywhere with λ_P^* except for the case of $\exp(-\lambda_A\delta)$. This additional equation for ^{187}Os/^{187}Re allows us to explore the parameter S_o, which describes the initial enrichment in metals.

4. R-PROCESS PRODUCTION RATIOS AND METEORITIC ABUNDANCE RATIOS

4.1 Production Ratios

The production ratios employed in the present study, will be mostly based on recent r-process calculations which include the important effect of β-delayed fission, but a short historical summary is given in table I (for more information see table 2 in Thielemann et al. 1983b).

TABLE I
Production Ratios

^{232}Th/^{238}U	^{235}U/^{238}U	^{244}Pu/^{238}U	Ref.
1.65	1.42	0.90	Seeger, Fowler, Clayton (1965)
			Fowler (1972)
1.90	1.89	0.96	Seeger, Schramm (1970)
1.80	1.42	0.90	Fowler (1978)
1.70	0.89	0.53	Wene, Johansson (1976)
$1.90^{+0.2}_{-0.4}$	$1.50^{+0.5}_{-0.6}$	$0.90^{+0.1}_{-0.2}$	Symbalisty, Schramm (1981)
1.50	1.10	0.40	Krumlinde et al. (1981)
1.40	1.24	0.12	Thielemann et al. (1983ab)

The pattern presented above can be understood in the following way. A rather "flat" mass formula (in terms of nuclear mass excess as a function of N-Z), such as the one by Myers and Swiatecky (1967) used in Seeger and Schramm (1970), yields a relatively flat abundance distribution. The reason is that, in such a case, neutron separation energies of 1.5-2.0 MeV, which determine the location of the r-process

path, occur relatively far from the stability line, where very short
β-decay half-lives are encountered. This leads to a "flat" (almost
constant) abundance distribution with nearly equal abundances for all
short-lived radioactive progenitors.

Those are ^{232}Th, ^{236}U, ^{240}Pu, ^{244}Pu, ^{248}Cm, ^{252}Cf for ^{232}Th; ^{235}U,
^{239}Pu, ^{243}Am, ^{247}Cm, ^{251}Cf, ^{255}Fm for ^{235}U; ^{238}U, ^{242}Pu, ^{246}Cm for
^{238}U; and ^{244}Pu, ^{248}Cm, and ^{252}Cf for ^{244}Pu. In the case of equal
abundances for all progenitors, ratios of ^{232}Th/^{238}U = ^{235}U/^{238}U = 2
and ^{244}Pu/^{238}U = 1 are expected, close to the values obtained by
Seeger and Schramm (1970). If a "steeper" mass formula is assumed,
the abundances are more sensitive to structure effects in β-decay
half-lives. In such a case, a slight maximum occurs at A ≅ 238-240.
This led to reduced production ratios for ^{232}Th/^{238}U and ^{235}U/^{238}U in
the original work of Seeger, Fowler, and Clayton (1965). Thielemann,
Metzinger, Klapdor (1983a) used a relatively "steep" mass formula
(Hilf, von Groote, Takahashi 1976) and the same effect was evident. A
recent overview by Haustein (1984), which compares experimental data
and mass formula predictions, comes to the conclusion that steep mass
formulae, like e.g. Liran, Zeldes (1976), are to be preferred over
"flat" formulae, e.g. Myers (1976). The Hilf et al. (1976) formula is
not listed in this comparison but behaves close to the one by Liran
and Zeldes (1976).

The additional inclusion of β-delayed fission (i.e. fissioning of
nuclei in excited states after population by β-decay) cuts down drastically the abundances for A ≳ 238. This leads to a strong reduction
of heavy progenitors, and thus reduces ^{232}Th/^{238}U further and has a
drastic effect on ^{244}Pu/^{238}U. The calculations of Krumlinde et al.
(1981) and Thielemann et al. (1983), both of which incorporate the
recent (and only available) fission barrier predictions by Howard and
Möller (1980) give similar predictions: $1.40 \leq {}^{232}$Th/^{238}U ≤ 1.50, $1.10
\leq {}^{235}$U/^{238}U ≤ 1.24, and $0.12 \leq {}^{244}$Pu/^{238}U ≤ 0.40. The largest uncer-
tainty is in ^{244}Pu/^{238}U which, however, changes drastically in both
cases in comparison to the previous value of 0.90.

There are uncertainties in the fission barrier and β strength function
predictions far from stability, without doubt, but the general tendency described above, should persist.

For the ^{129}I/^{127}I pair, no theoretical prediction is necessary, as it
is known from the empirical decomposition of solar abundances in r-
and s-components (^{129}I/^{127}I = 1.26, Käppeler et al. 1982). Because of
a different chronometric equation in case of the Re/Os pair, no production ratio is involved.

4.2 Meteoritic Abundance Ratios

All ratios which are given in this section are dated back to the time
of meteorite formation i.e. Δ+δ. The best known ratio is ^{235}U/^{238}U,
as it is not affected by chemical fractionation. The value is 0.317

(Anders and Ebihara 1982) or 0.315 (Cameron 1982). For a literature overview see Anders and Ebihara(1982). $^{232}Th/^{238}U$ came down from 2.48 (see Symbalisty and Schramm 1981) to present values 2.32 (Anders and Ebihara 1982) or 2.22 (Cameron 1982). This reduction balances somewhat the reduction in production ratios.

$^{244}Pu/^{238}U$, which can at present only be determined from decay products of ^{244}Pu, has undergone drastic changes: 0.035 (Wasserburg et al. 1969), 0.015 (Podosek 1972), 0.016 (Drozd et al. 1977), 0.0068 ∓ 0.0019 (Hudson et al. 1984). This reduction in meteoritic ratios again balances, in part, the strong reduction in the production ratios, when β-delayed fission is included. $^{129}I/^{127}I$ seems to be a weak chronometer, as the short ^{129}I half-life made it even vary over the condensation period of meteorites and only limits can be given: $0.9 \times 10^{-4} \leq {}^{129}I/^{127}I \leq 2.3 \times 10^{-4}$ (Jordan et al. 1980). These limit were slightly reduced by Crabb, Lewis, Anders (1982) to 1.09×10^{-4} and 1.60×10^{-4}, respectively.

In the case of the $^{187}Os/^{187}Re$ pair, ^{187}Re is produced primarily in the r-process. ^{187}Os is produced both in the s-process and by β-decay of the unstable ^{187}Re. In the framework of the chronometric equation, given in section 2, the value of $(^{187}Os/^{187}Re)^r_{\Delta+\delta}$ has to be determined, i.e. the pure r-component at the time of meteorite formation. This involves the knowledge of the ^{187}Re half-life for backdating and the subtraction of the ^{187}Os s-component to derive the "cosmoradiogenic" ^{187}Os.

$$(13) \quad r = \frac{^{187}Os_{cr}}{^{187}Re} = \frac{^{187}Os(\Delta+\delta) - F_\sigma \sigma(186)/\sigma(187) {}^{186}Os}{^{187}Re(\Delta+\delta)}$$

This equation is originally given in Fowler(1972), the ^{187}Os s-component is determined by assuming a constant nσ-curve in the s-process and can thus be expressed by $n(187) = \sigma(186)/\sigma(187) \, n(186)$. An additional complication is the 9.75 KeV excited state in ^{187}Os, which is thermally populated under s-process conditions and causes the stellar $\sigma^*(187)$ to be different from the laboratory value $\sigma(187)$. Hershberger et al. (1983) calculate a value for the correction factor F_σ to be 0.81-0.83, based on inelastic neutron scattering experiments. Fowler (1985), arguing by analogy with the ^{189}Os ground state (Browne and Berman 1981), finds 0.716.

The determination of $^{187}Os(\Delta+\delta)$ and $^{187}Re(\Delta+\delta)$ carries also uncertainties stemming from the uncertainty in the very long ^{187}Re half-life. Present compilations vary between $t_{1/2} = 5 \times 10^{10}$ y (Seelmann-Eggebert et al. 1981) and 4.35×10^{10} y (Lederer and Shirley 1978); Naldrett (1984) cites 3.5×10^{10} y. Using a half-life of 4.0×10^{10} y, an F_σ-value of 0.716, $\sigma(186)/\sigma(187)=0.478$ (Winters and Macklin 1982), and a meteoritic age of 4.55×10^9 y, we find r=0.144 with the abundances of Cameron (1982) and Anders and Ebihara (1982). A recommended change in the elemental Os/Re ratio from 14.14 to 12.7 (Anders 1984) changes r to 0.122. When using F_σ = 0.82 (see Winters and Macklin 1982 and also Woosley and Fowler 1979), r = 0.107 results. A longer half-life of 4.5×10^{10} y changes the two latter numbers into 0.131 and 0.116, respectively. Taking all these facts into account, an uncertainty range of $0.10 \leq r \leq 0.15$ seems not to be unreasonable.

Arnould at al. (1984; see also the numerical results of Truran and Iben 1977; Cosner, Iben, Truran 1980; and Cosner 1981) give arguments that a pulsed s-process could also give an s-process contribution for ^{187}Re and express this in terms of an effective $F = F_s F_\sigma$. It should be noted that the mean age inferred from Re-Os by the "model-independent" approach is not strongly sensitive to the s-process production of ^{187}Re.

5. RESULTS

When utilizing the r-process production ratios and meteoritic abundance ratios as given in section 4 in the context of the exponential model which was outlined in section 3, we obtained results for the time of nucleosynthesis in the Galaxy, prior to the formation of the solar system, in the range 8×10^9 y $\leq \Delta \leq 15 \times 10^9$ y. Adding the free decay period δ and the age of the meteorites, this gives galactic ages from 12.6 to 19.6 billion years. Besides the fact that such a range is not very satisfying, the assumption of a final spike introduces also a number of uncertainties.

First of all, we do not know if there is precisely one final spike, or rather a number of irregularities in the time dependence of the nucleosynthesis production close to the formation of the solar system (Reeves 1979, Clayton 1983). Besides this, the size of the required final spike is completely controlled by the ^{129}I/^{127}I chronometer, because of the short ^{129}I half-life ($t_{1/2} = 1.57 \times 10^7$ y). The ^{244}Pu/^{238}U chronometer is also mostly determined by the final spike; but because of the slightly longer half-life of ^{244}Pu (8.26×10^7 y), it is also sensitive to nucleosynthesis occurring over the last 10^8 y before the contribution of a final spike. Unfortunately both chronometers involve large uncertainties. The iodine pair shows a large uncertainty range in the meteoritic abundance ratio, while ^{244}Pu, being very sensitive to variations in fission barrier heights, is quite uncertain in its production ratio (see section 4). This assures that the use of these chronometric pairs introduces a large uncertainty in the time dependence of the nucleosynthesis production. If one of them fixes the size of the final spike, the other can enforce a too strongly decreasing or increasing production term in the exponential model, i.e. enforce an incorrect behavior of λ_R, which has a strong influence on the more important quantity Δ, the main aim of our study. Therefore, we will proceed to concentrate on the three long-lived chronometric pairs to determine the long-term behavior in galactic nucleosynthesis.

5.1 Models with S_o, λ_R, and Δ

Since we have chosen to drop the two short-lived chronometers ^{129}I and ^{244}Pu which are sensitive to the final spike S_Δ and the free decay period δ, those two parameters can also be omitted in our analysis. This leaves us with S_o, λ_R, and Δ. At first we only want to consider ^{232}Th/^{238}U and ^{235}U/^{238}U. Having only two chronometric equations, λ_R will be used as a free parameter; this will allow us to explore the possible range in λ_R, derived from astronomical observations (see section 2).

Fig. 1a shows the results for two cases:
(a) the production ratios are taken from Fowler (1978) with values of 1.80 and 1.42 for ^{232}Th/^{238}U and ^{235}U/^{238}U, respectively (see Table I), and
(b) the corresponding values 1.40 and 1.24 from Thielemann et al. (1983) are used.

Solutions are displayed as function of λ_R, while its absolute value is also given in the middle frame on a logarithmic scale, to enable a better reading. The S_o-frame contains only values with $S_o \leq 0.35$, i.e. less than 35% initial metal enrichment. The oldest disk populations show 10-30% of the solar metal content, therefore larger values of S_o are excluded on the basis of observations.

Solutions with $S_o = 0$ and the production ratios of 1.80 and 1.42 give a declining effective nucleosynthesis rate ($\lambda_R \geq 0$), $\Delta = 6.3 \times 10^9$ y, and a galactic age of 10.9×10^9 y, in accordance with the findings of Fowler (1978). It may be recognized that the inclusion of a non-zero initial enrichment even reduces the age further, an effect which was expected (Fowler and Meisl 1985). When applying the more recent production ratios 1.40 and 1.24, which result from the inclusion of β-delayed fission, larger ages result. In the vicinity of $S_o = 0$ a drastic increase is noticable. This strong dependence on S_o introduces a large uncertainty in Δ and values in the range 11×10^9 y $\leq \Delta \leq 20 \times 10^9$ y are possible.

There is one interesting point with regard to the discussion in section 2. The old production ratios allow only for a declining effective nucleosynthesis rate, somewhat in contrast to astronomical observations. The more recent values give a (slightly) increasing nucleosynthesis rate with -2×10^{-10} y$^{-1} \leq \lambda_R \leq -10^{-10}$ y^{-1}, in agreement with the discussion of section 2.

To avoid the danger of overinterpretating these results too much, we also performed a further sensitivity study with regard to intermediate values for both chronometers. Fig. 1b displays the results for ^{232}Th/^{238}U 1.50 and ^{235}U/^{238}U = 1.24 and 1.30, respectively. First of all we see a reduction in Δ, now allowing for values between 8 and 12 $\times 10^9$ y. In addition it can be recognized that λ_R is mainly determined by the ^{235}U/^{238}U ratio. The small change from 1.24 to 1.30 shifts the function $S_o(\lambda_R)$ to the right, which means that the time dependence of the nucleosynthesis rate is governed by the uranium isotopes while the Th/U ratio is mostly responsible for Δ. This also means that astronomical evidence alone only supports the change in the ^{235}U/^{238}U ratio.

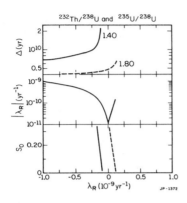 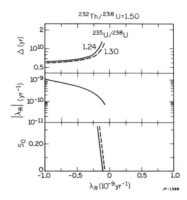

Fig. 1 Using the two chronomatic pairs ^{232}Th/^{238}U and ^{235}U/^{238}U, the values of Δ (duration of galactic nucleosynthesis before formation of the solar system) and S_o (initial metal enrichment) are presented as a function of λ_R, which governs the time dependence of the effective nucleosynthesis rate $\Psi_o \exp(-\lambda_R t) = \Psi(t)\exp(\omega t)$. The absolute value of λ_R is also displayed on a logarithmic scale. Fig. 1a shows two cases with production ratios ^{232}Th/^{238}U = 1.40 and ^{235}U/^{238}U 1.24 or 1.80 and 1.42, respectively. The latter ratios do not allow for solutions with $\lambda_R \leq 0$ in the range $S_o \leq 0.3$. Larger values of S_o result in shorter galactic ages. Fig. 1b displays results for ^{232}Th/^{238}U = 1.50 and ^{235}U/^{238}U = 1.24 or 1.30. In comparison with Fig. 1a it is seen that the shift of $S_o(\lambda_R)$ is mostly due to the ^{235}U/^{238}U ratio.

5.2 Models with the ^{187}Os/^{187}Re Chronometric Pair

As discussed in section 3, the additional equation for the ^{187}Os/^{187}Re pair allows the determination of all three remaining parameters and could help to find narrower ranges of possible solutions. Unfortunately, however, besides the fact of an uncertain stellar half-life, the "cosmoradiogenic" Os/Re ratio (r) also involves relatively large uncertainties. Therefore we choose ^{187}Os$_{cr}$/^{187}Re as a free parameter in the range from 0.1 to 0.15 and display Δ, λ_R, and S_o as a function of r. This is seen in Fig. 2 where an "effective" half-life of ^{187}Re (averaged over stellar and interstellar environments) of 4×10^{10} y is used. Two solutions are shown, for a ^{232}Th/^{238}U ratios of 1.40 and 1.50. While we find similar results for Δ and λ_R when $S_o \leq 0.35$ is considered, the values of the cosmoradiogenic Os/Re ratio r differ strongly. From present knowledge in meteoritic and s-process abundances the whole range from 0.1 to 0.15 seems to be possible and the hope for stronger conclusions for cosmochronology might not be fulfilled (see also Arnould, Takahashi, Yokoi 1984). It also seems, however, that ^{232}Th/^{238}U ratios in excess of 1.50 cause problems even for the present limits of r, if one does not allow for shorter effective half-lives of ^{187}Re.

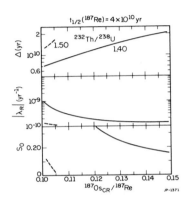

Fig. 2 Δ $|\lambda_R|$ and S_0 as functions of the cosmoradiogenic ^{187}Os/^{187}Re ratio, when employing ^{232}Th/^{238}U, ^{235}U/^{238}U, and ^{187}Os/^{187}Re in a chronometric study. Varying the ^{232}Th/^{238}U production ratio from 1.40 to 1.50 results in a drastic change of the solution. Within the uncertainties of ^{187}Os$_{CR}$/^{187}Re the ^{232}Th/^{238}U production ratio seems to have an upper limit close to 1.50, if ^{187}Re has an "effective" half-life of 4×10^{10} yr during galactic nucleosynthesis.

This leads to another idea. Suppose we know the amount of initial enrichment and the Os/Re ratio. Which "effective" half-life of ^{187}Re would be compatible with those values? Figs. 3a and 3b show the results of such tests, for a ^{232}Th/^{238}U ratio of 1.40 and 1.50, respectively. Assumed values of S_0 are 0.2 and 0.3 or 0.1 and 0.2. Allowing for the whole uncertainty range of r and S_0 we find values for the "effective" Re half-life which roughly range from 2.7 to 5.5 $\times 10^{10}$ y. The evaluation of complex models, dealing with a temperature dependent Re half-life (Yokoi, Arnould, Takahashi 1983) in terms of such an "effective" Re half-life over the period of galactic evolution, might provide clues and consistency checks.

While the present study has shown that all long-lived chronometers might give a consistent picture of galactic evolution, when using the production and meteoretic abundance ratios within their uncertainty limits, the hope for a precise age determination is not fulfilled. The present study gave results for Δ in the range from 8 to 20×10^9 y, and thus allowed for galactic ages from 12.6 to 24.6 billion years, but even these limits might broaden when functional dependences other than the present exponential model with its included extensions, are taken into account.

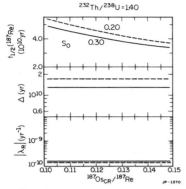 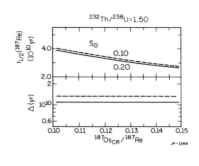

Fig. 3 The assumption of a certain value for S_o determines Δ and λ_R. The upper part shows which effective ^{187}Re half-life is required as a function of ^{187}Os$_{cr}$/^{187}Re, to make the ^{187}Os/^{187}Re pair consistent with such a solution. Fig. 3b shows that a ^{232}Th/^{238}U production ratio of 1.50 requires shorter effective ^{187}Re half-lives (and also results in smaller galactic ages, as discussed previously).

6. CONCLUSIONS

The results presented in the previous section showed that our present knowledge concerning cosmochronology is quite uncertain and allows one only to set limits on the age of the Galaxy, which can range from 12.6 - 24.6 billion years. These uncertainties have several origins
(a) nuclear physics: properties of nuclei far from stability (β-decay half-lives, fission barriers, mass formulae).
(b) meteoritic abundance ratios: most chronometer ratios seem to be quite well determined, but especially the back-dating of the Re/Os ratio is highly dependent on uncertainties associated with the Re half-life, s-process contributions, and the stellar neutron capture cross section of ^{187}Re.
(c) the interpretation relies on a model ansatz for the nucleosynthesis production as a function of time. This model ansatz can introduce additional uncertainties, as it can deviate substancially from the "real" galactic nucleosynthesis history.

There is hope, however. Improving galactic evolution models (Matteucci 1986) might make a model ansatz obsolete, if also (and hopefully) the site of the r-process can be identified.

In addition, Beer and Macklin(1985) have succeeded in determining the radiogenic ^{207}Pb abundance (resulting from U-decay). Thus, an additional and maybe more precise chronometer can be used, similar to the Re/Os pair, for age determinations.

One interesting outcome of the present study should also be mentioned.

Presently, the new inflationary universe scenario (Guth 1981; Linde 1984; Steinhardt 1984) can solve the cosmological horizon problem and explain the large scale homogeniety of the observable universe. In case of a vanishing cosmological constant (vacuum energy in the early universe), this scenario predicts $\Omega = 1$, i.e. a closed universe. Under those conditions the age of the universe is $t_U = 2/3H_0^{-1}$, an upper bound for t_G, the age of the Galaxy. With a Hubble constant of 50 ∓ 7 km s^{-1} Mp^{-1} (Sandage and Tammann 1984), this gives $t_U = (13 \mp 1.8) \times 10^9$ y; other evaluations would give even smaller values. Such an age is at the very lower edge of the uncertainty range, derived for galactic ages in the present work. It is also of interest that such a galactic age would be at the low side of ages determined for globular clusters $(14-19) \times 10^9$ y, James and Demarque 1983; $(15-18) \times 10^9$ y, VandenBerg 1982; $(17 \pm 2) \times 10^9$ y, Sandage 1982; $(14 \pm 3.5) \times 10^9$ y, Iben and Renzini 1984). This effect is discussed by Blome and Priester (1984) and by Fowler and Meisl (1985) in terms of a non-vanishing cosmological constant. Big Bang nucleosynthesis requires $\Omega \leq 0.14$-0.18 (Steigman 1986) and seems also to allow for such a conclusion (see also Turner, Steigman, and Krauss 1984). But in order to give a clear answer, cosmochronology has to come up with much more precise age predictions of galactic nucleosynthesis.

We want to thank W.A. Fowler, whose interest and suggestions initiated the present study, and W. Priester, for pointing out the cosmological implications. This study was supported in part by NSF Grant AST 83-14415.

REFERENCES

Anders, E. 1984, private communication
Anders, E., Ebihara, M. 1982, Geochim. Cosmochim. Acta 46, 2363
Arnould, M., Takahashi, K., Yokoi, K. 1984, Astron. Astrophys. 137, 51
Beer, H., Macklin, R.L. 1985, Kernforschungszentrum Karlsruhe, preprint
Blome, H.-J., Priester, W. 1984, Naturwissenschaften 71, 528
Browne, J.C., Berman, B.L. 1981, Phys. Rev. C23, 1434
Cameron, A.G.W. 1982, in Essays in Nuclear Astrophysics, eds. C.A. Barnes, D.D. Clayton, D.N. Schramm (Cambridge Univ. Press) p. 23
Clayton, D.D. 1983, Ap. J. 268, 381
Cosner, K.R. 1981, Ph.D. thesis, Univ. of Illinois, unpublished
Cosner, K.R., Iben I,Jr., Truran, J.W. 1980, Ap. J. (Letters) 238, L91
Crabb, J., Lewis, R.S., Anders, E. 1982, Geochim. Cosmochim. Acta 46, 2511
Drozd, R.J., Morgan, C.J., Podosek, F.A., Poupeau, G., Shirk, J.R., Taylor, G.J. 1977, Ap. J. 212, 567
Fowler, W.A. 1972, in Cosmology, Fusion and Other Matters, ed. F. Reines (Colorado Associated University, Boulder) p. 67
Fowler, W.A. 1978, Proceedings of the Welch Foundation Conferences on Chemical Research XXI, ed. W.D. Milligan (Robert A. Welch Foundation, Houston) p. 61

Fowler, W.A., Meisl, C.C. 1985, Symposium on Cosmogonical Processes, Boulder, Colorado, preprint OAP-660 and private communication
Fowler, W.A., Hoyle, F. 1960, Ann. Phys. 10, 280
Guth, A.H. 1981, Phys. Rev. D23, 347
Hainebach, K.L., Schramm, D.N. 1977, Ap.J. 212, 347
Haustein, P.E. 1984, Proc. 7th Intl. Conference on Atomic Masses (AMCO-7), ed. O. Klepper, THD Schriftenreihe für Wissenschaft und Technik 26, p. 413
Hershberger, R.L., Macklin, R.L., Balakrishnan, M., Hill, N.W., McEllistrem, M.T. 1983, Phys. Rev. C28, 2249
Hilf, E.R., von Groote, H., Takahashi, K. 1976, CERN 76-13, p.142
Howard, W.M., Möller, P. 1980, At. Data Nucl. Data Tables 25, 219
Hudson, B., Kennedy, B.M., Podosek, F.A., Hohenberg, C.M. 1984, Geochim. Cosmochim. Acta
Iben, I. Jr., Renzini, A. 1984, Phys. Rev. 105, 329
Jones, K., Demarque, P. 1983, Ap. J. 264, 206
Jordan, J., Kirsten, T., Richter, H., 1980, Z. Naturforsch. 35a, 145
Käppeler, F., Beer, H., Wisshak, K., Clayton, D.D., Macklin, R.L., Ward, R.D. 1982, Ap. J. 257, 821
Krumlinde, J., Möller, P., Wene, C.O., Howard, W.M. 1981, Proc. 4th Intl. Conference on Nuclei far from Stability, CERN 81-09, p.260
Lederer, C.M., Shirley, V.S. 1978, Table of Isotopes, 7th Edition (John Wiley Sons)
Linde, A.D, 1984, Rep. Prog. Phys. 47, 925
Liran, S., Zeldes, N. 1976, At. Data Nucl. Data Tables 17, 431
Matteucci, F. 1986, this conference
Meyer, B. 1986, this conference
Miller, G.E., Scalo, J.M. 1979, Ap. J. Suppl. 41, 513
Myers, W.D. 1976, At. Data Nucl. Data Tables 17, 411
Myers, W.D., Swiatecky, W.J. 1967, Ark. Fys. 36, 343
Naldrett, S.N. 1984, Can. J. Phys. 62, 15
Podosek, F.A. 1972, Geochim. Cosmochim. Acta 36, 755
Reeves, H. 1979, Ap. J. 237, 229
Sandage, A. 1982, Ap. J. 252, 553
Sandage, A., Tammann, G.A. 1984, Nature 307, 326
Schramm, D.N., Wasserburg, G.J. 1970, Ap. J. 162, 57
Seeger, P.A., Fowler, W.A., Clayton, D.D. 1965, Ap. J. Suppl. 11, 121
Seeger, P.A., Schramm, D.N. 1970, Ap. J. (Letters) 160, L157
Seelmann-Eggebert, W., Pfennig, G., Münzel, H., Klewe-Nebenius, H. 1981, Chart of the Nuclides, Kernforschungszentrum Karlsruhe
Sneden, C., Pilachowski, C.A. 1985, Ap. J. in press
Steigman, G. 1986, this conference
Steinhardt, P.J. 1984, Comments on Nucl. Part. Phys. A12, 273
Symbalisty, E.M.D., Schramm, D.N. 1981, Rep. Prog. Phys. 44, 293
Thielemann, F.-K. 1984, in Stellar Nucleosynthesis, eds. C. Chiosi, A. Renzini (Reidel, Dordrecht) p. 389
Thielemann, F.-K., Metzinger, J., Klapdor, H.V. 1983a, Z. Phys. A309, 301
Thielemann, F.-K., Metzinger, J., Klapdor, H.V. 1983b, Astron. Astrophys. 123, 162
Tinsley, B.M. 1977, Ap. J. 216, 548

Tinsley, B.M. 1980, Fund. Cosmic Phys. $\underline{5}$, 287
Truran, J.W. 1980, Astron. Astrophys. $\underline{97}$, 391
Truran, J.W., Cameron, A.G.W. 1971, Astrophys. Space Sci. $\underline{14}$, 179
Truran, J.W., Iben, I.Jr. 1977, Ap. J. $\underline{216}$, 797
Turner, M.S., Steigman, G., Krauss, L.M. 1984, Phys. Rev. Lett. $\underline{52}$, 2090
VandenBerg, D.A. 1983, Ap. J. Suppl. $\underline{51}$, 29
Wasserburg, G.J., Hunecke, J.C., Burnett, D.S. 1969, J. Geophys. Res. $\underline{74}$, 4221
Wene, C.O., Johannson, S.A.E. 1976, Proc. 3rd Intl. Conference on Nuclei far from Stability, CERN 76-13, p.584
Winters, R.R., Macklin, R.L. 1982, Phys. Rev. $\underline{C23}$, 1434
Woosley, S.E., Fowler, W.A., 1979, Ap. J. $\underline{233}$, 411
Yokoi, K., Takahashi, K., Arnould, M., Astron. Astrophys. $\underline{117}$, 65

A POSSIBLE UNIFIED INTERPRETATION OF THE SOLAR ^{48}Ca/^{46}Ca ABUNDANCE RATIO AND Ca-Ti ISOTOPIC ANOMALIES IN METEORITES

K.-L. Kratz[1], W. Hillebrandt[2], J. Krumlinde[3],
P. Möller[3], F.-K. Thielemann[2], M. Wiescher[1]
and W. Ziegert[1]

[1] Institut für Kernchemie, Univ. Mainz, Germany
[2] Lund Inst. of Technology, Lund Univ., Sweden
[3] MPI für Physik und Astrophysik, Garching, Germany

ABSTRACT. When taking into account nuclear structure effects in β-decay properties of neutron-rich S to K isotopes, neutron exposures similar to those occurring in explosive He-burning can explain the solar abundance ratio ^{48}Ca/^{46}Ca as well as the correlated Ca-Ti isotopic anomalies observed in the EK-1-4-1 Allende inclusion, assuming both to be the result of the same nucleosynthesis process.

1. INTRODUCTION

Progress in nuclear astrophysics is intimately related to continued experimental and theoretical efforts in nuclear physics. For example, the fact that up to now no astrophysical calculation in whatever nucleosynthetic environment was able to reproduce the solar ^{46,48}Ca abundances [1] or the correlated Ca-Ti isotopic anomalies observed in inclusions of the Allende meteorite [2,3], may - at least partly - be due to the poor knowledge of the relevant nuclear physics data for n-rich isotopes around ^{48}Ca. Therefore, we have investigated the influence of nuclear structure properties of short-lived S to K nuclides on the production of their Ca-Ti β-decay daughters in astrophysical n-capture processes.
    While the lighter Ca isotopes up to A=44 are suggested to be produced by explosive O- and Si-burning [4,5] with possible s-process contributions [6], the rare heavy isotopes ^{46,48}Ca are probably the result of a different, independent nucleosynthesis process. The burning products of this process may then later be mixed into presolar material to form the observed solar abundances, or - if remaining unmixed - may show up as premordial meteoritic inclusions with the original isotopic abundance composition which may be different from the solar one.

2. THE ^{48,49}Ca(n,γ) CROSS SECTIONS AND THE nβ-PROCESS

Among the numerous attempts to identify the exotic origin of the heavy Ca-Ti isotopes, the most successful is that of Sandler, Koonin and Fowler (hereafter referred to as SKF [7]). The authors suggest that these isotopes could be produced in a high n-density environment of 10^{-7}mol/cm^3 with a n-exposure time of 10^3 s, in which both n-capture and β-decay are effective (nβ-process). Assuming the initial abundances to be solar and applying statistical Hauser-Feshbach (HF) n-capture cross sections [8], SKF calculate a ^{48}Ca/^{46}Ca abundance ratio of 21.5, which is only a factor 2.5 smaller than the observed solar value of 53±11 [1]. Apart from this success, however, the predicted anomalies for ^{46}Ca and ^{49}Ti are too large compared to those in the Allende EK-1-4-1 inclusion [2,3]. In order to reduce the abundances of these two isotopes, SKF propose - using straightforward shell model arguments - a low-lying s-wave resonance in ^{46}K(n,γ) and ^{49}Ca(n,γ), respectively, which would enhance the global HF-rates by a factor of 10. This would increase considerably the depletion of the above progenitors of ^{46}Ca and ^{49}Ti.

In the case of ^{49}Ca(n,γ), indeed, such an s-wave n-capture resonance was observed experimentally in the 'inverse reaction' to n-absorption, i.e. β-delayed neutron (βdn)

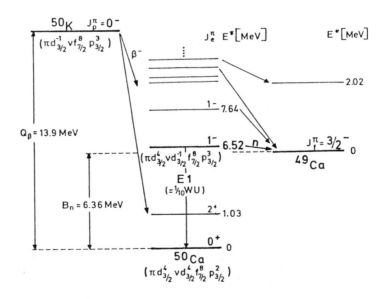

Figure 1. Principal characteristics of βdn-decay of ^{50}K. The $J^\pi=1^-$ state at 6.52 MeV in ^{50}Ca may have the properties of the s-wave n-capture state required by SKF [7].

emission of 740-ms ^{50}K [9]. As is indicated in Fig. 1, the corresponding $J^{\pi}= 1^-$ state in the compound nucleus (CN) ^{50}Ca lies 155 keV above the reaction threshold B_n and has the suggested particle-hole structure, which may result in the required large E1 radiation width of 1/10 Weisskopf unit [7]. However, subsequent measurements of the partial decay widths of this state of $\Gamma_n \lesssim 5.4$ keV and $\Gamma_\gamma \lesssim 30$ eV [10] have put constraints on the Breit-Wigner (BW) resonance n-capture rate. It will be smaller than the HF-rate for temperatures of $T_9 < 0.5$, but may be enhanced up to a factor six over the statistical cross section for $T_9 \gtrsim 1.0$. Nevertheless, this is not sufficient to support the explanation of the meteoritic ^{49}Ti abundance suggested by SKF [7] and to account for the remaining discrepancy in the solar abundance ratio of ^{48}Ca/^{46}Ca.

Another possible interpretation for the high ^{48}Ca/^{46}Ca abundance ratio and the relatively low ^{49}Ti abundance in EK-1-4-1 may be a non-statistical behaviour of n-capture on these two nuclides due to the low level densities in the respective CNs. In an nβ-process-like environment, successive n-capture in the Ca chain would allow to bridge to ^{48}Ca across the unstable isotopes ^{45,47}Ca. Assuming then a substabtially decreased (n,γ)-rate for the doubly-magic ^{48}Ca below the HF-estimate [8] would reduce its 'destruction' and thus enhance the ^{48}Ca abundance. With this, also the overflow of ^{49}Ti would be reduced. In an attempt to determine the n-capture rate of ^{48}Ca, we have investigated βdn-decay of 1.26-s ^{49}K [12]. This decay mode will - due to the GT selection rules - only populate s- and d-wave n-emitting states in the CN ^{49}Ca which will, however, contain the major components of the 'inverse' ^{48}Ca n-resonance capture. Our data are remarkable in that no low-energy s- and d-wave resonances were observed in the βdn-spectrum of ^{49}K. Whereas the lack of $l_n=2$ neutrons below about 150 keV is due to their small decay widths ($\Gamma_n < \Gamma_\gamma$), the absence of $l_n=0$ neutrons up to about 1 MeV can only be understood in terms of the specific particle-hole structure of ^{49}Ca [12,13]. With this, the BW resonance rate is estimated to be only ~25 μb compared to the HF-rate of 1.1 mb [8], apparently supporting the above speculation about a low (n,γ)-cross section for ^{48}Ca. However, a recent determination of the the total n-capture rate of ^{48}Ca yielded 0.95 mb [14], suggesting that in this case direct radiative capture must be the dominant reaction mechanism, which - more or less fortuitously - resembles the statistical HF-rate [8].

Nevertheless, these results on the ^{48,49}Ca(n,γ) cross sections exclude a straightforward explanation of the abundance ratios ^{48}Ca/^{46}Ca and ^{49}Ti/^{50}Ti via steady-state capture reactions. Hence, alternative interpretations of the above isotopic anomalies have to be found, without speculation of unknown resonances.

3. BDN-DECAY OF SHORT-LIVED S TO K ISOTOPES

Initiated by the unexpectedly high βdn-emission probability of P_n=86 % for ^{49}K [12], we have investigated possible implications of this decay mode on the final Ca-Ti isotopic abundances. The idea behind this attempt is the following: Assuming the nucleosynthesis path around A=48 to lie in the K chain with considerable production of ^{49}K, βdn-emission from this precursor will predominantly (by 86 %) form ^{48}Ca, whereas ^{49}Ti will only be populated weakly (by 14 %) by direct β-decay of ^{49}K. This would enhance the abundance of ^{48}Ca and - at the same time - reduce the ^{49}Ti abundance. Similarly, one may speculate that the low ^{46}Ca abundance could be due to βdn-emission of its A=46 progenitors.

In the A=48 mass region, experimental P_n-values exist only for K precursors [12], their a priori unexpected variations in the βdn-branching ratios (e.g. ^{48}K: 1.1 %, ^{49}K: 86 %) being due to the specific shell structure of the respective Ca emitters [12,13]. Since the commonly used sta-

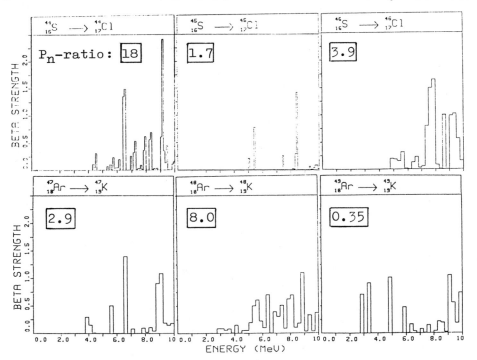

<u>Figure 2.</u> GT strength functions of $^{44-46}$S and $^{47-49}$Ar calculated with the RPA shell model of [16]. The influence of nuclear structure on the βdn-decay is reflected by the P_n ratio RPA/Gross Theory.

tistical 'Gross Theory' [15] fails badly to reproduce these P_n-values (e.g. ^{48}K: 0.17 %, ^{49}K: 8.9 %), we have used the recent RPA shell model of [16] to derive theoretical βdn-branching ratios for S to Ca precursors from the respective model Gamow-Teller (GT) strength functions. Since these calculations were quite successful in reproducing the known K P_n-values (e.g. ^{48}K: 0.65 %, ^{49}K: 95 %), we believe that the predicted variations in the P_n-values of neighbouring S to Ar precursors are reliable, too. Again, these variations directly reflect the rapid changes in the structure of the respective GT strength functions in this mass region. As examples, in Fig. 2 the shell model strength functions for $^{44-46}$S and $^{47-49}$Ar are shown; and a comparison of the RPA with the 'Gross Theory' P_n-values is included in order to demonstrate the influence of nuclear structure on βdn-decay of these isotopes.

4.   BDN-DECAY AND NEUTRON CAPTURE MODELS

The network calculations of SKF [7] do not include βdn-emission; but at the assumed n-density of 10^{-7}mol/cm^3 the nβ-process path lies rather close to the β-stability line (see Fig. 3), so that in this model βdn-branching anyhow will not cause substantial changes in the residual isotopic abundances. For example, in the nβ-process only 4 % of the ^{48}Ca production originates from ^{49}K βdn-decay, and ^{46}S or ^{46}Cl as possible progenitors of ^{46}Ca are not reached at all.

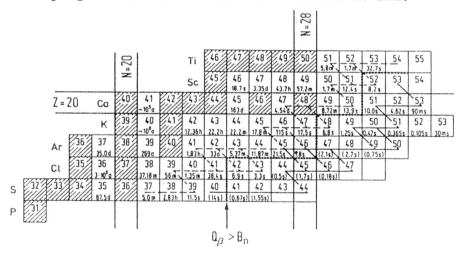

Figure 3. Reaction paths in the nβ-process [7] (dashed line) and in our n-process attempt for a neutron density of $6 \cdot 10^{-5}$mol/cm^3 (solid line). The n-process path lies significantly beyond the ($Q_β$-B_n) line where βdn-decay is possible.

At higher n-densities, however, the nucleosynthesis path will be shifted far-off stability, where the decay mode of βdn-emission may play a significant role in altering the decay back to the stability line after freeze out. Such an astrophysical scenario may be explosive He-burning (n-process) [17,18], which may be associated with a supernova shock-front in the He-burning shell of a massive presupernova star. With n-densities of $10^{-5}-10^{-3}$ mol/cm^3 at short time-scales of ≤1 s, fast n-capture will shift the process path to far-unstable isotopes with β-decay half-lives of ≤1 s. As an example, the solid line in Fig. 3 shows the likely n-process path for a 'typical' n-density of $6 \cdot 10^{-5}$ mol/cm^3. Starting from solar abundances, we have calculated the build-up of isotopic abundance distributions of short lived S to Ti nuclei as a function of n-exposure by using HF-rates [11] and the simplification of treating subsequent n-capture for each element separately (see Fig. 4), without considering competing β-decay and (γ,n)-reactions. This simplification is justified to a good approximation for n-process conditions.

From the calculated abundance distributions at freeze out, (a) without and (b) with additional corrections for estimated effects from β-decay of the shortest-lived members of each Z-chain during n-exposure, the decay back to the stability line was determined by taking into account βdn-branching. In this scenario, ^{48}Ca will predominantly be

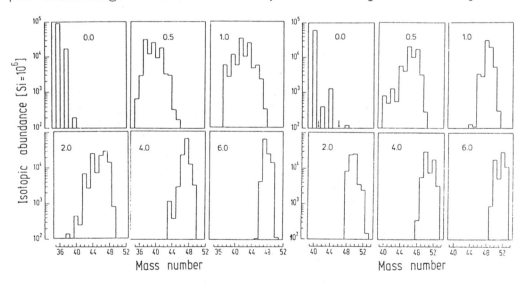

Figure 4. Abundance distributions of Ar (left) and Ca isotopes (right) as a function of n-exposure in units of 10^{-5} mol·cm^{3}s. The initial abundances are solar ones [1].

formed by $^{48}Ar(\beta-)^{48}K(\beta-)^{48}Ca$ and by $^{49}Ar(\beta-)^{49}K(\beta-n)^{48}Ca$, both Ar isotopic abundances peaking at n-exposures of $4-6 \cdot 10^{-5} mol \cdot cm^{-3} s$ (see Fig. 4). Just at these n-exposures, where the production of the ^{48}Ca progenitors is largely enhanced, the formation of the ^{46}Ca progenitors, $^{46,47}Cl$ and $^{46,47}Ar$, shows a minimum. For these conditions, the final abundance ratio of $^{48}Ca/^{46}Ca$ is calculated to be in the range 40-66, which is in excellent agreement with the observed solar abundance ratio of 53 ± 11 [1].

In order to derive meteoritic Ca-Ti abundances, the resulting n-process abundances at different n-exposures were mixed into solar material, as done by SKF [7]. From these mixed abundances, isotopic anomalies δ, as defined in [2,3], were derived. Fig. 5 compares the results from our approach with those from the nβ-process [7] and the observed Ca-Ti anomalies in the EK-1-4-1 Allende inclusion [2,3]. At a n-exposure of about $7 \cdot 10^{-5} mol \cdot cm^{-3} s$, optimum overall agreement for the isotopic anomalies can be derived, including the two 'problem-nuclei' of the nβ-process, ^{46}Ca and ^{49}Ti. With particular regard to the latter two nuclides, βdn-emission from the progenitors $^{46,47}Cl$, respectively

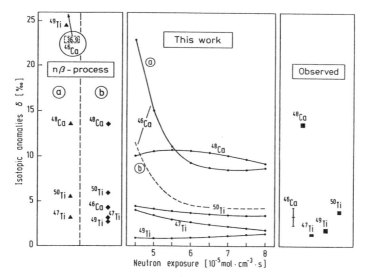

Figure 5. Comparison of the Ca-Ti isotopic anomalies predicted by the nβ-process [7] (ⓐ with HF-rates for all nuclei, and ⓑ with rates increased by a factor of 10 for ^{46}K and ^{49}Ca) with our calculations (ⓐ without, and ⓑ with estimated competition of β-decay of ^{46}Ca progenitors during n-exposure) and with the observed anomalies in EK-1-4-1 [2,3].

^{49,50}K, obviously explains the low isotopic abundances of ^{46}Ca and ^{49}Ti in the EK-1-4-1 Allende inclusion.

5. SUMMARY

We have demonstrated, that with only slighly different n-exposures the same nucleosynthesis process may yield both the correct solar ^{48}Ca/^{46}Ca abundance ratio and the meteoritic Ca-Ti isotopic anomalies of EK-1-4-1 without speculation of unknown n-capture resonances. Instead, the inclusion of nuclear structure effects in the β-decay properties of n-rich S to K isotopes into high-density n-capture models seems to be the nuclear physics clue to a solution of these astrophysical problems. A more quantitative confirmation of our ideas by complete network calculations is in progress.

REFERENCES

[1] E. Anders and M. Ebihara, Geochim. Cosmochim. Acta, 46, 2363 (1982)
[2] T. Lee et al., Ap. J., 220, L21 (1978); 228, L93 (1979)
[3] F.R. Niederer et al., Ap. J., 240, L73 (1980); and Geochim. Cosmochim. Acta, 45, 1017 (1981)
[4] S.E. Woosley and T.A. Weaver, in: Essays in Nuclear Astrophysics, ed. C.A. Barnes et al., Cambridge, Univ. Press, p. 337 (1982)
[5] D. Bodansky et al., Ap. J. Suppl., 148, 16 (1968)
[6] F. Käppeler et al., Ap. J., 257, 821 (1982)
[7] D.G. Sandler et al., Ap. J., 259, 908 (1982)
[8] S.E. Woosley et al., Nucl. Data Tables, 22, 371 (1978)
[9] K.-L. Kratz et al., Astr. Ap., 125, 381 (1983)
[10] W. Ziegert, Ph.D. Thesis, Univ. Mainz (1985); and to be published
[11] F.-K. Thielemann, Ph.D. Thesis, TH Darmstadt; and in Ref. [9]
[12] L.C. Carraz et al., Phys. Lett., 109B, 419 (1982)
[13] A. Dobado and A. Poves, Proc. Summer School on Heavy Ion Collisions, La Rabida, Spain (1982)
[14] F. Käppeler et al., Ap. J., in print
[15] K. Takahashi, Progr. Theor. Phys., 47, 1500 (1972)
[16] J. Krumlinde and P. Möller, Nucl.Phys., A417, 419 (1984)
[17] J.W. Truran et al., Ap. J., 222, L63 (1978)
[18] F.-K. Thielemann et al., Astr. Ap., 74, 175 (1979)

V – RELEVANT NUCLEAR PHYSICS PROBLEMS

THE SCREENING OF PHOTODISINTEGRATION REACTIONS

R. MOCHKOVITCH[1] and K. NOMOTO[2]

[1]Institut d'Astrophysique du CNRS
98 bis, Boulevard Arago, F-75014 PARIS
[2]Department of Earth Science and Astronomy
College of Arts and Sciences, University of Tokyo
Meguro-ku, Tokyo 153, Japan

ABSTRACT

We calculate the rates of photodisintegration reactions in a strongly coupled plasma for the non-resonant and resonant cases. The screening produces a decrease of the Coulomb barrier between two interacting nuclei together with an increase of the energy threshold of the reaction. We show that these two effects nearly cancel out so that the photodisintegration rates differ little from their values in a dilute medium. Some astrophysical consequences of this result are briefly discussed in conclusion.

1. INTRODUCTION

Photodisintegration reactions become important during neon burning when ^{20}Ne is disintegrated into ^{16}O+^{4}He at temperatures as high as T~10^9K. At still higher temperatures, during silicon burning photodisintegration reactions contribute to the rearrangement of nuclei toward nuclear statistical equilibrium. The physical parameters of the plasma often correspond to the strong or intermediate screening regime

$$\Gamma = \frac{Z_1 Z_2 e^2}{kT \, a_{12}} > 1 \qquad (1)$$

where Z_1 and Z_2 are the charges of the two interacting nuclei; T is the temperature and a_{12} the typical internuclear distance. The rate of the forward reaction is then enhanced by a large factor as first shown by Schatzman (1948) and Salpeter (1954); (see also Salpeter and Van Horn, 1969; Graboske et al., 1973; Mitler, 1977, Itoh et al., 1977, 1979; Alastuey and Jancovici, 1978). The situation is not so clear for the photodisintegration rate : will it be enhanced or not ? Different opposite answers have been given to this question in the literature (Weaver et al., 1984; Nomoto, 1984) and the present work will try to clarify the situation.

2. PHOTODISINTEGRATION REACTIONS : THE FORMALISM

Consider the two reactions,

$$1 + 2 \underset{p}{\overset{f}{\rightleftharpoons}} 3 + \gamma \qquad (2)$$

where f stands for forward and p for photodisintegration, 1, 2 and 3 being the three interacting nuclear species.
The photodisintegration lifetime of nucleus 3 is related to the reaction rate of the forward reaction by the reciprocity theorem (Blatt and Weisskopf, 1952),

$$\frac{1}{\tau_{3\gamma}} = \lambda_{3\gamma} = \frac{1}{1+\delta_{12}} \frac{G_1 G_2}{G_3} \left[\frac{M_1 M_2}{M_3}\right]^{3/2} \left[\frac{2\pi kT}{h^2}\right]^{3/2} \langle \sigma v \rangle_{12} \exp\left[-\frac{Q}{kT}\right] \qquad (3)$$

(Fowler et al., 1967). δ_{12} is the Kronecker delta function and the G_i and M_i are respectively the nuclear partition functions and masses; Q is the energy threshold of the phodisintegration reaction and $\langle \sigma v \rangle_{12}$ the rate of the forward reaction per second per pair of interacting particles,

$$\langle \sigma v \rangle_{12} = \frac{(8/\pi)^{1/2}}{M^{1/2}(kT)^{3/2}} \int_0^\infty \sigma(E) \, E \, \exp(-E/kT) \, dE \qquad (4)$$

where M is the reduced mass, E the kinetic energy in the center of mass system and $\sigma(E)$ the reaction cross section. In a strongly coupled plasma, due to the screening effects, the potential between two interacting nuclei is not simply the pure Coulomb term $Z_1 Z_2 e^2/r$ but becomes,

$$V_{12}(r) = \frac{Z_1 Z_2 e^2}{r} - \Delta\mu^c + w(r) \qquad (5)$$

The difference $\Delta\mu^c = \mu_1^c + \mu_2^c - \mu_3^c$, where μ_i^c is the contribution to the chemical potential of nucleus i resulting from the Coulomb interactions. It produces the main part of the enhancement factor,

$$EF \approx \exp\left(\frac{\Delta\mu^c}{kT}\right) \qquad (6)$$

which is the result obtained by Salpeter in 1954.

More detailed calculations involve the effective potential w(r) which is given by numerical Monte-Carlo simulations.
Another screening effect consists in a shift of the energy threshold of the reaction,

$$Q = Q^0 + \Delta\mu^c \tag{7}$$

where $Q^0 = \Delta M c^2$.
We then have,

$$\langle\sigma v\rangle_{12} = \langle\sigma v_{12}\rangle^0 \text{ (low density limit)} \times EF$$

$$\approx \langle\sigma v_{12}\rangle^0 \times \exp\left(\frac{\Delta\mu^c}{kT}\right) \tag{8}$$

together with

$$\lambda_3 \propto \langle\sigma v\rangle_{12} \times \exp\left(-\frac{Q}{kT}\right) \tag{9}$$

which leads to a cancelation of $\exp(\Delta\mu^c/kT)$ in $\langle\sigma v\rangle_{12}$ with $\exp(-\Delta\mu^c/kT)$ in $\exp(-Q/kT)$ and gives,

$$\lambda_{3\gamma} \approx \lambda_{3\gamma}^0 \text{ (low density limit)} \tag{10}$$

We conclude that the photodisintegration rate is not enhanced at least in first approximation. To go a little further we have to distinguish between non-resonant and resonant reactions.

3. NON-RESONANT REACTIONS

In non-resonant reactions the nuclear cross section factor S(E), defined by

$$\sigma(E) = \frac{S(E)}{E} P(E) \tag{11}$$

where P(E) is the barrier penetration factor, varies slowly with energy. Taking $S(E) \approx S(0)$ we get

$$\langle\sigma v\rangle_{12} = \frac{(8/\pi)^{1/2}}{M^{1/2}(kT)^{3/2}} S(0) \cdot g(r_n) \tag{12}$$

where $g(r_n) = \int_0^\infty P(E) \exp(-E/kT) dE$ is the value of the pair correlation function at a typical nuclear radius distance.
The enhancement factor for the forward reaction is given by,

$$EF = \frac{g(r_n)}{g_0(r_n)} \quad (13)$$

where the correlation function $g_0(r)$ corresponds to the limit of very low density.

The probability of tunneling through the potential $V_{12}(r)$ is obtained by the WKB method or by the path integral technique. The chemical potentials $\mu_i{}^c$ and the effective potential $w(r)$ have been computed by Hansen et al. (1977), De Witt et al. (1973) and Jancovici (1977). $w(r)$ is quadratic for small nuclear separation and then becomes quasi-linear. Finally, the enhancement factor yields,

$$EF \approx \exp\left[\frac{\Delta\mu^c}{kT} - \frac{5}{96}\tau\left(\frac{3\Gamma}{\tau}\right)^3\right] \quad (14)$$

Here, $\Gamma = Z_1 Z_2 e^2/kTa_{12}$ with $a_{12} = \left[\frac{3(Z_1+Z_2)/2}{4\pi n_e}\right]^{1/3}$, n_e being the electronic density and $\tau = 4.25\ A^{1/3}(Z_1 Z_2)^{2/3}/T_9^{1/3}$ where A is the reduced atomic mass and $T_9 = T/10^9 K$.

This formula (14) obtained by Alastuey and Jancovici (1978) is valid for $3\Gamma/\tau \ll 1$, which is fullfilled in the high temperature plasmas where photodisintegration reactions take place. Typical Γ and τ values are in the range $\Gamma \sim 1\text{-}5$ and $\tau \sim 10\text{-}100$. The $\exp(\Delta\mu^c/kT)$ term is large $\sim 2\text{-}100$ but cancels in the photodisintegration rate,

$$\lambda_{3\gamma} \approx \lambda_{3\gamma}^0 \times \exp\left[-\frac{5}{96}\tau\left(\frac{3\Gamma}{\tau}\right)^3\right] \quad (15)$$

which gives $\lambda_{3\gamma}^0/\lambda_{3\gamma} \sim 0.9\text{-}1$. The photodisintegration rate is not enhanced by the screening effects (notice that $\lambda_{3\gamma}$ is even smaller than $\lambda_{3\gamma}^0$!).

4. RESONANT REACTIONS

Photodisintegration reactions proceed in many cases through the resonant state. In that case the forward reaction rate (per second, per pair of interacting particles) is

$$\langle\sigma v\rangle_{12} = \hbar^2 \left(\frac{2\pi}{MkT}\right)^{3/2} \omega \frac{\Gamma_{12}(E_r)\Gamma_{3\gamma}(E_r)}{\Gamma(E_r)} \exp\left(-\frac{E_r}{kT}\right) \quad (16)$$

provided the resonance is sharp (e.g., Clayton, 1968). $\Gamma_{12}(E_r)$ and $\Gamma_{3\gamma}(E_r)$ are respectively the ingoing and outgoing channel widths at the resonance energy E_r and $\Gamma(E_r) = \Gamma_{12}(E_r)+\Gamma_{3\gamma}(E_r)$.
The factor ω is the statistical weight, $\omega = (2J+1)/((2I_1+1)(2I_2+1))$ where J is the resonance value of the angular momentum and I_1 and I_2 are the nuclear spins.
The resonance energy is shifted to $E'_r = E_r - \Delta\mu^C$, which leads to the following expression of the enhancement factor of the forward reaction,

$$EF = \frac{\Gamma_{12}(E'_r)\Gamma(E_r)}{\Gamma_{12}(E_r)\Gamma(E'_r)} \times \exp(\Delta\mu^C/kT) \qquad (17)$$

(Mitler, 1977), since the energy dependence of $\Gamma_{3\gamma}$ is weak.
The factor $\exp(\Delta\mu^C/kT)$ again cancels out in the photodisintegration rate, which leads to,

$$\lambda_{3\gamma} = \lambda_{3\gamma}^0 \times \frac{\Gamma_{12}(E'_r)\Gamma(E_r)}{\Gamma_{12}(E_r)\Gamma(E'_r)} \qquad (18)$$

If the partial width of the incident channel is a small fraction of the total width, Eq.(18) becomes,

$$\lambda_{3\gamma} \approx \lambda_{3\gamma}^0 \frac{\Gamma_{12}(E'_r)}{\Gamma_{12}(E_r)} \qquad (19)$$

Conversely, if the incident channel is dominant,

$$\lambda_{3\gamma} \approx \lambda_{3\gamma}^0 \qquad (20)$$

Since the partial width $\Gamma_{12}(E)$ is proportional to the Coulomb barrier penetrability,

$$\lambda_{3\gamma} \approx \lambda_{3\gamma}^0 \times \frac{P(E'_r)}{P(E_r)} \qquad (21)$$

for $\Gamma_{12} \ll \Gamma_{3\gamma}$. The penetrabilities are themselves proportional to the WKB integrals I and I'

$$P(E_r) \propto I = \int_0^{r_{tp}} \left[\frac{Z_1 Z_2 e^2}{r} - E_r\right]^{1/2} dr \qquad (22)$$

$$P(E'_r) \propto I' = \int_0^{r_{tp}} \left[\frac{Z_1 Z_2 e^2}{r} - \Delta\mu^c + w(r) - \underbrace{E_r + \Delta\mu^c}_{-E'_r} \right]^{1/2} dr \quad (23)$$

where r_{tp} is the classical turning point and $w(r)$ the effective potential. As long as $\Gamma/(E_r/kT) \leq 0.5$, r_{tp} belongs to the quadratic part of the effective potential $w(r)/kT = \frac{1}{4}(r/a_{12})^2$

Then,

$$P(E_r) \propto \int_0^{x_{tp}} \left[\frac{Z_1 Z_2 e^2}{a_{12}} \frac{1}{x} - E_r \right]^{1/2} dx \quad (24)$$

$$P(E'_r) \propto \int_0^{x_{tp}} \left[\frac{Z_1 Z_2 e^2}{a_{12}} \left(\frac{1}{x} + \frac{x^2}{4} \right) - E_r \right]^{1/2} dx \quad (25)$$

where $x = r/a_{12}$; $x_{tp} \sim \Gamma/(E_r/kT) \leq 0.5$. The two integrals differ only by a few percent and again $\lambda_{3\gamma}/\lambda_{3\gamma}^0$ is very close to unity.

5. THE CONCENTRATIONS AT EQUILIBRIUM

At equilibrium, the forward and reverse reaction rates $\lambda_{12} n_1 n_2$ and $\lambda_{3\gamma} n_3$ are equal, n_i ($i = 1,3$) being the number density of nucleus i. The reciprocity theorem gives,

$$\left(\frac{n_1 n_2}{n_3} \right) = \frac{\lambda_{3\gamma}}{\lambda_{12}} = \frac{\lambda_{3\gamma}}{\langle \sigma v \rangle_{12}} = \frac{1}{1+\delta_{12}} \frac{G_1 G_2}{G_3} \left(\frac{M_1 M_2}{M_3} \right)^{3/2} \left(\frac{2\pi kT}{h^2} \right)^{3/2} \exp\left(-\frac{Q}{kT} \right)$$

and, with $Q = Q^0 + \Delta\mu^c$ \quad (26)

$$\left(\frac{n_1 n_2}{n_3} \right) = \left(\frac{n_1 n_2}{n_3} \right)_0 (\text{low density limit}) \times \exp(-\Delta\mu^c/kT) \quad (27)$$

The equilibrium is shifted to the right side of (1), which is consistent with the thermodynamical consideration.

6. DISCUSSION AND CONCLUSION

We have shown that even in a strongly coupled plasma, photodisintegration reactions are not enhanced by the screening effects. Their rates are in fact slightly reduced compared to the low density limit, $\lambda_{3\gamma}/\lambda^0_{3\gamma}$ being in the range 0.9-1. This may be specially important in theoretical studies of the final evolution of massive stars in relation with presupernova models (see Woosley, 1985 - this workshop). Finally, our result can also be generalized to reactions with four charged particles such $(p,\alpha) - (\alpha,p)$. It can be shown, in a similar way, that only one of these two reactions, corresponding to the exothermic channel, will be enhanced.

7. REFERENCES

Alastuey, A., Jancovici, B. : 1978, Astrophys. J. 226, 1034.
Blatt, J.M., Weisskopf, V.F. : 1952, Theoretical Nuclear Physics (New-York : Wiley).
Clayton, D.D. : 1968, Principles of Stellar Evolution and Nucleosynthesis (New-York : McGraw-Hill).
De Witt, H.E., Graboske, G.R., Cooper, M.S. : 1973, Astrophys. J. 181, 439.
Fowler, W.A., Caughlan, G.R., Zimmermann, B.A. : 1967, Ann. Rev. Astron. Astrophys. 5, 525.
Graboske, H.C., De Witt, H.E., Grossman, A.S., Cooper, M.S. : 1973, Astrophys. J. 181, 457.
Hansen, J.P., Torrie, G.M., Vieillefosse, P.: 1977, Phys. Rev. A16, 2153
Itoh, N., Totsuji, H., Ichimaru, S. : 1977, Astrophys. J. 218, 477.
Itoh, N., Totsuji, H., Ichimaru, S., De Witt, H.E. : 1979, Astrophys. J. 234, 1079.
Jancovici, B. : 1977, J. Stat. Phys. 17, 357.
Mitler, H.E. : 1977, Astrophys. J. 212, 513.
Nomoto, K. : 1984, Astrophys. J. 277, 791.
Salpeter, E.E. : 1954, Australian J. Phys. 7, 373.
Salpeter, E.E., Van Horn, H.M. : 1979, Astrophys. J. 155, 183.
Schatzman, E. : 1948, J. Phys. Rad. 9, 46.
Weaver, T.A., Woosley, S.E., Fuller, G.M. : 1985, in Numerical Astrophysics, ed. J. Centrella, J. Leblanc and R. Bowers (Boston : Jones and Bartlett).

ELECTRON POLARIZATION IN NUCLEAR REACTIONS AT HIGH DENSITY

R. MOCHKOVITCH[1] and M. HERNANZ[2]

[1] Institut d'Astrophysique du CNRS
 98 bis, Boulevard Arago, 75014 Paris, France
[2] Departamento de Fisica, ETSEIB
 Universidad Politecnica de Cataluna, Barcelona, Spain

ABSTRACT

In most calculations of the enhancement factor of nuclear reactions at high density, the degenerate electrons have been treated as a uniform non-polarized background. In this paper, we check the validity of this assumption by computing the correction for finite electron polarizability. We compare our result with the recent work of Ichimaru and Utsumi (1983, 1984). We confirm that polarization effects are indeed small, in agreement with their 1984 paper. Moreover, we show that the uncertainty in the main contribution to the enhancement factor is of same order than the polarization correction. Electron polarization can then be safely neglected for example in deflagration scenarios for type I supernovae or in X-γ-ray burst models.

1. INTRODUCTION

Following the pioneering works of Schatzman (1948) and Salpeter (1954) it has been recognized that the rate of nuclear reactions taking place in a high density plasma can be very much increased. This enhancement is naturally due to the screening of the Coulomb potential between two interacting nuclei by the electrons and other nuclei. The potential barrier is lowered and its penetrability becomes easier.
Two rather similar techniques have been used to compute the enhancement factor : the WKB approximation (Salpeter and Van Horn, 1969; Graboske et al., 1973; Mitler, 1977; Itoh et al., 1977, 1979) and the path integral formalism (Alastuey and Jancovici, 1978). The many-body interactions in the plasma are generally accounted by an "effective potential" $w(r)$, which is added to the pure Coulomb term. Depending on the screening regime - weak, strong or pycnonuclear - the potential $w(r)$ can be obtained analytically or must be deduced from numerical Monte-Carlo simulations.
In this paper, we shall be interested by the strong screening regime where computations based on the WKB approximation and the path integral

formalism are in good agreement. However, until the work of Ichimaru and Utsumi (1983, 1984), the electrons were treated as a uniform, rigid background. In their first paper, these authors found that the effect of electron polarization was to produce a large increase of the enhancement factor, but finally reached the opposite conclusion in their second paper where they made a more careful analysis of the effective potential contribution. Our computations, based on Helmhotz free energy considerations confirm this last result. Its relevance to astrophysical problems involving strongly coupled plasmas is briefly discussed in conclusion.

2. SOME IMPORTANT QUANTITIES

We shall use the example of a plasma of pure carbon at a density of $10^9 g.cm^{-3}$ and a temperature of $10^8 K$ to introduce some important problem parameters. First of all, the electrons are strongly degenerate and relativistic as can be seen, by computing the two ratios,

$$\frac{\varepsilon_F^{kin}}{kT} \approx 420 \quad \text{and} \quad \frac{\varepsilon_F^{kin}}{m_e c^2} \approx 8 \tag{1}$$

where ε_F^{kin} is the kinetic energy of the electrons at the Fermi level and m_e is the electron rest mass.
The plasma coupling constant Γ is defined by,

$$\Gamma = \frac{(Ze)^2}{kTa} = 36 \tag{2}$$

where Ze is the electric charge of the nuclei and a, the typical internuclear distance ($4\pi/3\ a^3 n = 1$, n being the nucleus space density).
The ratio of the Gamow peak energy to the Coulomb energy is of the order of 0.6. According to the definitions of Salpeter and Van Horn (1969), this value, together with a Γ substancially larger than 1, corresponds to the strong screening regime with pycnonuclear corrections. The enhancement factor EF, can be obtained for example from Alastuey and Jancovici (1978). They give,

$$EF = \exp\left\{-C + \left(\frac{\tau}{3}\right)\left[\frac{5}{32}\left(\frac{3\Gamma}{\tau}\right)^3 - 0.014\left(\frac{3\Gamma}{\tau}\right)^4 - 0.0128\left(\frac{3\Gamma}{\tau}\right)^5\right]\right\} \tag{3}$$

$C = \Delta F/kT$ where ΔF is the free energy difference between two nuclei of charge Z and one nucleus of charge 2Z; $\exp(-\tau)$ is the penetration probability of the Coulomb barrier at the Gamow peak energy. The C factor can be extracted from the Monte-Carlo data of Hansen et al.(1977),

$$C = 1.0531\ \Gamma + 2.2931\ \Gamma^{1/4} - 0.5551\ \text{Ln}\Gamma - 2.350 \tag{4}$$

and τ, in a plasma of pure carbon is a function of temperature only. At $T = 10^8 K$, $\tau \approx 181$.

Then EF ≃ $1.6 \cdot 10^{16}$, the residual error in expression (3) being less than 1% in Ln EF which represents a factor of about 2, in rather good agreement with the Itoh et al. formula,

$$EF = \exp\left[1.25\ \Gamma - 0.11\ \tau\ \left(\frac{3\Gamma}{\tau}\right)^2\right] \quad (5)$$

which yields EF ≃ $2.9 \cdot 10^{16}$.

These calculations however, are made with the rigid electron background assumption, considering that the effects of electron polarization are negligible. It seems to be justified by the large λ_{TF}/a ratio where λ_{TF} is the Thomas-Fermi screening length

$$\lambda_{TF}/a = (\pi/12\ Z)^{1/3}\ r_s^{-1/2} \simeq 8.4 \quad (6)$$

The dimensionless parameter $r_s = a/(a_0 Z^{1/3})$ where a_0 is the electronic Bohr radius ($a_0 = 0.529$ Å). For $Z = 6$ and $\rho = 10^9 \text{g.cm}^{-3}$, $r_s = 1.75 \cdot 10^{-3}$.

This qualitative argument must naturally be checked by a more detailed analysis. This was done first by Ichimaru and Utsumi (1983, 1984) who obtained the corrections due to electron polarization to the pure Coulomb and averaged potentials $(Ze)^2/r$ and $w(r)$. In this paper, we shall adopt a (slightly) different point of view by computing the corrections to the Helmholtz free energy.

3. THE EFFECTS OF ELECTRON POLARIZATION

When the response of the electron background is taken into account, an additional contribution F^{pol} must be added to the Helmholtz free energy of the plasma. It can be computed by perturbation, following the method used by Galam and Hansen (1976). The correction to the enhancement factor will then take the form EF = EF (rigid background) × A_e with

$$A_e = \exp\left(\frac{\Delta F^{pol}}{kT}\right) \quad (7)$$

Here, ΔF^{pol} = F^{pol}(N nuclei of charge Z) - F^{pol}(N-2 nuclei of charge Z and 1 nucleus of charge 2Z) = $F^{pol}(N,0) - F^{pol}(N-2,1)$.
The expansion of $F^{pol}(N,0)/kT$ to first order yields,

$$\frac{F^{pol}(N,0)}{kT} = N\frac{\Gamma}{\pi} \int_0^\infty S_0(q)\left[\frac{1}{\varepsilon(q)} - 1\right] dq \quad (8)$$

$\varepsilon(q)$ is the dielectric function in dimensionless Fourrier space (q=ak) and $S_0(q)$ is the structure factor corresponding to the pure Coulomb unperturbed potential.

$S_0(q)$ depends on Γ only; it has been tabulated by Galam and Hansen (1976). The dielectric function for relativistic electrons reads,

$$\varepsilon(q) = 1 + (q_{TF}^2/q^2) \, F(x,y) \tag{9}$$

where $q_{TF} = a/\lambda_{TF}$, $x = \dfrac{\hbar k_F}{m_e c} = \dfrac{1}{137} \left(\dfrac{9\pi}{4}\right)^{1/3} r_s^{-1}$ and $y = q/q_F$ with

$q_F = ak_F = (9\pi Z/4)^{1/3}$. In our example $q_{TF} = 0.119$, $x = 8.0$ and $q_F = 3.49$. $F(x,y)$ is a complicated function of x and y which has been worked out by Jancovici (1962).
We now expand $F^{pol}(N,0)/kT$ in powers to q_{TF}, a natural small parameter in the high density limit. We obtain,

$$\frac{F^{pol}(N,0)}{NkT} = - (q_{TF}^2/\pi) \, \Gamma \, I(\Gamma,x) + O(q_{TF}^3) \tag{10}$$

with

$$I(\Gamma,x) = \int_0^\infty \frac{S_0(q) \, F(x,y)}{q^2} \, dq \tag{11}$$

This integral has been computed numerically. For $\Gamma = 36$ and $x = 8$, we get $I = 1.56$
We use the linear behaviour of F^{pol} noticed by Hansen et al. (1977) to obtain $F^{pol}(N-2,1)$

$$\frac{F^{pol}(N-2,1)}{kT} = (N-2)\frac{F^{pol}(N,0)}{NkT} + \frac{F^{pol}(0,N)}{NkT} \tag{12}$$

The two terms must be evaluated at the same electronic density so that $F^{pol}(0,N)/NkT$ is given by expression (10) with Γ and q_{TF} respectively replaced by $2^{5/3}\Gamma$ and $2^{1/3}q_{TF}$.
Then,

$$\frac{\Delta F^{pol}}{kT} = \frac{2}{\pi} \, q_{TF}^2 \Gamma \, [2^{4/3} I(2^{5/3}\Gamma,x) - I(\Gamma,x)] \tag{13}$$

Considering again our example, $2^{5/3}\Gamma \simeq 114$ for $\Gamma = 36$ and $I(114,8) \simeq 1.21$ which gives $\Delta F^{pol}/kT \approx 0.55$ and $A_e \approx 1.7$.
For larger Γ values, A_e will increase but never become much larger than unity.

4. DISCUSSION AND CONCLUSION

We now compare our result to that of Ichimaru and Utsumi (1983, 1984). We used the same plasma parameters - $Z=6$, $\rho=10^9 \text{gcm}^{-3}$ and $T=10^8 \text{K}$ - they considered in their 1983 paper where they find the formula,

$$A_e = \exp\left[(\alpha_0-\beta_0)Z^{1/3}\Gamma\right] \qquad (14)$$

They express the difference $(\alpha_0-\beta_0)$ in powers of r_s and finally obtain $A_e \simeq 600$, in contradiction with the present work. Moreover, A_e as given by (14), would also increase rapidly with Γ.
However, the treatment they made of the effect of electron polarization on the effective potential was uncorrect as stated in their 1984 paper. Their new calculation gives $A_e = 1.5$ in excellent agreement with us.
A 50-70% increase of the enhancement factor may appear to be rather important but it is in fact comparable to the uncertainty in EF due to other effects. In the context of astrophysical applications (type I supernova - X-γ-ray burst models) this turns out to be unimportant. The position of the ignition line for a given nuclear fuel (defined by $\varepsilon_{nuc} = \varepsilon_\nu$ where ε_ν represents the neutrino losses) remains nearly unchanged for such small A_e values.

Acknowledgments
The authors thank Jean-Pierre Hansen for many helpful discussions.

5. REFERENCES

Alastuey, A., Jancovici, B. : 1978, Astrophys. J. 226, 1034.
Galam, S., Hansen, J.P. : 1976, Phys. Rev. A14, 816.
Graboske, H.C., De Witt, H.E., Grossman, A.S., Cooper, M.S. : 1973, Astrophys. J. 181, 457.
Hansen, J.P., Torrie, G.M., Vieillefosse, P. : 1977, Phys. Rev. A16, 2153.
Ichimaru, S., Utsumi, K. : 1983, Astrophys. J. Lett. 269, L51.
: 1984, Astrophys. J. 286, 363.
Itoh, N., Totsuji, H., Ichimaru, S. : 1977, Astrophys. J. 218, 477.
Itoh, N., Totsuji, H., Ichimaru, S., De Witt, H.E. : 1979, Astrophys. J. 234, 1079.
Jancovici, B. : 1962, Nuovo Cimento, 25, 428.
Mitler, H.E. : 1977, Astrophys. J. 212, 513.
Salpeter, E.E. : 1954, Australian J. Phys. 7, 373.
Salpeter, E.E., Van Horn, H.M. : 1969, Astrophys. J. 155, 183.
Schatzman, E. : 1948, J. Phys. Rad. 9, 46.

LARGE-BASIS SHELL-MODEL TECHNOLOGY IN NUCLEOSYNTHESIS AND COSMOLOGY

G. J. Mathews, S. D. Bloom, K. Takahashi, G. M. Fuller
University of California
Lawrence Livermore National Laboratory
Livermore, CA 94550

and

R. F. Hausman, Jr.
Los Alamos National Laboratory
Los Alamos, NM

ABSTRACT We discuss various applications of the Lanczos method to describe properties of many-body microscopic systems in nucleosynthesis and cosmology. These calculations include: solar neutrino detectors; beta-decay of excited nuclear states; electron-capture rates during a core-bounce supernova; exotic quarked nuclei as a catalyst for hydrogen burning; and the quark-hadron phase transition during the early universe.

1. INTRODUCTION

Often in astrophysics one is faced, at a computational level, with a need to know the eigenstates of a microscopic many-body system. Most often this appears as a need to know properties of unstable or unknown nuclear states. Similar needs can occur, however, for atomic states (e.g. opacities) or even high-temperature properties of the vacuum during the early universe. In this paper we briefly overview one particular technology (the Lanczos method) which has evolved in recent years as a means to calculate such non-relativistic Schrödinger-equation problems. We discuss the application of this algorithm to several current problems in nucleosynthesis and cosmology including; 1) The response of the $^{71}Ga(\nu,e^-)^{71}Ge$ solar-neutrino detector; 2) Beta decay rates for excited nuclear states; 3) The dynamics of neutronization during core-bounce supernovae; 4) The possibility of quarked nuclei as a catalyst for the p-p chain; and 5) The quark-hadron phase transition during the early universe.

2. THE LANCZOS METHOD

The Lanczos method (Whitehead, et al. 1977; Hausman 1976) is an iterative scheme to generate eigenstates and eigenvalues of a large basis without having to store the entire Hamiltonian matrix or even compute the entire spectrum of eigenstates. The procedure begins from an initial vector, $|v_1\rangle$. Next, a new vector, orthogonal to the first, is constructed;

$$|v_2\rangle = [H|v_1\rangle - h_{11}|v_1\rangle]/h_{12} , \tag{1}$$

where $h_{ij} = \langle v_i|H|v_j\rangle$. This vector is then normalized and a third vector can be similarly constructed,

$$|v_3\rangle = [H|v_2\rangle - h_{21}|v_1\rangle - h_{22}|v_2\rangle]/h_{23} .$$

This algorithm can be rearranged as a matrix equation,

$$H \begin{pmatrix} v_1 \\ v_2 \\ v_3 \\ \vdots \end{pmatrix} = \begin{pmatrix} h_{11} & h_{12} & & \\ h_{12} & h_{22} & h_{23} & \\ & h_{23} & h_{33} & \cdot \\ & & \cdot & \cdot \end{pmatrix} \begin{pmatrix} v_1 \\ v_2 \\ v_3 \\ \vdots \end{pmatrix} \tag{3}$$

By diagonalizing this tridiagonal matrix after n iterations, one obtains n approximate eigenstates for the system. When n equals the dimension of the system this approach is an exact diagonalization. The power of this technique comes in, however, from the fact that the extrema of the spectrum converge first. Thus, if one is only interested in detailed properties for low-lying states and (or) average properties for high-lying states, adequate information can be obtained after a few Lanczos operations and the size of the matrix to diagonalize is miniscule.

This is the technique which we exploit. We are able to accommodate very large bases by using what we call the internal occupation-number representation, whereby the physical bits in the computer become the second-quantized occupation numbers. Thus, high-speed machine-language operations such as logical "or" and logical "and" become equivalent to the second-quantization creation and annihilation operators. This gives the code great speed and flexibility.

3. THE ^{71}Ga(ν,e$^-$) ^{71}Ge SOLAR NEUTRINO DETECTOR

As a first example, we consider the response of the proposed ^{71}Ga neutrino detector. The ^{71}Ga(ν,e$^-$)^{71}Ge reaction has been much discussed as an experimental way to resolve the solar neutrino problem. The main nuclear physics reason that ^{71}Ga is a good solar-neutrino detector is that the capture reaction has a low Q-value for the allowed ground-state to ground-state transition (about 233 keV). This low Q-value implies (Bahcall 1978) that a ^{71}Ga detector should be sensitive mostly to neutrinos from the primary p+p→d+ν+e$^+$ reaction. The existing ^{37}Cl detector (Davis 1978), on the other hand, involves a bigger Q-value (814 keV) and is not sensitive to p-p neutrinos. The chlorine experiment is sensitive mainly to neutrinos from the decay of ^{8}B.

Recently, there has been considerable discussion (Bahcall 1978; 1984; Orihara et al. 1983; Baltz et al. 1984; Grotz et al. 1984) concerning the contributions from neutrino captures which lead to excited states in ^{71}Ge (see Fig. 1). In order to use ^{71}Ga as a solar neutrino detector, one must know the cross sections for neutrino captures to excited states in ^{71}Ge or show that they are unimportant. The larger Q-values for excited state transitions imply that these states are predominantly populated by neutrinos from the decay of ^{7}Be and ^{8}B.

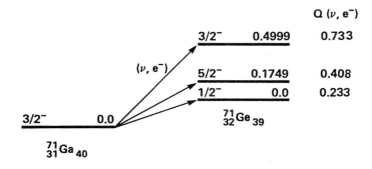

Figure 1 Transitions of interest for the ^{71}Ga(ν,e$^-$)^{71}Ge neutrino detector.

Our calculated (Mathews et al. 1985) GT strength to excited states in ^{71}Ge is lower than the value inferred by Orihara et al. (1983) from the low-energy (p,n) data. Thus, our results support the analysis of Baltz et al. (1984) which indicates that the inferred ℓ = 0, zero-degree Gamow-Teller strength is overestimated in the 35 MeV experiment because of contributions from higher ℓ-waves. Our results for the total neutron capture rate to the low lying levels in ^{71}Ge, which are most important for solar neutrino experiments, (Bahcall 1978) are in generally good agreement with the previous phenomelogical estimates (see Table 1). For the dominant sources, p-p and ^{7}Be neutrinos, the total cross sections are practically identical when they are calculated with the nuclear model and with the phenomenological estimate. We do, however, calculate a somewhat larger capture rate to the lowest $5/2^-$ state of ^{71}Ge than was inferred from the beta-decay systematics and a somewhat smaller rate to the lowest $3/2^-$ excited state. In addition, our nuclear model calculations suggest that there may be a large increase in the capture rate for ^{8}B neutrinos (see Table 1) that is caused by Gamow-Teller (GT) transitions to excited states with energies between 2-6 MeV above the ground state of ^{71}Ge.

Table 1. Calculated (Mathews et al. 1985) solar-neutrino capture rates (in SNU) compared with previous estimates.

Neutrino Source	Standard Solar Model Excited State Cross Sections:		Low Z Solar Model Excited State Cross Sections:		Mixed Solar Model Excited State Cross Sections:	
	This Work	Bahcall (1978)	This Work	Bahcall (1978)	This Work	Bahcall (1978)
p-p	70.2	70.2	72.5	72.5	74.2	74.2
pep	3.0	3.6	3.2	3.9	3.2	3.9
^{7}Be	31.2	31.7	10.9	11.1	11.2	11.4
^{8}B	11.6	1.2	1.6	0.3	2.4	0.4
^{13}N	3.3	3.1	0.1	0.1	0.6	0.6
^{15}O	4.6	4.7	0.0	0.1	1.0	1.0
Total	124	115	88	88	93	92

4. BETA DECAY OF THERMALLY EXCITED NUCLEAR STATES

Thermally populated nuclear excited states occur in most realistic scenarios for neutron-capture nucleosynthesis (Mathews and Ward 1985). This thermal population of nuclear excited states can lead to drastically different beta-decay rates which affect the nucleosynthesis.

For example, the terrestrial beta-decay half life of ^{99}Tc is long ($t_{1/2} \sim 2 \times 10^5$y) due to the second-forbidden decay of the $9/2^+$ ground state to the $5/2^+$ ground state of ^{99}Ru. There are, however, excited states at 140 and 181 keV in ^{99}Tc which can have GT

allowed transitions to the ground and first excited state of ^{99}Ru.
If typical GT-allowed log(ft) values are assumed (Cosner et al. 1984)
for these excited states the half life of ^{99}Tc reduces to about 1
yr. at T_9 = 0.35 (kT=30 keV).

To understand ^{99}Tc nucleosynthesis it is important to know its
thermally-enhanced beta-decay rate. In Fig. 2 we show our
calculations (Takahashi, Bloom, and Mathews 1985) of low-lying states
in ^{99}Tc, ^{99}Ru and the neighboring isotonic pair, ^{97}Nb-^{97}Mo.
The energies are difficult to reproduce since the low-lying states are
undoubtedly intruder states pushed down by the effects of coupling the
$\pi(1g_{9/2})^3$ configuration to collective excitations of the core.
On the other hand, the GT transition strength is probably dominated by
the one-body components of these states and is probably adequately
described by the limited model space employed here. Our calculations
indicate that the stellar lifetime for ^{99}Tc may be considerably
longer than previous estimates based on systematics (Cosner et al.
1984; Takahashi and Yokoi 1984).

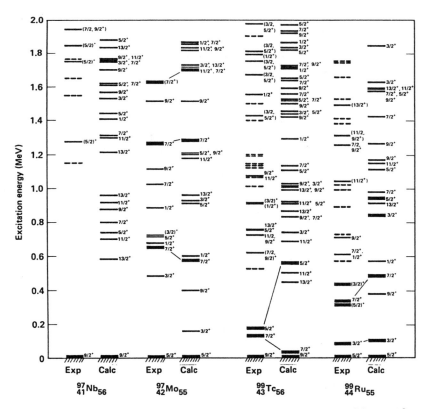

Fig. 2. Calculated (Takahashi et al. 1985) levels in ^{99}Tc, ^{99}Ru,
and the neighboring isotonic pair, ^{97}Nb-^{97}Mo.

5. GAMOW-TELLER TRANSITIONS IN A PRESUPERNOVA STAR

Shell-model calculations of electron-capture transitions in iron-group nuclei are particularly important for determining the structure of presupernova massive stars. This structure then, in turn, determines whether a core-bounce supernova mechanism can occur. The role of the GT resonance in presupernova electron capture rates and the physics of stellar collapse have been described by a number of authors (Arnett 1977; Bethe et al. 1979; Weaver et al 1984; Fuller 1982; Burrows and Lattimer 1983).

A presupernova $\sim 25 M_\odot$ star (Woosley and Weaver 1982) at the end of silicon burning contains about a solar-mass of ^{54}Fe and neutron-rich isotopes such as ^{48}Ca, ^{50}Ti, ^{54}Cr and ^{58}Fe. As the core grows in mass it eventually becomes unstable to collapse when it reaches the Chandrasekhar mass ($M \sim 5.8(Y_e)^2$), where Y_e is the ratio of free electrons to baryons). Electron capture rates will have an important effect in determining Y_e and therefore the Chandrasekhar mass.

Figure 3 is an example from some recent calculations (Bloom and Fuller 1984) of electron-capture GT strength for iron-group nuclei. This figure shows a calculation of the ^{56}Fe strength calculated in a two-particle two-hole configuration space for both the ground state and the 2^+ first excited state. The GT resonance lies fairly low in energy, $\sim 2-5$ MeV, and will participate in the neutronization. This resonance will speed the electron capture rate, and therefore reduce Y_e and the size of the Chandrasekhar mass relative to a calculation which has not included this resonance strength.

This is an important result since it makes the core-bounce mechanism more viable. The reason for this is that the core-bounce is actually experienced by an inner homologous ($v \propto r$) core which then must photodisintegrate the outer core before impinging on outer envelopes of the star. To prevent the complete dissipation of the shock due to photodisintegration, the size of the outer core must be as small as possible, and the size of the inner homologous core must be large. The GT strength function (and the amount of GT quenching) will be important in determining both of these parameters. The total core mass will be small because the presupernova electron-capture rates are fast. On the other hand, the inner homologous core will be large due to the fact that the neutronization process will lead to Pauli blocking of the GT transitions as the core collapses. Thus, the inner homologous core does not benefit from the rapid electron capture rates which minimized the Chandrasekhar mass.

6. QUARKED NUCLEI IN THE SUN

Recently (Boyd et al. 1983) it has been suggested that free quarks (if present in an abundance of 10^{-15} and bound to nuclei) could act as a catalyst to the p-p chain in the sun. They would do this by making A = 5 nuclei stable and thus vastly decreasing the time it takes to build up to the ^{7}Li^Q(p,α)^{4}He reaction which completes the

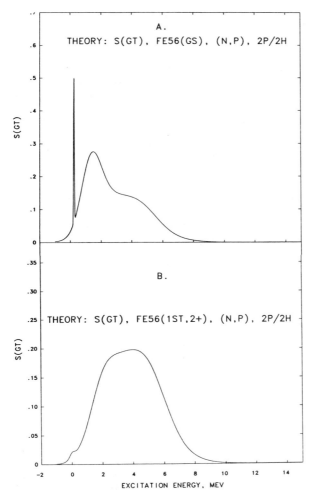

Fig. 3. Calculated (Bloom and Fuller 1984) GT strength for electron capture from the ^{56}Fe ground and first-excited state.

cycle. This hypothesis is contingent, however, on several assumptions. These are that ^{5}He^Q is stable, that ^{7}Be^Q is unstable to electron capture, and that the ^{7}Li^Q(p,α)^{4}He^Q reaction is energetically favored.

We were skeptical of these conditions and have attempted (Hughes et al. 1984) to at least qualitatively test them by calculating binding energies for quarked nuclei as best we could. We begin with a nuclear two-body force (Stone, et al. 1983) which reproduces the correct binding energies and charge radii for nuclei in this mass range. The quark nucleon potential is expected to behave as a one-pion-exchange potential at large distances, so we use this interaction.

To our surprise we found that the necessary conditions for the quarked-nucleus p-p cycle could be obtained with "reasonable" values of a bare quark mass which is 4 times the nucleon mass and a potential strength which is 3 times the nucleon one-pion-exchange potential. We do not claim that this proves the validity of the quarked-nucleus hypothesis, but it does appear that we have been unable to disprove it.

7. THE QUARK-HADRON PHASE TRANSITION

At $t \sim 10^{-5}$ sec, $T \sim 100$ MeV, the universive is expected to have undergone a phase transition from a quark-gluon plasma to hadrons (mostly mesons). It has been speculated (Crawford and Schramm 1982) that this transition may introduce long-range fluctuations which could produce primordial black holes to act as seeds for galaxy formation. To date, however, this transition has not been studied in sufficient detail because of the complexity of the QCD vacuum although many efforts are underway. The most promising approach to understanding this problem is via a discretization of space-time on a lattice (lattice guage theory).

At a computational level the problem in lattice gauge theory is very similar to the many-body nuclear shell-model problem. We have already applied our code to solve for the phases of simple Z_2 and $O(3)$ lattice guage theories. We are now in the process of upgrading the architecture of the code to accommodate the more challenging $SU(3)$ QCD problem.

8. CONCLUSION

In this brief overview we have attempted to indicate a sampling of some of the kinds of problems in astrophysics which demand a solution to a non-relativistic quantum-mechanical many-body system. Some of the problems are difficult, but perhaps not impossible with technology currently available.

9. ACKNOWLEDGEMENT

Work performed under the auspices of the U.S. Department of Energy by the Lawrence Livermore National Laboratory under contract number W-7405-ENG-48.

10. REFERENCES

Arnett, W. D. 1977, Astrophys. J., 218, 815.
Bahcall, J. N. 1978, Rev. Mod. Phys., 50, 881.

Bahcall, J. N., Huebner, W. F., Lubow, S. H., Parker, P. D., and Ulrich, R. K. 1982, Rev. Mod. Phys., 54, 767.

Bahcall, J. N. December 1, 1984, "Status of the ^{71}Ga Solar Neutrino Experiment," written report presented to the Indiana University Cyclotron Facility in support of experiment 181, (unpulished); J. Rappaport and E. Sugarbaker, (private communication).

Baltz, A. J., Weneser, J., Brown, B. A., and Rapaport, J. 1984, Phys. Rev. Lett., 53, 2078.

Bethe, H. A., Brown, G. E., Applegate, J. and Lattimer, L. 1979, Nucl. Phys., A234, 487 (1979).

Bloom, S. D. and Fuller, G. M., 1984 (Submitted to Nucl. Phys.).

Boyd, R. N., et al. 1983, Phys. Rev. Lett., 51, 609.

Brown, G. E., Bethe, H. A., Baym, G. 1981, Nucl. Phys., A375, 481.

Burrows, A. and Lattimer, L. 1983, Astrophys. J., 278, 735.

Cosner, K. R., Despain, K. H., and Truran, J. W. 1984, Atrophys. J., 283, 313.

Crawford, M. and Schramm, D. N. 1982, Nature, 298, 538.

Davis, R. 1978, "Proceedings of Informal Conference on the Status and Future of Solar Neutrino Research," ed. by G. Friedlander (Brookhaven National Laboratory Report No. BNL 50879) V. 1, p. 1.

Fuller, G. M. 1982, Astrophys. J. 252, 741.

Grotz, K., Klapdor, H. V. and Metzinger, J. September 1984, "Proceedings of the Fifth International Symposium on Capture Gamma-Ray Spectroscopy and Related Topics," (to be published).

Hausman, R. F., Jr. 1976, Ph.D. thesis, University of California Radiation
    Laboratory report No. UCRL-52178.

Hughes, C. A., Bloom, S. D., Mathews, G. J., and Fuller, G. M. 1984, BAPS, 29, 1503.

Mathews, G. J., Bloom, S. D., Fuller, G. M., and Bahcall, J. N. 1985 (Submitted to Phys. Rev. C).

Mathews, G. J., and Ward, R. A. 1985, Rep. Prog. Phys. (in press).

Orihara, H. et al. 1983, Phys. Rev. Lett., 51B, 1328.

Stone, C. A., et al. 1984, BAPS, 29, 1503.

Takahashi, K., Mathews, G. J., and Bloom, S. D. 1985, (Submitted to Phys. Rev. C).

Takahashi, K. and Yokoi, K. 1984 (submitted to Atomic and Nucl. Data Tables.

Weaver, T. A., Woosley, S. E., and Fuller, G. M., in "Numerical Astrophysics: In Honor of J. R. Wilson", ed. J. Centrella, J. Le Blanc, and R. Bower, (Science Books International; Portola, Calif.).

Whitehead, R. R., et al., 1977, Adv. Nucl. Phys., 9, 123.

Woosley, S. E. and Weaver, T. A. 1982, in "Essays in Nuclear Astrophysics", C. A. Barnes, et al. (ed.) (Cambridge Univ. Press; London) p. 377.

PROPERTIES OF ANOMALOUS NUCLEI AND THEIR POSSIBLE EFFECTS ON STELLAR BURNING

R.N. Boyd,
Department of Physics, Department of Astronomy
The Ohio State University, Columbus, OH 43210

Abstract

The effects of the possible existence of anomalous nuclei, Q-nuclei, on stellar burning are investigated. In particular, with properties only slightly different from those of normal nuclei they could catalyze a hydrogen-burning cycle, similar to the CNO cycle, which could contribute appreciably to solar energy generation with an abundance of about 1×10^{-15} Q-nuclei per normal nucleus. Furthermore, they are found to be able to solve many of the problems associated with the standard solar model, including the solar neutrino problem. Possible experiments to test for their existence are discussed.

1. Introduction

The possibility that anomalous nuclei, Q-nuclei, might exist was first suggested [1-3] about two years ago, and the studies of their effects on stellar energy generation, nucleosynthesis, and stellar evolution are continuing. Basically these particles were assumed to have properties which were both plausible and would make them very efficient for stellar burning. The requisite properties are threefold: The Q-nuclei must (1) have slightly more binding energy per nucleon than normal nuclei have, (2) have densities somewhat greater than those of normal nuclei, and (3) be stable, or nearly so, with respect to baryon emission.

Several candidates exist for these Q-nuclei, all of which have been suggested by nuclear or particle theorists. One possibility would be that of a nucleus with an extra quark embedded in it. Such quarks might have been left unconfined after the big bang at a small, but nonzero, abundance level. Such Q-nuclei have been investigated in detail by deRujula, Giles and Jaffe[4] and by Chapline[5]. Hadronic rare particles of the type suggested by Cahn and Glashow[6], if they could produce stable entities upon being embedded in a nucleus, could also serve as Q-nuclei. Superdense nuclei[7], although generally thought to have somewhat greater baryon numbers than the Q-nuclei discussed below, might be another possibility. While this list is far from complete, it does give some of the candidates for the Q-nuclei the possible manifestations of which are discussed below.

2. Q-Nuclear Properties

If the Q-nuclei do exist, and if they have the properties discussed above, they could catalyze the H-burning cycle indicated in Fig. 1. In that cycle, the Q-nucleus $^4\text{He}^Q$ would capture a proton to form $^5\text{Li}^Q$, a step which does not occur for normal nuclei. However, the hypothesized additional binding energy for Q-nuclei allows that capture to occur. A beta decay and two more (p,γ) reactions bring the cycle to $^7\text{Be}^Q$. At this point a critical beta decay must occur in the Q-nuclear cycle, i.e., $^7\text{Be}^Q$ must decay to $^7\text{Li}^Q$ much more rapidly than the next (p,γ) reaction to $^8\text{B}^Q$ occurs. Indeed, if this is not the case, the Q-nuclear abundance required to impact stellar burning might have to be much larger than is plausible, and the production rate of high-energy neutrinos would be large. However it is also possible that $^8\text{Be}^Q$ would decay into ^4He and $^4\text{He}^Q$, in which case only the question of high-energy neutrino production would be of concern.

If, however, the beta decay to $^7\text{Li}^Q$ does occur rapidly, then a (p,α) reaction on that Q-nucleus would complete the production of a normal ^4He and return the catalyst $^4\text{He}^Q$ for the next cycle. The speed of the beta decay of $^7\text{Be}^Q$ was found[1], from comparison of the rate of high-energy neutrino production anticipated from decay of $^8\text{B}^Q$ to that observed[8], to re-

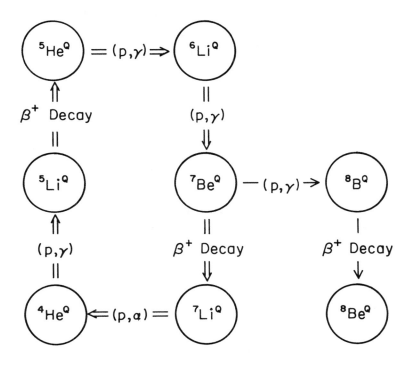

Figure 1. The Q-nuclear cycle for H-burning, consisting of three proton captures and a (p,α) reaction. Proton capture on ^{7}Be^Q would give leakage from the cycle.

quire a half-life for $^7\text{Be}^Q$ of roughly 1000 seconds or less, which corresponds[2] to a positron end-point energy of about 1.0 MeV. A beta-decay energy that large would require some enhancement of the Q-nuclear density from that of normal nuclei; roughly a factor of 2 would suffice[2].

In order to describe the participation of Q-nuclei in stellar burning, their thermonuclear reaction rates had to be determined. The prescription of Woosley et al.[9], who parameterized reaction rates in terms of a statistical model for the reaction cross section, was used. While the accuracy of this model is questionable for light nuclei (or Q-nuclei), the dominant feature of low-energy charged particle reactions, namely the Coulomb barrier, is correctly handled. Thus it would be expected to give at least a reasonable description of the Q-nuclear thermonuclear reaction rates.

Since $^6\text{Li}^Q$ would be expected to be burned most slowly in the Q-nuclear cycle, its abundance is essentially that of Q-nuclei. Some estimate of that abundance which would be required for the Q-nuclei to compete on even terms with the normal solar burning modes can be determined by equating the rate of alpha-particle production in the Q-nuclear cycle, which will be essentially the rate at which the $^6\text{Li}^Q(p,\gamma)^7\text{Be}^Q$ reaction occurs, to that for the p-p chain. It was found[1] to be about 1×10^{-15} Q-nuclei per normal nucleus.

3. Stellar Modelling Calculations

A considerably more detailed account of the effects of Q-nuclei on stellar burning was obtained[2] by including their thermonuclear reaction rates in the stellar modelling code of Paczynski[10], as modified by Joseph[11]. A series of calculations with varying assumed quantities of $^6\text{Li}^Q$ was performed, varying the zero-ago ^4He abundance and convection parameter to produce the sun's size and luminosity after 4.6 billion years. The fraction of the sun's energy produced by the Q-nuclear cycle and the high-energy solar neutrino production rate were examined. If it is assumed that the Q-nuclear cycle produces no high-energy neutrinos, then, at the Q-nuclear abundance at which two-thirds of the sun's energy is produced by the Q-nuclei, the predicted solar neutrino rate will be in agreement with that detected in the ^{37}Cl detector of Davis[8]. At that Q-nuclear abundance,

2×10^{-15} Q-nuclei per normal nucleus, the density at the core of the sun drops by about 25% from its standard solar model (SSM) value, while the temperature is just about what it is predicted to be in the SSM.

4. Catalog of Problems with the SSM

Although the solar neutrino problem[12] is the best known problem associated with the SSM, it is not the only one. Several others were noted by Joseph[13], including (1) the 5 minute oscillations, (2) the Li abundance, (3) the zero ago He abundance, (4) the ages of globular clusters and (5) the solar variability.

The 5 minute oscillations observed in solar spectra have been identified by Ulrich et al.[14] as pressure-mode oscillations in the solar convective envelope. Their study demonstrated that the depth of the convective envelope resulting from the SSM was too small by about 30% to yield oscillations of the observed frequency. At the Q-nuclear abundance which solves the solar neutrino problem, 2×10^{-15} Q-nuclei per normal nucleus, the depth of the convective envelope is increased by about 20% from the SSM value. While this does not remove the problem completely, it does reduce it to the point at which small effects from, e.g., rotation could remove the theoretical-observational discrepancy altogether.

The solar Li abundance is considerably lower than can be explained by the SSM[15]. The usual explanation for this discrepancy is that Li is burned at the bottom of the convective envelope, thus requiring a temperature at that point of $T_6=2.5$. The SSM predicts that value to be $T_6=2.0$. With a Q-nuclear abundance of 2×10^{-15} Q-nuclei per normal nucleus, the temperature at the bottom of the convective envelope is predicted to be $T_6=2.5$ thus eliminating the Li abundance problem.

The zero-age abundance of ^{4}He in SSM calculations generally is required[16] to be 0.25+.01, in excellent agreement with the primordial abundance[17]. Unfortunately, this value is well below that typical of stars which have evolved through previous generations of stellar burning. A value of around 0.30 seems to be characteristic of stars having the sun's metallicity[13]. At the same Q-nuclear abundance which solves the solar neutrino problem, the zero-age He abundance required is increased 0.04 from the value obtained without Q-nuclei, in good agreement with the value for stars having the sun's metallicity.

Three methods have been used to determine the age of the universe or of galaxies. One of those involves examination of the stars on the main sequence of the Hertzsprung-Russell diagram in distant globular clusters[18]; the mass of the most massive stars remaining on the main sequence, together with a SSM calculation to determine the age at which stars leave the main sequence, thus determines the ages of the stars in that cluster. We find that inclusion of the Q-nuclei at a level which solves the other problems of the SSM reduces the predicted lifetimes of stars on the main sequence by about 20%. While this still leaves the results from this technique for age determination in agreement with the results of the other techniques, primarily because of the large uncertainties involved, it does suggest that the standard globular cluster age determination may give a value which is too large.

Finally, there are indications[19] that life existed on earth roughly a billion years ago. The variability of the energy output from the sun, however, predicted[20] to be about 44% over that time span by the SSM, would not allow for that possibility. With the inclusion of the Q-nuclei, however, that variability is reduced[13] to 28%, thus allowing for, e.g., greenhouse effects to reduce the temperature variation on earth enough to accomodate life for that time span.

5. Other Predictions from Q-Nuclei

But do Q-nuclei exist? The variety of possible candidates makes that question difficult to answer. While searches for fractionally charged objects have attained apparent abundance limits as low as 1×10^{-18} fractional charges per baryon, it is generally felt[21], mainly because of chemical caveats, that there could well be many orders of magnitude difference between the implied abundance limits from those experiments and those which actually exist in nature. Thus more searches for fractionally charged entities in nature need to be performed. One such search[22] has attempted to use an enriched source sample to circumvent at least some of the chemical caveats associated with previous experiments.

Other types of searches have attempted to search for integrally charged nuclei having anomalous masses[23]. While the limits in some cases begin to approach the level needed to test the Q-nuclear hypothesis, the general situation is far from adequate at the present time. However more such searches are in progress.

A new generation of solar neutrino detectors, some of which will have a considerably lower threshold energy than that of ^{37}Cl, can also produce new tests of the Q-nuclear hypothesis. We have performed an estimate of the rates which would be anticipated for several of those detectors[24], and have found that, for the ^{71}Ga and ^{115}In detectors, the predicted neutrino detection rate, if the Q-nuclei do exist at the abundance required to solve the Cl detector solar neutrino problem, will be at least 50% higher than that predicted by the standard solar model. For the ^{81}Br detector, the predicted rate increases by at least 30%. For higher threshold detectors, such as ^{98}Mo and ^{205}Tl, the neutrino detection rates will, as in the case of the Cl detector, decrease if Q-nuclei are generating a significant fraction of the sun's energy.

While the Q-nuclei are only hypothetical particles at present, the remarkably large amount of circumstantial evidence in support of their existence encourages considerably more exhaustive searches for them.

Acknowledgements

My collaborators in these studies of Q-nuclei have been R. Turner, C. Joseph, B. Sur, M. Wiescher, and L. Rybarcyk. Their efforts are gratefully acknowledged. This work was supported in part by National Science Foundation Grant 82-03699.

1. R.N. Boyd, R.E. Turner, M. Wiescher and L. Rybarcyk, Phys. Rev. Letters 51, 609 (1983).
2. R.N. Boyd, R.E. Turner, L. Rybarcyk and C. Joseph, Astrophys. J. 289, 155 (1985).
3. R.N. Boyd, R. Turner, B. Sur, L. Rybarcyk and C. Joseph, Proc. Conf. on Solar Neutrinos and Neutrino Astronomy, ed. by M.L. Cherry, W.A. Fowler

and K. Lande, AIP Conf. Proc. No. 126, (New York, American Institute of Physics), p. 145.
4. A. deRujula, R.C. Giles and R.L. Jaffe, Phys. Rev. D17, 285 (1978).
5. G.F. Chapline, Phys. Rev. D25, 911 (1982).
6. R.N. Cahn and S.L. Glashow, Science 213, 607 (1981).
7. T. Ohnishi, Nucl. Phys. A362, 480 (1981).
8. R. Davis, Jr., Science Underground, ed. by M. Nieto et al., AIP Conf. Proc. No. 96, (New York, American Institute of Physics), p. 2.
9. S. Woosley, W.A. Fowler, J.A. Holmes and B.A. Zimmerman, Caltech preprint OAP-422, 1975.
10. B. Paczynski, Acta Astr. 20, 47 (1970).
11. C. Joseph, M.S. Thesis, Ohio State University, 1984.
12. W.A. Fowler, Science Underground, ed. by M. Nieto et al., AIP Conf. Proc. No. 96, (New York, American Institute of Physics), p. 80.
13. C. Joseph, Nature 311, 254 (1984).
14. R.K. Ulrich and E.J. Rhodes, Astrophys. J. 265, 551 (1983).
15. R. Weymann and R.L. Sears, Astrophys. J. 142, 174 (1965).
16. J.N. Bahcall, Highlights in Astrophysics: Concepts and Controversies, ed. by S. Shapiro and S. Teukolsky (New York, John Wiley) 1985.
17. J. Yang, M.S. Turner, G. Steigman, D.N. Schramm and K. Olive, Astrophys. J. 281, 493 (1984).
18. A. Sandage, Astr. J. 88, 1159 (1983); D.A. VandenBerg, Astrophys. J. Suppl. 51, 29 (1983).
19. L.P. Knauth and S. Epstein, Geochim. Cosmochim. Acta 40, 1095 (1976); G. Newkirk, Jr., Ann. Rev. Astr. Astrophys. 21, 429 (1983).
20. J.N. Bahcall, W.J. Huebner, S.H. Lubow, P.D. Parker and R.K. Ulrich, Rev. Mod. Phys. 54, 767 (1982).
21. L. Lyons, Oxford University Report No. 80-77, 1982.
22. R.E. Turner, Ph.D. Thesis, Ohio State University, 1984.
23. W.J. Dick, G.W. Greenlees and S.L. Kaufman, Phys. Rev. Letters 53, 431 (1984); R. Middleton, R.W. Zurmuhle, J. Klein and R.V. Kollarets, Phys. Rev. Letters 43, 429 (1979).
24. B. Sur and R.N. Boyd, Phys. Rev. Letters 54, 485 (1985).

STELLAR REACTION RATE OF $^{14}N(p,\gamma)^{15}O^+$

U. Schröder, H.W. Becker, J. Görres, C. Rolfs and
H.P. Trautvetter, Universität Münster, Münster, W. Germany
and
R.E. Azuma and J. King
University of Toronto, Toronto, Canada

The $^{14}N(p,\gamma)^{15}O$ reaction is a member of the chain of reactions in the hydrogen burning CNO cycles. It is the slowest reaction in the main CN cycle and thus controls the energy generation in the CNO cycles. Due to its importance for nuclear astrophysics the reaction has been studied by several investigators in the years 1951 - 1963. The earliest investigations[1-3] used nitrited solid targets and the β^+-activity of the residual nuclei ^{15}O to arrive at extrapolated S-factors of $S(0) = 1.5$ to 3.0 keV-b. The prompt capture γ-ray transitions were studied in the work of Bailey and Hebbard[4] using NaI(Tl) detectors and TiN solid targets. These studies[4] were carried out at $E_p = 0.2 - 1.1$ MeV and indicated that the direct capture process into the 6.18 and 6.79 MeV states contribute predominantly to the S-factor at stellar energies ($S_{6.18}(0) = 1.0$ keV-b, $S_{6.79}(0) = 1.4$ keV-b) with $S_{tot}(0) = 2.75 \pm 0.50$ keV-b. This result has been used in stellar model calculations.

The above results of the direct capture processes lead to spectroscopic factors of $C^2S = 0.6$ and 0.7 for the 6.18 and 6.79 MeV states, while stripping reactions reveal[5] discordant values of 0.04 - 0.16 and 0.31 - 0.47, respectively. It should be noted that the studies of Hebbard and Bailey[4] were hampered by the poor energy resolution of the NaI(Tl) detectors and the high γ-ray background yields due to contaminant reactions such as $^{15}N(p,\alpha\gamma)^{12}C$ and $^{19}F(p,\alpha\gamma)^{16}O$. As a consequence, the capture processes into excited states of ^{15}O could only be detected via their high-energy secondary γ-ray transitions. In view of the above discrepancies and the astrophysical importance of this reaction, the reaction has been reinvestigated recently.

As a first step, the reaction has been investigated[6] at the 350 kV Münster accelerator in the energy range of $E_p = 170 - 350$ keV using Ge(Li) detectors and TiN solid targets as well as N_2 gas targets of the extended and quasi- point supersonic jet type.[7] The investigations involved the measurement of excitation functions, γ-ray angular distributions and absolute yields. The resulting properties[6-8] of the $E_p = 278$ keV resonance are in excellent agreement with previous work[5] execpt for the total width (previous[5]: $\Gamma_{lab} = 1.7 \pm 0.5$ keV, present[5,6]: $\Gamma_{lab} = 1.1 \pm 0.1$ keV). With these resonance properties the observed yields at the tails of the 278 keV resonance have been analysed in terms of resonant and direct capture contributions.

The yields of the R/DC→6.79 MeV γ-ray transition are consistent with an incoherent suerposition of resonant and direct capture amplitudes leading to $S_{6.79}(0) \approx 1.4 \pm 0.2$ keV-b, in excellent agreement with the work of Bailey and Hebbard[4]. The R/DC→6.18 MeV transition reveals constructive and destructive interference features at energies below and above the 278 keV resonance, respectively, which can be explained by the presence of a direct capture process of $S_{6.18}(0) \approx 0.2 \pm 0.1$ keV-b. The results are in poor agreement with Bailey and Hebbard[4] but would resolve the above discrepancies with stripping data. The data for the transitions into the groundstate and 5.18 MeV state allowed only to extract upper limits of $S_0(0) \lesssim 0.3$ keV-b and $S_{5.18}(0) \lesssim 0.14$ keV-b, which are not in disagreement with the reported values[4] of $S_0(0) \approx 0.27$ kev-b and $S_{5.18}(0) \approx 0.08$ keV-b.

In order to substantiate these results as well as to clarify the reaction mechanisms involved in all capture transitions, the studies have been extended, as a second step, to the energy range of $E_p = 0.2 - 1.1$ MeV at the 1MV JN Van de Graaff accelerator of the University of Toronto. Critical to this program was the availability of ^{14}N targets depleted in ^{15}N, since the $^{15}N(p,\alpha\gamma)^{12}C$ reaction is a prolific source[4,5] of 4.4 MeV γ-rays in this energy range of $E_p \gtrsim 0.4$ MeV. Since ^{14}N gas depleted in ^{15}N cannot be obtained commercially, solid targets had to be produced by ^{14}N ion implantation[9]. These targets had isotopic ratios of $^{15}N/^{14}N \approx 10^{-4}$ and withstood high beam loads without significant deterioration. The experimental program was similar to that at the Münster accelerator. The results are shown in Fig. 1 in form of the S(E)-factor. The excitation functions and γ-ray angular distributions (not shown) of the transitions into the 6.18, 6.79, 6.86 and 7.28 MeV states are consistent with a direct capture process leading to S(0)-values of about 0.027, 1.40, 0.036 and 0.033 keV-b, respectively. The results confirm the basic conclusions drawn from the Münster data. The observed feature of the transitions into the groundstate, 5.18 and 5.24 MeV excited states could not be understood easily and demanded experimental information at higher proton energies.

Such measurements have been carried out at the 4MV Dynamitron Tandem at Bochum in the energy range of $E_p = 0.8 - 3.5$ MeV. The analysis of the available data is not yet completed, but the results indicate the following features: (i) the groundstate transition is dominated at low energies by the tail of a $\Gamma = 0.6$ MeV broad resonance at $E_p = 2.4$ MeV, leading to $S_0(0) \approx 0.030$ keV-b; (ii) the transition into the 5.24 MeV state follows at all energies the direct capture process except for interference effects near the 1.06 MeV resonance: $S_{5.24}(0) \approx 0.035$ keV-b; and (iii) the capture process into the 5.18 MeV state is dominated by the tails of the 0.28 and 1.06 MeV resonances with the inclusion of interference effects : $S_{5.18}(0) \approx 0.010$ keV-b.

The analyses of the new data indicate a astrophysical S(0) factor of $S(0) \approx 1.6 \pm 0.3$ keV-b, which arises to 90 % from the direct capture process into the 6.79 MeV state. The observed direct capture amplitudes to the 5.24, 6.18, 6.79, 6.86 and 7.28 MeV states (Fig. 1) are all consistent with the respective spectroscopie factors reported form stripping reactions[5]. If final analyses of the data confirm the above total S(0) -

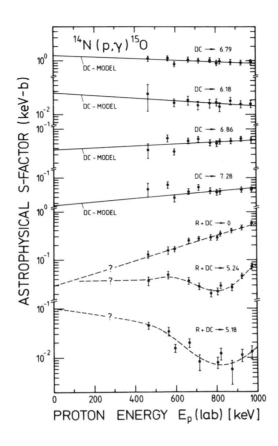

Fig. 1: Data obtained for the ^{14}N(p,γ)^{15}O reaction at the 1MV Van de Graaff of the University of Toronto. Analyses of the results for the transitions R/DC →0, 5.18 and 5.24 MeV require data at higher proton energies.

value, the $^{14}N(p,\gamma)^{15}O$ reaction burns nearly a factor of 1.8 slower at stellar energies. It might be interesting to note that the unpublished work of Pixely[4] in 1957 suggested a value of $S(0) \approx 1.5$ keV-b. The astrophysical consequences of a reduced state for the $^{14}N(p,\gamma)^{15}O$ reaction must await stellar model calculations.

References
1) D.B. Duncan and J.E. Perry, Phys. Rev. 82 (1951) 809
2) W.A.S. Lamb and R.E. Hester, Phys. Rev. 108 (1957) 1304
3) R.E. Pixely, Thesis, California Institute of Technology (1957)
4) G.M. Bailey and D.F. Hebbard, Nucl. Phys. 46 (1963) 529 and 49 (1963) 666
5) F. Ajzenberg-Selove, Nucl. Phys. A 360 (1981) 1
6) U. Schröder, Diplomarbeit, Universität Münster (1984)
7) H.W. Becker et al., Nucl. Instr. Meth. 198 (1982) 277
8) H.W. Becker, W.E. Kieser, C. Rolfs, H.P. Trautvetter and M. Wiescher, Zeit. Phys. A 305 (1982) 319
9) S. Seuthe, Diplomarbeit, Universität Münster (1985)

* Supported by the Deutsche Forschungsgemeinschaft and the National Sciences and Engineering Research Council of Canada

Recent progress in experimental determination of the
$^{12}C(\alpha,\gamma)^{16}O$-reaction rate

H.P. Trautvetter, A. Redder and C. Rolfs
Inst. f. Kernphysik der Univ. Münster, West-Germany

I. INTRODUCTION

The importance of a change in the $^{12}C(\alpha,\gamma)^{16}O$ reaction rate on nucleosynthesis during hydrostatic shell burning has been shown by Arnett and Thielemann [Ar 84] and was also pointed out many times during this conference. Woosley [Wo 85] has shown that an increase of the $^{12}C(\alpha,\gamma)^{16}O$-reaction rate by a factor of 3 as compared with [FCZ 75] improved the nucleosynthesis pattern considerably for a 25 $M_\odot$ star and for elements up to the iron group and he also shows that such a change has large effects on stellar structure [Wo 86].

This presentation will illustrate first the experimental situation for obtaining this important $^{12}C(\alpha,\gamma)^{16}O$ reaction rate, then new experimental results are presented, which will finally be discussed and compared with previously and presently available theoretical extrapolations for this rate to stellar energies.

II. REVIEW OF EXPERIMENTAL SITUATION

Typical situations and problems in experimental nuclear astrophysics are dicussed in [Ro 78, Tr 84]. The problems involved in determining the tails of subthreshold states, which often can totaly determine the reaction rate at stellar energies and can not be obtained by a direct measurement, has been discussed at the example of the $^{12}C(\alpha,\gamma)^{16}O$ reaction [Dy 74, Ke 82].

On hand of fig. 1 the relevant arguments can be briefly summarized. Stellar burning temperatures in units of T_9 are given on the right hand side of this fig. 1 which also indicates that for He-burning temperatures there is no prominent feature in the compound nucleus ^{16}O at the stellar enery E_0

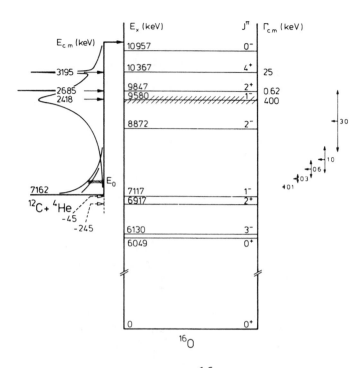

Fig. 1: Level diagram of the ^{16}O-compound nucleus from the ^{12}C + ^{4}He fusion reaction. Stellar temperatures in units of T_9 are shown on the right and the expected S-factor curve on the left.

(fig. 1). The broad J^π = 1$^-$ resonance at 2418 keV can contribute by its low energy tail to the reaction rate, but there is also a J^π = 1$^-$ state bound by just 45 keV against α-particle decay. These two J^π = 1$^-$ states give rise to an electric dipole radiation (E1) which can in the energy region between both states interfer either constructively or destructively. By careful experimental investigation of the low energy tail of the broad J^π = 1$^-$, 2418 keV resonance one can therefore study the contribution of the subthreshold J^π = 1$^-$ state.

In fig. 2 the cross section data of [Ja 70, Dy 74 and Ke 82] are compared. From this fig. 2 it is obvious that the data of [Ja 70] and [Ke 82] are in good agreement at the important low energy tail of the 2418 keV resonance but are substantially higher than the data of [Dy 74]. It has been argued by [Ke 82] that the additional yield stems from the J^π = 2$^+$ state in ^{16}O which is bound by 245 keV (fig. 1).

This state gives rise to an electric quadrupole radiation
(E2) which can tail out to the astrophysical interesting
energy region and can there interfere with the non resonant
E2-direct capture amplitude [Ke 82]. A direct experimental
test for such an E2-contribution would be angular distribution measurements as already suggested by [Dy 74].

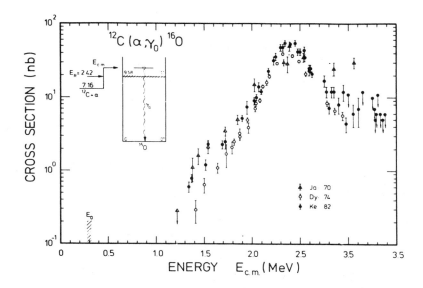

Fig. 2: Cross section measurement results for
$^{12}C(\alpha,\gamma)^{16}O$ as obtained by different
groups [Ja 70, Dy 74, Ke 82].

III. DIRECT EXPERIMENTAL INVESTIGATION OF THE E2-CONTRIBUTION

The principle of the direct determination for the cross
section ratio $\sigma(E1)/\sigma(E2)$ is illustrated in fig. 3. With the
z-axis being in the direction of the beam the resulting spin
in the compound nucleus ^{16}O can only come from the relative
angular momentum ℓ of the two spinless particles α and ^{12}C,
which is perpendicular to the z-axis, hence the system is
aligned and the destinct angular distribution patterns result for the E1-transition (fig. 3a) and the E2-transition
(fig.3b). From this picture it is clear that with positio-
ning a γ-detector at 90° one would only detect the E1-
radiation since the E2-part vanishes at this angle. (In the
work of [Ke 82] angle integrated data were obtained which
might explain the difference to [Dy 74] where the detector
was placed at 90°). Fig. 3c shows the incoherent sum of the

E1 and E2 radiation while in general interference effects between the two amplitudes are expected (fig. 3d). Depending on the ratio of $\sigma(E1)/\sigma(E2)$ (two examples are shown in fig. 3d) and the relative phase between both radiation types a destinct foreward-backward asymetry results which can be used to measure the $\sigma(E2)/\sigma(E1)$-ratio.

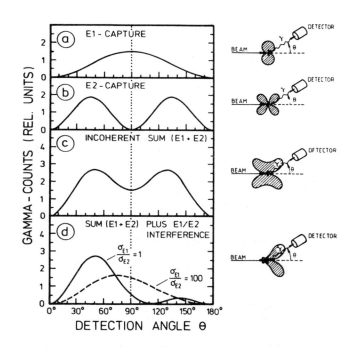

Fig. 3: Expected γ-angular distributions for the $^{12}C(\alpha,\gamma)^{16}O$ reaction (for details see text).

This fact has been realized already by [Dy 74] and their results are shown in fig. 4 as experimental points at four different energies. The fit through the data in [Dy 74] (dashed curve) neglected contribution from the subthreshold $J^\pi = 2^+$ state which is bound by 245 keV (see fig. 1) and resulted therefore in a negligible contribution of an E2-component at stellar energies. The analyses of the yield curve measurements of [Ke 82] resulted in a curve which also fits the [Dy 74] data (solid curve in fig. 4) but shows

quite different behaviour below E = 2 MeV compard to
[Dy 74]. Subsequent theoretical predictions of [La 84] for
the ratio σ(E2)/σ(E1) are shown by the dashed-dotted curve
in fig. 4 and also predict a sharp increase below E = 2 MeV.

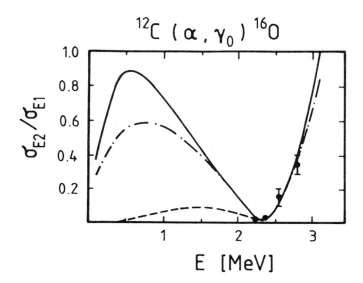

Fig. 4: Ratio of the cross section for the
$^{12}C(\alpha,\gamma)^{16}O$ reaction of the E2- to the
E1-contribution. The data points and
dashed curve are from [Dy 74], the solid
curve comes from the analysis of the
data by [Ke 82] and the dashed dotted
curve is a theoretical prediction by
[La 84].

The critical point for angular distribution measurements
below E = 2 MeV is the expected very low cross section of
the order of n-barns and smaller. Such measurements are
also always hampered by the ever present $^{13}C(\alpha,n)^{16}O$
reaction which produces a large neutron induced background.
These problems where overcome by producing ^{12}C-targets by
implanting ^{12}C with E = 110 keV in gold-backings using a
small accelerator (Münster) as a mass-seperator. An array of
6 Ge(Li)-detectors were placed at 15°, 40°, 60°, 90°, 120°
and 150° relative to the beam axis to measure simultaneously
the γ-yield from the $^{12}C(\alpha,\gamma)^{16}O$ reaction where beam
currents of up to 500 μA were used from the Stuttgart Dynamitron accelerator (for details see [Re 85]).

Typical angular distributions obtained at four different
energies are shown in fig. 5 together with a two parameter

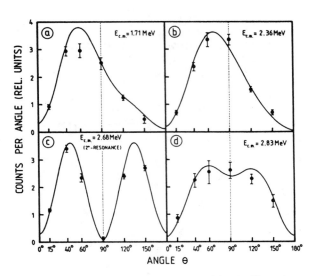

Fig. 5: Examples for angular distribution results shown together with a two parameter fit (for details see text).

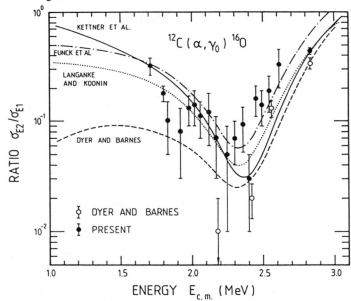

Fig. 6: Present results for the cross section ratio (solid points) compared with the results of [Dy 74]. Also shown are curves as obtained from the analyses of [Ke 82] and theoretical predictions of [Fu 84] dashed dotted curve, [La 84] dotted curve and [Dy 74] dashed curve.

χ^2-fit where the ratio $\sigma(E2)/\sigma(E1)$ and the relative phase were varied. In the $J^\pi = 2^+$ resonance at $E = 2.68$ MeV (fig. 5c) the well known pattern for E2-radiation (fig. 3b) was observed. The extracted $\sigma(E2)/\sigma(E)$-ratios from these patterns are shown in fig. 6 together with the previous results of [Dy 74], the analysis of [Ke 82] and the theoretical analysis of [La 84] and [Fu 84]. The destinct rise of the $\sigma(E2)/\sigma(E1)$ ratio below $E = 2$ MeV clearly indicates that the E2 component of the $J^\pi = 2^+$ subthreshold state bound by 245 keV can not be neglected.

IV. CONCLUSION

The data of [Dy 74] have been analysed by different authors and the S-factor was extrapolated to the astrophysical interesting energy $E_0 = 0.3$ MeV [Ba 76, Dy 74, Hu 76, Ko 74, La 84, We 74]. One also finds in literature the result of [Be 78] where the reduced particle width $\theta_\alpha^2(7.12)$ was derived from α-transfer data using the $^{12}C(^7Li,t)^{16}O$ reaction, together with an R-matrix parametrisation for the extrapolation. The data of [Ke 82] have been extrapolated by a simple single level Breit-Wigner formalism [Ke 82] as well as with a R-matrix formalism [La 84]. All extrapolation procedures for the E1-contribution seem to agree within their respective quoted errors and would yield an average value which is about a factor two higher then the previously accepted value [FCZ 75]. However, in addition to the E1 component the E2-contribution can not be neglegted and has to be added to the E1-part. The latest theoretical analyses [La 84, Fu 84, De 84] for the E2 component are based on the data of [Dy 74] as well as on the data of [Ke 82] and contribute up to 50 % to the S-factor at $E = 0.3$ MeV. From this survey it is concluded that the total S(0.3 MeV)-value for the $^{12}C(\alpha,\gamma)^{16}O$ reaction has to be increased by at least a factor of 3 and possibly by a factor of 5. Efforts are presently under way both at Münster as well as in CalTech to improve the accuracy in the extrapolated S-factor value.

REFERENCES

[Ar 84] W.D. Arnett and F.K. Thielemann: Stellar Nucleosynthesis", eds. Chiosi and Renzini (Reidel Publ., Holland) (1984) 145

[Ba 76] F.C. Barker (1976): Private communication

[Be 78] F.D. Becchetti, E.R. Flynn, D.L. Hanson and J.W. Sunier: Nucl. Phys. A305 (1978) 293

[De 84]   P. Descouvemont, D. Baye and P.-H. Heenen
          Nucl. Phys. A 430 (1984) 426

[Dy 74]   P. Dyer and C.A. Barnes: Nucl. Phys. A 233 (1974)
          495

[FCZ 75] W.A. Fowler, G.R. Gaughlan and B.A. Zimmerman:
          An. Rev. Astron. Astrophys. 13 (1975) 69

[FU 84]   C. Funck, K. Langanke and A. Weiguny:
          Phys. Lett. 152 (1985) 11

[Hu 76]   J. Humblet, P. Dyer and B.A. Zimmerman:
          Nucl. Phys. A 271 (1976) 210

[Ja 70]   R.S. Jaszcak, J.H. Gibbons and R.L. Macklin:
          Phys. Rev. C2 (1970) 63 and 2452

[Ke 82]   K.W. Kettner, H.W. Becker, L. Buchmann, J. Görres,
          H. Kräwinkel, C. Rolfs, P. Schmalbrock, H.P. Trautvetter and A. Vlieks: Z. Phys. A 308 (1982) 73

[Ko 74]   S.E. Koonin, T.A. Tombrello and G. Fox:
          Nucl. Phys. A 220 (1974) 221

[La 84]   K. Langanke and S.E. Koonin: in print

[Re 85]   A. Redder: Ph. D. Thesis, to appear in Inst. f.
          Kernphysik Münster (1985)

[Ro 78]   C. Rolfs and H.P. Trautvetter:
          Ann. Rev. Nucl. Part. Sci. 28 (1978) 115

[Tr 84]   H.P. Trautvetter, J. Görres, K.U. Kettner and C.
          Rolfs: "Stellar Nucleosynthesis", eds. Chiosi and
          Rinzini (Reidel Publ. Holland) (1984) 79

[We 74]   D.C. Weisser, J.F. Morgan and D.R. Thompson:
          Nucl. Phys. A 235 (1974) 460

[Wo 86]   S.E. Woosley: these proceedings (1986)

RADIOACTIVE ION BEAMS:
RESEARCH MOTIVATION AND METHODS OF PRODUCTION

Richard N. Boyd
Department of Physics, Department of Astronomy
The Ohio State University, Columbus, OH 43210

Abstract

Radioactive ion beams can provide research opportunities not available with ordinary ion beams. In particular, radioactive beams allow investigation of nuclear reactions important to the stellar burning and nucleosynthesis which occur in high temperature and/or density environments in stars. General design considerations are presented which apply to any such facility. The facility presently being built at Ohio State University is described.

1. Introduction

The recent surge of interest[1,2,3] in radioacive ion beam (RIB) facilities has been motivated primarily by their use in examining reactions of interest to astrophysics. In particular, the conditions which occur in stellar burning in the periphery of an accreting neutron star[4], or in internal boundary regions of stars which exist at very high temperatures[5], produce nuclear burning which will be fast enough that it will involve nuclear reactions on short-lived radioactive nuclei. Characterization of such burning thus requires detailed knowledge of the proton and alpha-particle induced reactions which can occur on these short-lived nuclei.

Some of the nuclear reactions encountered in such situations have been described in detail by Wallace and Woosley[5], in their investigations

of the r-p process, and are indicated in Fig. 1. The normal CNO cycle
burning requires a beta-decay from ^{13}N to ^{13}C, a decay which always has
time to occur in normal stellar conditions since ^{13}N has a half-life of 10
minutes. However, at temperatures around T9=0.6, this decay almost never
has time to occur before proton radiative capture to ^{14}O occurs. What
happens next depends on the details of the ^{14}O$(\alpha,p)^{17}$F reaction, but that
reaction probably occurs frequently at temperatures like T9=0.6 in competition with the beta-decay of ^{14}O. Indeed, as can be seen from Fig. 1,
most of the reactions needed to characterize the stellar burning under
these conditions, i.e., all those with arrows originating at nuclei above
and to the left of the double line, require knowledge of nuclear reactions
on radioactive nuclei. While some of these reaction strengths can be
inferred[6] from existing nuclear data, most cannot.

Another situation which occurs occasionally in nucleosynthesis considerations involves isomeric nuclei. The details of the nucleosynthetic
processes associated with creation and annihilation of ^{26}Al, a nucleus
thought to be responsible for one of the most interesting isotopic anomalies observed[7], have been discussed previously[8]. There are several radioactive nuclei involved in the relevant nuclear reactions, e.g., ^{25}Al,
^{26}Al, and ^{27}Si, but an isomeric state of ^{26}Al, ^{26}Al^m is involved, as well.
At high, but plausible, stellar temperatures, these states would be in
thermal equilibrium with their photon bath. Thus information about nuclear reactions initiated on a nucleus in an excited state is required for
characterization of the processes involved. This situation not unique, as
a similar one exists around mass 34 amu as well[9].

Since most of the nuclei involved in fast stellar burning have half-lives of seconds or minutes, cross sections involving them must be measured by creating beams of them, and observing the reactions resulting from
bombardment of hydrogen and helium targets with those beams.

2. General Design Considerations

While the concept of producing beams of radioactive ions for research
purposes is not new, the idea of using those secondary beams to produce
observable nuclear reactions is. It is this feature of the Ohio State

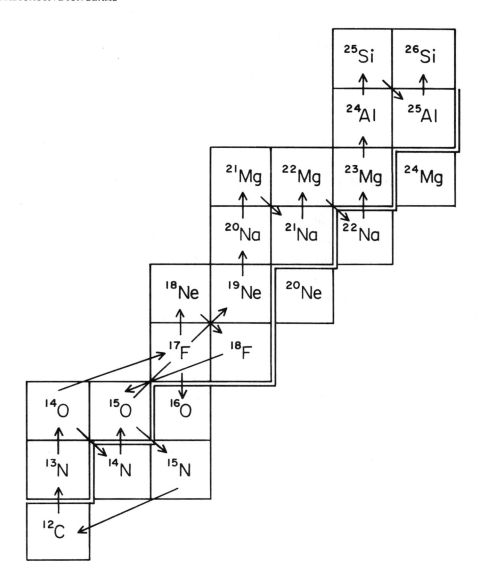

Figure 1. Some of the reactions and beta decays which occur in the r-p process. All the nuclei above and to the left of the double line are radioactive.

University (OSU) RIB facility which distinguishes it from the isotope separators which have existed for many years.

The basic components of the OSU RIB facility are indicated in Fig. 2. An intense beam of stable ions is first produced from an accelerator, with energy slightly in excess of 1 MeV per amu. That stable beam is focussed on to a target of light ions, e.g., ^{3}He, at which many of the stable ions are converted to the radioactive ions of interest. All the ions emerging from the first target are then passed through a filter, which selects and focusses the radioactive ions of interest, and eliminates all others. The resulting secondary beam may then pass through an energy measuring and tagging system, and then into a second target. Detectors in, or around, the second target then observe the products from the reaction of interest.

It should be noted that another RIB design exists[10] in which the incident beam is brought to bear on a target located in an ion source. The radioactive ions produced by reactions in that target are then extracted, analyzed, and accelerated, thus producing the RIB of interest.

3. Description of the OSU RIB Facility

The features of a RIB facility are most easily described by selecting a particular reaction. I will use the $^{15}O(\alpha,\gamma)^{19}$Ne as an example. The reaction which would be used to produce the ^{15}O beam, the ^{3}He(^{16}O,^{15}O)^{4}He reaction, yields reaction products, at appropriate ^{16}O incident energies back to about 24°. Of course, elastically scattered and unscattered ^{16}O ions exist in intensities many orders of magnitude greater than those of the reaction products. Fortunately, the elastically scattered beam extends back in typical cases only about half as far in angle as do the radioactive reaction products. Thus, in the example reaction, blocking of the forward angle region of the entrance to the filter removes virtually all the ^{16}O ions, and allows only the ^{15}O ions to be transmitted.

The target at which the radioactive ions are produced, Target I, necessarily operates at a fairly high pressure, about half atmospheric pressure, in order to produce as high a yield of radioactive ions as possible. Because any entrance window into this gas cell presents insurmountable difficulties, this cell has been designed[6] to be differentially

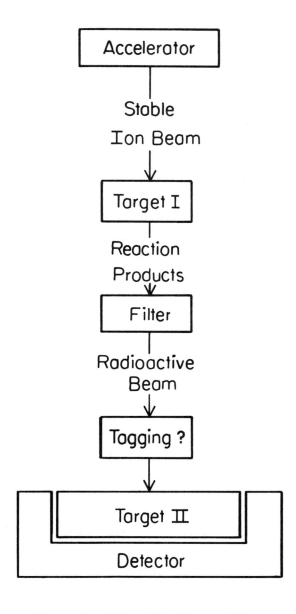

Figure 2. Schematic diagram of a radioactive beam facility which could be used to investigate reactions on short-lived nuclei.

pumped over several stages. A coupled system of Roots blower pumps handles the gas load on the high pressure regions, and a gas purification system allows recycling of the ^{3}He gas back into the target cell.

The OSU RIB facility presently uses a dipole magnet as the filter. While this device does produce the separation of the desired reaction products from the incident beam, it also has an extremely small acceptance solid angle, thus producing an intolerably small count rate for most reactions. Thus we are presently considering different filter configurations, e.g., a quadrupole system or a solenoid system. Either would allow an increase in the acceptance solid angle of the filter of about two orders of magnitude from that of the dipole magnet.

The large entrance solid angle to the filter, however, results in a large energy spread of the radioactive ions emerging from it, about 30% of the maximum energy in typical cases. Thus, if precise measurements of excitation functions are to be obtained, the energy of each ion which enters the second gas cell must be measured. This is accomplished by a time-of-flight system, located after the filter, consisting of two large (4 cm x 8 cm) thin foils separated by 30cm. As the radioactive ions pass through each foil, they knock out electrons which are then accelerated and focussed, by a curved electrostatic mirror, into a channel plate detector. The timing signal between the detectors at each of the two foils then determines the time-of-flight, hence energy, of the ions passing through the system. This system has been tested with an alpha-particle source, and has been found to give timing resolution of about 450 psec.

Following the time-of-flight system is Target II which, for the $^{15}O(\alpha,\gamma)^{19}Ne$ reaction study, contains ^{4}He gas. Surrounding this cell will be large NaI detectors which will detect the gamma-rays produced by the reaction being studied. Since the ^{15}O ions will lose energy as they pass through the gas in this cell, their energies can be made to span a significant fraction of the excitation function for the reaction of interest. A time measurement between the second foil in the time-of-flight system and the fast signals from the gamma-ray detectors, together with knowledge of each ion's energy as it enters Target II, determines its location in the cell where the reaction occured. Then an energy loss algorithm determines the energy at which the radiative capture event occurred. Slow signals from the gamma-ray detectors are also analyzed so as to provide information about gamma-ray decay schemes of the nuclei produced and to allow discrimination between events of interest and spurious background events. In this way approximately one-third of the excitation function for any reaction can be observed at a single accelerator setting; energy straggling limits the observable range to that fraction.

4. Conclusions

It is anticipated that the facility described above will, at least in favorable cases, produce radioactive beam intensities of at least 10^6 ions per second at energies slightly in excess of 1 MeV per amu. These beams would be expected to yield several hundred events per day for the radiative capture reactions of interest. It should be noted that the energies of interest to the astrophysical situations described in Section 1 would be around 0.5 MeV for proton radiative capture reactions, and around 2 MeV for alpha-particle radiative capture. Because the experiments to be performed with the OSU RIB facility have the heavier ion as the incident particle, the center-of-mass energies which will exist in these reactions will be close to the energies of interest. Thus this facility is ideally suited to study of reactions of importance to astrophysics.

This facility and other facilities similar to it should provide information to nuclear astrophysicists which was heretofore unobtainable. This information should be crucial to the understanding of the processes of nucleosynthesis which have provided a significant fraction of the elements in the universe.

Acknowledgements

It is a pleasure to acknowledge the numerous people who have contributed to the design and development of the OSU RIB facility. In this context, L. Rybarcyk deserves special mention. Others who have contributed significantly are M. Wiescher, C. Rolfs, W. Kim, H.J. Hausman, P. Schmalbrock, P. Corn and M.J.A. deVoigt. This work was supported in part by NSF Grant 82-03699.

REFERENCES

1. R.N. Boyd, Proc. of the Workshop on Rad. Ion Beams and Small Cross Section Measurements, ed. by R.N. Boyd, 1981, p. 30.
2. R.N. Boyd, L. Rybarcyk, M. Wiescher and H.J. Hausman, IEEE Trans. on Nucl. Sci. NS-30, 1387 (1983).
3. R. Haight, G.J. Mathews, R.M. White, L.A. Aviles and S.E. Woodard, IEEE Trans. on Nucl. Sci. NS-30, 1160 (1983).
4. C.J. Hansen and H.J. van Horn, Astrophys. J. 195, 735 (1975); R.E. Tamm and R.E. Picklum, Astrophys. J. 224, 210 (1978).
5. R.K. Wallace and S.E. Woosley, Astrophys. J. Suppl. 45, 389 (1981).
6. G.J. Mathews and F.S. Dietrich, Astrophys. J. 287, 969 (1984).
7. G.J. Wasserburg and D.A. Papanastassiou, Essays in Nuclear Astrophysics, ed. by C.A. Barnes, D.D. Clayton and D.N. Schramm, Cambridge Press, London, 1982, p. 77.
8. R.A. Ward and W.A. Fowler, Astr. Astrophys. 238, 266 (1980); L. Buchmann, M. Hilgemeier, A. Krauss, A. Redder, C. Rolfs, H.P. Trautvetter and T.R. Donoghue, Nucl. Phys. A415, 93 (1984).
9. R.N. Boyd, L. Rybarcyk, W. Kim, P. Schmalbrock and H.J. Hausman, to be published in Nucl. Instr. Methods, 1985.
10. J. D'Auria, L. Buchmann, M. Arnould, R. Azuma, C. Barnes, R. Boyd, C. Davids, J. King, C. Rolfs, T. Ward and M. Wiescher, TRIUMF Research Proposal, 1984, unpublished.

VI – NEUTRINOS AND MONOPOLES

THE GALLIUM SOLAR NEUTRINO EXPERIMENT GALLEX

Heidelberg-Karlsruhe-Milano-Munchen-Nice-Rehovot-Roma-Saclay Collaboration

D.Vignaud
DPhPE , CEN Saclay
91191 Gif/Yvette

Abstract

The low threshold (233 keV) of the neutrino capture reaction $^{71}Ga(\nu_e,e^-)^{71}Ge$ is a real chance for the detection of solar neutrinos , particularly those coming from the pp reaction. The GALLEX experiment will measure the solar neutrino flux by counting the ^{71}Ge atoms produced in a 30 ton gallium target in the form of a $GaCl_3$ solution. A calibration will be done with a ^{51}Cr source which emits 751 keV neutrinos. The detector will be set up in the Gran Sasso Underground Laboratory (Italy) . The expected result will allow a better understanding of solar models and to discriminate between solar models problems and possible neutrino oscillations .

1. Introduction.

The so-called solar neutrino puzzle consists in the discrepancy between experimentally observed solar neutrino flux and solar models theoretical predictions . The latest result of the only experiment which has ever been performed to detect solar neutrinos , the chlorine experiment [1] , is now 2.0 ± 0.3 SNU (1 SNU corresponds to 10^{-36} capture / atom / second) while the theoretical expectations of the standard solar model give [2] 5.8 ± 2.2 SNU (3σ limit) . However the main criticism to the chlorine experiment is that the threshold for ν_e capture by ^{37}Cl in the reaction $^{37}Cl (\nu_e , e^-) ^{37}Ar$ is 814 keV which corresponds only to the tail of the solar ν_e spectrum . Two major reasons are generally invoked to explain the chlorine experiment results : either there are some problems with the solar models or there are neutrino oscillations .

In order to try to solve this puzzle a new radiochemical experiment has been proposed some time ago, involving the gallium as a target [3,4]. The ^{71}Ga is sensitive to solar neutrinos by the capture reaction

$$\nu_e + {}^{71}Ga \rightarrow {}^{71}Ge + e^- \qquad (1)$$

The threshold (233 keV) is very convenient to detect a large part of the solar ν_e spectrum, particularly the ν_e coming from the pp fusion reaction. The newly proposed experiment (GALLEX) needs 30 tons of gallium in the form of $GaCl_3$. In the frame of the standard solar model 1 atom of ^{71}Ge should be produced each day. The produced ^{71}Ge atoms have to be extracted and counted each two weeks ($T_{1/2}$ = 11.43 days). Such techniques have already been developed and the experiment has been proved to be feasible. Due to the requirement of a low level background site it is intended to set up the detector in the Gran Sasso Underground Laboratory (120 km from Rome in the Apennine mountains). Moreover it is planned to calibrate our detector with a 800000 Ci ^{51}Cr source which provide monochromatic ν_e (751 keV).

In this paper we first recall briefly the standard solar model predictions and the neutrino oscillation problem. We then describe our proposed gallium experiment : extraction, counting, background, calibration with a ^{51}Cr source, schedule, with some details. Finally we give what would be the implications of the results on solar physics and/or neutrino oscillation physics. More details on this experiment can be found in reference [5].

2. The solar neutrino flux in the standard solar model (SSM).

The standard solar model is described elsewhere [6]. We just recall here the origin of the neutrinos coming from the sun, the predictions for the corresponding flux and for the neutrino capture cross sections.

Most of the energy produced in the sun comes from the pp fusion reaction chain :

$$p + p \rightarrow {}^2H + e^+ + \nu_e \quad (2 \text{ times}) \qquad \nu_{pp}$$
$$^2H + p \rightarrow {}^3He + \gamma \quad (2 \text{ times})$$
$$^3He + {}^3He \rightarrow {}^4He + p + p$$
$$\text{or} \quad {}^3He + {}^4He \rightarrow {}^7Be + \gamma$$

Electron or proton capture on the ^{7}Be may also give :

$$e^- + {}^7Be \rightarrow {}^7Li + \nu_e \qquad \nu_{Be}$$
$$p + {}^7Be \rightarrow {}^8B + \gamma$$
$$\hookrightarrow {}^8Be^* + e^+ + \nu_e \qquad \nu_B$$

A small fraction comes from the pep reaction :

$$p + e^- + p \rightarrow {}^2H + \nu_e \qquad \nu_{pep}$$

There is also a small contribution (1.5 %) of the CNO cycle in the total fusion energy liberated.

The predictions of the standard solar model concerning the neutrino flux at the earth level are given in Fig. 1a where the six main contributions are represented. The integrated flux value is given for each neutrino source in Table 1. In Table 1 are also shown the predictions for the neutrino capture cross sections in ^{71}Ga, in SNU's (see for example ref. [7] for neutrino capture cross sections). The values are the ones recently published by Bahcall et al. [2] and take into account recent calculations of ^{71}Ge halflife (11.43 days) and Q_{EC} value (233.2 keV) [8]. These predictions do not take into account the contributions for ^{71}Ge* excited states which have been estimated to be of the order of 10 % of the contribution for the ^{71}Ge ground state [9]. (The decay scheme of ^{71}Ge is represented in Fig. 1b where energies of ν_{pp}, ν_{Be} and ν from ^{51}Cr are also shown).

Table 1: Solar neutrinos flux (in $\nu/cm^2/s$) and gallium capture rates (in SNU) for the different sources in the sun.

Source	Flux [6]	Energy spectrum (MeV)	Capture rate
pp	6.07×10^{10}	0.-0.420	67.1
pep	1.50×10^{8}	1.440	2.4
^{7}Be	4.3×10^{9}	0.862 (90%), 0.383 (10%)	30.3
^{8}B	5.6×10^{6}	1.-14.	1.5
^{13}N	5.0×10^{8}	0.-1.20	2.6
^{15}O	4.0×10^{8}	0.-1.73	3.7
Total			107.6

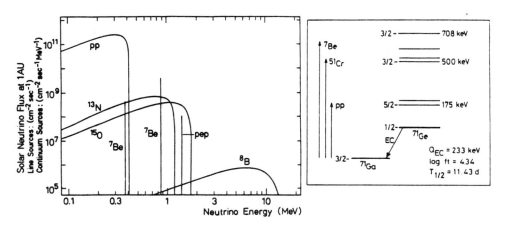

Figure 1: a) Solar neutrino spectrum in the SSM - b) ^{71}Ge decay scheme.

3. Neutrino oscillations.

The hypothesis of neutrino oscillation between their different flavours (ν_e, ν_μ, ν_τ) is now well known (see for example ref. [10]). It is based on the idea that neutrinos, eigenstates of the weak interaction, are not eigenstates of the mass but linear combinations of ν_1, ν_2 and ν_3 which are the eigenstates of the mass :

$$|\nu_l\rangle = \Sigma\, U_{li}\, |\nu_i\rangle \qquad l=e,\mu,\tau \quad i=1,2,3$$

The fraction of flavour m neutrinos observed at a distance x from the flavour l neutrinos source can be written :

$$P(\nu_l \to \nu_m) = \Sigma\, U_{li}^2 \cdot U_{mi}^2 + \Sigma\, U_{li} \cdot U_{mi} \cdot U_{lj} \cdot U_{mj}\, \cos(2\pi x/l_{ij})$$

where $l_{ij}(m) = 4\pi p_\nu / |m_i^2 - m_j^2| = 2.48\, p(\text{MeV}/c) / |\Delta m^2|\, (\text{eV}^2)$
is the oscillation length.

The distance between the sun and the earth is $L = 1.5\, 10^{11}$ m and we may hope to be sensitive to Δm^2 until about 10^{-11} eV^2 which is much more than all known results on limits to neutrino oscillations (for recent results see [11]).

If we consider a mixture of two flavours only the U matrix is a rotation matrix (angle θ) and we have :

$$P(\nu_e \to \nu_e) = 1 - \sin^2 2\theta \cdot \sin^2(1.27\, \Delta m^2\, L / p_\nu) \qquad (2)$$

4. The gallium experiment.

a) history.

The first idea for using gallium for the detection of solar neutrinos was suggested by Kuzmin [3]. The first proposal for such an experiment is due to Bahcall et al. [4] in 1978 who requested 50 tons of gallium. A pilot experiment with 1.3 ton of gallium was done at Brookhaven to show the feasability of the extraction of small quantities of ^{71}Ge. A proposal by a Brookhaven-Heidelberg-Rehovot-Philadelphia-Princeton Collaboration was written in 1981 [12]. This collaboration was dismantled in 1983 due to lack of money. A new collaboration was built in 1984, with mainly European Laboratories [13].

b) design.

The 30 tons of gallium (39.6 % of ^{71}Ga and 60.4 % of ^{69}Ga) in form of a concentrated GaCl$_3$-HCl solution are placed in a large tank. The ^{71}Ge produced in the neutrino capture reaction (1) form the volatile GeCl$_4$ compound which is swept out of the solution by circulating air through the tank. The gas flow is passed through gas scrubbers where GeCl$_4$ is absorbed in water. The GeCl$_4$ is then extracted into CCl$_4$, back extracted into tritium-free water and finally reduced to the germane GeH_4 gas by using KBH_4. The germane is introduced with xenon in a small proportionnal counter where the number of ^{71}Ge atoms is determined by observing their radioactive decay.

c) extraction of the ^{71}Ge.

The feasability of the extraction has been proved in a pilot experiment done at Brookhaven with 1.3 ton of gallium (see fig. 2a). More than 30 runs have been performed by introducing small amounts (a few mg down to 100 μg) of stable Ge carrier, as well as runs in which the ^{71}Ge was produced in the $GaCl_3$ by cosmic rays or by a neutron source, or by decay of ^{71}As.

An example of germanium extraction from the pilot tank is shown in fig. 2b where more than 99 % of the 2 mg added to the $GaCl_3$ solution before sweeping have been extracted in 1.2 day. The conclusion is that the entire chemical process (extraction and conversion into GeH_4) can be carried out with more than 95 % overall yield.

In the final experiment it is planned to extract the produced ^{71}Ge every two weeks, taking one day for extraction.

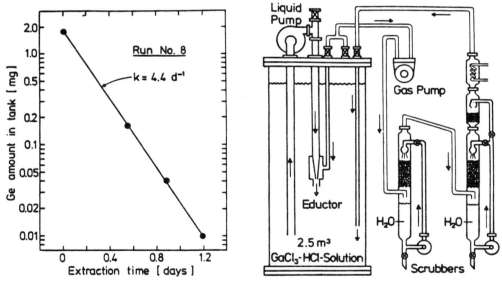

Figure 2: Extraction system in the pilot experiment and result of an extraction run.

d) counting of the ^{71}Ge.

In the frame of the standard solar model and with the expected capture rate about 1 ^{71}Ge atom is produced each day in the tank (1 SNU corresponds to $8.7\ 10^{-3}$ capture/day in the 30 t detector). A counting system able to measure such low decay rates has been developed by the Heidelberg group. ^{71}Ge decays by electronic capture and the energy deposition from Auger electrons and X rays emitted results mainly in a spectrum with an L peak at 1.2 keV and a K peak at 10.4 keV. The miniaturized proportional counters are built with ultrapure materials. They are placed in an anticoincidence shield with NaI and plastic scintillation detectors, completed by heavy passive shielding with lead and iron. The counter background is then of the order of 1 count per day. But the pulse shape of this background (materials radioactivity, external γ rays, electronic noise) is different of that of the ^{71}Ge decay [5] and a pulse shape analysis using a transient digitizer allows to reduce the background level. Fig. 3 displays the results of a 51.5 day counter background measurement obtained with one of the best counters. G*I, a quantity characterizing the pulse shape, is plotted versus the energy for each count. The boxes represent the regions where 95 % of all ^{71}Ge events with energies in the L and K peak windows are located. The background obtained is .08 (.04) count per day in the L (K) peak and the corresponding counting efficiencies are respectively 28 and 43 % for the L and K peaks.

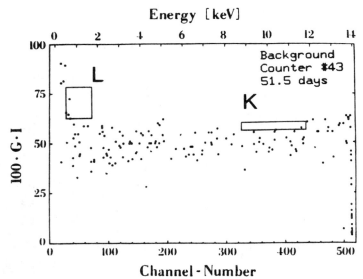

Figure 3 : Proportional counter background measurement.

e) background due to ^{71}Ge produced by other sources than solar neutrinos.

^{71}Ge atoms may be produced in the 30 t detector by other sources than solar neutrinos. This background consists mainly in the reaction ^{71}Ga(p,n)^{71}Ge. The interacting protons come from :

i) cosmic ray muons interactions. In the Gran Sasso Laboratory (shielding depth $\simeq$ 3500 m water equivalent) we expect .01 atom per day.

ii) (α,p) reactions where α come from radioactive decays of U, Th and Ra present in the $GaCl_3$ solution. Low level of these radioactive atoms can be obtained for large $GaCl_3$ quantities and give less than .01 ^{71}Ge atom per day.

iii) (n,p) reactions induced by neutrons from the surrounding rock. 0.8 n/cm^2/day would give .006 ^{71}Ge atom per day. Preliminary measurements [14] give an upper limit of 2.6 n/cm^2/day. It can be measured more precisely by using the reaction $^{40}Ca(n,\alpha)^{37}Ar$ where ^{37}Ar atoms can be counted. If the level was too high an additional 25 cm water shielding around the detector would reduce this background by an order of magnitude.

The total background is then less than a few percent and can be monitored by the reaction $^{69}Ga(p,n)^{69}Ge$ which has a large cross section and which can be counted in the same way as the ^{71}Ge.

f) calibration with a ^{51}Cr source.

There are still uncertainties about the cross section contribution from the two excited states in ^{71}Ge at 175 and 500 keV (which can be populated by allowed Gamow-Teller transitions) whose contribution is about 10 % of the ground state one [9] (see Fig. 1b). In order to complete information about the ^{71}Ga neutrino capture cross section and to have an overall consistency check of the detector a calibration experiment using an artificial ^{51}Cr source is planned as a full part of the experiment. ^{51}Cr decays ($T_{1/2}$ = 27.7 days) by the electronic capture reaction $^{51}Cr(e^-,\nu_e)^{51}V$. It emits monoenergetic neutrinos : 90 % at 751 keV and 10 % at 431 keV. The 751 keV ν_e can populate the two excited states at 175 and 500 keV (see Fig. 1b). An excess of the measured ^{51}Cr signal above the ground state contribution could be attributed to these excited states.

The characteristics of the source are the following : 100 kg of chromium powder are irradiated in Siloé (Grenoble) and Osiris (Saclay) during 2 months, giving an activity of 800000 Ci, activity necessary to perform a significant experiment. The source is then placed inside the $GaCl_3$ tank during 2 months, with a ^{71}Ge extraction each 2 weeks. The expected sensitivity is 18-20 % after one such run and 10 % after 4 runs (1 year).

g) schedule.

In 1985 the proposals are being submitted to respective German, French and Italian authorities and to Gran Sasso committee. The approvals are expected before the end of the year and the installation could begin immediately. The completion for gallium acquisition could then be at the end of 1986. 2 years of running (50 extractions) are necessary to obtain 10 % statistical error on the number of ^{71}Ge produced atoms. These 2 years and 1 year of calibration could be planned in 1987-1988-1989 and a first result could be given at the end of 1989.

5. Interpretation of the forthcoming result.

The interpretation of the forthcoming result depends on the result itself :
- between 100 and 130 SNU the standard solar model is probably correct and we can exclude neutrino oscillations with a large mixing.
- below 70 SNU the neutrino oscillation hypothesis is the most probable since no actual solar model predicts a value lower than 70 SNU which corresponds to "extreme models" i.e. models with energy generation in sun by fundamental pp reactions only.
- between 70 and 100 SNU the chlorine result and the gallium result should be combined to give a reliable explanation.
- above 130 SNU the field is open for models such the model with quarked nuclei which can predict until 250 SNU [15].

To illustrate this we display in fig. 4 the ^{37}Ar production rate in the chlorine experiment versus the ^{71}Ge production rate in the gallium experiment. The two horizontal lines correspond to the chlorine result with 2 σ [1]. The circle corresponds to the standard solar model (SSM). The dashed line corresponds to different solar models and particularly the turbulent diffusion mixing model which predicts a very small flux of ν_B [16]. The cross corresponds to the minimum rate (neutrino production by pp reaction only). The full line gives the prediction of the SSM in case of neutrino oscillation (with $\Delta m^2 > 10^{-8}$ eV^2) with a mixing varying along the line. The two marked points along the line correspond respectively to the two neutrino mixing and three neutrino mixing with the maximum value of the mixing parameter. If Δm^2 is below 10^{-8} eV^2 the situation is a little more complicated since the average value of the $\sin^2(1.27 \Delta m^2 L / p_\nu)$ term in equation (2) depends on Δm^2 but definite conclusions can still be drawn (see [17]).

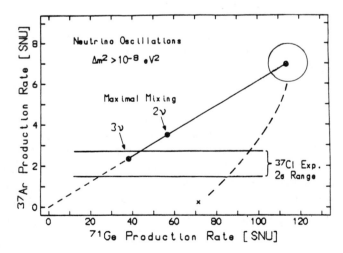

Figure 4 : ^{37}Ar observed production rate versus ^{71}Ge production rate (see text).

References

[1] J.K.Rowley, B.T.Cleveland and R.Davis, in "Solar neutrinos and neutrino astronomy", Homestake (1984), AIP Conf. Proc. n° 126, p.1

[2] J.N.Bahcall, B.T.Cleveland, R.Davis and J.K.Rowley Astrophysical J. Lett. (1985)

[3] V.A.Kuzmin, Sov. Phys. JETP 22 (1966) 1051

[4] J.N.Bahcall et al., Phys. Rev. Lett. 40 (1968) 1351

[5] T.Kirsten, "The gallium solar neutrino experiment", in Proc. of the Resonance Ionization Spectroscopy meeting, Knoxville, Tennessee, April 1984
W.Hampel, "The gallium solar neutrino detector", in "Solar neutrinos and neutrino astronomy", Homestake (1984), AIP Conf. Proc. n° 126, p.162

[6] J.N.Bahcall et al., Rev. Mod. Phys. 54 (1982) 767

[7] J.N.Bahcall, Rev. Mod. Phys. 50 (1978) 881

[8] W.Hampel and R.Schlotz, "The ^{71}Ge Q_{EC} value and the neutrino capture cross section for the gallium solar neutrino detector", Paper presented at the 7th Int. Conf. on Atomic Masses and Fundamental Constants, Sept. 1984, Darmstadt-Seeheim

[9] G.J.Mathews, S.D.Bloom, G.M.Fuller and J.N.Bahcall "A shell model study of the ^{71}Ga$(\nu,e^-)^{71}$Ge solar neutrino detector", February 1985, submitted to Phys. Rev. C

[10] S.M.Bilenky and B.Pontecorvo, Phys. Rep. 41 (1978) 225

[11] V.A.Lubimov, in Proc. of the XXII Int. Conf. on High Energy Physics, Leipzig, July 1984, vol. 2, p.108

[12] W.Hampel, "The gallium solar neutrino experiment", in Proc. of the Int. Conf. on Neutrino Physics and Astrophysics, Hawai, July 1981, R.J.Cence et al. editors, p.6

[13] Members of the collaboration : T.Kirsten (spokesman), W.Hampel, G.Eymann, G.Heusser, J.Kiko, E.Pernicka, B.Povh, M.Schneller, K.Schneider, H.Volk (**Heidelberg**) - K.Ebert, E.Henrich, R.Schlotz (**Karlsruhe**) - R.L.Mossbauer (**Munchen**) - I.Dostrovsky (**Rehovot**) - M.Cribier, G.Dupont, B.Pichard, J.Rich, M.Spiro, D.Vignaud (**Saclay**) - G.Berthomieu, E.Schatzman (**Nice**) - E.Fiorini, E.Bellotti, O.Cremonesci, C.Liguori, S.Ragazzi, L.Zanotti (**Milano**) - L.Paoluzi, S.D'Angelo, R.Bernabei, R.Santonico (**Roma**) -

[14] E.Fiorini, C.Liguori and A.Rindi, "Preliminary measurements of the γ and neutron background in the Gran Sasso tunnel", Frascati report LNF85/7 (15.3.85)

[15] B.Sur and R.N.Boyd, Phys. Rev. Lett. 54 (1985) 485

[16] E.Schatzman, in "Solar neutrinos and neutrino astronomy", Homestake (1984), AIP Conf. Proc. n° 126, p.69

[17] W.Hampel, "The gallium solar neutrino detector and neutrino oscillations", in Proc. of the Int. Conf. on Neutrino Physics and Astrophysics, Nordkirchen, June 1984, K.Kleinknecht and E.A.Paschos editors, p.530

REAL-TIME DETECTION OF LOW ENERGY SOLAR NEUTRINOS WITH INDIUM AND SUPERHEATED SUPERCONDUCTIVITY

Georges WAYSAND
Groupe de Physique des Solides de l'Ecole Normale Supérieure
Tour 23 - Université Paris VII
75251 Paris Cedex 05, France

ABSTRACT

The principle of detection in real-time of low energy solar neutrinos with indium is recalled. Because four tons of indium are necessary to get one event per day, an elegant way to realize this experiment is to use indium as the target and the detector of solar neutrinos. Superheated superconductivity provides such an opportunity at the expense of the division of the metal in microspheres embedded in a dielectric. The properties of this detecting material are discussed in relation with the specific goal of solar neutrinos detection.

This talk is dedicated to Y. Orlov, presently in exile in Siberia.

1. INTRODUCTION

The importance of neutrinos for early nucleosynthesis is underlined by many contributions in this volume. So far however, not even a single neutrino coming from the p-p reaction in a star have been detected. Neutrino astronomy is much more a conceptual frame than an experimental field. Only a single device has been built : the chloride experiment of R. Davis looking at neutrinos coming from the sun. The Davis device is based on the radio chemical reaction in which Cl^{37} (which represents 25 % of the natural chloride) captures one neutrino and is therefore transformed in Ar^{37}. The isotope 37 of Argon is detected by the release of one electron when it desexcites. The energy threshold of this reactions is 814 keV. What is then detectable ?

The solar neutrino flux consists of three components :
1. pp neutrinos with energy in the range $0 < E_\nu < 420$ keV and a flux of 6×10^{10} ν_e/cm^2 sec.
2. Be^7 neutrinos which are monochromatic with $E_\nu = 861$ keV and have a flux of 4×10^9 ν_e/cm^2 sec.
3. finally : B^8 neutrinos with energy E_ν between 0 and 14 MeV and a flux of 3×10^6 ν_e/cm^2 sec.

Therefore, the chloride device looks essentially to this component which is 4 order of magnitude smaller than the pp one. This pioneer experiment has detected a flux of 2 ± 0.3 SNU whereas Bahcall has predicted a value of 6 ± 2 SNU.

As it has often been discussed, this discrepancy can be assigned either to neutrinos oscillations or to a deficient solar model. From this puzzle comes the need for solar neutrinos detectors with lower energy threshold. Since the flux of pp neutrinos is proportional to the luminosity of the sun, it is subject to no more than a 10% uncertainty, therefore if the low energy solar neutrino flux is not in agreement with the luminosity evaluation, this will prove the existence of neutrinos oscillations. The gallium experiment discussed in these proceedings is also a radio chemical experiment with a threshold of 236 keV and can provide an integrated value of the flux above value. Therefore, it can provide a partial answer to the problem. The indium reaction is a real-time detection one, its energy threshold is 128 keV, and can detect an expected flux for pp neutrinos of 571 SNU to be compared with 71 for the gallium. However, starting from the constraints defined by the indium reaction it is not easy to design a realistic experiment. Nevertheless, after one and a half year of reflexion and discussion an ad-hoc french committee has decided to recommand the launching of a feasibility study of a solar neutrino detector based on the indium reaction and using the rupture of a superheated superconducting state in indium granules to detect the ionising events associated with the naturalization of the neutrino. The principle of such an experiment is discussed below.

2. THE NEUTRINO-INDIUM^{115} INTERACTION : THE INVERSE β DECAY

All over the range of atomic masses from 111 to 131, R.S. Raghavan[4] has noticed that one observes a β decay from the $9/2^+$ state of the indium isotopes to the $7/2^+$ state of indium. At the lower atomic masses tin is transmuted into indium, whereas at the high atomic masses the reverse process occurs. At the atomic mass 115, the two levels are separated only by 128 keV ; the $9/2^+$ state is the state of ^{115}In whereas the $7/2^+$ state just above is an excited state of ^{115}Sn with a lifetime of 3.26 μsec. It is the shortness of this lifetime which makes this reaction completely radioelectrical and not radiochemical as in all the other situations :

$$^{115}_{49}In_{66} + \nu_o \longrightarrow {}^{115}Sn\,7/2^+ + e_\nu \quad (\sim 200 \text{ keV from pp})$$

then : $\downarrow$ 3.26 μsec

$$^{115}Sn\,3/2^+ + e_1 \quad (90 \text{ keV})$$

$$\downarrow$$

$$^{115}_{50}Sn_{65}\,1/2^+ + \gamma_2 \quad (498 \text{ keV})$$

In this process, the energy E_ν of the incoming solar neutrinos is known if one measures the energy e_ν of the capture electron :

$e_\nu = E_\nu - 128$ keV.

The e_ν pulse itself is identified by the triple delayed coïncidence

$$e_\nu \xrightarrow{3.26 \,\mu s} e_1 + \gamma_2$$

associated with a spatial localisation of e_ν and e_1.

It is the association of the triple delayed coïncidence with the localization of e_ν and e_1 which constitutes the signature of the neutrino. To get one event per day four tons of indium are necessary which is considerably less than with other techniques.

The main drawback of the indium reaction is, since it is a realtime phenomenon, that the background noise must be carefully treated. This general consideration is of special importance in the present project because the main source of noise comes from ^{115} Indium itself. Though it is the most abondant isotope (95,7%) with a large lifetime : 5.1×10^{14} years, this means that the spontaneous β decay of indium into tin creates 225 electrons β (with a maximum energy of 490 keV) per kilogram of indium and per second. Some of these β can produce γ ray by bremstrahlung. In such a case, there is a chance to receive a fake event. A fake event can also be produced by a β, from indium, in coïncidence with a γ ray coming from the environment. This second source of noise is reduced if the effective concentration in indium of the experiment is large. The effective signal to noise ratio relies on the actual structure of the detecting system.

The first type of solution that comes to mind is to seek for various combinations of conventional techniques such as multiwire proportional chambers, light collecting devices, etc... In such devices if one wants to be efficient, the indium mass be divided, at least in thin foils, to allow the electrons to escape from the metal. Studied by a Bell-MIT collaboration and again recently surveyed by a Saclay team, this type of solutions has been left over.

Non conventional techniques involving proved principles have been proposed. In these techniques, the leading idea is to use indium not only as the target but also as the detecting element itself. This requires that indium must be either :
i) a semiconductor (but the production of InP by the ton is not predictible and even if it is produced, the packaging problem will make the device poorly efficient).
ii) a part of an organic liquid scintillator with long mean free path : such a compound does not exist at the present time.

iii) a superconductor : this is the only bulk property of indium which can be used for detection.

In this later case, two solutions are proposed :
a) the realization of superconducting tunnel junction with one electrode made of bulk indium (a few cc. for each elementary cell). Since, the energy gap of the superconducting state with respect to the normal (resistive) state is of the order of 1 meV, such a device will have a very good resolution. This was proved with junctions made of thin film. Therefore, one can disregard any localization of the neutrino and rely only on the energy resolution to identify any event. This proposition of N. Booth which is also studied in our program has, for the moment, a major weakness : fabrication of a junction with bulk electrodes of indium has not yet been achieved. For a practical neutrino detector at least 10^4 devices have to be produced. This is a difficult task but the potentially very high energy resolution of such a device makes it attractive.
b) the other proposition for the neutrino detector of indium is to use superheated superconducting granules. The properties of these granules are discussed hereafter.

3. SUPERHEATED SUPERCONDUCTING GRANULES AS PARTICLE DETECTORS

3.1. Superheated superconductivity for a sphere

Bubble chambers, Wilson chambers are metastable systems where the particle is detected by rupture of metastability resulting in the creation of respectively bubble or droplets. Metastability occurs because the liquid-vapor phase transition is of the first order (there is a latent heat of transformation).

Some superconductors, among them indium, exhibits also a first order phase transition in magnetic field at temperatures $T<T_c$ (T_c : the critical temperature of apparition of the superconductivity in zero magnetic field) the transition for the superconducting state to the normal (resistive state) occurs abruptly at a field $H = H_c(T)$. Like in other cases of first order phase transition metastable states can be created : the existence of superheated (the superconducting state

persists above H_c when one sweeps up the magnetic field from O) and supercooled state (the sample remains resistive even when the magnetic field is decreased below H_c).

The superconducting state in type I superconductors is characterized by a complete expulsion of the magnetic field from the volume of the granule due to the flow of supercurrents in a very thin layer of thickness λ_L (called the Landau penetration depth) just under the surface. The supercurrent obeys the Maxwell-Landau equation characteristic of a superconductor :

$$\vec{J} = - \frac{1}{\mu_o \lambda_L^2} \vec{A}$$

and not the usual Maxwell relation $\vec{J} = \sigma \vec{E}$ for a normal metal. The Landau equation together with the Maxwell relation $\vec{\nabla} \wedge \vec{B} = \mu_o \vec{J}$ gives for the magnetic field.

$$\vec{\nabla}^2 \vec{B} = \frac{1}{\lambda_L^2} \vec{B}$$

which has no uniform solution in the superconductor except O : at a distance x from the surface

$$B(x) = B(o) e^{(-x/\lambda_L)}$$

Therefore, perturbations of the surface such as spikes, defects, irregularities, etc... over λ strongly affect the screening supercurrent and therefore weakens the probability of observation of the superheated state. Practically the theoretical limit of metastability H_{SH} is easily observed with microspheres under 40 microns. Suspensions of such spheres in wax constitutes the detecting divided material.

An hysteresis cycle of an isolated granule is represented in Figure 2 and the corresponding phase diagram in Fig. 1. Please note that for an isolated sphere, the maximum superheated magnetic field is 2/3 H_{SH} ; H_{SH} = superheated critical field. The demagnetization of the sphere is responsible of that effect. Our isolated superheated sphere of radius R in an external field B can be sketched by a dipolar magnetic moment $\vec{m}$ given by :

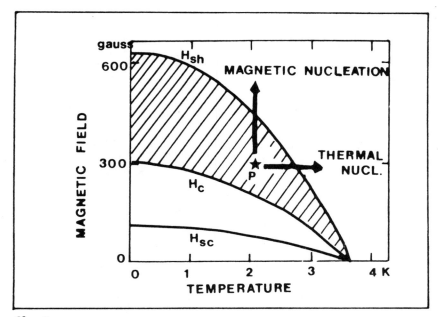

fig 1

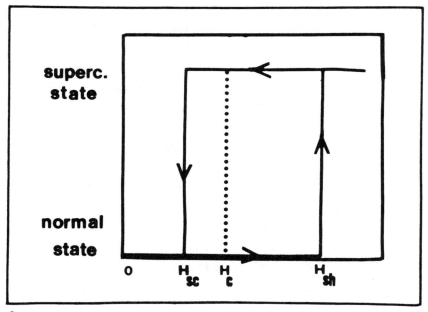

fig 2

$$\vec{m} = \alpha \vec{B}_o$$

α, the polarisability is given by :

$$\alpha = -\frac{1}{2} R^3$$

and the effective magnetic field $B(\vec{r})$ is :

$$B(r) = B_o + H_o \frac{R^3}{2} \nabla \left(\frac{\cos \theta}{r^2} \right) \quad ; \quad \text{with } \vec{r} = (r) e^{i\theta}$$
$$\theta = 0 \text{ for } \vec{r} \,//\, \vec{B}$$

The component of B_o parallel to the surface is :

$$B_o \bigg|_{r=R} = \frac{3}{2} B_o \sin \theta$$

at the equator $B_\theta = \frac{3}{2} B_o$

Therefore, the superheated critical H_{SH} at the equation is reached when $B_o = 2/3\ H_{SH}$.

For a suspension of spheres, the situation is much more complicated because of N-body diamagnetic interactions. The essential experimental fact is that the magnetic field inside the suspension varies from point to point. Therefore, each granule changes of state for a given value of the applied magnetic field. A crude approach of this situation has been examined in [15] and [18]. A much more correct treatment will be soon provided by U. Eigenmuller and P. Mazur [19]. If the granules have not a regular surface shape they will not reach the theoretical limit for the superheating. This also strongly contributes to the shape of the hysteresis cycle of a collection of granules.

3.2. Energy threshold

Let us restrict ourselves to a single microsphere. The change of state can happen following two main paths :
- at $T<T_c$ kept constant the magnetic field is increased until H_{SH} is reached at one point of the surface of the granule : this is a magnetic nucleation provided by sweeping up the magnetic field or by the arrival of a magnetic monopole

- at $H < H_{SH}(T)$ kept constant the granule is heated up to the temperature $T+\Delta T = \Delta H$ applied. This is how the granule changes of state under irradiation by ionizing particle. This is a thermal nucleation.

A microscopic description of the interaction between the ionizing particle and the granule is underway by N. Perrin but a simple calorimetric model is not irrelevant in the present situation. Effectively, during the transition to the normal state the fraction of the granule which is in this final state can be considered as isothermal : since, we want to heat the granule only up to the moment where at least one point of the surface has reached $H_{SH}(T+\Delta T)$, it suffices to heat one half of the volume.

If $\int_0^l \frac{\partial E}{\partial r} dr$ represents the total energy deposited by the particle on the length l of its trajectory in the sphere. The normal zone has a temperature increased by ΔT.

$$\Delta T = \frac{\int_0^l \frac{\partial E}{\partial r} dr}{2/3 \, \pi R^3 C_p}$$

$H_{SH}(T)$ is decreased down to :

$$H_{SH} + \frac{\partial H_{SH}}{\partial T} \Delta T$$

If this new value is lower than the applied field the change of state will be irreversible. This is the case for all the granules with a local magnetic field in the range $[H, H+\Delta H]$ defined by

$$\Delta H = \frac{\partial H_{SH}}{\partial T} \Delta T$$

which is a measurable quantity.

It is important to note that ΔH is inversely proportional to the volume of the granule. Since indium granule as small as a few thousand angströms in diameter will remain metastable, it turns out that the energy threshold of one granule in this range will be of a few electron-volts.

It is an additional argument that justifies to turn to low temperature techniques for the solar neutrino detection by indium.

However, we are left with a naïve but decisive question : how do we read the change of state of those granules ?

3.3. Real time read out techniques

3.3.1. Charge preamplifier

As it has already been said above where a granule is in the superheated state, it expels completely the magnetic field from its volume. When the same granule is switched to the normal state, this field comes in with a characteristic time dt proportional to R^2 ; R the radius of the granule. During this time, if the granule is within a pick up coil of width L, this coil senses a flux variation

$$d\phi \sim \frac{\pi R^3}{L} H.$$

Up to now for real time read out, we have used charge preamplifiers. As of March 85, the best estimated performance is the detection of one 10μ granule in a pick up coil of 200 microns width, center to center. (We use U shaped pick up coil so that each of them is the equivalent of a single wire for a wire chamber).

What can we hope with this level of performance of our charge preamplifier ? One 10μ granules produces the same signal as 8 granules of 5μ diameter with the same pick up coil. For the indium experiment, this size of granule is probably good but the pick up coil must be more larger if we do not want to have a weight of printed circuitry larger than the weight of the granules. Let us suppose 2 mm width for a pick up coil is a good compromise. In that case, our present charge amplifier requires the similtaneous flipping of 80 granules of 5 microns to detect something. If we estimate the energy threshold of a 10 microns to be around 30 keV the energy threshold for a 5μ granules will be eight times less 30/8 keV ; and for 80 granules, we will have to deposit :

$\frac{30 \times 80}{8} \sim 300$ keV which is, by the way, the present energy threshold for the indium experiment.

Needless to say this is a poor performance, however, if the background was not so severe, one could notice that we have already a real-time solar neutrino counter, sensitive to the higher part of the pp neutrino spectrum which is spreaded up to 420 keV.

Let us notice however that if we use a pick up coil with many turns (q the energy resolution is almost improved by the number of turns : for a 5 turns pick up coil, it will be around 60 keV in the previous example.

Therefore, in a sense, it seems that the recent improvements have created a situation where one cannot say a priori that we have to turn to cryoelectronics to achieve the full scale experiment. (Indeed, Josephson devices have already a level of performance many orders of magnitude better than our preamplifier but it is not easy to handle thousands of them in a single experiment).

3.3.2. Linear amplifier

Up to now this type of preamplifier has not used for the flippling of the granules because its performance have been less sensitive. However, one should note that in this case, the signal is $(d\Phi/dt)$ αR and not R^3 like with a charge preamplifier.

We can deduce from this remark that, in spite of the fact that the initial performance is poorer, it will require less progress to reach the size of interest than may be with charge preamplifier.

R. Bruere-Dawson (Collège de France) is presently giving a renewal of interest at this approach.

3.4. Sketch of the detector :

The pick-up coils are U shaped therefore the localization is provided by families of X and Y loops. When a granule changes of state a pulse voltage appears on one loop of each families and the time coïncidence validates the event.

The main problem for a practical realization is to find a compromise between the number of wires and the localization requirement for a convenient background rejection. With respect to this, the actual filling factor in indium must be as high as possible, with the actual mode of preparation of the granule it roill be over 20%. In that case, the total volume of the detector will be less than 8 cubic meters, requiring therefore a small space in the tunnel when this experiment has to be set up.

4. Conclusion

This brief survey shows that the real-time detection of solar neutrinos by superconducting techniques is presently a promising direction of development. With respect to this goal, the constitution of a devoted team in France for a feasibility study is encouraging. Recent results seem to indicate that the use of a cryoelectronics technique may be not necessary in which case, the building of small-scale prototype will be a further step ahead. Needless to say in such a cross disciplinary project is it hard to give credit to each physicist on the basis of a summary report. This is why I would like to mention the colleagues involved in one or in all the aspects of the program as of March 85 : A. de Bellefon (Collège de France), R. Bruere Dawson (Collège de France), M. Cribier (Saclay), P. Espigat (Collège de France), B. Escoubès (CRN Strasbourg), L. Gonzales Mestres (LAPP Annecy), R. Kuenzler (U. de Strasbourg), F.R. Ladan (G.P.S.), D. Limagne (G.P.S), J.P. Maneval (G.P.S.), D. Perret-Gallix (LAPP Annecy), N. Perrin (G.P.S.), Pichard (Saclay), J. Rich (Saclay), M. Spiro (Saclay).

REFERENCES

SOLAR NEUTRINOS
(1) J.N. BAHCALL, Nucl. Phys. 71, 267 (1965).
(2) J.N. BAHCALL, Rev. Mod. Phys. 50, 881 (1978).
^{115}In REACTION
(4) R.S. RAGHAVAN, Phys. Rev. Lett. 37, 259 (1976).

(5) M. DEUTSCH et R.S. RAGHAVAN, Proposal to develop a Detector for Solar Neutrinos. M.I.T. - 17 mai 1979.

(6) R.S. RAGHAVAN - Direct detection and spectroscopy of solar neutrinos using indium. Talk given at the University in Miniconference on the Neutrino Mas - Oct. 2-5, 1980

SUPERCONDUCTIVITY (main references).

(7) C. VALETTE, Thèse série A n° 846 - Orsay- 1971 (non publiée).

(8) C. VALETTE and J.P. BURGER - Journal de Physique 30, 562 (1969).

(9) H. BERNAS, J.P. BRUGER, G. DEUTSCHER, C. VALETTE, S.J. WILLIAMSON, Phys. Lett. 24A, 721 (1967).

(10) A. K. DRUKIER, C. VALETTE, G. WAYSAND, L.C.L. YUAN, F. PETERS. Lettere al Nuovo Cimento, 14, 300, (1975).

(11) R. L. CHASE, CH. GRUHN, A. HRISOHO, C. VALETTE, G. WAYSAND. Fast electronics for a metastable superconducting detector.

(12) C. VALETTE, G. WAYSAND - Proceedings XXth Cospar 38 (1977), North Holland.

(13) D. HUEBER, C. VALETTE, G. WAYSAND - Cryogenics, 21, 387 (1981).

(14) D. HUEBER, C. VALETTE, G. WAYSAND - Proceedings LT 16, Physica, 1088, 1229 (1981).

(15) C. VALETTE, G. WAYSAND, D. STAUFFER - Solid State Com. 41, 305 (1982).

(16) A. HRISOHO, G. WAYSAND, D. STAUFFER - Instrumental limit and number of effective granules in superheated superconducting detectors. Nucl. Instr. & Methods 214, 415 (1983).

(17) M. KOUIRITI, Experimental Study of Superheating in Suspension of Superconducting Tin Granules with High Filling Factors (in French, unpublished) - Thèse de 3ème cycle, Université Paris VII, December 1984.

(18) M. HILLEN and D. STAUFFER, Solid. State Com. 43, 487 (1982).

(19) P. MAZUR, Com. Journ. Phys. 63, 24 (1985). See also P. MAZUR and U. EIGENMULLER (to be published).

CAN NEUTRINOS FROM CYGNUS x-3 BE SEEN BY PROTON DECAY DETECTORS?

Arnon Dar
Department of Physics
Technion-Israel Institute of Technology
Haifa
Israel

ABSTRACT. The flux of high energy neutrinos from Cygnus x-3 is estimated directly from the observed light curve of high energy γ-rays from this source. These neutrinos induce an underground muon flux which can be distinguished by its characteristic direction and duration from the muon background induced by atmospheric neutrinos. For zenith angles ≥84° the expected rate of muon events in the IMB detector due to Cygnus x-3 is between 1 to 10 events/year, depending on the actual duration of the γ ray pulses from this source.

It has been suggested independently by a few authors that close binary systems which are composed of a pulsar or a black hole and of a nearby large companion may produce large fluxes of neutrinos[1]. These neutrinos are produced when high energy cosmic rays from the compact object are dumped into the companion's atmosphere and produce there mesons that decay into neutrinos. The best known candidate in our galaxy for such an "astronomical beam dump" which produces a large flux of ultrahigh energy neutrinos is the binary system Cygnus x-3, believed to consist of a neutron star (or a black hole) and a main sequence companion star. Cygnus x-3 has been detected[2] as a radio, infrared, X-ray and γ-ray source displaying a period P ~ 4.8h associated with the eclipsing of the pulsar by the companion. The 4.8h orbital period and the duration of the eclipse ~0.4P imply that the orbital radius of the pulsar is ~1.05R, where R is the radius of the companion, and that the mass of the companion is ≤4.5$M_\odot$ assuming R ~ $M^{0.6}$ and a ≥1.4$M_\odot$ neutron star Cygnus x-3 has also been detected as a source of ultrahigh energy (UHE) γ's. The highest energy photons from Cygnus x-3 that have been detected[3] are in the energy range 2×10^6 GeV to 2×10^7 GeV. In Fig. 1 we show a compilation[3] of the experimental results on the integral flux of high energy photons from Cygnus x-3. This integral flux can be well represented by[3]

$$N_\gamma(>E) = (6.4\pm3.6)\times10^{-7} E^{-1.108\pm0.021} \text{ cm}^{-2}\text{sec}^{-1}, \qquad (1)$$

where E is expressed in GeV.

The UHE photons from Cygnus x-3 seem to be emitted in short pulses, $\Delta t_\gamma \le 0.05P$, just at the beginning and just at the end of the eclipse. This has led Vestrand and Eichler[4] to propose that these UHE photons are produced by the decay of π^0's (and η^0's) in the cascades which are generated in the companion's atmosphere by high energy cosmic rays from the compact object. The high energy photons in the cascade are collinear with the incident cosmic rays. Because of

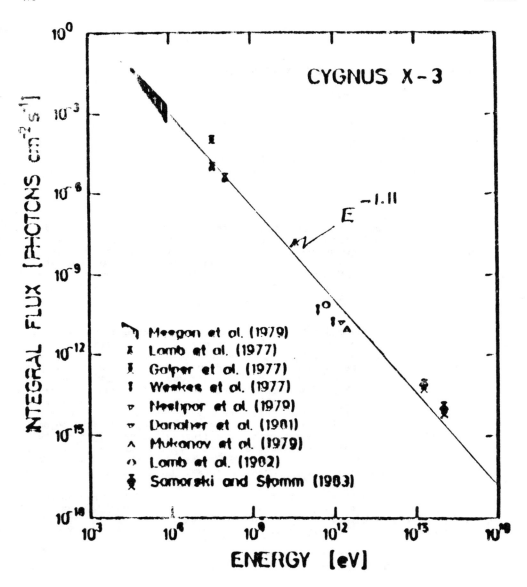

Fig. 1. Compilation[3] of the integral flux of photons with E > 100 MeV from Cygnus x-3.

their attenuation in the companion's atmosphere the compact object is visible in the UHE γ-ray region only when the line of sight to it passes not too deep in the companion's atmosphere (see Fig. 2). This may explain the two short pulses of UHE photons just at the beginning and just at the end of the eclipse.

In addition to the production of π^o's and n^o's which decay into photons, the cosmic rays from the compact object also produce $\pi^{\pm}$'s and $K^{\pm}$'s when they collide in the companion's atmosphere. The number of π^o's, π^+'s and π^-'s which are produced in these collisions are approximately equal. If a large fraction of these π^+'s and K^+'s decay into neutrinos and if these neutrinos penetrate the companion then Cygnus x-3 produces a neutrino flux which is comparable to the high energy γ flux, but which lasts during the whole eclipse[5], $\Delta t_\nu \sim 0.4P$. Thus Cygnus x-3 may be a much brighter source of high energy neutrinos, than that of high energy γ's.[5] Can these neutrinos be detected by the massive underground proton decay detectors?

The search of high energy neutrinos from Cygnus x-3 (as well as from other extraterrestrial sources) is based on the search of muons which are induced by interaction of ν_μ's from this source in the material surrounding the underground detector. The main background of atmospheric muons can be eliminated by limiting the search to sufficiently large angles[6], where the only remaining background is due to muons induced by atmospheric ν_μ's. In Figure 3 we show the expected background at two typical depths, 1570 mwe of the IMB detector and 4200 mwe of the Homestake detector. Thus the Homestake detector can search extraterrestrial sources only at zenith angles $67° \leq \theta \leq 180°$ while the IMB detector can search such sources only at zenith angles $84° \leq \theta \leq 180°$. The zenith angle of Cygnus x-3 varies like:

$$\cos\theta = \sin\phi\sin\delta + \cos\phi\cos\delta\sin(2\pi t/p_s), \qquad (2)$$

where p_s = 23.93h is the sidereal period, δ = 40.9° is the declination of Cygnus x-3 and ϕ is the latitude of the detector. At IMB ϕ = 43.5°N while at Homestake ϕ=44.5°N. Consequently IMB can look for Cygnus x-3 during ~7 hours while Homestake can look for it during ~12 hours of the sidereal period.

The neutrinos flux at Earth from Cygnus x-3 and the induced muon flux can be calculated[5] following the simple analytical methods that were developed in ref. 6. The neutrino flux depends on:

(i) The cosmic ray flux from the compact object that impinges on the companion's atmosphere.

(ii) The density distribution in the companion's envelope where the meson production takes place.

(iii) The column density through the companion which must be penetrated by the high energy neutrinos in order to reach Earth.

The power law dependence of the high energy γ ray spectrum (Eq. (1)) from Cygnus x-3 implies that the cosmic ray flux has the same power law dependence, $dN/dE \sim cE^{-p}$ where $p \sim 2$. For a power law spectrum the γ rays flux from $\pi^o \to 2\gamma$ decays, and the ν_μ flux from $\pi^{\pm} \to \mu\nu$ and $K^{\pm} \to \mu\nu$ decays, can be related directly using only general information on the density distribution in the system. For instance, if the scale height of the density distribution in the companion's envelope where the production takes place is H_a(km) then $\pi^{\pm}$'s and $K^{\pm}$'s, which are produced along a cosmic ray trajectory with zenith angle θ^*, decay before they undergo inelastic collisions if their energies are smaller than ~18 H_a secθ^* GeV and ~1.33x10^2 H_a secθ^* GeV, respectively. (The decay probabi-

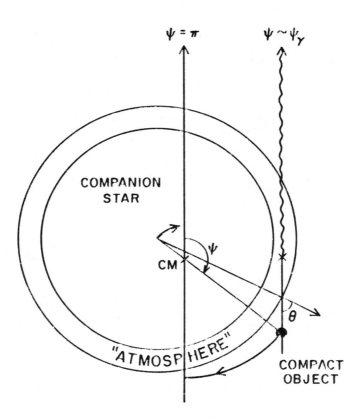

Fig. 2. The expected flux of underground muons at depths of 4200 mwe and 1570 mwe, due to cosmic ray interactions in the Earth's atmosphere.

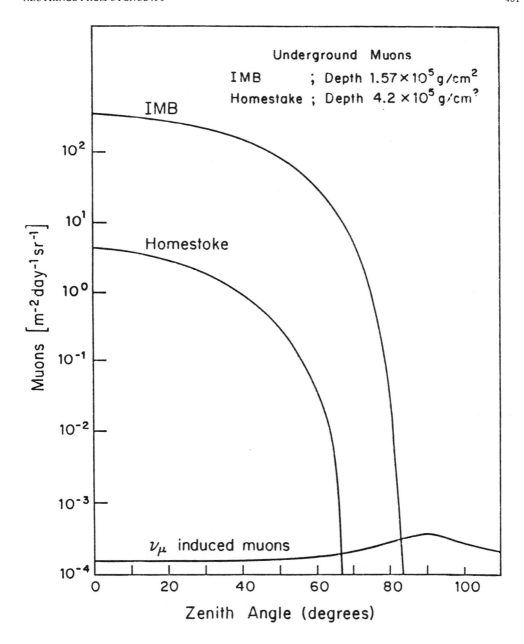

Fig. 3. The assumed geometry of Cygnus x-3 during the UHE γ-pulse.

lity of mesons of energy E and life time τ is given approximately by[6] $D(E) = 1/(1+\gamma E)$ where $\gamma^{-1} = mc^2(H_a \sec\theta^*/c\tau))$. For such energies the γ and ν_μ fluxes <u>before attenuation</u> are related through[6]

$$\frac{dn_\nu}{dE} \equiv \frac{1}{2} \sum_{M=\pi^\pm, K^\pm} (\alpha_M^{1-P} B_M g_M/g_\pi 0) \frac{dn_\gamma}{dE} \cong 0.50 \frac{dn_\gamma}{dE} , \qquad (3)$$

where $\alpha_M = m_M^2/(m_M^2 - m_\mu^2)$, B_M is the branching ratio for $M \to \mu\nu$ decays and[7] $g_M = \int_0^1 x^{P-1} (dn_M/dx) dx$ with dn_M/dx being the cross section for inclusive production of M with scaled momentum $x = p/p_{max}$. If muons (from $\pi \to \mu\nu$ and $K \to \mu\nu$ decays) also decay in flight ($E < 0.16\, H_a \sec\theta^*$) and produce ν_μs through $\mu \to e\nu_e\nu_\mu$ then their contribution modifies Eq. (3) as follows[6]:

$$\frac{dn_\nu}{dE} \cong 1.16 \frac{dn_\gamma}{dE} . \qquad (4)$$

Note however that H_a cannot be taken from calculations of stellar envelopes of normal $M \sim 4.5\, M_\odot$ main sequence stars, because the companion's outer layers which face the compact object are highly inflated by strong tidal forces and large fluxes of radiation from the nearby (~0.05R) compact object. A rough estimate of the scale height of the "inflated envelope" can be obtained as follows: A high energy proton beam that collides with a target at impact parameter b produces collinear beams of HE γ's and ν's which emerge from the target with intensities proportional to

$$n_i \sim \mu_p T(b) \left[\frac{e^{-\mu_i T(b)} - e^{-\mu_p T(b)}}{(\mu_p - \mu_i) T(b)}\right] , \qquad (5)$$

where $i = \gamma, \nu$, $T(b) = \int \rho(b,z) dz$ is the target thickness (column density) along impact parameter b, and μ is the absorption coefficient given by $\mu \equiv N_A \sum X_i \sigma_i / W_i$ where N_A is Avogadro's number and X_i is the fraction (by weight) of "atoms" of atomic weight W_i and cross section σ_i. Maximal fluxes emerge from the target at impact parameters that satisfy

$$T(b) = \frac{1}{\mu_i - \mu_p} \ln \left(\frac{\mu_i}{\mu_p}\right) . \qquad (6)$$

For standard stellar composition (~75% H + 25% He) $\mu_\gamma^{-1} \sim 100$ g/cm^2 and $\mu_p^{-1} < 40$ g/cm^2 [8] and the optimal thickness for production of HE γ's is ~60 gm/cm^2. From Eq. (5) it can be seen also that significant production of HE γ ray fluxes occur when T(b) changes between ~few g/cm^2 and ~few x 100 g/cm^2 during which $\ln T(b)$ changes by about a factor of 5. During the UHE γ ray pulse the line of sight to the compact star moves from the "surface" of the companion, $b \cong R$, into a depth of $\Delta b = (R+\Delta R)(\cos 18° - \cos 36°) \sim 0.15R$ below the surface. The radius R of a main sequence star of standard composition and zero age behaves like $\sim M^{0.6}$ for $M > 1.8\, M_\odot$, i.e. for $M \sim 4.5\, M_\odot$ the radius is $R \sim 2.5\, R_\odot$ and then

$$H_a \sim 0.15R/5 \sim 0.07\, R_\odot \sim 5 \times 10^4 \text{ km} . \qquad (7)$$

Consequently μ's, $\pi^{\pm}$'s and $K^{\pm}$'s which are produced in the "companion's envelope" with energies below $10^4 \sec\theta^*$, $10^6 \sec\theta^*$ and $10^7 \sec\theta^*$ GeV, respectively, decay before suffering inelastic collisions. Below we shall show that only ν_μ's with $E \leq M_W^2/2m_p \sim 3.5 \times 10^3$ GeV contribute significantly to the muon counting rate in the underground detectors. Hence we shall base our estimates on Eq. (4). Since $\mu_\nu \gg \mu_\gamma$ it follows from Eqs. (4) and (5) that for $T(b) \gtrsim 100 \text{gm/cm}^2$

$$(dn_\nu/dE \cong 2.14 \max(dn_\gamma/dE). \tag{8}$$

If one sets $\mu_\nu = 0$ and uses the values $\Delta t_\gamma = 0.05P$ and $\Delta t_\nu = 0.40P$ for the duration of the UHE γ and ν pulses[4], respectively, one finds that the time averaged fluxes satisfy

$$\frac{d\bar{n}_\nu}{dE} \simeq 4.28 \frac{\Delta t_\nu}{\Delta t_\gamma} \frac{d\bar{n}_\gamma}{dE} \simeq 34 \frac{d\bar{n}_\gamma}{dE}. \tag{9}$$

Note that Eq. 9 is valid only if the absorption of γ rays between source and Earth can be neglected and $\mu_\nu = 0$. However, γ rays with energies above the threshold energy for the process $\gamma + \gamma \to e^+ + e^-$ are attenuated by collisions with galactic and cosmological photons.[9] If we denote the mean free path of HE photons in the background radiation by λ and the distance of Cygnus X-3 by D ($D \sim 12.4$ kpc $= 3.86 \times 10^{22}$ cm [10]) then $d\bar{n}_\gamma/dE$ in Eq. (9) should be replaced by $\exp(D/\lambda)(d\bar{n}_\gamma/dE)$. Note in particular that in the energy range $10^6 \leq E_\gamma \leq 10^7$ GeV the mean free path of photons in the cosmological 3°K black body radiation is[9] $\lambda \sim 3.5 \times 10^{22}$ cm and $\exp(D/\lambda) \sim 3$. When one substitutes Eq. (1) into Eq. (9) and includes the effects of photon attenuation in the background radiation, one obtains in the limit $\mu_\nu \to 0$ that

$$\frac{d\bar{n}_\nu}{dE} \simeq 2.4 \times 10^{-5} E^{-2} \text{ cm}^{-2} \cdot \sec^{-1}. \tag{10}$$

For sufficiently large energies and trajectories near the center of the companion, the absorption of neutrinos is quite significant. This absorption is due to both charged current interactions and neutral current interactions with both nucleons and electrons. Their cross sections can be calculated[11] from the standard Glashow-Salam-Weinberg theory of electroweak interactions, using for nucleons the quark structure functions that were measured at accelerators and using QCD to correct for scaling violations. For a standard composition ($n_p/n_{He} \sim 12$) the attenuation of ν_μ's and $\bar{\nu}_\mu$'s is dominated by interactions with protons. Their total cross sections can be approximated by the following interpolating formula

$$\sigma_{\nu p}^t \sim \sigma_{\bar{\nu} p}^t \sim 0.62 \times 10^{-38} E(\text{GeV}) \frac{\ln(E/E_t)}{\ln(E/E_t) + 6 \times 10^{-2}(E/E_W)} \text{ cm}^2, \tag{11}$$

where $E_t \sim$ is the threshold energy for $\bar{\nu}_\mu p \to \mu^+ n$. The ν-flux during the eclipse can be related then to the γ-flux as follows:

$$\frac{dn_\nu}{dE} \cong 2.14 \, e^{-\mu_\nu T(b)} e^{\lambda/D} \max(\frac{dn_\gamma}{dE}) \cong 4.28 \, e^{-\mu_\nu T(b)} \frac{P}{\Delta t_\gamma} e^{\lambda/D} \frac{d\bar{n}_\gamma}{dE}, \tag{12}$$

where $\mu_\nu T(b) \sim N_A \sigma_\nu(E) T(b)$. Since the distortion of the companion's surface due to its proximity to the compact object has very little effect on $T(b)$ for trajectories near its center, and since ν attenuation is significant only for such trajectories, therefore $T(b)$ can be estimated from standard stellar structure calculations of spherical symmetric stars. Such calculations for a main sequence star of a mass $4.5~M_\odot$, a standard composition and a zero age yield a density distribution which can be described approximately by $\rho(r) \sim \rho_c \exp(-6.75 r/R)$ where $\rho_c \sim 20~g/cm^3$ is the central density in the star and $R \sim 2.46 R_\odot$ is its radius. Consequently $T(b) \sim 2\rho_c b K_1(b/H)$, where $H \equiv R/6.75 \cong 2.5 \times 10^{10}$ cm, is the scale height of the density distribution. The neutrino flux as given by Eq. (12) can be folded with the probability for producing muons which reach the underground detector, in the material surrounding it.[6] We have found that our numerical calculations can be well approximated by the following analytical procedure: The attenuation factor $\exp(-\mu_\nu T(b))$ can be replaced by a step function $\theta(E-E_c(b))$ where the cutoff energy E_c satisfies $\mu_\nu(E_c) T(b) \sim 1$. Following the methods of Ref. 6 one can show then that the underground flux of muons induced by a ν-flux of the form $dn_\nu/dE \sim cE^{-2} \theta(E-\bar{E}_c(b))$ is given approximately by

$$\frac{dN_\mu}{dE} = \frac{cN_A}{\alpha} (A+\frac{B}{3}) \left[\ln(\frac{E_c}{E}) - (\frac{1}{\epsilon}+\frac{1}{E_c}) \ln(\frac{E_c+\epsilon}{E+\epsilon}) \right], \quad (13a)$$

$$N_\mu(>E) = \frac{cN_A}{\alpha} (A+\frac{B}{3}) \left[\epsilon(1+\frac{\epsilon}{E_c} + \frac{E}{E_c}) \ln(\frac{E_c+\epsilon}{E+\epsilon}) - E \ln(\frac{E_c}{E}) - \epsilon(1-\frac{E}{E_c}) \right], \quad (13b)$$

where $\bar{E}_c \equiv \min(\bar{E}_c, 10~E_W)$ and $\epsilon = \alpha/\beta$ where α and β are the coefficients in the energy-range relation $dE/dx = -\alpha - \beta E$. QED calculation of μ cross sections in standard rock ($\rho = 2.6~g/cm^3$, $Z=11$ and $A=26$) yield $\alpha \cong 1.86 + 0.077~\ln(E_\mu/m_\mu)$ MeV cm^2/g and $\beta = [1.78 + 0.2~\ln(E_\mu/m_\mu)] \times 10^{-6}$ cm^2/g. A and B are coefficients in the cross section for the charged current reaction $\nu_\mu N \to \mu X$ on isoscalar nucleon. For $E_\nu \le E_W$ $d\sigma_\nu/E_\mu \cong A + B(E_\mu/E_\nu)^2$. This functional form follows from the Glashow-Salam-Weinberg theory of electroweak interactions. Experimental results yield $A \cong B \cong 0.36 \times 10^{-38}$ cm^2 (for $n_\nu = n_{\bar{\nu}}$). With these values ($cN_A/\alpha) \times (A+B/3) \sim 4.8 \times 10^{-18}$ cm^{-2} sec^{-1} GeV^{-1}. If $\Delta t_\gamma/P \sim 5\%$ then during the eclipse the flux of ν-induced muons with $E > 2$ GeV changes from its maximum value $N_\mu \sim 2 \times 10^{-13}$ cm^{-2} sec^{-1} when $b \sim R$ to a minimum value $N_\mu \sim 1 \times 10^{-14}$ cm^{-2} sec^{-1} when $b \sim 0$ and back to $N_\mu \sim 2 \times 10^{-13}$ cm^{-2} sec^{-1} when b increases back to R. The time averaged μ-flux (over the whole period) is $N_\mu \sim 5 \times 10^{-14}$ cm^{-2} sec^{-1}. The expected number of underground muons induced by ν_μ's from Cygnus x-3 which penetrate the IMB detector (23m x 18m x 17m) at zenith angles larger than 85° (about 7h out of 24h) is ~ 1 per year if $\Delta t_\gamma/P \sim 5\%$ while it is ~ 10 per year if $\Delta t_\gamma/P \sim 0.5\%$. The "light curve" of these muons is shown in Fig 4.

The background of underground muons which are induced by atmospheric ν_μ's in the same angular range $84° \le \theta \le 96°$ is[6] $N_\mu(E \ge 2~GeV) \sim 4 \times 10^{-13}$ cm^{-2} sec^{-1} sr^{-1}. Therefore during the ν-pulse from Cygnus x-3 the signal to background ratio for zenith angles $\ge 84°$ is about 5 to 1, assuming an angular resolution of $\pm 7°$ and $\Delta t_\gamma/P \sim 5\%$. However if Δt_γ, the "width" of the UHE γ-pulse reflects mainly the experimental resolution and not the "true" width, and if Δt_γ is much shorter, $\Delta t_\gamma/P \le 0.5\%$, then the expected rate of underground muons induced by ν_μ's from Cygnus x-3 which penetrate the IMB detector at zenith angles $\ge 84°$ is ≥ 10 events·year^{-1}. Such a rate is well above the background and in that case IMB could see ν_μ's from Cygnus x-3.

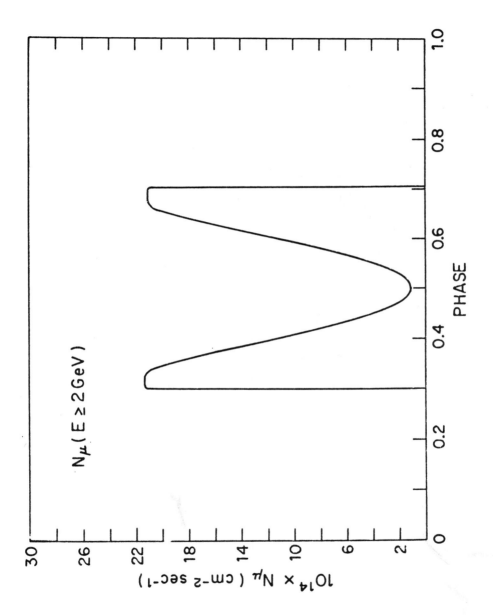

Fig. 4. The "light curve" of underground muons induced by ν_μ's from Cygnus x-3.

ACKNOWLEDGEMENTS

The author gratefully acknowledges useful discussions with M. Livio, O. Regev and G. Shaviv.

REFERENCES AND FOOTNOTES

1. D. Eichler and D.N. Schramm, Nature 275, 709 (1978);
   D. Eichler, Ap. J. 222, 1109 (1978); Nature 275,725 (1978);
   A. Dar, PRL, 51, 227 (1983) and references therein.
2. R. Giaconni et al., Ap. J. 148, L119 (1967).
   B.M. Vladimirsky et al., Proc. 13th Int. Conf. on Cosmic Rays 1, 456 (1973);
   D.R. Parsignault et al., Ap. J. 218, 232 (1977);
   Yu I. Neshpor et al., Ap. Space Sci. 61, 349 (1979);
   S. Danaher et al., Nature 289, 568 (1981);
   R.C. Lamb et al. Nature 296, 543 (1982), and references therein.
3. M. Samorski and W. Stamm, Ap. J. 268 L17 (1983);
   J. Lloyd-Evans et al., Nature 305, 784 (1983).
4. W.T. Vestrand and D. Eichler, Ap.J. 261, 251 (1982).
5. A preliminary version of this work was presented at a CERN Colloquium in memory of Roger Van Royen, CERN, Geneva, 28 July, 1984.
6. A. Dar, Technion Preprint PHYS-84-41, submitted for publication in Phys. Rev. D. (August, 1984).
7. For $p \cong 2$, $g_{\pi^+} = 3.5 \times 10^{-3}$, $g_{\pi^-} = 2.4 \times 10^{-3}$, $g_{\pi^0} = \frac{1}{2}(g_{\pi^+}+g_{\pi^-})$, $g_{K^+} = 5.0 \times 10^{-4}$, $g_{K^-} = 1.5 \times 10^{-4}$.
8. $\sigma_{in}(pp)$ changes slowly with energy. The quoted value of μ_p is for $\sigma \sim 45$ mb which is its average value in the relevant energy range $E < E_W^p = M_W^2/2m_p$.
9. J. Wdowczyk et al., J. Phys. A5, 1419 (1972).
10. J.M. Dickey Ap. J. 273, L71 (1983).
11. Y.M. Andreev et al., PL 84B, 247 (1979)
    A. Dar and P. Langacker, unpublished.

DETECTION OF MAGNETIC MONOPOLES WITH METASTABLE TYPE I SUPERCONDUCTORS

L.Gonzalez-Mestres and D.Perret-Gallix
Laboratoire d'Annecy-le-Vieux de Physique des Particules
Chemin de Bellevue, B.P. 909
74019 Annecy-le-Vieux Cedex
France

ABSTRACT. Metastable type I superconductors provide a natural way to detect magnetic monopoles. The method is independent of the speed of the monopole, and provides tracking and timing information. The electronic signal exceeds by two orders of magnitude that of induction experiments. As a consequence, large area detectors can be built at a reasonable cost. The detector consists of a colloid of superheated tin microspheres (30 to 100 μm diameter). Background rejection is very high, due to the large size of granules. Time resolution is in the range 100 nsec - 1 μsec and space resolution is $\sim$ 1 cm. Therefore, the speed and direction of the monopole can be determined with good accuracy. We discuss the present status of the project, as well as some of the main technical problems. In particular, we study the role of impurities in the superconductive material, and propose a way of improving the performance (time resolution, signal in voltage) of superconducting granules detectors.

1. INTRODUCTION

Type I superconductors are characterized by values of K smaller than $1/\sqrt{2}$. K is the Ginzburg-Landau parameter [1], $K = \lambda/\xi$. λ is the London penetration depth and ξ is the temperature dependent coherence length. Metastable states exist whenever there is a positive surface energy, which is the case of type I superconductors in the presence of an external magnetic field. In the present case, we mean by surface energy the energy of a normal-superconducting interphase at the transition point. Let H be the value of the external magnetic field, H_c the critical field, T the temperature and T_c the critical temperature. The critical line is usually drawn as $H_c = H_c(T)$, and by T_c one means the critical temperature at H = 0.
    A superconductive sample can remain superconducting for $H > H_c(T)$. This situation is called superheating. For each material, there is a value of the magnetic field, H_{sh} (the superheated critical field) above which the superheated state can no longer exist. Conversely, a superconductive specimen can remain in the normal state for $H < H_c(T)$. It was

the first metastable phase of type I superconductors to be experimentally observed, and is called supercooling. Supercooling cannot exist below a certain value of H, H_{sc} (the supercooled critical field).

A small deposit of energy in a superheated sample can create a nucleation center of the normal state, that propagates to the whole specimen. In this way, a microscopic phenomenon may lead to a macroscopic effect. The principle is similar to that of bubble chambers or emulsions. Therefore, it is not surprising that the use of superheated type I superconductors for particle detection was proposed as early as 1967 [2].

β-rays destroy the superheated state for 2 μm diameter mercury spheres [2]. Similar results were obtained for Sn and In granules [3] using 425 and 930 keV electrons. Tin granules of 10 μm diameter have been proven to be sensitive to γ-rays of 30 keV, which allows for X-ray imaging [4] and transition radiation detection [5]. Small In granules (about 6 μm diameter) are expected to be sensitive to ionizing particles, thus allowing for solar neutrino detection [6,7]. Also, small tin granules may be used to detect solar neutrinos [8] or supersymmetric galactic dark matter [9] (scalar neutrinos, photinos). Efforts to improve the electronic read out and to incorporate fast electronics have been made in the last ten years [10].

More recently [11,12], we have proposed the use of superheated granules detectors to search for magnetic monopoles with all values of β (β = v/c, v = monopole speed). Since the basic mechanism is independent of the size and shape of the specimen, large granules can be used (30 to 100 μm diameter) producing a signal that can be detected with conventional electronics. Time and space resolution (100 nsec to 1 μsec, ~1 cm) are rather confortable and allow to determine the speed and direction of the monopole. Furthermore, large granules are rather insensitive to any thermal deposit of energy [11] and provide a high background rejection. A large area detector based on superconducting granules should therefore be much cheaper and manageable than any equivalent induction experiment, where SQUIDs are used and sophisticated shielding techniques are required [13]. Note, however, that induction experiments provide a direct measure of the magnetic charge of the monopole.

2. HOW A MAGNETIC MONOPOLE DESTROYS THE SUPERHEATED STATE

A magnetic monopole traversing a type I superconductor leaves behind [14] a tube of magnetic flux $\phi = 2\phi_o$ trapped into the sample (ϕ_o = 2.067·10^{-7} Gauss cm^2). Inside the flux tube, the trapped magnetic field breaks Cooper pairs [11,12,15] and lowers the value of the Ginzburg-Landau order parameter [1,12,15]. In cylindrical coordinates around the axis of the flux tube, the real Ginzburg-Landau order parameter f [1] varies from 0 to 1 in a distance of about a coherence length [11,12,14,15]. Figure 1 exhibits the variation of f as a fonction of the distance to the axis of the flux tube. The calculation [12] was made for K = 0.16.

Based on the experimental fact [16] that surface defects of size $\gtrsim \xi$ destroy the superheated state, our claim [11] was that, if the monopole traverses a superheated granule, the ends of the flux tube will

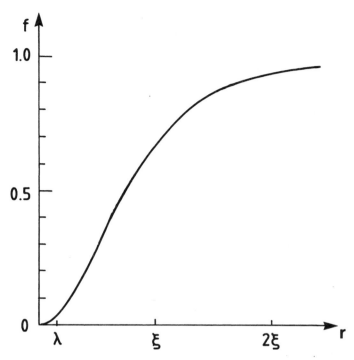

Figure 1. Structure of the flux tube ($\phi = 2\phi_o$). The real order parameter f is plotted as a function of the distance r to the axis.

originate nucleation centers on the surface of the grain and the whole granule will become normal. More detailed theoretical calculations [12, 15] confirm this principle. In particular:

a) We have computed [12,17] the superheating energy barrier per unit surface, that in the low K limit, and using the Ginzburg-Landau approximation, turns out to be

$$\Delta F \simeq \frac{4\sqrt{2}}{3} (1-\sqrt{1-x^2})(\sqrt{1+x}-\sqrt{1-x}) \frac{H_{sh}^2}{8\pi} \lambda \qquad (1)$$

where $x = (1 - H^2/H_{sh}^2)^{1/2}$ and H is the value of the magnetic field on the surface of the superconductor.

Numerical calculations for $K = 0.16$ ($H_{sh} = 2.3 H_c$) agree with (1) within 10% in a wide range of values of H. We have also performed calculations in the extreme anomalous limit [18] and the results agree with (1) within 30%. For a point on the surface of a spherical grain submitted to an applied magnetic field $H_o \simeq 2/3 H_{sh}$, one has $H \simeq H_{sh} \sin\theta$, where θ is the angle between H_o (the applied field) and the position vector taken from the center of the granule ($0 \leq \theta \leq \pi$, Fig.2). Then, $x \simeq \sigma(\theta) = |\cos\theta|$, where $\sigma(\theta)$ is the fraction of the grain surface lying in the region $|\theta'-\pi/2| \leq |\theta-\pi/2|$ (the equatorial zone from θ).

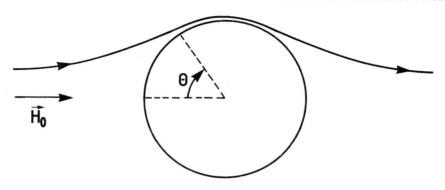

Figure 2. The polar angle between the applied magnetic field H_0 and the position vector. The line going close to the equator of the granule is a field line deviated by Meissner effect.

This can be compared with the energy of the flux tube per unit length computed in ref. [14] . The vortex energy, when calculated for $K = 0.16$ and integrated over a depth of ξ, turns out to exceed by an order of magnitude the value of ΔF for $H = 0.7\ H_{sh}$ ($H = 1.6\ H_c$, $\sigma(\theta) \simeq 70\%$), integrated over a surface of $\pi \xi^2$.

b) Pressure balance arguments also go in the same direction. For $K = 0.16, \sigma(\theta) \simeq 70\%$, we have been able to prove the following statement: Starting from a configuration with a flux tube perpendicular to the surface and a constant external field $H \geqslant 0.7\ H_{sh}$, there exists a a continuous path in configuration space such that: 1) it leads to a complete flip of the sample to normal state 2) the total free energy is always decreasing.

This does not by itself provide a complete mathematical proof of the nonexistence of equilibrium configurations other than the normal state. But it clearly exhibits the instability of the superheated state in the presence of a vortex line. The pressure of the external magnetic field, that tends to penetrate into the specimen and widen the normal zone, turns out to be stronger than the rigidity of the flux tube, that tends to stay close to its equilibrium configuration.

c) We neglect ohmic energy losses. However, they can heat locally the granule and help nucleation. For $\beta = 3 \cdot 10^{-5}$, ohmic energy losses in tin, when integrated over a depth of ξ, exceed by an order of magnitude the value of the energy barrier ΔF integrated over a disk of radius ξ.

If the mechanism works for $H \geqslant 0.7\ H_{sh}$, the probability that none of the ends of a vortex line flips a single spherical granule is less than 10% (90% efficiency). In a practical detector, because of diamagnetic interactions between grains and clustering effects, it is reasonable to estimate that the monopole flips at least 50% of the grains it crosses.

Finally, we cannot elude the question of whether the principle can be tested experimentally. This is obviously necessary before a large area detector will actually be built. Drawing a flux tube near the surface of a superheated granule can fake to some extent the effect of a monopole interacting with the surface of the grain, but it does not a priori reproduce the injection of the vortex line into the specimen.

In order to check the incompatibility between the superheated state and the presence of a fluxoid with $\emptyset = 2\emptyset_o$, we have imagined an alternative test [19]. We use the fact that macroscopic specimens can reach the superheated state with sizeable lifetimes (10 sec to 1 min for tin cylinders 12 cm long, 3 mm diameter [20]). It therefore should be possible to prepare a thin sample ($\sim 4 \mu$m thick, ~ 30 mm^2 of surface) able to reach the superheated state with ~ 1 min lifetime. After having checked the metastability properties of the specimen, we inject into it a flux tube with $\emptyset = 2\emptyset_o$. This may be done, for instance, heating locally near the border of the sample, and in the presence of a magnetic field $\vec{H}_\perp$ perpendicular to the specimen ($H_\perp \sim 1$ Gauss), a region of $\sim 4 \mu$m radius with a laser beam introduced inside the cryostat by an optical fiber. After removal of $\vec{H}_\perp$, the presence of the vortex can be checked by scanning.

Having obtained a sample with a trapped flux tube of $\emptyset = 2\emptyset_o$, a magnetic field $H_{\parallel} > H_o$ parallel to the surface can be turned on. The presence of the fluxoid should prevent the formation of the superheated state for $H_{\parallel}$ larger than some value, H_m, in the range $H_c < H_m < H_{sh}$.

3. THE DETECTOR

The superheated grain monopole detector would be a colloid of spherical granules (30 to 100 μm diameter) into paraffin, or some other insulating material. Dilution factor in practical detectors is usually taken to be 10 to 20% in volume, but using grains coated with insulating material it is likely that clustering effects can be reduced and the dilution factor increased. Granules are often produced in the industry, just because they are less volatile than powder and fill the volume better. One can imagine that, if granules can be coated with a well suited dielectric, the use of paraffin would no longer be necessary. Grains can probably be just piled up. Increasing the filling factor would then increase the signal.

In the case of the monopole detector, the material would be pure tin and the colloid would be spread in foils ~ 1 cm thick, forming several parallel planes. An applied magnetic field $\vec{H}_o$, perpendicular to the plane of the detector, is increased until the granules reach the superheated state. Because of diamagnetic interactions, H_o would be substantially below 2/3 H_{sh}. An X-Y coordinate system of current loops (~ 1 cm wide, ~ 1m long) connected to fast amplifiers, would provide position information on each plane of detector (Fig.3a). With at least three parallel planes, the signal for a monopole would be one point per plane, all of them aligned and the timing consistent with the flight of a single particle (Fig.3b).

The electronic read out is based on the disappearence of the Meissner effect, when one or several metastable granules become normal. The local variation of magnetic field induces currents in the loops surrounding the region where the event takes place. In this way, spatial resolution of 1 cm is reached. Taking time resolution to be ~ 100 nsec, and with at least three parallel planes of granules spaced by 30 cm, the speed of a monopole with $\beta \sim 10^{-4}$ would be determined with $\sim 1\%$ accuracy. We have in

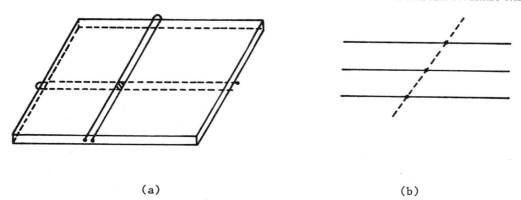

(a)   (b)

Fig.3 a) The position of an event given by a coincidence between two perpendicular current loops. b) The signature for a monopole crossing three parallel planes of granules.

this way a <u>real time, track detector</u>.

For tin (T_c = 3.72 K), the detector would work in the range 1.5 K $<$ T $<$ 2.5 K. The He^4 λ-point (T = 2.18 K) lies in this interval. Working near T = T_c appears to be better for large specimens, since ξ becomes larger and surface deffects play a less important role, but has the drawback that both H_c and H_{sh} become smaller.

The signal in magnetic flux due to the disappearence of the Meissner effect for a granule of radius R sitting on the axis of a current loop of width d and length $\ell \gg$ d is [21]:

$$\Delta \phi = \frac{4 H_o R^3}{d} \quad (2)$$

and remains of the same order for grains lying at a distance a $\sim$ d of the induction loop plane. We therefore have the proportionality law :

$$\Delta \phi \text{ (1 granule)} \sim \frac{R^3}{d} \quad (3)$$

On the other hand, taking foils of thickness A = d/2 , the number of grains traversed by the monopole will be proportional to d/R , giving:

$$\Delta \phi \sim R^2 \quad (4)$$

Therefore, the monopole experiment must be performed with large grains in order to increase the electronic signal. Then, <u>conventional electronics</u> can be used. As an example, with: R = 15 μm, $\overline{2A = d = 1 \text{ cm}}$, ℓ = 1 m (L $\simeq$ 2 μH); assuming that the monopole trajectory is perpendicular to the loop plane and goes through the loop axis ; and taking 50% efficiency for the flipping mechanism with a 10% filling factor in volume (about 12 grains change state)), we get: $\Delta \phi \simeq$ 500 n ϕ_o , where n is the number of turns of the current loop. Several turns of coil are often used [22] , which may be convenient for voltage amplifiers, but also if one wants to increase the ratio L/Z (Z = impedance of the circuit). For n=1, the signal in current is : $I \simeq 0.5 \mu A$. Induction experiments give : $\Delta \phi = 2 n \phi_0$, which is two orders of magnitude lower than in our case.

To discuss the signal in real time, one must start by a careful study of the time associated to the transition to normal state of a superheated granule. According to Faber [23] and to Valette and Waysand [21], the flipping time τ for a spherical granule follows the law :

$$\tau \propto \frac{R^2}{\rho} \qquad (5)$$

where ρ is the normal state resistivity of the material. For tin Valette and Waysand use the value : $\rho = 10^{-9} \Omega$ cm (high purity) and give the estimate: $\tau \sim 0.8 \mu$sec for 10 μm diameter grains. According to (5), for R = 15 μm one would have $\tau \sim 7 \mu$sec and, for $\Delta \phi = 500 \phi_0$, the average signal in voltage would be : $\mathcal{E} \simeq \Delta \phi / \tau \sim 0.15 \mu V$. For $n \neq 1$, $\mathcal{E}$ is proportional to n and independent of R.

It seems, however, that the performance of the detector can be considerably improved working with <u>less pure materials</u>. At fixed n, one has:

$$\mathcal{E} \simeq \frac{\Delta \phi}{\tau} \propto \rho \propto \frac{1}{\Lambda} \qquad (6)$$

and, according to our theoretical estimate [15,17], the superheating properties are preserved taking $\Lambda \gtrsim 2 \xi_0$ (Λ is the mean free path of conduction electrons). For tin, $\Lambda = 2 \xi_0$ means $\Lambda \simeq 4600$ Å and, in the free electron approximation, $\rho \simeq 10^{-7} \Omega$ cm. More refined estimates give somewhat higher values of ρ. Using 100 μm diameter grains, one would get : $\mathcal{E}$ (n=1) $\sim$ 15 μV, τ ($\rho = 10^{-7} \Omega$ cm, R=50 μm) $\sim$ 0.8 μsec ; $\Delta \phi$(n=1) $\simeq$ 6000 ϕ_0 and I (n=1) $\simeq$ 5 μA. ξ_0 is the Pippard coherence length for pure material [1].

Preliminary specific heat measurements made in Strasbourg [24] suggest that it is indeed possible to preserve the superconductivity properties for low purity materials (type I properties, value of K). For In, it has been shown experimentally [25] that In-Bi alloys (0.2% Bi, $\rho = 3 \cdot 10^{-7} \Omega$ cm) have $H_{sh} > 2 H_c$ for $T > 0.5 T_c$. Then, a 6 μm diameter grain made with such an alloy would be expected to flip in less than 1 nsec. This is certainly <u>important for solar neutrino experiments</u>. If similar numbers work for tin, one would have: $\tau \sim 1 \mu$sec, $\mathcal{E} > 50 \mu V$ for a monopole detector made with 200 μm diameter grains.

<u>A large area monopole experiment</u> with superconducting granules can be performed in coincidence with conventional detectors (e.g. Gran Sasso experiment). Scintillators around the granules detector allow to directly identify charged particles. Flat cryostats could be required if the detector is to be installed between planes of scintillators. Otherwise, cylindric configurations look more appropriate [15]. With current loops $\sim$ 1 m long and $\sim$ 1 cm wide, the number of electronic channels for 3×100 m^2 would be $\sim$ 60.000. The final choice for the amplifiers would be conditioned by the optimized size and flipping time of grains. With foils $\sim$ 5 mm thick, the amount of pure Sn required is $\sim$ 1 ton, for 10% dilution factor in volume. It is likely that the optimized detector will use several tons of tin. Very high purity Sn (10^{-6} impurity level or less) is normally used, but according to our estimates, it may be better to work with less pure material. The program for a large area experiment can be as follows:

- Optimize grains and material (size,purity,resistivity,coating).
- Given the optimized size of grains,flipping time and filling factor, choose the best suited amplifier.
- Test prototypes with a few electronic channels.
- Optimize the size, shape and design of the elementary unit for a large area experiment (10 m^2 ?). build and test this unit, run it for $\sim$ 1 year in an underground laboratory (Gran Sasso ?).
- If the elementary unit works correctly, reproduce it several times ($\longrightarrow$ 100 m^2 or more?).

References

1. See, for instance,Superconductivity, Ed.R.D.Parks,Marcel Dekker Inc. 1969
2. H.Bernas,J.P.Burger,G.Deutscher,C.Valette,S.J.Williamson Phys.Lett. 24A, 721 (1967) ; C.Valette,Thesis,Orsay 1971.
3. J.Blot,Y.Pelan,J.C.Pineau and J.Rosenblatt, J.Appl.Phys. 45 ,1429 (1974)
4. J.Behar,J.M.Cardi,B.Herszberg,D.Hueber,C.Valette and G.Waysand, J.Phys. Coll.39 C6-1201 (1978).
5 A.K.Drukier,C.Valette,G.Waysand,L.C.L.Yuan,F.Peters,Lett. al Nuovo Cimento 14, 300 (1975)
6. R.S.Raghavan, Phys.Rev.Lett. 37 ,259 (1976)
7. G.Waysand, Proceedings of the Moriond Meeting on Massive Neutrinos,January 1984, Ed.J.Tran Thanh Van
8. A.K.Drukier and L.Stodolsky, Phys.Rev.D30,2295 (1984)
9. M.W.Goodman and E.Witten,Princeton preprint November 1984.
10.R.L.Chase,C.Grühn,A.Hrisoho,C.Valette and G.Waysand,Proceedings of the Second ISPRA Nuclear Electronics Symposium, Published by C.E.C. Luxembourg 1975
A.Hrisoho and G.Waysand,Nucl.Instr.and Meth. 214 , 415 (1983).
11. L.Gonzalez-Mestres and D.Perret-Gallix,Proceedings of the Moriond Meeting on Massive Neutrinos,January 1984,Ed.J.Tran Thanh Van.
12 L.Gonzalez-Mestres and D.Perret-Gallix,LAPP-Preprint TH-117 (1984),Contribution to the XVII International Conference on Low Temperature Physics.
13. D.Fryberger, invited Talk at the Applied Superconductivity Conference,San Diego September 1984.Published by IEEE Transactions on Magnetics (1985).
14. C.Bernard,A.De Rujula and B.Lautrup, Nucl.Phys. B242 ,93 (1984)
15. L.Gonzalez-Mestres and D.Perret-Gallix,Preprint LAPP-EXP-85-02 (1985), Contribution to Underground Physics 85.
16. See,for instance,J.P.Burger and D.Saint-James in [1], and references therein. Also, J.P.Burger in Superconductivity ,Ed. P.R.Wallace,Gordon and Breach 1969.
17. L.Gonzalez-Mestres and D.Perret-Gallix, LAPP Preprint EXP-84-05-TH-112 (April 1985)
18. See,for instance,A.Baratoff in Superconductivity , Ed. P.R.Wallace 1969
19. This idea follows from discussions with M.Ocio (CEA,Saclay) .
22. G.V.Ermakov and N.L.Sorokin, Soviet Physics Solid State 24 , 2086 (1982)
21. See,for instance,C.Valette and G.Waysand, "Détecteur supraconducteur de rayons gamma à resolution intrinsèque submillimétrique" Orsay 1976.
22. A.Schlawlow and G.Devlin, Phys.Rev. 113 , 120 (1959)
23. T.E.Faber, Proc.Roy.Soc. 219 A ,75 (1953).
24. B.Escoubès and R.Kuentzler, in preparation.
25. H.Parr,Phys.Rev. B 14,2842 and 2849 (1976)

LIST OF AUTHORS

Arcoragi J.P.	293
Arnould M.	189, 197, 293, 303, 363
Audouze J.	27, 57, 65
Azuma R.E.	431
Barbuy B.	325
Becker H.W.	431
Becker S.A.	285
Beer H.	263
Bloom S.D.	413
Boyd R.N.	423, 443
Cahen S.	243
Canal R.	121
Carr B.	87
Cassé M.	243, 339
Danziger I.J.	233
Dar A.	477
De Loore C.	189, 197
Dearborn D.	37
Delbourgo-Salvador P.	27, 65
Doom C.	189, 197
Durouchoux Ph.	331
El Eid M.F.	167, 177
Eriguchi Y.	143
Fricke K.J.	167, 177
Fuller G.M.	413
Glatzel W.	87, 167
Gonzales-Mestres L.	487
Görres J.	105, 431
Greggio L.	315
Hausman R.F.	413
Hernanz M.	407
Hillebrandt W.	389
Howard W.M.	271, 277
Isern J.	121
Jorissen A.	303
Käppeler F.	253
King J.	431
Kolb E.W.	71
Kratz K.L.	389
Krumlinde J.	389
Labay J.	121
Langer N.	177

Lindley D.	57
López R.	121
Luminet J.P.	215
Maeder A.	207
Malinie G.	27
Mathews G.J.	271, 277, 285, 413
Matteucci F.	315
Meyer B.S.	355
Mochkovitch R.	399, 407
Möller P.	389
Müller E.	143
Nomoto K.	131, 399
Perret-Gallix D.	487
Pichon B.	223
Prantzos N.	189, 197, 293, 339, 363
Redder A.	435
Reeves H.	3, 13, 23
Ritter H.	105
Rolfs C.	431, 435
Schaeffer R.	65, 243
Schramm D.N.	37, 79, 355
Schröder U.	431
Silk J.	57
Steigman G.	37, 45
Takahashi K.	271, 277, 285, 413
Thielemann F.K.	105, 131, 373, 389
Trautvetter H.P.	347, 431, 435
Truran J.W.	97, 373
Turner M.S.	71
Vignaud D.	453
Walker T.P.	71
Ward R.A.	271, 277, 285
Waysand G.	463
Weaver T.A.	145
Wheeler J.C.	113
Wiescher M.	105, 389
Woosley S.E.	145
Yokoi K.	131
Ziegert W.	389

LIST OF PARTICIPANTS

AGUER Pierre   CSNSM - Batiment 104
                           Campus
                           91406   ORSAY
                           FRANCE

ARCORAGI Jean-Pierre   Physics Dept - Montréal Univ.
                           C.P. 6128 SUCC "A"
                           H32 3J7 MONTREAL
                           CANADA

ARNETT David   Astronom. & Astrophys. Center
                           5640 South Ellis Ave
                           60637   CHICAGO IL
                           USA

ARNOULD Marcel   Inst d'Astronomie & Astrophys.
                           U.L.B. - C.P. 165
                           1050   BRUXELLES
                           BELGIUM

AUDOUZE Jean   Inst. d'Astrophysique
                           98 Bis Bd Arago
                           75014   PARIS
                           FRANCE

BARBUY Béatrix   Dept de Astronomia
                           Univ. Sao Paulo - CP 30627
                           01051   SAO PAULO
                           BRASIL

BEER Hermann   Kernforschungszentrum
                           IK III - Postfach 3640
                           7500   KARLSRUHE 1
                           FEDERAL REP. GERMANY

BORENSTEIN Samuel   LPNHE - Ecole Polytechnique
                           Route de saclay
                           91128   PALAISEAU Cedex
                           FRANCE

BOYD Richard   Physics Dept.-Ohio State Univ.
                           174 West 18th Ave.
                           43210   COLUMBUS OH
                           USA

CAHEN Sebastien	Service d'Astrophysique CEN Saclay 91191 GIF/YVETTE Cedex FRANCE
CANAL Ramon	Dept di Fisica de la Tierra y del Cosmos, Univ Granada 8001 GRANADA SPAIN
CASSE Michel	Service d'Astrophysique CEN Saclay 91191 GIF/YVETTE Cedex FRANCE
CESARSKY Catherine	Service d'Astrophysique CEN Saclay 91191 GIF/YVETTE Cedex FRANCE
CRANE Philippe	European Southern Observatory Karl Schwarzschildstr 2 8046 GARCHING FEDERAL REP. GERMANY
DANZIGER John	ESO Karl-Schwarzschild Str 2 8046 GARCHING Bei MUNCHEN FEDERAL REP. GERMANY
DE LOORE Camiel	Astrophysical Institute V.U.B. - Pleinlaan 2 1050 BRUXELLES BELGIUM
DEARBORN David	Lawrence Livermore Lab. P.O. Box 808 94550 LIVERMORE CA USA
DELBOURGO SALVADOR Pascale	Inst. d'Astrophysique 98 Bis Bd Arago 75014 PARIS FRANCE
DUROUCHOUX Philippe	DPHG/SAP CEN Saclay 91191 GIF/YVETTE Cedex FRANCE

LIST OF PARTICIPANTS

EL EID Mounib	Dept of Physics American Univ. of Beirut BEIRUT LIBAN
GLATZEL W.	Math. Dept- Queen Mary College Mile End Road E1 4NS LONDON UNITED KINGDOM
HAOUAT Gerard	Physique Neutronique & Nucl. C.E. Bruyères - BP 12 91680 BRUYERES LE CHATEL FRANCE
HEGYI Dennis	Physics Dept.- Randall Lab. University of Michigan 48109 ANN ARBOR MI USA
HOWARD Michael W.	L-297 Lawrence Livermore Lab. 94550 LIVERMORE CA USA
ISERN Jordi	Dept Fisica Terra y Cosmos Diagonal 645 08028 BARCELONA SPAIN
KAPPELER Franz	Kernforschungszentrum Karlsruhe IK - Postf. 3640 7500 KARLSRUHE FEDERAL REP. GERMANY
KRATZ Karl-Ludwig	Inst f Kernchemie- Univ. Mainz Joh. Gutenberg- Postf 3980 6500 MAINZ FEDERAL REP. GERMANY
LABAY Javier	Dept de Fisica de la Tierra y del Cosmos-Diagonal 647 08028 BARCELONA SPAIN
LANGER Norbert	Universitaetssternwarte Geismarlandstrasse 11 3400 GOTTINGEN FEDERAL REP. GERMANY

LUMINET Jean-Pierre	Groupe d'Astrophys. Relativist Observatoire de Paris 92195 MEUDON PRINCIPAL Cede FRANCE
MAEDER André	Geneva Observatory 1290 SAUVERNY SWITZERLAND
MATHEWS Grant	Lawrence Livermore Nat. Lab. L-405 - California Univ. 94550 LIVERMORE CA USA
MATTEUCCI Francesca	European Southern Observatory Karl Schwarzschilde Str 2 8046 GARCHING Bei MUNCHEN FEDERAL REP. GERMANY
MEYER Bradley	Dept of Astronomy & Astrophys. University of Chicago 60637 CHICAGO IL USA
MOCHKOVITCH Robert	Inst. d'Astrophysique 98 Bis Bd Arago 75014 PARIS FRANCE
MULLER Ewald	Max-Planck-Inst. f Astrophysik Karl-Schwarzschild Str 1 8046 GARCHING Bei MUNCHEN FEDERAL REP. GERMANY
PICHON Bernard	Groupe d'Astrophys. Relativiste DAPHE/LAM-Obs. Meudon 92195 MEUDON PRINCIPAL Cedex FRANCE
PRANTZOS Nicolas	Section d'Astrophysique CEN Saclay 91191 GIF/YVETTE Cedex FRANCE
REEVES Hubert	DPhG/SAP Bat. 28 91191 GIF/YVETTE Cedex FRANCE

LIST OF PARTICIPANTS

RODNEY William	NSF 1800 George St-Nord West 20550 WASHINGTON DC USA
ROLFS Claus	Inst. Kernphysik Univ. Munster-Domagkstr 71 4400 MUNSTER FEDERAL REP. GERMANY
SALATI Pierre	L.A.P.P. Boite Postale 909 75014 ANNECY LE VIEUX Cedex FRANCE
SCHAEFFER Richard	Dept. de Physique Théorique CEN-Saclay 91191 GIF/YVETTE Cedex FRANCE
SCHRAMM David N.	AAC-100 - Univ. of Chicago 5640 S. Ellis 60637 CHICAGO IL USA
SHAPIRO Maurice	MPI f Physik & Astrophysik Karl-Schawarzschild Str 1 8046 GARCHING Bei MUNCHEN FEDERAL REP. GERMANY
SIGNORE Monique	ENS Radio Astronomie 24 Rue Lhomond 75231 PARIS Cedex 05 FRANCE
STEIGMAN Gary	Bartol Research Foundation Univ. of Delaware 19716 NEWARK DE USA
TAKAHASHI Kohji	Lawrence Livermore Nat. Lab. University of California 94550 LIVERMORE CA USA
THIBAUD Jean-Pierre	CSNSM - Batiment 104 Campus 91406 ORSAY CAMPUS FRANCE

THIELEMANN Fiedrich K.	MPI f Physik & Astrophysik Karl-Schwarzschild Str 2 8046 GARCHING Bei MUNCHEN FEDERAL REP. GERMANY
TRAUTVETTER Hanns-Peter	Inst. Kernphysik Univ. Munster-Domagkstr 71 4400 MUNSTER FEDERAL REP. GERMANY
TRURAN James W.	Astronomy Dept.- Illinois Univ 1011 West Springfield Ave 61801 URBANA IL USA
VANGIONI-FLAM Elisabeth	Inst. d'Astrophysique 98 Bis Bd Arago 75014 PARIS FRANCE
VIGNAUD Daniel	D.Ph.P.E CEN Saclay 91191 GIF/YVETTE Cedex FRANCE
WALKER Terry	Theoretical Astrophysics Group MS 209 - Fermilab, Box 500 60510 BATAVIA IL USA
WAYSAND Georges	Groupe de Phys. des Solides ENS Tour 32- Univ. Paris 7 752311 PARIS Cedex 05 FRANCE
WHEELER Craig J.	Dept. of Astronomy University of Texas 78712 AUSTIN TX USA
WIESCHER Michael	Inst. f. Kernchemie Univ. Mainz - Postf. 3980 6500 MAINZ FEDERAL REP. GERMANY
WOOSLEY Stan	Lick Observatory University of California 95064 SANTA CRUZ CA USA